World Geomorphological Landscapes

Series editor

Piotr Migoń, Wroclaw, Poland

More information about this series at http://www.springer.com/series/10852

Nabil Sayed Embabi

Landscapes and Landforms of Egypt

Landforms and Evolution

Springer

Nabil Sayed Embabi
Department of Geography,
Faculty of Arts
Ain Shams University
Cairo
Egypt

ISSN 2213-2090 ISSN 2213-2104 (electronic)
World Geomorphological Landscapes
ISBN 978-3-319-65659-5 ISBN 978-3-319-65661-8 (eBook)
https://doi.org/10.1007/978-3-319-65661-8

Library of Congress Control Number: 2017949508

Printed on acid-free paper

This Springer imprint is published by Springer Nature
The registered company is Springer International Publishing AG
The registered company address is: Gewerbestrasse 11, 6330 Cham, Switzerland

This book is dedicated to all Earth Sciences Students and Researchers

Series Editor Preface

Landforms and landscapes vary enormously across the Earth, from high mountains to endless plains. At a smaller scale, Nature often surprises us creating shapes which look improbable. Many physical landscapes are so immensely beautiful that they received the highest possible recognition—they hold the status of World Heritage properties. Apart from often being immensely scenic, landscapes tell stories which not uncommonly can be traced back in time for tens of million years and include unique events. In addition, many landscapes owe their appearance and harmony not solely to the natural forces. Since centuries, or even millennia, they have been shaped by humans who modified hillslopes, river courses, coastlines, and erected structures which often blend with the natural landforms to form inseparable entities.

These landscapes are studied by geomorphology—"the Science of Scenery"—a part of Earth Sciences that focuses on landforms, their assemblages, surface and subsurface processes that molded them in the past and that change them today. Shapes of landforms and regularities of their spatial distribution, their origin, evolution, and ages are the subject of research. Geomorphology is also a science of considerable practical importance since many geomorphic processes occur so suddenly and unexpectedly, and with such a force, that they pose significant hazards to human populations and not uncommonly result in considerable damage or even casualties.

To show the importance of geomorphology in understanding the landscape, and to present the beauty and diversity of the geomorphological sceneries across the world, we have launched a new book series *World Geomorphological Landscapes*. It aims to be a scientific library of monographs that present and explain physical landscapes, focusing on both representative and uniquely spectacular examples. Each book will contain details on geomorphology of a particular country or a geographically coherent region. This volume presents geomorphology of Egypt—a country widely known for their archaeological heritage but much less for its outstanding geomorphology. The physical environment of Egypt is usually seen from the perspective of the Nile, its floodplain and its delta, whereas the rest is "just desert." This volume, while inevitably considering the Nile and the Delta in detail, also demonstrates the huge diversity of desert landscapes on both sides of the Nile and in the Sinai Peninsula. It also shows the variety of coastal features along both long coastlines that Egypt has—along the Mediterranean Sea and the Red Sea. More importantly, however, the reader will have the chance to virtually visit remote desert landscapes which are hardly accessible and few geomorphologists enjoyed being there in person.

The World Geomorphological Landscapes series is produced under the scientific patronage of the International Association of Geomorphologists—a society that brings together geomorphologists from all around the world. The IAG was established in 1989 and is an independent scientific association affiliated with the International Geographical Union and the International Union of Geological Sciences. Among its main aims are to promote geomorphology and to foster dissemination of geomorphological knowledge. I believe that this lavishly illustrated series, which sticks to the scientific rigor, is the most appropriate means to fulfill these aims and to serve the geoscientific community. To this end, my great thanks go to Professor Nabil Sayed Embabi for undertaking this enormous task, to produce the geomorphological synthesis of his country as a single-author volume. His many years of experience in researching Egyptian landscapes, from the coast to the heart of the Sahara, culminated in this comprehensive account of the fascinating morphology of Egypt that is now presented to the international geomorphological community.

Wroclaw, Poland Piotr Migoń

Acknowledgements

I owe special thanks to many people who helped me in different ways to bring this book into being. Special thanks go to Prof. Poitr Migon, the Series Editor for his careful editing of the manuscript of this book and for his valuable suggestions, which added to the scientific value of this book. I thank him also for writing such a magnificent preface to this book. Thanks also is due to Dr. Robert Doe, Senior Publisher, Earth Sciences and Geography, Springer, whom I met in Paris, during the IAG Conference, August 2013, for helping me approach Springer to contribute this book about Egypt in the series of World Geomorphology. I would like also to express my gratitude to Dr. Moawad Badawy Moawad, Department of Geography, Faculty of Arts, Ain Shams University, Cairo, Egypt, who assisted me in so many ways, provided me with papers which were not available in Egypt, reviewed the chapters of this book, and suggested valuable comments on various issues. Thanks also goes to Dr. Ashraf Abul-Fetooh of Suez University, Suez, Egypt, for providing me with some references on Karst in Egypt and reviewing some chapters of this book as well as lists of references. I thank Dr. Ahmed Said of The Library and Information Department, Faculty of Arts, Ain Shams University, Cairo, Egypt, and Dr. Samira Adam of the Geological Survey of Egypt for providing me with some papers from Egypt or abroad. I thank also my daughter Dr. Doaa Embabi, for revising the language of the manuscript. Acknowledgment is due to all publishers and institutions for permissions to reprint some of the figures appearing in this book. I acknowledge also Google Earth for providing high-resolution space images on the Internet free of charge, without which so many forms could not have been studied. Thanks also is due to my Ph.D student Miss Wafaa Khalaf for providing me with somefield photographs for karst forms of El-Galalah El-Qibliyah. Finally, my deep and sincere gratitude is due to my late wife, Dr. Faiza Salem, whose support and kindness was the key to my success.

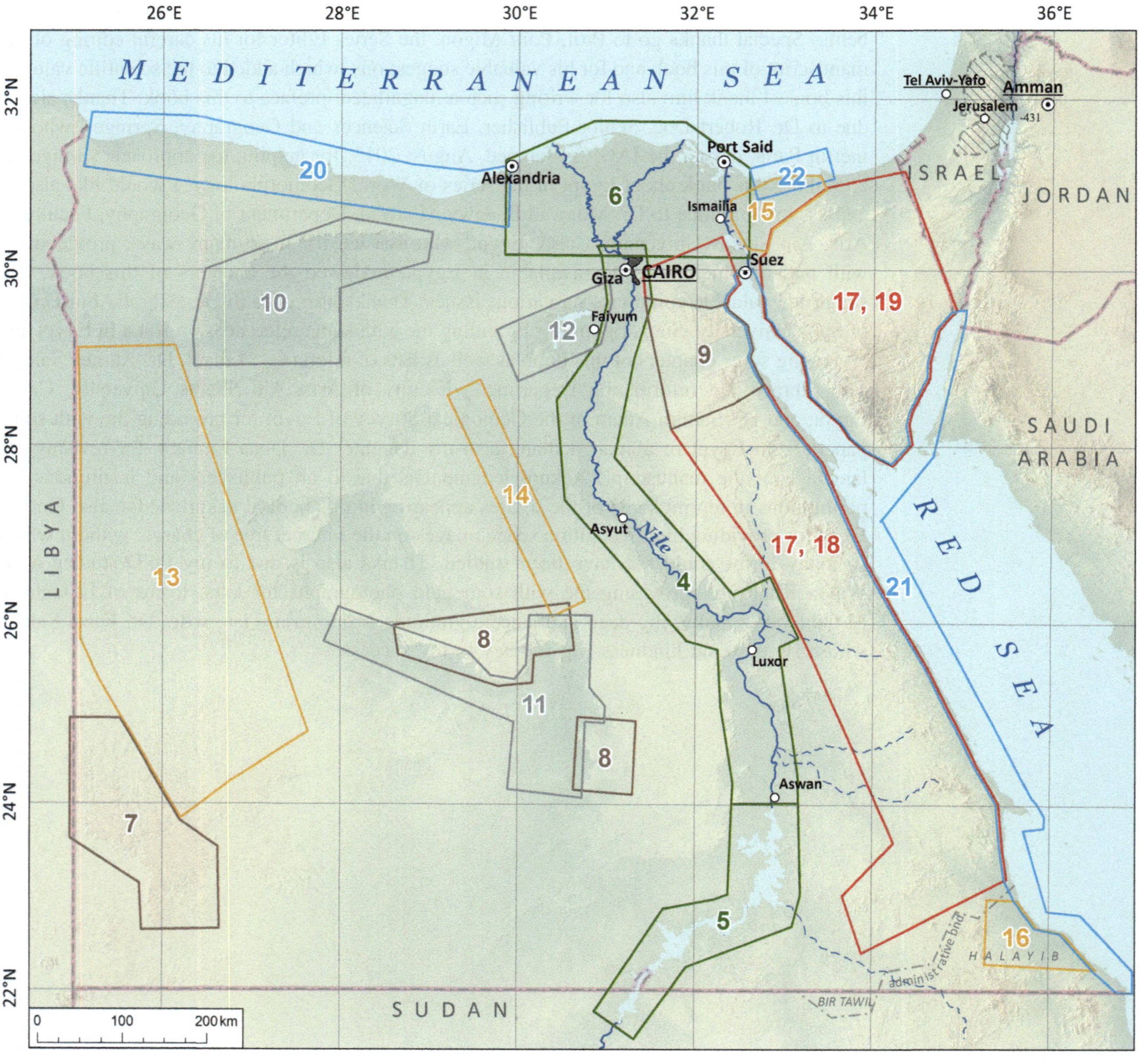

Fig. 1 Location of areas covered by each chapter (numbers of chapters are indicated in each box)

Contents

Part II The Nile

Part III The Plateaus

About the Author

Nabil Sayed Embabi has been a Professor of Desert Geomorphology in the Department of Geography, Faculty of Arts, Ain Shams University, Cairo, Egypt since 1979. He obtained his Ph.D. from Bristol University, UK in 1967. Dr. Embabi has taught at several Arab Universities: King Abdel-Aziz University at Jeddah (1972), Libyan University at Tripoli (1973), Qatar University at Doha (1979-1983), and the United Arab Emirates University at Al-Ain (1988–1992). During his research spanning over 50 years, he participated in or led a number of expeditions or research projects in Egypt/Arabian Deserts. He has also been involved in scientific collaboration with professors at universities in the United States of America, Finland, and the United Kingdom. He has authored and coauthored more than 50 papers and four books in English/Arabic, and attended and presented papers at several international conferences such as those of the International Geographical Congress and the International Association of Geomorphologists. His research papers were cited 87 times in previously published scientific work according to Academia.edu.

Introduction

This book is an attempt to introduce the landscapes and landforms of Egypt to the Earth Scientists. This country has a huge wealth of landforms and landscapes which were inherited from the past geological periods. Some of these inherited forms developed more than 500 Ma ago, influenced by the past (arid and wet) climates. This huge legacy deserves to be presented to scholars of Earth Sciences. For limitations of space and scope, this book does not claim to cover every landform. However, it deals with all types of landscapes and landforms, selecting representative examples or, which have never been studied. Landscapes and forms in this book depend on three sources of data. First, field observations and studies conducted over a span of more than fifty years; some of the field photographs included in this book are taken in the late fifties and early sixties of the twentieth century. Second, previous geologic and geomorphologic studies published in scientific journals/books, as well as unpublished scientific dissertations (both on the masters and on the doctoral level). The third source of data is the space images of Google Earth and Landsat-TM images, aerial photographs, and topographic and geologic maps. The availability of high-resolution images on the Internet through Google Earth made it possible to discover and study some forms—which would have otherwise been an impossible task without such images.

This book is divided into 24 chapters, grouped into eight parts. **Part I** presents the general characteristics of Egypt: Chap. 1 covers the geographical regions of Egypt, Chap. 2 presents the geologic setting of Egypt, and Chap. 3 addresses the past and present climatic conditions. **Part II** deals with the Nile in Egypt and is divided into three chapters: Chap. 4 presents major and remarkable events which occurred during the life of the Nile, Chap. 5 deals with Lake Nasser, and Chap. 6 analyzes the geomorphic aspects of the Nile Delta. **Part III** deals with some selected representative plateaus that are either typical or unique in their geologic and geomorphic aspects. Here, the sandstone plateaus in the SW of the Western Desert of Abu Ras and the Gif Kebir (Chap. 7), the vast karstified plateau between the Nile Valley and the mega-depressions (Chap. 8), and the two Galalah plateaus in the Eastern Desert (Chap. 9) are selected. **Part IV** tackles the depressions in the Western Desert, of which three are selected. These three depressions are the Qattara, which is the largest and deepest one (Chap. 10), the Kharga-Dakhla representing such depressions developed in the leeward side of cuesta escarpments (Chap. 11), and the Fayum Depression due to its unique connection to the Nile Valley (Chap. 12).

In **Part V**, four sand seas and three dune fields are selected out of 16 seas and fields in Egypt. The Great Sand Sea is the largest (>100,000 km^2) in Egypt (Chap. 13). Contrary to it, dune fields in SE of Eastern Desert are the smallest fields in Egypt (a few hundred square kilometers) which are presented in Chap. 16, and have been examined for the first time in detail with the aid of Google Earth images. In fact, study of dune fields in SE of the Eastern Desert would not have been possible without the high-resolution images of Google Earth. The Ghard Abu Moharik Sand Sea, however (Chap. 14), is an exceptional sand sea—not only in Egypt, but also in other deserts—because of its linearity. The North Sinai Sand Sea (Chap. 15) covers the north Sinai plains between the coastal strip and the Syrian Arcs. It is one of the most complex sand seas in Egypt. **Part VI** (Mountains and drainage systems) is divided into three chapters. Chapter 17 deals with the Red Sea and Sinai Mountains; Chap. 18 addresses the drainage system of the Red Sea Mountains, whereas Chap. 19 analyzes the Sinai drainage system. **Part VII** covers the coastal forms: It is divided into three chapters. Although they are small forms, the carbonate ridges of the northwest coast of the Mediterranean Sea make a characteristic rolling landscape in this coastal region and are tackled in Chap. 20. The Red Sea and Sinai coasts have a wealth of forms, because they are the location where several continental and marine processes meet together along the coastal fringes (Chap. 21). The Bardawil is one of the unique lagoons on the Mediterranean coasts since it has a tectonic origin and it has no connection to the Nile water unlike other large lagoons, and it is dealt with in Chap. 22. **Part VIII** examines Geo-Parks and geomorphologic hazards. Selected Geo-Parks and some specific forms are presented in Chap. 23, because of their characteristic forms/landscapes. Finally, it was seen appropriate to conclude this book with a presentation of the most recurring geomorphologic hazards in Egypt (Chap. 24).

Part I
General

Geographic Regions of Egypt

1

Abstract

This chapter presents the main characteristics of the four geographic regions of Egypt. Since it runs in a north–south direction, the Nile Valley divides the Egyptian landmass into two vast regions: one in the west, and the other one in the east. The one in the west is called the Western Desert, and the one in the east (except for the Sinai Peninsula) is called the Eastern Desert. Although some landforms are repeated in more than one region, every region has its own landscape that makes it different from others. This means that Egypt is divided into four geographic regions: the Nile Valley, the Western Desert, the Eastern Desert, and Sinai Peninsula. Characteristic forms of each region are presented to draw a clear and accurate picture for each one of them, but also of the whole of Egypt. Furthermore, this analysis will serve as an introduction to the detailed geomorphic studies that will follow in this book.

Keywords

The Nile Valley • The Nile Delta • The Western Desert • The Eastern Desert • Sinai Peninsula • The Mediterranean Sea • The Red Sea

1.1 Introduction

Egypt, or *Misr*, occupies the north–eastern corner of Africa and extends beyond the Isthmus and Gulf of Suez into Asia to the Sinai Peninsula. It lies approximately between latitudes 22° and 32° N, and longitudes 25° and 37° E. Its maximum width from east to west is about 1230 km, and its maximum length from north to south is about 1100 km. Egypt covers an area of approximately 1 million km^2 and takes, to a large extent, the neat form of a square. Egypt is bound by the Mediterranean Sea to the north, the Republic of the Sudan to the south, the Republic of Libya to the west, and the Red Sea, the Gulf of Aqaba, and Palestine/Israel to the east. Since the Tropic of Cancer crosses the southern part of the country, Egypt lies in the tropical and subtropical arid climate. It forms the eastern part of the Sahara and forms a part of the great desert belt, which extends from the African Atlantic coast in the west, to Arabia and Iran in Asia in the east.

Egypt is divided into four geographic regions. These regions are the Nile Valley and the Delta, the Western Desert, the Eastern Desert, and the Sinai Peninsula (Fig. 1.1). **The Nile Valley and the Delta** represent a unique feature, not only in Egypt but also in the north of Africa. The **Western Desert** is dominated by plateaus, sand dunes, and depressions, whereas the **Eastern Desert** is the region of mountains, which dominate the landscape from north to south. **Sinai** is a characteristic triangular-shaped region and

N.S. Embabi, *Landscapes and Landforms of Egypt*, World Geomorphological Landscapes,
https://doi.org/10.1007/978-3-319-65661-8_1

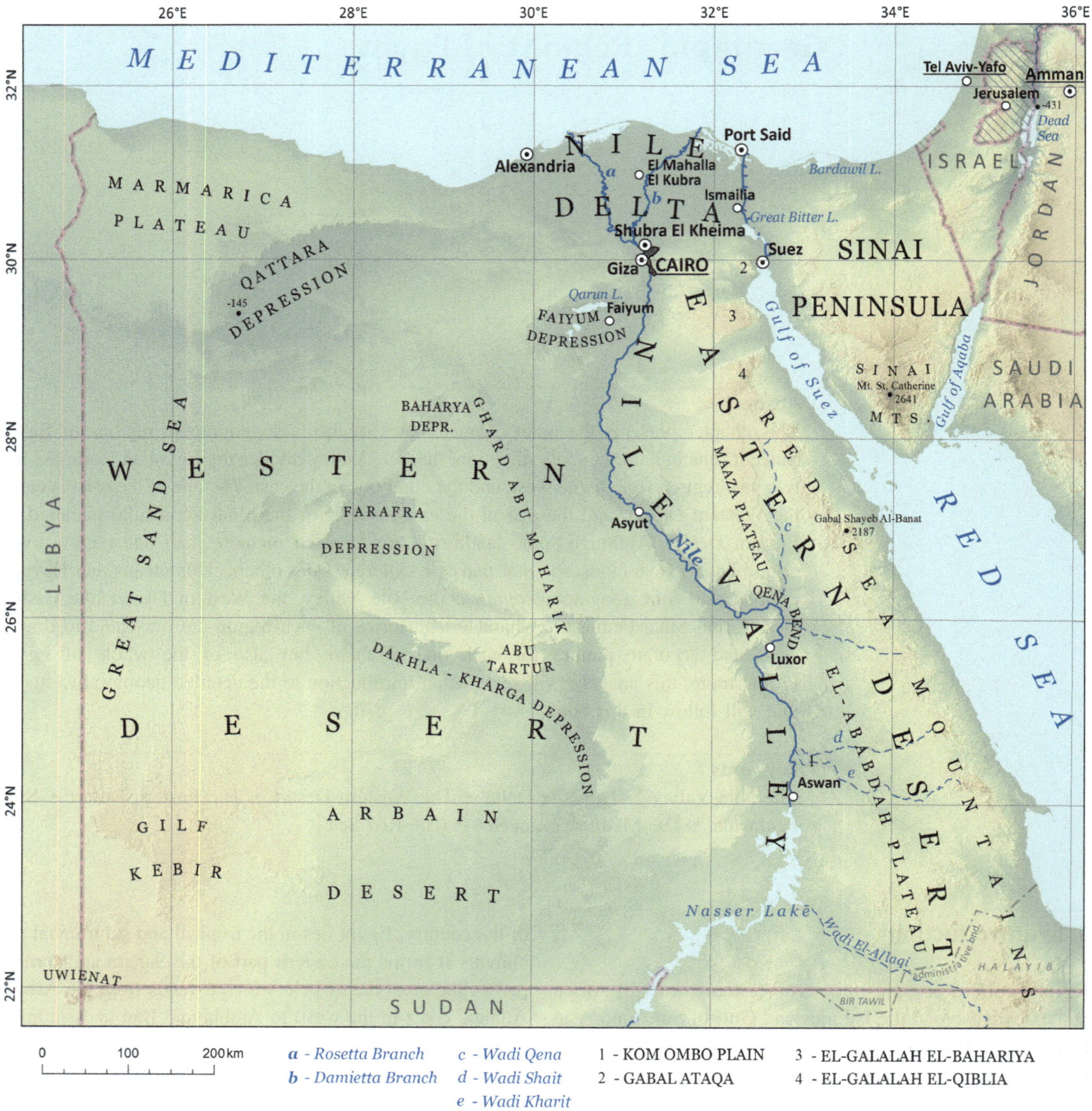

Fig. 1.1 Map of Egypt showing the geographical names mentioned in the text

is bound by water bodies along long stretches in all directions. Furthermore, in order to understand the geomorphology of Egypt, it is useful to note that the Mediterranean and the Red Seas are as important as those four regions. This chapter summarizes the main geomorphologic aspects of each region.

1.2 The Nile Valley and the Nile Delta

The **Nile Valley** and the **Nile Delta** constitute the smallest region in Egypt. They cover about 3.5% (35,000 km^2) of the total area of the country. As a perennial stream running in a desert region, the **River Nile** is a salient feature in the geomorphology of Egypt. Not only does it divide the country into two distinct physical regions (the Western and Eastern Deserts), but it has also shaped Egypt's history and human occupations. The Nile in Egypt represents the terminal 1530 km of the River Nile, which extends for about 6825 km. Its level decreases from 114 m above sea level (asl) at the Egyptian-Sudanese borders, to 84 m asl at Aswan, to 18 m asl at Cairo. Along this distance, it receives no water from any of its tributaries. During its geological history, the Nile developed a valley and formed a delta.

The Nile Valley widens gradually northwards from several hundreds of meters in the south to 23 km at the latitude of Beni Suef city ~122 km to the south of Cairo. To the south of the First Cataract, about 60 km south of Aswan, the Nubian Nile used to pass through a very narrow valley, which was surrounded by cliffs from both sides. At present, this reach is drowned by the waters of Lake Nasser, which came into being after the construction of the High Dam in the late 1960s of the twentieth century. The total length of **Lake Nasser** at 180 m is about 500 km from Aswan to Dal Cataract in the Sudan. Due to previous topography, the width of the lake varies greatly from place to place with a general mean of 12.5 km. The mean gradient of the Nubian sector is 1/11,000. In the north of Aswan, the flood plain broadens gradually except for certain sections where the plain disappears such as the case with the Silsila gorge (60 km north of Aswan). At about 300 km north of Aswan, the Nile makes a large and abrupt bend known as the **Qena Bend**, which is bound by high and steep limestone sides rising up to 300 m asl. The eastern side of the Nile Valley from Aswan to Cairo is generally higher and steeper than the western side (Fig. 1.2). Most of the important wadis, which were former tributaries to the Nile, come from the eastern side of the Valley. However, the sides of the Nile Valley rise up to the level of the surrounding plateaus, which are 200–400 m a.s.l. Along this distance (1100 km), the Nile occupies the eastern side of its floodplain. In some instances, the Nile abuts on the base of the eastern side. The width of the Nile channel varies between 300 m in El-Silsila Gorge and 4–5 km in various parts from the Qena Bend to the apex of the Delta to the north of Cairo.

Parallel to the River Nile, at the western margins of the floodplain, a meandering channel called **Bahr Youssef**, branching at Dyrut (a town 300 km to the south of Cairo) from the Ibrahimiya irrigation canal, flows northward. At the latitude of Beni Suef city, this channel changes its

Fig. 1.2 Nile floodplain and the eastern side of the Nile Valley in the vicinity of Asyut city

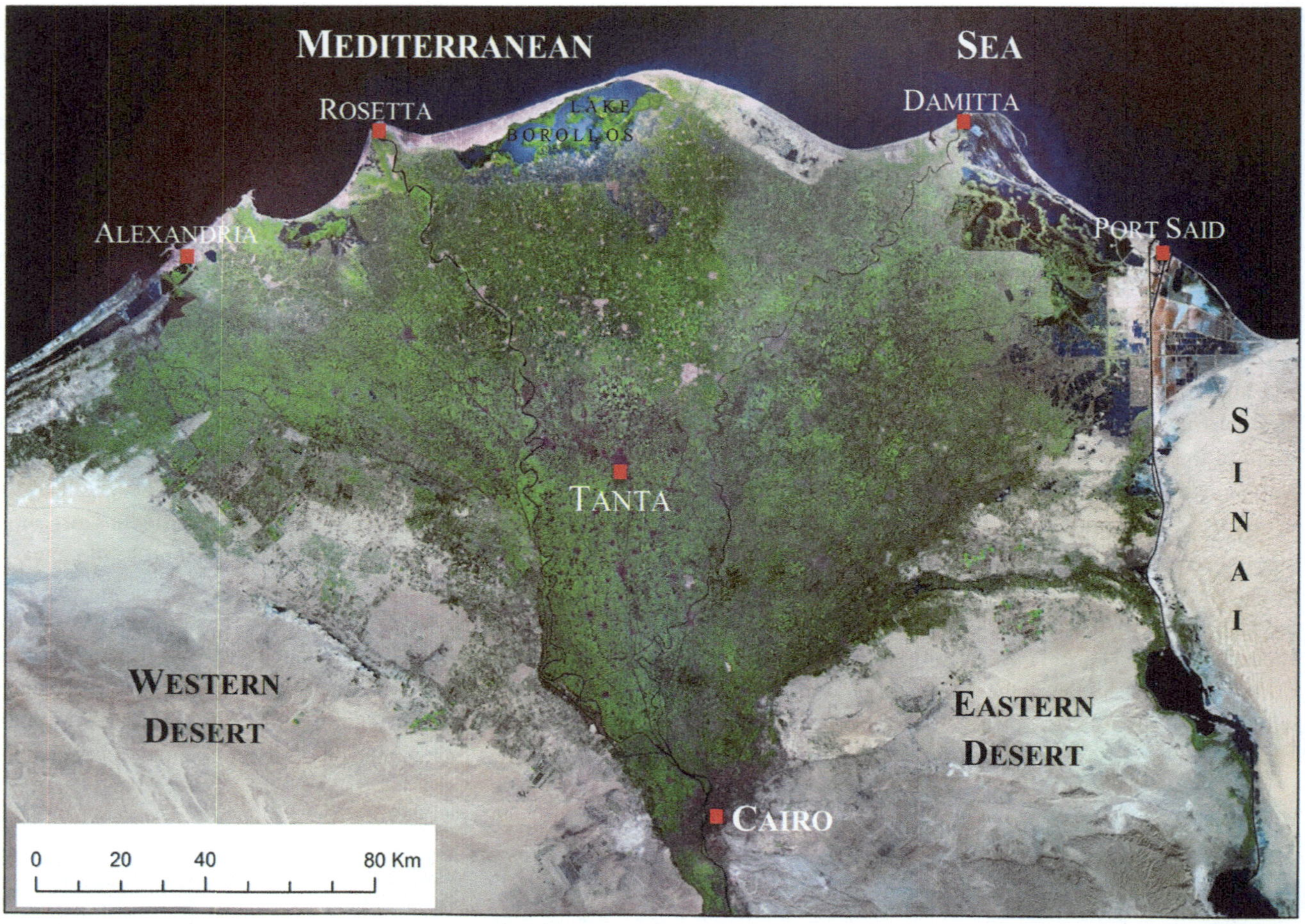

Fig. 1.3 Nile Delta with its triangle shape. (*Source* Google Earth 2012)

destination to the west and cuts through the divide separating the **Fayum Depression** from the Nile Valley. When entering the Depression, Bahr Youssef branches into various canals to irrigate the fertile lands at its floor. Excess irrigation water drains into two internal basins. The first is the **Qarun Lake**, which occupies the lowest part of the Fayum Depression (45 m below sea level), whereas the second is the **Wadi El-Rayan Depression**, lying to the west of Fayum. After the artificial connection with the Fayum drainage network in 1973 (Hamdan 1980), two small lakes were formed in Wadi El-Rayan.

In the north of Cairo, the **Nile Valley** opens up to the Nile Delta (Fig. 1.3). Due to its shape, the **Nile Delta** was the first of its kind to be named as such by Herodotus in 450 BC, when he visited Egypt and realized the great similarity between the shape of the Greek letter Δ and the shape of the Nile Delta. Before splitting into two branches, the Nile flows for about 20 km in the direction of north–west. The eastern branch of the delta is called **Damietta Branch**, whereas the western one is called **Rosetta Branch**. Those two branches are the last of several others, which have recently disappeared. The channels of these two branches, like that of the main Nile, are all characterized by meandering and the presence of silt islands, which total more than 400 islands in both the main Nile and its two branches (Taha 1997). The Delta is a vast depositional plain formed by the Nile a long time ago. The gradient of this deltaic plain is about 1/10,000. Along the Mediterranean coast, several shallow lagoons (Maryut, Idku, El-Brollos, and Manzala) were developed. These lagoons are connected to the sea by small openings (locally known as Bogaz) in the sand barriers, which separate them from the sea. They are kept open by artificial dredging since they represent good fisheries.

1.3 The Western Desert

The **Western Desert** is the largest geographic region in Egypt covering more than two-thirds of the total area of Egypt (681,000 km^2). The Western Desert is the region of plateaus, depressions, sand dunes, and plains. It is also a region of desert oases, which depends on the underground water of the Nubian aquifers. Except for the south–western part of the **Gilf Kebir** and **Uwienat**, which rises in some places to more than 1000 m a.s.l, most of the Western Desert does not rise more than 300–400 m a.s.l (Fig. 1.1). Meanwhile, there are several localities in this desert lying below sea level (Fig. 1.1). The deepest point is −134 m (Ball 1927) to −145 m (IGN/SFS 1977) in Qattara Depression. Other areas are **Siwa** (−18.5 m), **Fayum** (−45 m), **Wadi El-Natrun** (−24 m), and **Wadi El-Rayan** (−64 m).

The southern part of the Western Desert is a vast plain bound to the north by the Kharga-Dakhla Depression, in the west by the Gilf Kebir Plateau, in the east by the Eocene plateau and the Nile Valley, and extends southwards into northern Sudan. This plain is called the **Arbain Desert** (or the Forty's Desert by Haynes 1982) after the well-known caravan route Darb El-Arbain, which crosses the middle area of this plain. Fig. (1.4) shows a camel caravan on this Darb El-Arbain dating to September 1978, photographed by the author while participating in a joint expedition of Ain Shams University, the Geological Survey of Egypt and the Smithsonian Institution of Washington DC. Most of this plain rises 150–200 m asl, but generally gets higher westward until it reaches 600 m asl on average at the foot slopes of the Gilf Kebir Plateau. Vast areas of this plain extend northward and merge into the floors of Abu Minqar, Kharga-Dakhla Depression. Toward these depressions, the level of the ground decreases gradually, where in some localities in Kharga, it reaches just a few meters above sea level. Hills in the depression, such as Gabal Edminston in Dakhla, and Gabal Taref, Gabal El-Ter, Gabal Um El-Ghanaiem, and Gabal El-Qarn in Kharga, may rise up to 100–300 m asl. Dunes spreading in these depressions are organized as dune belts (Fig. 1.5). Barchans moving southward under the effect of northern winds represent a permanent hazard for highways, farms, wells, and villages (Embabi 1982, 1990). Recently, buried wadis and channels were discovered by Radar Imaging by Columbia Shuttle in 1981. This paleo-drainage is known in previous studies as **Radar Rivers**, Subsurface Valleys, and Paleochannels (McCauley et al. 1982). The detailed picture of these wadis is still incomplete, since imaging has not covered the whole buried drainage net(s).

The escarpments of the depressions of Kharga-Dakhla and Abu Minqar represent the southern edges of the vast limestone and chalk plateaus, known as the karstified Eocene plateaus dominating the central part of the Western Desert. These plateaus extend eastward to the Nile Valley, westward to the Great Sand Sea, and northward to Siwa, Qattara, and Fayum depressions. Only the small plateaus of **Abu Tartur** and **El-Quss Abu Said** carry names. The maximum height attained by these plateaus is about 500 m asl in Abu Tartur Plateau, whereas the dominant height is 300–400 m asl. However, height decreases westward toward the Great Sand Sea, indicating the presence of a great basin occupied by the **Great Sand Sea**, since the same happens in north of the Gilf Kebir Plateau. A similar basin occupied by a small sand sea is present in the eastern part of **Farafra Depression**. The dominant features on the surfaces of these plateaus are sand seas and dune belts, small closed depressions, and relict karst forms. **Ghard Abu Moharik Sand Sea**, which extends for about 300 km from the northeast of **Bahariya Depression** to the northern escarpment of Kharga, represents a

Fig. 1.4 Camel caravan on Darb El-Arbain in the southern part of the Western Desert. (*Photo* taken in Sept. 1978)

Fig. 1.5 Dunes descending the Dakhla escarpment, forming a barchan belt in the Dakhla lowland

characteristic feature on this plateau. Bahariya and Farafra Depressions also characterize this central part of the Western Desert.

The Great Sand Sea is a salient feature in the Western Desert, extending for about 600 km from the northern escarpments of the Gilf Kebir and Abu Ras plateaus in the south to the southern margins of Siwa Depression in the north (Fig. 1.1). In addition to the evidence mentioned in the previous paragraph, which shows that this sand sea occupies a huge basin, other evidence indicates that the floor of its northern part approaches sea level or lies below it (Embabi 1998). Although the linear dune form is predominant in the Great Sand Sea (Fig. 1.6), its pattern changes from network to parallel and *Y*-junction from north to south. This sand sea represents the main source of sand, which forms the small dune field that extends from the south of the Gilf Kebir Plateau to Uwienat and north–west Sudan. Several kilometers to the south of Gilf Kebir, there is a small and unique plateau (47 km N–S, 25 km E–W maximum width). Its uniqueness comes from it being a peneplained plateau and is composed of igneous and metamorphic rocks. Heights range between 850 and 950 m a.s.l, with a single peak reaching 1114 m a.s.l.

The northern part of the Western Desert is dominated by the five major depressions of Siwa, Qattara, Wadi El-Natrun, Wadi El-Rayan and Fayum, and by the Miocene (Marmarica) plateau. As mentioned in Sect. 1.2, Bahr Youssef connects the Fayum Depression with the Nile Valley. This is why it is the only depression that has soils similar to those of the Nile Valley and is irrigated and cultivated by the Nile water. The floors of these five northern depressions are below sea level. The **Qattara Depression** is the largest (45,000 km^2) and the deepest (−145 m) erosional depression not only in Egypt but in Africa as well. The north–eastern escarpment of Qattara is only 60 km away from the Mediterranean coast, and this is why it was suggested by Ball (1927) to utilize this depression in the production of hydroelectric power by connecting it to the Mediterranean Sea through an open canal and a tunnel.

Fig. 1.6 Parallel linear dunes of the central section of the Great Sand Sea. (*Source* Google Earth 2012)

To the north of Qattara and Siwa, a vast Miocene limestone plateau extends to the coastal plain along the Mediterranean coastline, locally known as **Marmarica** or **El-Diffa Plateau**. This plateau rises 200–250 m a.s.l and is dotted by numerous small closed basins. A narrow coastal plain with a sequence of low **carbonate ridges** (bars) which are roughly parallel to the present coastline separates this plateau from the Mediterranean coast. This coastal plain opens up at a point to the south of Alamain and widens gradually eastward until it merges with the sandy plains of the western margins of the Nile Delta.

1.4 The Eastern Desert

This region occupies about one-fourth (225,000 km^2) of the total area of Egypt. It is bound by the Nile Valley and the Delta from the west, the Red Sea, the Gulf and Isthmus of Suez from the east, and the Egyptian-Sudanese border from the south. This is the region of mountains, plateaus, and large wadis. In this region, the **Red Sea Mountains** and the fringing plateaus are all dissected by drainage nets, which drain either toward the **Red Sea** in the east or toward the **River Nile** in the west. The heights of some mountainous peaks in this region may reach more than 2000 m a.s.l, whereas those of the plateaus vary greatly between more than 1000 and several hundreds of meters above sea level. Very few localities at the southern end of the **Gulf of Suez** are a few meters below sea level.

The Red Sea Mountains range extends southward from **Gabal Um Tenasib** (latitude 28° 30′ N) parallel to the coastline of the Red Sea up to and beyond the Egyptian borders within the Sudan. The mountains do not in fact form a continuous range, but are rather disconnected blocks separated by basins and deep wadis. Along this range, about 30 peaks rise more than 1000 m a.s.l, such as Gabal Qattar (1963 m), Gabal Abu Abid (1900 m), **Gabal Hamata** (1977 m), Gabal Shendib (1911 m), and **Gabal Shayeb Al-Banat** (2187 m). These mountains are formed of igneous and metamorphic rocks, with a thin sedimentary cover along the eastern flanks. Due to the effect of tectonics, the eastern slopes of the mountains are steeper than the western flanks, leaving a narrow coastal plain along most of the Red Sea coast.

To the north and west of the mountains, vast plateaus extend to the Nile Valley. The southern sandstone plateau is called **El-Ababdah Plateau**, extending northward from the Egyptian-Sudanese borders to the Qena Bend of the Nile Valley. This plateau is highly dissected by wadis that reach the Nile Valley, of which wadis **El-Allaqi**, **Shait**, and **Kharit** are the longest. This plateau is characterized by the presence of some large basins, such as the one lying to the east of Kom Ombo city in the Nile Valley. This basin is filled with fluvial sediments originated from the two wadis of Kharit and Shait draining from the Red Sea Mountains. These materials are covered by the silt of the Nile and that is why the surface of the basin has the appearance of a vast plain. It is called **Kom Ombo Plain**. To the east of this plain, another basin is developed and is called **Atmur Noqra**. The northern extensive limestone tablelands carry the names (from north to south) of **Gabal Ataqa**, **El-Galalah El-Bahariyah**, **El-Galalah El-Qibliyah**, and **El-Maaza**. Only the two Galalahs attain heights that reach more than 1000 m a.s.l. These two Galalahs are separated by the very wide (30 km) wadi named Wadi Araba that represents a reversed topography feature, since it is originally a structural swell. In addition to the wadis dissecting the plateaus, inliers, outliers, and gravel sheets characterize the surface of the plateaus. Due to the effect of tectonics, the northern margins of the Eastern Desert are divided into several separate hills, which rise several hundreds of meters above sea level. Examples of these hills are Gabal Uweibed, Gabal El-Nasuri, and Gabal El-Anqabia. To the north of these hills, ancient fluvial deposits cover the plains extending to the east of the Nile Delta. These plains decrease in height from about 200 m asl in the south to a few meters above sea level in the north.

Along the coasts of the Gulf of Suez and the Red Sea, a narrow coastal plain developed between the coastline and the foot slopes of the mountains. The plain is covered with fluvial sediments that were deposited during the past pluvial periods. Ancient coral reefs, raised beaches, and marine terraces are characteristic features in this plain. The width of the plain varies from one place to another, where it widens in the area of **Aish El-Mallaha** in the north and in the area of **Halayib**, and **Shalatin** in the extreme south. The coastline of this plain is generally straight due to the tectonic origin of the Red Sea. Fringing coral reefs developed along most of the coastline of the Gulf of Suez and the Red Sea.

The Eastern Desert is highly dissected by drainage networks, which can be divided into two opposite systems. The first is the Red Sea drainage system, with short wadis and small basins in general, whereas the second may be called the Nile-Mediterranean drainage system, with relatively long wadis and large basins. Examples of the Red Sea wadis are Araba, Abu Had, El-Mallaha, Safaga, Gasus, El-Hammamat, El-Gimal, Hodain, and Kraf. Examples of the wadis of the second system are Sannur, Tarfa, Assyuti, Qena, Shait, Kharit, and Allaqi. **Wadi Qena** is distinguished from other wadis in that its main course runs for about 200 km in a north–south direction, which is almost opposite to that of the Nile Valley. **Wadi El-Allaqi** with its extension in the Sudan is by far the largest in the Eastern Desert.

1.5 The Sinai Peninsula

Though considered a subregion of the Eastern Desert, the **Sinai Peninsula** with a distinct triangular shape possesses its own characteristics. It is bound from the north by the Mediterranean Sea, from the west by the Gulf and Isthmus of Suez, and from the east by the **Gulf of Aqaba** and the Palestinian-Israeli boundary. Sinai covers about 6% (61,000 km^2) of the total area of Egypt. Its coasts extend for about 700 km, a characteristic which makes Sinai less continental compared to other regions of Egypt. In Sinai, nearly all the geologic formations, structures, and landforms of Egypt are represented. Moreover, regional climatic variations are quite similar to those of Egypt. Due to these characteristics and its distinctive location in Egypt, Hamdan (1980) called Sinai "Egypt Minor."

The southern part of Sinai is the mountainous Sinai, whereas the central part is a tableland region. The northern part is divided into two subdivisions: isolated domal hills and mountains in the southern subdivision, whereas the northern part is mainly covered by sand dunes. Southern Sinai consists of high, complex, and rugged igneous and metamorphic mountains. Most of these mountains rise more than 1000 m asl. However, several peaks rise above 2000 m asl, of which **Mount St. Catherine** (2641 m a.s.l) is the highest not only in Sinai, but also nationwide (Fig. 1.1). Other conspicuous peaks are Gabal Um Shomar (2586 m), **Gabal Mousa** (2285 m), and Gabal Serbal (2070 m). The mountains are dissected by incised drainage lines which drain either into the Gulf of Aqaba in the east or the Gulf of Suez in the west. Examples of wadis draining into the Gulf of Aqaba are (from north to south) Watir, Dahab, and Kid, whereas wadis Baabaa, Sudr, and Feiran are examples of those draining into the Gulf of Suez. Due to this dissection and the effect of lithological variations, the peaks have the appearance of a jigsaw (Fig. 1.7). However, some of the high areas are bevelled and appear as high plains. These are the Haute Surfaces of Awad (1951). They are locally known as farsh, which means flat surface, such as Farsh Zebir, Farsh Itla, and Farsh El-Qasab. Because they are sites for fluvial deposition, intermountain basins lying between peaks have the appearance of small isolated depositional plains.

The eastern slopes of the **Sinai Mountains** directly overlook the waters of the Gulf of Aqaba, except for the

Fig. 1.7 Sinai Mountains, with its steep slopes and jigsaw crests, accommodating Saint Catherine Monastery

areas where small deltas developed along the coastline by the wadis, which drain into the Gulf of Aqaba. In contrast, the western slopes of the mountains are separated from the coastline of the Gulf of Suez by a broad depositional plain (10–30 km) called **El-Qaa plain**. This plain is featureless, except for the small and faulted sedimentary chains of Gabal Qabilyat, Gabal Abu Dorba, and Gabal Araba, which extend parallel to the coastline. The most common coastal features are coastal inlets (drowned wadis), small deltas, and coral reefs along the Gulf of Aqaba, and tidal flats and coastal lagoons along the Gulf of Suez.

The sedimentary tablelands of central Sinai lie to the north of the mountains. They consist of two immense structural plateaus. The first is called El-Tih in the south and the second is El-Egma in the north. Both plateaus are characterized by the whitish color of their calcareous rocks, which is in contrast to the dark color of the igneous and metamorphic mountains of southern Sinai. Steep escarpments bound both plateaus from the south, west, and east. The escarpments are bold and not dissected, except for a few localities where wadis were able to cut through its high walls (Fig. 1.8). The relative heights of the escarpment of El-Tih and that of El-Egma are 300–700 m and 300–500 m, respectively. Both escarpments rise more than 1000 m a.s.l (**Gabal El-Tih** 1187 m; **Gabal El-Gunna** 1583 m; **Gabal El-Egma** 1626 m). Drainage lines of El-Tih form the tributaries of the main wadis which drain either into the Gulf of Aqaba or the Gulf of Suez. Wadis draining the surface of El-Egma form the upper tributaries of **Wadi El-Arish**, which crosses the northern region of Sinai and debouches into the Mediterranean Sea.

The northern part of Sinai is dominated by sand plains in the north and folded hills and mountains in the south. The folded forms are organized into parallel lines trending ENE-WSW and rise between 1090 m asl (**Gabal Yelleq**) and 368 m a.s.l (**Gabal Risan Ainaza**). They are elliptical in shape and exhibit structural domal morphology. Their length varies between 10 and 50 km, and width between 3 and 20 km. Some of the drainage lines dissecting the slopes of these hills and mountains drain internally forming fans,

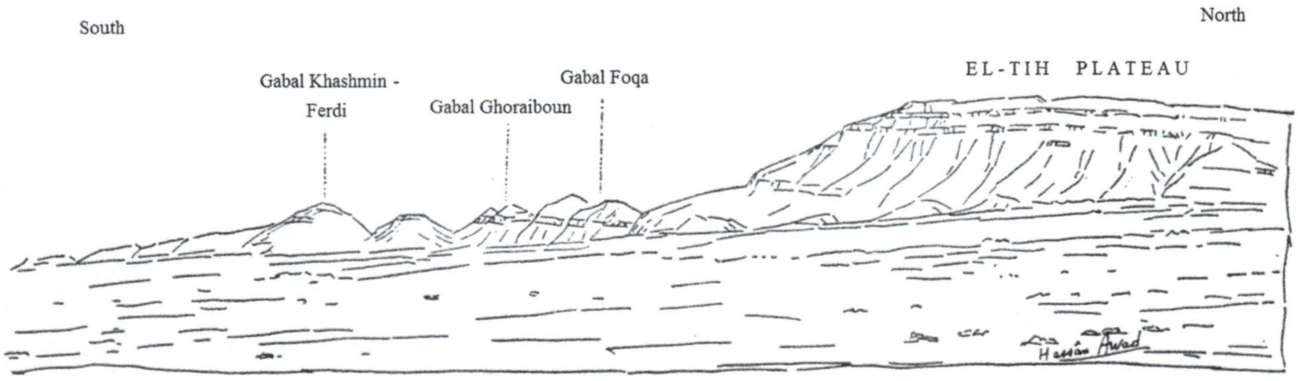

Fig. 1.8 Block diagram for El-Tih Cuesta in central Sinai. (*Source* Awad 1951)

bajadas, and playa plains. To the north of these domal forms, the sandy plains extend to the Mediterranean coastline. Dunes of different types, patterns, and age spread from the eastern borders of Egypt to the Suez Canal, forming a sand sea, which is called the **North Sinai Sand Sea** (Embabi 1998). In this plain, there are other two significant forms. The first is the lower reaches of Wadi El-Arish, and the second is the **Bardawil Lake**. To the west of this lake, the sandy plain changes to muddy flats that are composed of ancient Nile silt and locally known as **El-Tinah plain**.

1.6 The Mediterranean Sea

The **Mediterranean Sea** covers an area of about 250,000 km^2. It has an east–west maximum length of about 3550 km between the coast of the Levant and the **Strait of Gibraltar**. The widest north–south breadth between Libya and the coasts of the Adriatic basin is about 1100 km. The sea is divided into two major basins by an underwater sill between Sicily and Tunisia. The eastern basin is larger and deeper than the western one. The deepest point lies in the Hellenic Trough (5100 m) in the eastern basin. Most of the waters of the Mediterranean come from the Atlantic Ocean via the Gibraltar Strait. Only a few rivers (the Nile, the Po, the Rhone, and the Ebro) debouch into the Mediterranean. High evaporation rates due to warm dry climate result in relatively high surface saline water. Ancient climatic and sea level changes were of great importance to the geologic and geomorphologic development of the Mediterranean and the surrounding lands. Eustatic lowering of the sea level disconnected the Mediterranean from the oceanic system of the Atlantic and converted it into a dry basin during the Late Miocene period (Said 1990). During the lower Pliocene period, the rise of sea level resulted in the submergence of the Nile Valley in Egypt up to Aswan, converting it into a marine arm (Ball 1939).

The total length of the Egypt Mediterranean coast is about 1055 km from the Egyptian-Libyan border in the west to the Egyptian-Palestinian border in the East, as measured on Landsat-TM Mosaic. Bathymetric contours indicate that the continental shelf is wide and shallow in front of the Nile Delta and Sinai Peninsula and becomes narrow and ends abruptly along the coast to the west of Alamain (100 km west Alexandria). The shape and pattern of bathymetric lines show that the Nile Delta extends below the waters of the Mediterranean for at least 50–70 km. Until recently, the Nile sediments were transported eastward by the longshore current along the shores of Sinai and as far as Gaza in Palestine. Black sands of the Nile sediments can be seen clearly on the beaches of Sinai. At present, and after the completion of the **High Dam**, the supply of Nile sediments has been cut off, and the coasts of the **Nile Delta** have been subjected to severe coastal erosion. This is especially clear at points where the northwest wind occurs at right angles to the coastline such as the mouths of the two branches of Damietta and Rosetta, and the Borollos promontory.

1.7 The Red Sea

The **Red Sea** is an elongate, narrow body of water, which extends in a south–southeast direction from Suez and Aqaba to the **Strait of Bab El-Mandab**, separating northeast Africa from the **Arabian Peninsula**. Its total length is around 2000 km, with a variable breadth between 210 km at the latitude of **Ras Mohamed** where it is divided into the Gulf of Suez and the Gulf of Aqaba, and 400 km at the latitude of Mossawaa. Egypt is one of seven countries which have shorelines on the Red Sea (Egypt, Saudi Arabia, Sudan, Eriteria, Yemen, Jordan, and Israel). Egypt occupies the western coasts of the Red Sea down to latitude 22° N, including the coasts of the Gulf of Suez and the western coasts of the Gulf of Aqaba.

The total length of the Egyptian Red Sea coast is about 982.8 km from the northern limb of El-Zeit Bay to the Egyptian-Sudanese border, whereas the total length of the coast of the Gulf of Suez is about 627.8 km, and the length of the Egyptian coast along the Gulf of Aqaba is about 191.4 km from **Taba** to the latitude of **Tiran Island**.

The bathymetry of the Red Sea shows that it has an irregular floor with a maximum depth of 2850 m (Edwards and Head 1987). It has a characteristic wide continental shelf (about 30 km) along the Egyptian coast. This shelf drops away rapidly to sea bottom (1000–1500 m) in a series of stepped cliffs at a depth of about 500 m (Edwards and Head 1987). Over much of the Egyptian coasts, the continental shelf is covered with coral reefs. Bathymetry lines also show that nearly all the small deltas of the wadis, which debouch in the Red Sea, extend below present-day sea level. Large-scale topographic maps show the presence of several coastal inlets along the lower section of some wadis, locally called sharms. These sharms developed due to the **sea level rise** during the last 5000 bp (Moawad 2013). On the contrary, the presence of ancient coral reefs inland along the Red Sea coasts indicates higher sea level in the past. The **Gulf of Suez** and the **Gulf of Aqaba** exhibit a great difference in bathymetry. The Gulf of Suez has a shallow and flat bottom basin with a depth of 55–57 m, deepening toward the entrance of the Red Sea. The Gulf of Aqaba is a steep-sided, deep, and narrow (30 km) trough, with an average depth of 1250 m and maximum depth of 1850 m. Due to these characteristics, the continental shelf which characterizes the Red Sea is absent in the Gulf of Aqaba except in the southern part. The Gulf of Aqaba is separated from the Red Sea at its entrance by the sill of Tiran, which is about 250–300 m deep. Similar to the Red Sea, actively growing reefs continue to develop along the coasts of both Gulfs, Aqaba and Suez.

The Red Sea and both Gulfs are located in a region of arid climate with high evaporation rates. Therefore, surface salinity of the water is higher than the normal salinity of the oceans, varying between 40 and 41‰. Salinity is higher in the Gulfs than in the Red Sea. By and large, rain is very sparse over the Red Sea as a whole. It falls rather sporadically and is mostly localized (Edwards and Head 1987). These characteristics of surface water control water movement and sediment transport along the shores of the Red Sea and the two Gulfs of Suez and Aqaba.

References

Awad H (1951) La Montagne du Sinai Central Étude Morphologique. Société Royale de Géographie d'Égypte, Le Caire

Ball J (1927) Problems of the Libyan desert. Geograph J 70:21-38, 105–28, 209–224

Ball J (1939) Contributions to the geography of Egypt. Geol, Survey of Egypt, Cairo

Edwards A, Head S (eds) (1987) Red Sea. Pergamon Press, Oxford

Embabi N (1982) Barchans of the Kharga Depression. In: El-Baz F, Maxwell T (eds) Desert landforms of Southwest Egypt: a basis for comparison with Mars. NASA, Washington DC, pp 141–155

Embabi N (1990) Dune movement in the Kharga and Dakhla Oases Depressions, the Western Desert, Egypt. In: El-Baz F, El-Tayeb I, Hassan M (eds) Proceedings of the international workshop on sand transport and desertification in Arid lands, Khartoum, Singapore, 1985

Embabi N (1998) Sand Seas of the Western Desert of Egypt. In: Al-Sharhan A, Glennie K, Whittle G & Kendall C (eds) Proceedings of the International Conference on Quaternary Deserts and Climatic Change, Al-Ain, Rotterdam, Balkema, 1995

Hamdan G (1980) The personality of Egypt, part I: the physical personality, (in Arabic) Aalm El-Kotob, Cairo

Haynes V (1982) The Darb El-Arbain Desert: a product of Quaternary climatic change. In: El-Baz F, Maxwell T (eds) Desert landforms of Southwest Egypt: a basis for comparison with Mars. NASA, Washington D C, pp 91–118

IGN/SFS (Institut Géographique National and Société Française Stréotopographié) (1977) Survey of the Qattara Depression, surface and volume. Qattara Project Authority, Cairo

McCauley J, Schaber G, Breed C, Grolier M, Haynes C, Issawi B, Elachi C, Blom R (1982) Subsurface valleys and geoarcheology of the Eastern Sahara revealed by shuttle radar. Science 218:1004–1019

Moawad M (2013) Geomorphological characteristics of the submerged topography along the Egyptian Red Sea Coast: In Scalon L, Ranieri J (eds) Continental shelf: geographical distribution, biota and ecological significance, (Earth Sciences in the 21st Century), Nova Science Publishers, Inc., New York

Said R (1990) (ed) The geology of Egypt, Balkema, Rotterdam

Taha M (1997) The Geomorphology of the Nile depositional islands. (in Arabic). Arab Geograph J 29:223–229

The Geologic Setting

2

Abstract

This chapter presents the main geological aspects of Egypt, which are relevant to its geomorphology. It depends on data collected from previous geological work, the interpretation of Geological Maps (scales 1:2,000,000 and 1:500,000 and 1:250,000), Landsat-TM and Google Earth images, and aerial photographs of different scales for parts of the various regions. Geologic aspects of Egypt are divided into two sections. The first section deals with rock formations, whereas the second deals with the tectonic framework.

Keywords

Egypt • Basement rocks • Sedimentary cover • Tectonic framework

2.1 Rock Formations

Rock formations are classified into two groups according to their mode of development. The first group is the **Crystalline Basement Complex rocks**, composed of igneous and metamorphic rocks and developed during Precambrian times. The second group comprises all the sedimentary rocks which developed during the Phanerozoic period and cover the basement complex.

2.1.1 The Basement Complex (Precambrian)

The crystalline basement is primarily Precambrian in age. It covers about one-tenth of the total land surface of Egypt. It is exposed in the Red Sea Mountains of the Eastern Desert, the southern part of Sinai and Gabal Uweinat area (Fig. 2.1). Small exposures also occur in the southern part of the Western Desert and the Cataract region of Aswan and Nubia. Basement rocks are also found in the depths as they extend below the sedimentary cover throughout Egypt. The basement complex of Egypt, Ethiopia, and Somalia together shape the Nubian Shield that had formed as a contiguous part of the Arabian Shield prior to the opening of the Red Sea (Hermina et al. 1989).

The oldest dated basement rocks are found in **Uweinat Mountains** in the southwest corner of Egypt and which belong to the Late Archean period, about 2.7 Ga (List et al. 1989). These rocks are known as the **Karkur Murr** Series and consist of charnokitic gneisses metamorphosed to granulite facies. Newly obtained ages from **Gabal Kamil** and **Gabal Al-Asr** areas range from 2300 to 1900 Ma. An age of about 1800 Ma is recorded in Ain Daw Series, indicating the formation of a new crust in the Early-to-Middle Proterozoic period and adds new evidence to the proposal of Harris et al. in (1984) that the Late Archean nucleus of Gabal Uweinat is rimmed by a Middle Proterozoic fold belt. This nucleus was named by Gass (1977) the Nile Craton (Morgan 1990). A wide range of younger ages of basement rocks have also been determined for other areas ranging from about 1200 to about 500 Ma.

2.1.2 The Sedimentary Cover (Palaeozoic, Mesozoic, Cenozoic)

The sedimentary cover, which overlies the basement complex, is composed of three major lithological divisions (Fig. 2.1): a **Lower Clastic Division** (Palaeozoic to Late Cretaceous), a **Middle Calcareous Division** (Cenomanian to

N.S. Embabi, *Landscapes and Landforms of Egypt*, World Geomorphological Landscapes,
https://doi.org/10.1007/978-3-319-65661-8_2

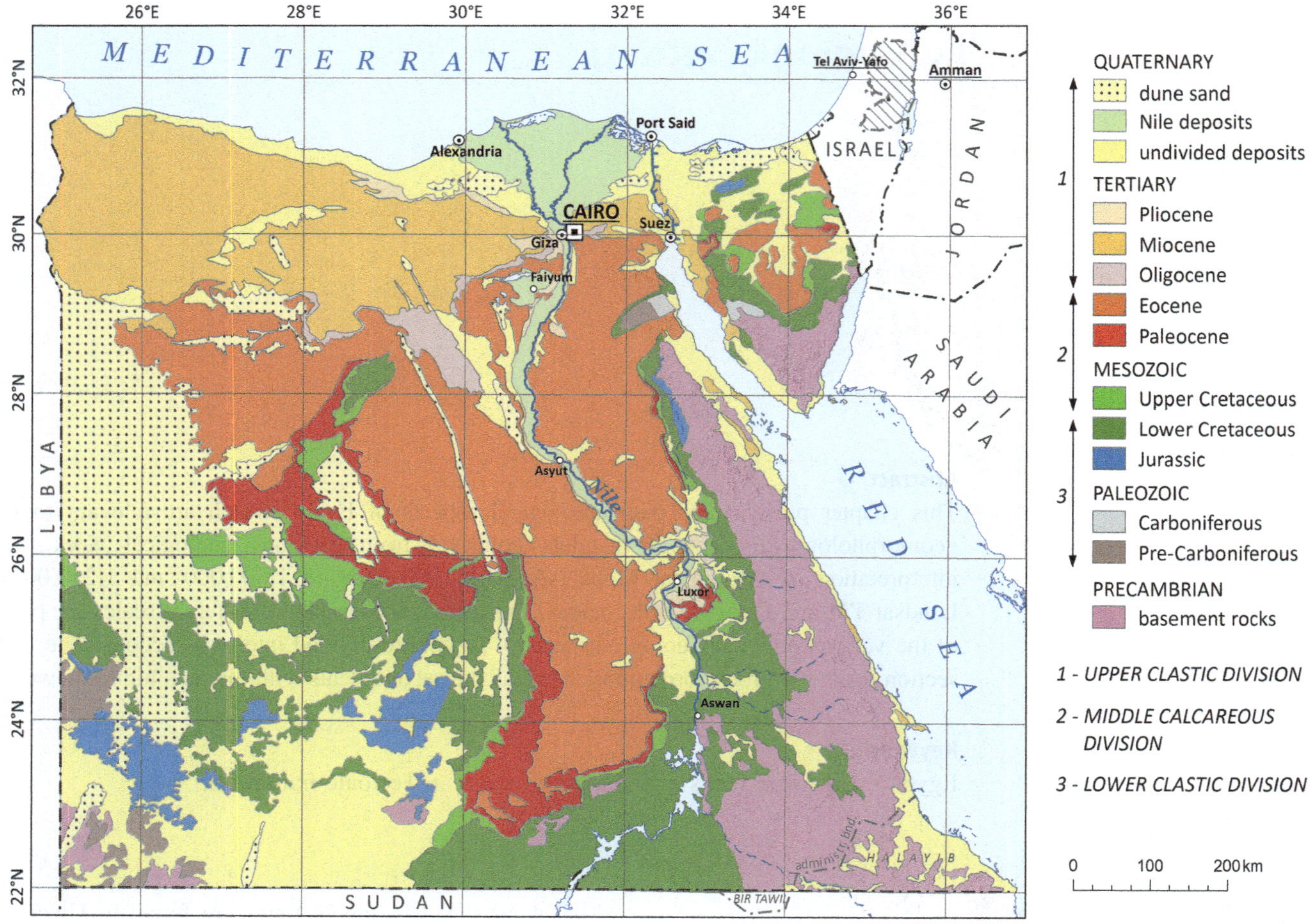

Fig. 2.1 Simplified Geological Map of Egypt. (*Source* The Geological Map of Egypt, scale 1: 2,000,000)

the end of Eocene), and an **Upper Clastic Division** from Oligocene to Holocene (Said 1962; Morgan 1990).

2.1.2.1 The Lower Clastic Division

The strata of the Lower Clastic Division are primarily clastic with a few calcareous beds. Although Said (1962) excluded the group of beds, which are known as Nubia formations from this division, it is preferable to include them, since this formation is all clastics. In this case, the age of this division will be from Palaeozoic to Upper Cretaceous. According to the area covered by exposures, the strata of this division can be divided into two groups: Pre-Jurassic, and Jurassic to Upper Cretaceous.

The Pre-Jurassic strata cover small areas in the southwest corner of Egypt around Uweinat and Gilf Kebir, **Wadi Araba** between **El-Galalah El-**Bahariyah and **El-Galalah El-Qibliyah** to the west of the Gulf of Suez, in various parts of Sinai at the eastern side of this Gulf, and in Arif El-Naqa in northern Sinai. Until recently, Pre-Carboniferous clastics were undifferentiated (Geological Map of Egypt 1981); however, recent work made it possible to classify them according to their age and environment of deposition (Klitzsch 1989, 1990). In the northern part of the Western Desert, Cambrian to Permian strata of up to several 1000 m in thickness have been identified in the subsurface. Palaeozoic beds are missing over most of the remaining parts of Egypt. Areas free of Palaeozoic beds today were probably covered by some of these formations, but due to intensive erosion following a regional up-arching of southern and central Egypt during a Late Palaeozoic–Early Mesozoic event, pre-Carboniferous strata were stripped from large parts of Egypt.

Strata of this subdivision are composed mainly of sandstone of fluvial, marine, and beach origin with a few shale and calcareous beds. However, the most interesting units are the chaotic glacial deposits of **Wadi Abd El-Malik** (**Abu Ras Plateau**) and Uweinat. These beds are composed of unsorted tillite of irregular boulders and blocks that can reach the size of a house placed in a clay sandy and conglomerate matrix intertwined with fluvial conglomerate and sandstone of the Middle and/or Late Carboniferous period (Klitzsch 1983, 1989). They are described by Klitzsch as

Northern Wadi Malik Formation. Their maximum thickness is 50–60 m near the northern end of Wadi Abdel-Malik, which cuts through the Abu Ras Plateau (Klitzsch 1989).

During the period from Jurassic to Upper Cretaceous, clastic sediments with over a kilometer thickness were deposited in the southern and eastern parts of Egypt. Their deposition was controlled primarily by marine transgressions and regressions. They are known collectively in the geological literature as **Nubia /Nubian Sandstone** since they are mainly composed of sandstone strata (Said 1962, 1990a). This sandstone was deposited during various cycles of sedimentation under different environments such as fluvial, deltaic, and lacustrine. Due to these cycles of deposition, the strata were divided into different formations (Klitzsch 1989). Examples of these formations are the **Gilf Kebir Formation**, the **Six Hills Formation**, and the **Sabaya Formation**. However, the Nubia Sandstone Formation will continue to be used as a regional facies term and is recommended to be used to these Late Jurassic to Upper Cretaceous sediments of similar origin. Strata of this formation are well exposed in the southern part of Egypt to the south of Kharga-Dakhla Depression and further south into the Sudan, in Nubia around **Lake Nasser** and Aswan, and in Bahariya Depression. They are also exposed in Wadi Qena, central and northern Sinai such as Gabal Maghara and other folded hills, and Wadi Araba between the two Galalah plateaus (Fig. 2.1).

Volcanics are inter-bedded at the base of the Nubia strata in southern Egypt. These, together with a group of intrusions cutting the basement rocks, represent a period of volcanic activity during the Upper Cretaceous period (Said 1962). There is another type of intrusions that characterize the Nubian strata. This type is represented by the sedimentary **sandstone ferruginous dykes**, which cut the Nubian strata and extend for several kilometers. These dykes appear on the surface of the ground as micro-ridges rising several tens of centimeters in the southern plain of the Western Desert (Fig. 2.2).

2.1.2.2 The Middle Calcareous Division

From Upper Cretaceous (Senonian) to the end of the Eocene, a series of beds, which are dominantly calcareous, was deposited and covered wide areas in the central parts of the Western and Eastern Deserts on both sides of the Nile Valley and central and northern parts of Sinai. They also extend to the southernmost parts of the Western Desert and along the Red Sea coast. The Upper Cretaceous strata show considerable variations in lithology. In the southern parts of Egypt, they are represented by **phosphate rocks** (Duwi Formation), which are best seen at the escarpment of the Abu Tartur Plateau, the Nile Valley sides, and along the Red Sea coastal zone, and overlie the Nubia Series. These in turn are overlain by a unit of grayish-green shales, which are best developed along the escarpments of Dakhla Depression (Dakhla Formation).

The **Dakhla Formation** is separated from the overlying **Esna Shale** by a chalky limestone unit known as **Tarawan Formation**. The age of these three formations dates back to the Late Cretaceous/Paleocene period. Toward the top, the Esna Shale becomes more calcareous and is replaced by Eocene rock units made up of carbonates and clastics, which form the cap rock of the wide plateaus of the Western and Eastern Deserts and central Sinai. The Eocene succession is divided into several groups of formations, since they were deposited in variable sedimentary sub-environments. Lower Eocene carbonates are known as **Thebes Group**, whereas those of Middle Eocene strata are known as the **Mokattam Group**, and the Upper Eocene formations, which are more clastics, are classified into various formations such as Maadi, and Qasr El-Sagha Formations (Hermina and Lindenberg 1989). Lower Eocene limestone (Thebes Group) is characterized by the presence of silica bands and concretions. Exposed concretions reach large sizes and frequently occur lying in crowds on the surface of the plateaus in the Western Desert recalling melon fields. This Lower Eocene limestone is

Fig. 2.2 Sandstone dyke in the southern plain of the Western Desert

also characterized by the presence of fossil remains of the earliest, and extinct, suborder of whales, Archaeoceti. This fossil was found in Wadi El-Rayan to the west of Fayum Depression. Due the concentration of the skeletons of the whales in a small area in the north of Wadi El-Rayan, this area is locally named **Wadi El-Hitan**, i.e., Valley of the Whales. Because of its scientific and international value, Wadi El-Hitan was recognized as a UNESCO World Heritage Site in 2005.

2.1.2.3 The Upper Clastic Division

The sedimentary rocks of the **Upper Clastic Division** are divided into two subdivisions: the Tertiary (Oligocene–Pliocene) rocks and the Quaternary deposits. Oligocene to Pliocene units occupy vast areas in the northern part of the Western Desert, small areas in the northern parts of the Eastern Desert, along the Red Sea coast, both sides of the Gulf of Suez, and several patches along the Nile Valley.

Oligocene Sediments are of fluviatile origin and are composed of sand and gravel. They cover a small narrow belt from the Cairo-Suez region to northern Fayum and Bahariya Depression, with quite a wide area extending southward between Bahariya and the Nile Valley. Due to geyser activity, the Oligocene sands and gravel of the **Gabal El-Ahmar Formation** are known for their distinguished red color and silicified tree trunks (Said 1962; Hermina and Lindenberg 1989). The Oligocene fluvial sediments were connected according to some previous studies with a river system that flowed consequent to the uplift of North Africa at the end of the Eocene period and was called by Blankenhorn the **Ur-Nil** (Said 1962). In some other studies, the Oligocene gravel was correlated with another regional drainage system draining northward (Said 1990b), or draining southwestward and westward and then northward after being connected with tributaries from the Gilf highlands. This system is called the **Gilf System** (Issawi and McCauley 1993). Volcanic activity during the Mid-Tertiary period resulted in the development of a broad belt of basaltic intrusions to the north of Fayum, which extends to the Nile Valley in Cairo and continues eastward along the Cairo-Suez road. Similar volcanic outcrops of the same age are present in various parts in the Western and Eastern Deserts, Sinai, and the Gulf of Suez.

The **Miocene rocks** exhibit local variations due to their deposition in various environments. However, the Lower Miocene deltaic to estuarine sand and gravel clastics, which are exposed at the southern margins of Qattara Depression, are known as **Moghra Formation**. The top of this formation intertwines with the **Marmarica carbonates** toward Siwa in the west, carrying a mixed fluvio-marine fauna (Hermina and Lindenberg 1989; Said 1990b). These sediments seem to have been deposited under one regime belonging to a river system, which must have flowed into the sea from the north of Fayum (Said 1990b). Along with other evidence, it was postulated that during the Early and Middle Miocene period, a drainage system developed in Egypt, of which the main river was called the **Qena River** (Issawi and McCauley 1993). In some other studies, however, this drainage system is considered to be a continuation of the **Ur-Nil system** (Salem 1976). The Middle Miocene reefal limestone, **Marmarica Formation**, is the most extensive exposure, covering the entire Marmarica (El-Diffa) Plateau of the northern Western Desert, and Cairo-Suez stretch (Said 1962; Hermina et al. 1989). Other Miocene sediments occupy small areas in the Western and Eastern Deserts, the Red Sea coast, and Sinai. Upper Miocene marine strata are not exposed in any part of Egypt.

The **Pliocene sediments** are scattered as relatively small outcrops along the margins of the Mediterranean coastal plain, in and around Wadi El-Natrun in the northern part of the Western Desert, and in the Nile Valley from Aswan to Cairo. They also occur in the lower sections of some Nile tributaries in the Eastern and Western Deserts, along the western coast of the Gulf of Suez and along the Red Sea coastal belt. Alongside the Nile Valley and its tributaries, there are two main facies: marine limestone, marls and coquina beds of Kom El-Shelul Formation on the surface in the north, and in the subsurface in the south (Hermina and Lindenberg, 1989). The second is conglomerate sandy facies in the south and in the mouths of several large wadis that debouch at the Nile Valley (Said 1962). The marine facies along the Nile valley and its tributaries are taken as an indication of the rise of the Mediterranean Sea level forming a Marine Gulf that submerged the Nile Valley and the mouths of the tributaries. This Marine Gulf is known as the Pliocene Gulf of the Nile Valley (Said 1993).-

The **Quaternary deposits** are spread all over Egypt and lie unconformably over the Pliocene or older rocks. These deposits are closely connected with climatic changes and sea level fluctuations that occurred during the Pleistocene period. They are classified according to their mode of origin into eolian, fluvial, and marine sediments.

Eolian sands are the most widespread and most recent, covering more than 20% of the total area of Egypt (Embabi 1998). They occupy vast areas in the south and west of the Western Desert and North Sinai, and small areas in the north and east of the Western Desert, and the north and south of the Eastern Desert. They are formed as dunes of different types sand sheets shaped as sand seas and dune fields.

Fluvial deposits include the following types:

(a) Nile sediments (silt, clay, gravel) on the sides and floor of the Nile Valley, the surface and margins of the Delta, and the floor of Fayum Depression.
(b) Wadi gravel in the form of fills, terraces, fans, bajadas, and sheets in various localities in the Western and Eastern Deserts and Sinai.

(c) Playa silt, sand, and gravel clastics in the form of low-lying plains occupying the floors of internal basins or different sections of some wadis in the Western and Eastern Deserts, and Sinai.
(d) Calcareous tufa and travertine developed in the form of thin laminae terraces or sheets on the faces of the escarpments of the plateaus in the southern part of the Western Desert.

Marine and coastal deposits include the following types:

(a) The Mediterranean coastal oolitic limestone in the form of parallel linear ridges separated by longitudinal Depressions covered by evaporites, marl, and gypsum. These ridges were formed as offshore bars in the receding Mediterranean Sea during the Pleistocene (Shukri et al. 1956; Selim 1974).
(b) Sabkha deposits along the Mediterranean coasts of the Nile Delta, various stretches to the west of Alexandria, Sinai and in Siwa, Qattara, Wadi El-Natrun and small patches in other Depressions and along the Red Sea coast.
(c) Shelly-coralline sandy and silty sediments along the Red Sea coast, which indicate a shallow reefal marine environment during the Pleistocene epoch.

2.2 Tectonic Framework

Studies of Said (1962) have divided Egypt tectonically into four units as follows (Fig. 2.3).

(1) The Arabo-Nubian Shield
(2) The Stable Shelf
(3) The Unstable Shelf
(4) The Gulf of Suez-Red Sea Graben

2.2.1 The Arabo-Nubian Shield

The rocks of the Arabo-Nubian unit (the basement rocks) are exposed, as previously mentioned, over large areas in the Eastern Desert and Sinai and small patches in the extreme

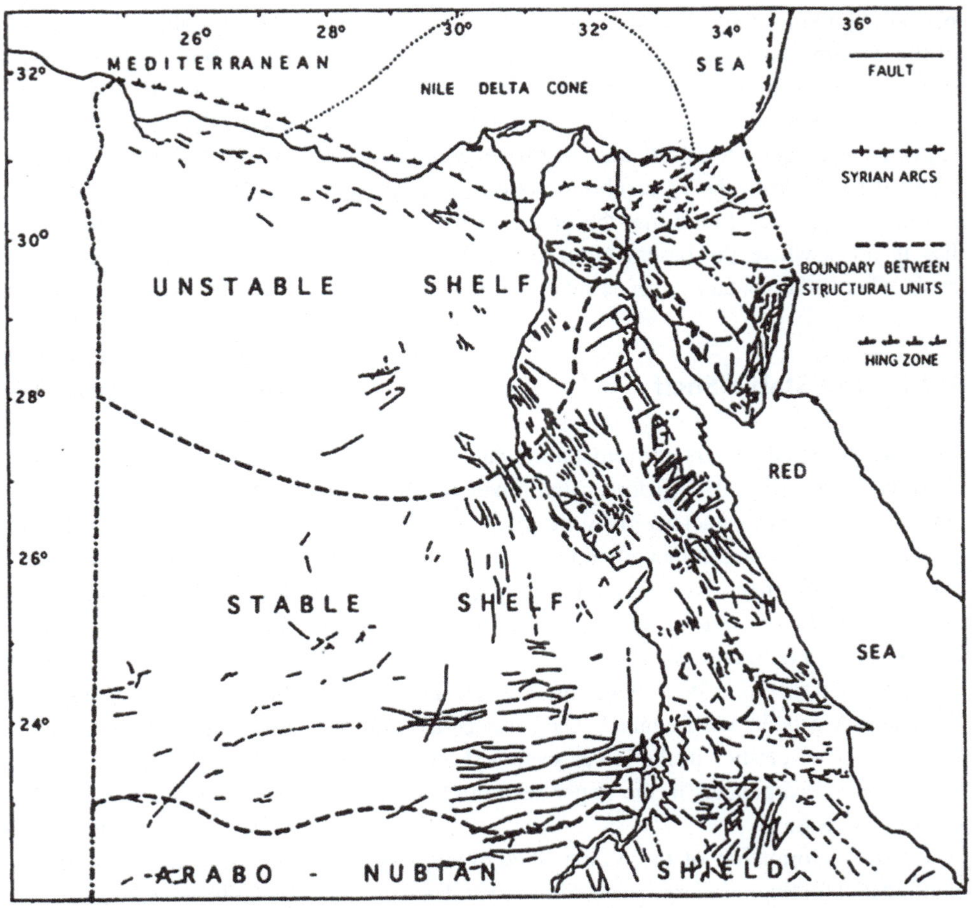

Fig. 2.3 Main structural features of Egypt (Compiled from various geological sources)

southwestern part of the Western Desert and along both sides of the Nile Valley in the Aswan environs. Also, evidence indicates that these rocks extend northward beneath the sedimentary cover of the Stable and Unstable Shelves. It is now well established that this shield was developed during several stages around an Archean Continental Nucleus (The Uweinat Massif) onto which a more recent crust was accreted (Morgan 1990). The shield was peneplained, and an extensive peneplain forms the base of the clastic sedimentary cover of the Nubia rocks. The **Arabo-Nubian Shield** remnants of this peneplain can be seen in central Sinai and is called by Awad (1951) "Pre-Nubian Erosion Surface." It is also found in the southwest corner of the Western Desert to the north of the famous hills of Peter and Paul, and in the central part of the Red Sea Mountains (Philobbos and Hassan 1975). Due to this peneplanation, the **Arabo-Nubian Shield** has been the source of clastics of the sedimentary cover.

At the end of the Precambrian (520 Ma), the eastern Mediterranean had not yet been formed and the Nubian plate was not separated from the Arabian plate. Therefore, Egypt, the Arabian Peninsula, and Turkey were a continental mass on the southwest corner of the Gondwanaland. At the beginning of the Palaeozoic period, Egypt was drifting south and had probably reached its most southerly latitude with the location of Cairo at almost 70° south during the Ordovician period (Morgan 1990). This location explains the presence of the Carboniferous glacial deposits in southwest Egypt that was mentioned in the previous section.

Since basement rocks are the oldest formations, it is expected that they record all the tectonic events that have affected Egypt from the Precambrian period to the present. Furthermore, it was found that the greater density of faults lies in areas where the basement rocks are exposed or in areas where sedimentary cover is relatively thin.

2.2.2 The Stable Shelf

The **Stable Shelf** covers the area north and west of the **Nubian Shield** (Fig. 2.3). It looks like a belt of ill-defined boundaries and is characterized by the relatively thin sedimentary sequence of continental and epicontinental deposits such as the Mesozoic Nubian sandstone. Thickness of the sedimentary succession increases northward. It is about 400 m near the Shield nucleus, 1100 m in Dakhla and Kharga, and 2500 m in Bahariya in the Western Desert near the Stable–Unstable Shelf contact. The sedimentary column is composed of clastics in the lower section (Lower Clastic Division) and carbonates in the upper section (Middle Calcareous Division).

Crustal deformation of the Stable Shelf is expressed in faulting and folding on a large scale (Said 1962). It seems that folding played a minor role in the structure of this Shelf. Some of these folds are mainly flexured blocks of which Ataqa and the two limestone plateaus of El-Galalah El-Bahariya and El-Qiblia are examples. Several of these folds (anticlines and synclines) are noted in different areas, such as those present along the Nile Valley banks (Sandford 1934), Wadi Qena in the Eastern Desert (The Geological Map of Wadi Qena Quadrangle NG 36-6), and El-Egma Plateau in central Sinai (Awad 1952). Many of the small folds are bounded partially or completely by faults, such as those in the northern part of Kharga Depression which extends along a north–south line. According to orientation, the faults of the Stable Shelf were classified by Said (1962) into four groups: north–south (East African) faults, northwest (African) faults, east–west (Tethyan) faults, and northeast–southwest faults. The effect of these tectonic lines on local and regional features will be discussed in the following sections.

2.2.3 The Unstable Shelf

The **Unstable Shelf** occupies most of northern Egypt between the transition line that separates the two Stable and Unstable Shelves and the Hinge Zone along the Mediterranean geosyncline (Fig. 2.3). The sedimentary column of the Unstable Shelf is thick. Thickness is estimated to be more than 7000 m along the Mediterranean coast (Schlumberger 1984). The sedimentary column is composed of clastics in the lower part, calcareous series in the middle, and an upper clastic cover topped with carbonates (Said 1962).

The Unstable Shelf shows signs of tectonic disturbances. This structural deformation is related to the **Laramide Phase** of the **Alpine Orogeny** which began by the end of the Lower Cretaceous period (Schlumberger 1984). The result of this deformation can be seen in the numerous anticlines that distinguish the northern parts of Egypt. These folds are known as the **Syrian Arcs**. The tectonic disturbances also resulted in faulting parallel to the lines of folding. The anticlines of these folds are all asymmetrical, double plunging and are organized into numerous parallel lines with northeast and east-northeast orientation. They were classified into three groups according to size (Mostafa and Khalil 1990). Large folds include **Gabal El-Maghara**, **Gabal Yelleq**, and **Gabal El-Halal**. Medium folds comprise **Gabal El-Minsherah**, **Gabal Kherim**, **Gabal Arif El-Naga**, and some others. Small folds are many and are less than 2 km long. The largest two anticlines are Gabal El-Maghara and Gabal Yelleq. Gabal El-Maghara is 40 km long and 24 km wide and has a height of 735 m asl. Gabal Yelleq is 45 km long and 20 km wide and has a height of 1090 m asl. Most of these structures form a distinctive tectonic province in northern Sinai. However, they extend along an arcuate trend from North Sinai and north of the Gulf of Suez across the southern part of the Nile Delta to the Western Desert to include **Gabal Abu Rawash** and the **Bahariya** and **Farafra** swells (Shata 1953; Shukri 1954). They also extend eastward toward the Negev Desert and the Mediterranean Basin to the north of Sinai. The geomorphologic characteristics of these structural hills and mountains will be discussed in Chap. 23 on **Geoparks**.

The **Unstable Shelf** is separated from the Mio-geosyncline of the Mediterranean Basin by a **Hinge Zone**, which practically coincides with the present coastal area of the Mediterranean. According to Sigaev (1959) and Schlumberger (1984), the Nile Delta occupies a large trough in the Hinge Zone. This trough which is called by Said (1981) the **North Delta Embayment** is present within the continental slope of the Mediterranean. It is limited in the south by a fault running along the northern boundary of the Cairo-Suez horst and in the north by structures consisting of parallel, elongated tilted blocks. This embayment has been filled with the sediments of the **Nile Delta Cone** since the Upper Miocene times. This sedimentary cone extends below the Mediterranean and covers an area which equals three times that of the present Nile Delta.

2.2.4 The Gulf of Suez—Red Sea Graben

The Gulf of Suez and the Red Sea are structurally and genetically closely related. They form the northern section of the **East African Rift System**. According to Said (1962), the **Gulf of Suez**, though situated within the Stable Shelf, has been an area of subsidence and a site of immense accumulation of sediments. It was formed during the Early Palaeozoic era as a narrow embayment of the Tethys and rejuvenated during the rifting phase of the Great East African Rift system in Oligocene–Miocene times. The Gulf is bordered by, and made up of, a large number of faulted blocks. Its connection with the Mediterranean in the north and with the Red Sea to the south was established during the Early Miocene period (Schlumberger 1984).

The Red Sea and the **Gulf of Aqaba** originated during the Oligocene period after the arching and crustal thinning in the area of the Nubian-Arabian Shield and subsequent collapse as a part of the East African Rifting. The spreading of the Red Sea was due to the northeastward motion of the **Arabian Plate**. The left lateral displacement of the Arabian Plate along the Gulf of Aqaba was also established then (Schlumberger 1984). In Egypt, it is likely that the resulting topography from the rising margins of the Red Sea controlled the development of the ancestral rivers of the Nile which have their origin at about the same time as the initiation of the Red Sea Rifting (Said 1981).

2.2.5 Volcanicity

Different volcanic activities took place during the Phanerozoic period in Egypt. Such tectonic activity was mainly controlled by the fracture system, which originated in the Late Precambrian. This relationship is clear in Fig. 2.3, which shows the main tectonic features of Egypt. **Volcanic rocks** of the Palaeozoic and Mesozoic eras are diversified in terms of their chemical characteristics, size, and mode of eruption, whereas Cainozoic eruptions produced basalt volcanites (Meneisy 1990).

The earliest Palaeozoic igneous activity resulted in the emplacement of minor alkaline intrusions, granite and what is known as "**Dyke swarms,**" which is common within basement rocks (Meneisy 1990). The oldest known alkaline intrusions in the Eastern Desert and Sinai have an average age of 550 Ma. A number of intrusions and extrusions yielded Ordovician-Devonian ages (Meneisy 1990). Dykes are a characteristic feature of the basement rocks of the Eastern Desert and Sinai, and they are among the most intensely dyke-intruded regions known. The dykes are narrow, steeply dipping bodies, a few meters thick, and several kilometers long (Fig. 2.4). Dykes north of the Qena-Safaga highway in the Eastern Desert are estimated to be 530–480 Ma old (Meneisy 1990). Because they vary greatly in composition from acidic to basic and alkaline, dykes respond differentially to erosion so they form positive or negative landforms relative to the host basement rocks.

Age determination revealed three main volcanic episodes in the Late Palaeozoic era: Late Carboniferous, Permian, and Permo-Triassic. The rocks are widely intruded as ring complexes, sills, dykes, plugs, sheets, and flows into the basement of the Eastern Desert and the sedimentary cover in the Western Desert (Meneisy 1990).

The Mesozoic volcanics include basaltic rocks, alkaline ring complexes, and minor granite intrusions. Basaltic dykes, sills, fills, and plugs scattered in the Eastern and Western Deserts, and Sinai were intruded or extruded in the Mesozoic, of which those of Wadi Natash in the southern part of the Eastern Desert (Late Cretaceous) are the largest (600 km^2) Phanerozoic volcanic associations in Egypt. Early Cretaceous volcanics occur in the Wadi Araba and Wadi Darag areas in the northern part of the Eastern Desert.

The Cainozoic era witnessed several episodes of volcanic activity. The earliest started in Late Eocene, and the latest occurred in Lower-Middle Miocene. However, Mid-Tertiary (Oligocene–Miocene) volcanism was the most widespread and was related to the opening of the Red Sea. Quaternary volcanism is limited to Zabargad Island in the Red Sea and to Uweinat, Kamil peneplain, and Bir Safsaf in the south of the Western Desert.

Fig. 2.4 Two dykes projecting in Sinai Mountains

Tertiary volcanics are mainly basaltic, and most of them are distributed north of latitude 28° N. They take the form of sheets, dykes, sills, cinder cones, or small ridges. Significant outcrops occur in the Fayum (Gabal Qatrani)-Abu Rawash area, Cairo-Suez stretch, El-Bahnasa environs to the west of the Nile valley and Bahariya Depression. Minor exposures are present in the southern parts of the Western and Eastern Deserts, Red Sea coast, and Sinai. Tertiary basaltic rocks are also present beneath the Nile Delta and adjacent areas of the Western Desert (Said 1981).

Hydrothermal activities accompanied volcanic eruptions in Egypt. The most widespread and pronounced are those connected with Tertiary volcanism. The mineralization effects include ubiquitous ferrugination, silicification, and dolomitization. These activities affected several localities, of which El-Gabal El-Ahmar to the east of Cairo is a classical example with the red coloration of its sandstone. Silicified tree trunks are another classical example for silicification during hydrothermal activity.

2.3 Conclusion

In Egypt, it can be said that nearly all types of rocks and structures are well-represented and exposed on the surface of the ground. Also, ages extend from the Precambrian to Holocene indicating a long geological history. Geomorphic regions vary greatly in the distribution of rock formations, in that the Western Desert is dominated by sedimentary rocks, while the basement rocks are concentrated in the Eastern Desert and southern Sinai. In general, rock formations get younger from south to north. Thickness of the sedimentary cover increases on the whole northward, since the basement rocks get deeper in the same direction. This is why strata are on the whole-dipping northward. Two of the sedimentary formations are of relative significance, which are the carbonates and Nubian sandstone. These two types of rocks cover wide areas in the Western Desert, the Eastern Desert, and central Sinai. They form all the plateaus in these regions except for the small peneplained plateau in SW Western Desert. Of the Quaternary sediments—eolian sands are the most widespread, especially in the Western Desert.

Faults represent the most dominant type of structure, especially the eastern part of the Eastern Desert and the southern part of Sinai Peninsula. These mountainous are the most affected areas in Egypt by tectonics that accompanied the opening of the Red Sea. Another region which is tectonically significant is that one which is dominated by fold that lies in northern Sinai and is known as the Syrian Arcs. Other folds are present in several other localities in other regions as separate ones and not clustered as those of Sinai. Volcanic activities occurred in all regions and in all periods of the geological history of Egypt. Their materials take several forms, such volcanic cones, lava sheets, dykes, and ring complexes.

References

Awad H (1951) La Montagne du Sinai Central. Étude Morphologique. Société Royale de Géographie d'Égypte, Le Caire

Awad H (1952) A new type of desert cuesta in central Sinai. In: Proceedings Of The XVII International Geographical Congress, Washington

Embabi NS (1998) Sand seas of the Western Desert of Egypt. In: Alsharhan AS, Glennie KW, Whittle GL, St. C. Kendall CG (eds) Quaternary Deserts & Climatic Change, Proceedings of Intranational Conference on Quaternary Deserts and Climatic Change, Al-Ain, United Arab Emirates, 9-11 Dec. 1995, Balkema, Rotterdam, pp 495–509

Gass I (1977) The evolution of the Pan-African crystalline basement in NE Africa and Arabia. J Geolog Soc London 134:129–138

Harris NBW, Hawkesworth CJ, Ries AC (1984) Crustal evolution in northeast and east Africa from model Nd ages. Nature 309:773–776

Hermina M, Lindenberg HG (1989) The tertiary. In: Hermina M, Klitzsch E, List F (eds) Stratigraphic lexicon and explanatory notes to the geological map of Egypt, scale 1:500,000. Conoco Inc, Cairo, pp 141–218

Hermina M, Klitzsch E, List F (eds) (1989) Stratigraphic Lexcon and explanatory notes to the geological map of Egypt, Scale 1:500,000. Conoco Inc, Cairo

Issawi B, McCauley J (1993) The Cenozoic landscape of Egypt and its river systems. Ann Geol Sur Egypt 19:357–384

Klitzsch E (1983) Paleozoic Formations and a Carboniferous Glaciation from Gilf Kebir-Abu Ras area in southwestern Egypt. J Afr Earth Sci 1:17–19

Klitzsch E (1989) The Paleozoic. In: Hermina M, Klitzsch E, List F (eds) Stratigraphic lexicon and explanatory notes to the geological map of Egypt, scale 1:500,000. Conoco Inc, Cairo, pp 57–76

Klitszch E (1990) Paleozoic. In: Said (ed) The Geology of Egypt, Balkema, Rotterdam, pp 393–406

List F, El-Gaby K, Tehrani R (1989) The basement rocks in the Eastern and Western Deserts and Sinai. In: Hermina M, Klitzsch E, List F (eds) Stratigraphic Lexicon and explanatory notes to the geological map of Egypt 1:500,000. Conoco Inc, Cairo, pp 33–56

Meneisy M (1990) Volcanicity. In: Said (ed) The Geology of Egypt, Balkema, Rotterdam, pp 157–172

Morgan P (1990) Egypt in the framework of global tectonics. In: Said (ed) The Geology of Egypt, Balkema, Rotterdam, pp 91–111

Mostafa AR, Khalil MH (1990) Structural characteristics and tectonic evolution of north Sinai fold belts. In: Said R (ed) The geology of Egypt. Balkema, Rotterdam, pp 381–389

Philobbos ER, Hassan KK (1975) The contribution of palaeosoil to Egyptian lithistratigraphy. Nature 253:33

Said R (1962(The geology of Egypt. Elsevier Publishing Co, Amesterdam

Said R (1981) The geological evolution of the River Nile. Springer-Verlag, New York

Said R (1990a) Red Sea coastal plain. In: Said R (ed) The geology of Egypt. Balkema, Rotterdam, pp 345–359

Said R (1990b) Cenozoic. In: Said R (ed) The geology of Egypt. Balkema, Rotterdan, pp 451–486

Said R (1993) The River Nile: Geology, Hydrology and Utilization. Pergamon Press, Oxford

Salem R (1976) Evolution of Eocene-Miocene sedimentation patterns in parts of northern Egypt. Amer Assoc Petrol Geol Bull 60:34–64

Sandford K (1934) Paleolithic man and the Nile valley in upper and middle Egypt: a study of the region during Pliocene and Pleistocene times. Oriental Institute Publication No 3, Chicago, Illionis

Schlumberger (1984) Well Evaluation Conference. Egypt, Cairo

Selim A (1974) Origin and lithification of the Pleistocene carbonates of the Sallum Area, western coastal plain of Egypt. J Sedimentary Petrology 44:70–78

Shata A (1953) New light on the structural development of the Western Desert of Egypt. Bull Institut du Desert d'Égypte 3:101–106

Shukri NM (1954) Remarks on the geological structure of Egypt. Bull. Soc. Géogr. d'Égypte, 27: 65–82

Shukri NM, Philip G, Said R (1956) The geology of the Mediterranean coast between Rosetta and Bardia, II Pleistocene sediments: geomorphology and microfacies. Bull Institute d'Égypte 37:395–427

Sigaev N (1959) The main tectonic features of Egypt: an explanatory note to the tectonic map of Egypt, scale 1:2,000,000. Geological Survey of Egypt, Cairo, Paper, p 39

Climatic Conditions: Present and Past

3

Abstract

Analysis of present Egyptian climatic data indicates that Egypt is dominated by tropical and subtropical arid and hyper-arid climates. In the meantime, available evidence (geological, geomorphological, and biological) indicates that Egypt enjoyed wetter climates in the past than at present. All data indicates that from the end of the Eocene till the end of the Miocene periods, the climate was tropical wet, and from the Pliocene period, rainfall decreased gradually and aridity increased until the Mid-Holocene period when the tropical/subtropical arid climate prevailed all over Egypt and the Eastern Sahara. The main purpose of the study of present and past climates is to show their effect on the landscapes and landforms of Egypt.

Keywords

Egypt • Aridity index • Temperature • Rainfall • Wind conditions • Past climates

3.1 Present-Day Climatic Conditions

Although present-day arid conditions are known to prevail over the whole of Egypt, variations in the degree of aridity are largely unknown. To investigate such aspects, the UNESCO **Index of Aridity** (1987) as a ratio of (Pmm) annual precipitation (mm)/(Eptmm) annual evapotranspiration (mm) is applied to all meteorological stations (60 stations) shown in the published Climatological Normals of Egypt (Meteorological Authority 1979).

3.1.1 Degree of Aridity

To investigate degree of aridity all over Egypt, the calculated indexes were classified according to the UNESCO (1977) into four categories as follows:

A Hyper-arid: >0.03
B Arid: 0.03–0.2
C Semiarid: 0.2–0.5
D Sub-humid: 0.5–0.75

The calculated indexes of all meteorological stations (60 stations) were plotted according to the designated degree of aridity in Fig. 3.1. It can be seen from this figure that Egypt is dominated by hyper-arid and arid climates. Only one station (Rafah) in the extreme northeast corner belongs to the semiarid category. The map also shows that the arid climate prevails only along a narrow strip parallel to the Mediterranean coast from Sallum to El Arish. If Rafah, which belongs to category C (semiarid), is added, it can be said that present-day climatic conditions along the whole Mediterranean coast are arid to semiarid.

Although Fig. (3.1) shows that hyper-arid conditions prevail to the south of the arid Mediterranean coast, the actual indexes indicate that the southern half of the Western Desert (south of the latitude of Qena-Farafra) is extremely arid, since their indexes are zero. This suggests a new category of aridity, which can be called extremely arid, the index of which is zero. This is why the southern part of the Western Desert is known as the driest part in the Sahara, and probably the driest region on Earth (Kehl and Bornkamm 1993).

N.S. Embabi, *Landscapes and Landforms of Egypt*, World Geomorphological Landscapes,
https://doi.org/10.1007/978-3-319-65661-8_3

Fig. 3.1 Aridity index map of Egypt (calculated according to UNESCO (1987) and based on the Egyptian Meteorological Authority data)

The arid climates prevailing in Egypt are the result of the interaction of climatic elements, especially temperature and precipitation. The relevant characteristics of these climatic elements will be analyzed in the following section.

3.1.2 Temperature

3.1.2.1 Mean Monthly Minimum and Maximum

Climatological records of Egypt (Meteorological Authority 1979) show that winter months (December, January, February) are cold to warm, whereas summer months (June, July, August) are hot (Table 3.1).

The mean monthly minimum of January (the coldest month) varies between 2.1 °C for Nekhel in central Sinai and 13.8 °C for Quseir on the Red Sea Coast. In general, coastal stations are much warmer than continental ones. The mean monthly maximum of July/August (the hottest months) varies between 41.8 °C for Aswan and 29.6 °C for Alexandria. Similarly, in this part, coastal stations have lower records compared to continental ones. Except for highlands, distribution of maximum and minimum

Table 3.1 Minimum and maximum temperature in some meteorological stations

Station		Winter			Spring			Summer			Autumn		
		Months			Months			Months			Months		
		D	J	F	M	A	M	J	J	A	S	O	N
Cairo	Min.	10.3	8.8	9.5	11.6	14.2	17.4	20.2	21.8	21.8	20.1	17.8	13.8
		Average = 9.5			Average = 14.4			Average = 21.3			Average = 17.2		
	Max.	20.5	19.0	20.7	23.9	28.3	32.1	34.7	34.9	34.5	32.5	29.5	25.0
		Average = 20.1			Average = 28.1			Average = 34.7			Average = 29.0		
Alexandria	Min.	10.8	9.1	9.3	10.8	13.1	16.4	20.2	22.0	22.7	21.1	17.6	14.4
		Average = 9.7			Average = 13.4			Average = 21.6			Average = 17.7		
	Max.	20.1	18.4	19.3	21.3	23.5	26.6	28.6	29.7	30.6	29.6	27.6	24.2
		Average = 19.3			Average = 23.8			Average = 29.6			Average = 27.1		
Nekhel	Min.	3.7	**2.1**	2.9	4.9	8.8	11.5	14.5	16.4	17.1	15.2	12.3	8.4
		Average = 2.9			Average = 8.4			Average = 16.0			Average = 12.0		
	Max.	19.5	17.7	19.3	22.4	27.4	30.6	34.4	35.4	35.3	31.8	29.2	23.6
		Average = 18.8			Average = 26.8			Average = 35.0			Average = 28.2		
Quseir	Min.	15.7	**13.8**	14.	16.4	19.5	22.7	25.4	26.2	26.7	25.1	22.9	19.3
		Average = 14.6			Average = 19.5			Average = 26.1			Average = 22.4		
	Max.	28.2	22.5	22.9	24.7	27.2	30.1	32.2	32.9	33.4	31.7	29.9	27.0
		Average = 24.5			Average = 27.3			Average = 32.8			Average = 29.5		
Aswan	Min.	9.7	8.1	9.6	13.0	17.9	21.4	24.3	24.8	22.8	22.6	19.6	14.6
		Average = 8.9			Average = 17.4			Average = 24.0			Average = 18,9		
	Max.	25.0	23.5	26.2	30.3	35.3	38.7	**41.8**	**41.1**	**41.0**	39.5	36.4	29.8
		Average = 24.9			Average = 34.8			**Average = 41.3**			Average = 35.2		
Dakhla	Min	5.8	4.0	5.5	9.0	13.8	19.1	22.2	22.6	22.4	20.2	16.7	11.1
		Average = 5.1			Average = 14.0			Average = 22.4			Average = 16.0		
	Max.	23.0	21.4	23.8	27.8	32.8	36.7	38.7	38.4	38.4	35.7	32.3	27.4
		Average = 22.7			Average = 32.4			Average = 38.5			Average = 31.8		

Source Meteorological Authority of Egypt, Climatological Normals of Egypt up to 1979

temperatures (Figs. 3.2 and 3.3) and other temperature means shows that there is a general increase southward, toward the Tropic of Cancer, which crosses the southern section of Egypt.

3.1.2.2 Absolute Maximum and Minimum Temperature

Climatic records also show that high absolute temperature maximums are more than 40 °C and occur during late spring and early summer months, especially in June and May. Very high maximums (reaching 49°–50 °C) were recorded in Kharga, Dakhla, and Aswan (Fig. 3.2). Absolute minimums fall below freezing point in a few days only in Sinai, the Central Western Desert, and middle Egypt (Fig. 3.3). The lowest recorded minimum is −7.2 °C in Nekhel in central Sinai. In nearly all stations, drop of temperature below freezing point occurred once during the period of records (20–35 years). Only in very few stations did this drop in temperature below zero occur several times, such as Nekhel (4 times), El-Kontella (3 times) Sinai, Fayum (3 times), Minya (4 times), and Siwa, Farafra, and Dakhla (4 times). This is why Egypt witnesses no frost season.

3.1.3 Rainfall

3.1.3.1 Annual Rainfall

Mean annual rainfall does not exceed 200 mm along the Mediterranean coast as in Alexandria and is practically nil (less than 2.0 mm) in the southern half to the south of 28°sN (Fig. 3.4), that is, except for the southeast corner where little summer rainfall is recorded and the northeast corner where mean annual rainfall in Rafah is slightly higher than 300 mm. Rain falls as showers and varies considerably from year to year and from one locality to another. Rain is almost the only form of precipitation, but **snowfall** is recorded on

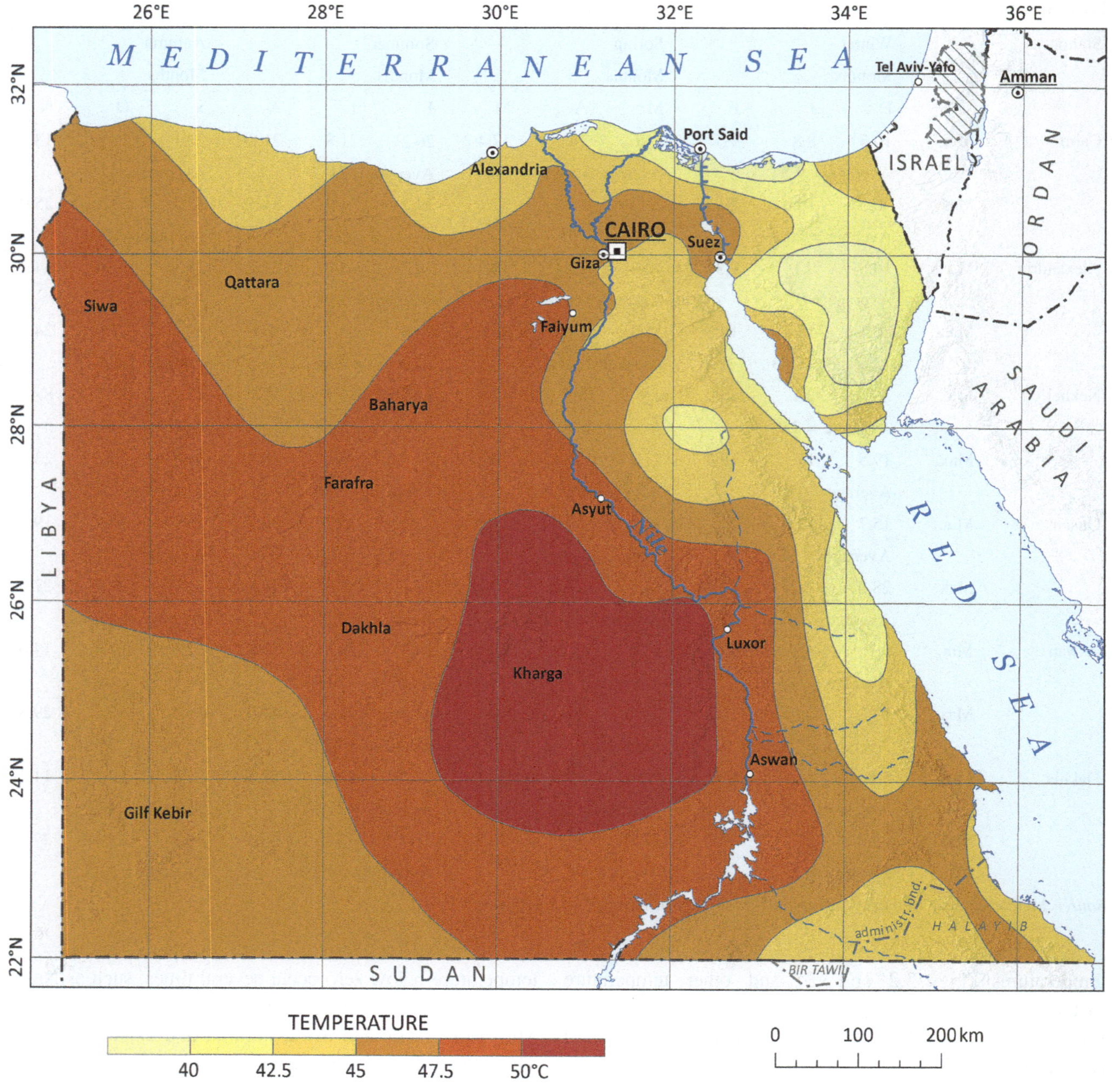

Fig. 3.2 Absolute maximum temperatures (°C) in June (based on the Egyptian Meteorological Authority data)

the summits of Sinai Mountains nearly every year during the months of November–April (Soliman 1972).

Hail occurs occasionally when favorable climatic conditions prevail. Normally, rainwater is absorbed in surface soils or is rapidly evaporated. Only during heavy rainstorms, rainwater might flow for short distances in wadis or as sheet floods, resulting in minor erosion effects. Some of this occasional runoff causes damage on highways, in marginal villages in the Nile Valley, and in cultivated fields, representing one of the most significant geomorphological hazards in Egypt.

3.1.3.2 Seasonal and Regional Variations

To investigate seasonal and regional variations in rainfall, relevant data was extracted from the climatic records for all stations. The examined data was as follows: rainy months (with rain not less than 1 mm), their number in each station and maximum fall in one day during the period of data recording, which varies between 20 and 40 years.

Rainy months cover winter and spring seasons and extend to the autumn months of November and October. Their number varies between 11 days and one month.

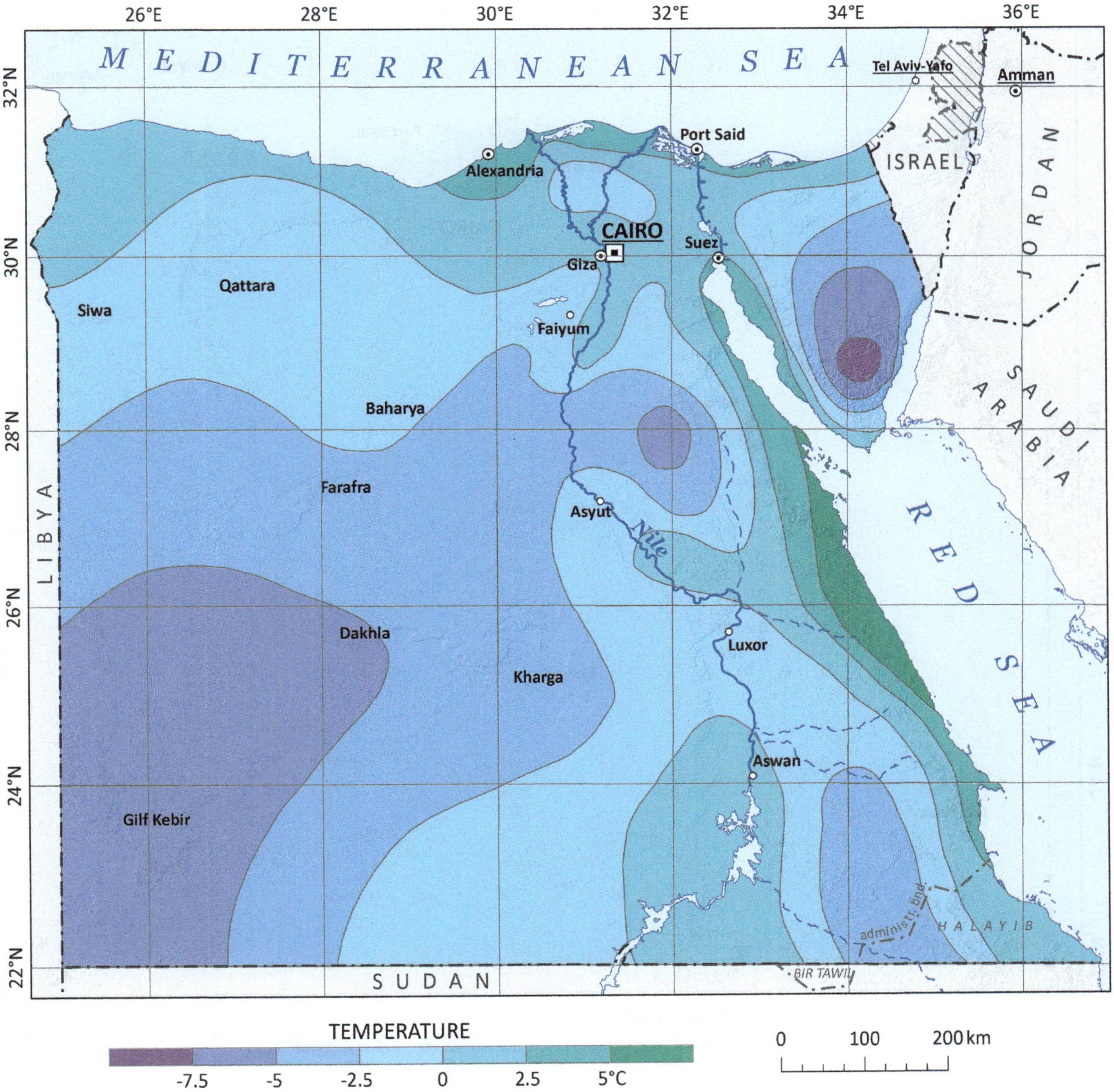

Fig. 3.3 Absolute minimum temperature (°C) in January (based on the Egyptian Meteorological Authority data)

In general, the number of rainy months decreases toward south and inland, a trend which is conformable with that of the index of aridity. Along the northern coast of the Mediterranean, the number of rainy months ranges between 10 and 11 at Sallum, Sidi Barrani, Matruh; and 8–9 for Dabaa, Dekheila, Alexandria, Balteem, Damiatta, Port Said, El Arish, and Rafah. In the Nile Delta, in the northern part of the Nile Valley and the adjacent Suez Canal region, the number of rainy months ranges between 6 and 8, decreasing to 3 months in Fayum and Beni Suef. From Minya at a latitude of approximately 28° N, all months are dry, except for a few locations that have only one rainy month, such as El-Minya, Mallawy, Bahariya, Hurghada, and Quseir. In the stations of Sinai, rainy months range between 6 and 7 for El-Kontella and Nekhel in central Sinai and are reduced to one month in the case of Sharm El-Sheikh, 3 months for Abu Redeis, 4 months for El-Qosiema, and 5 months for El-Tor. This picture of Sinai's rainy months is due to the effect of several local climatic and topographic factors.

The climatic records show an extraordinary phenomenon that happens over relatively long periods of time (10–30 years). This is the fall of unusually large quantities of

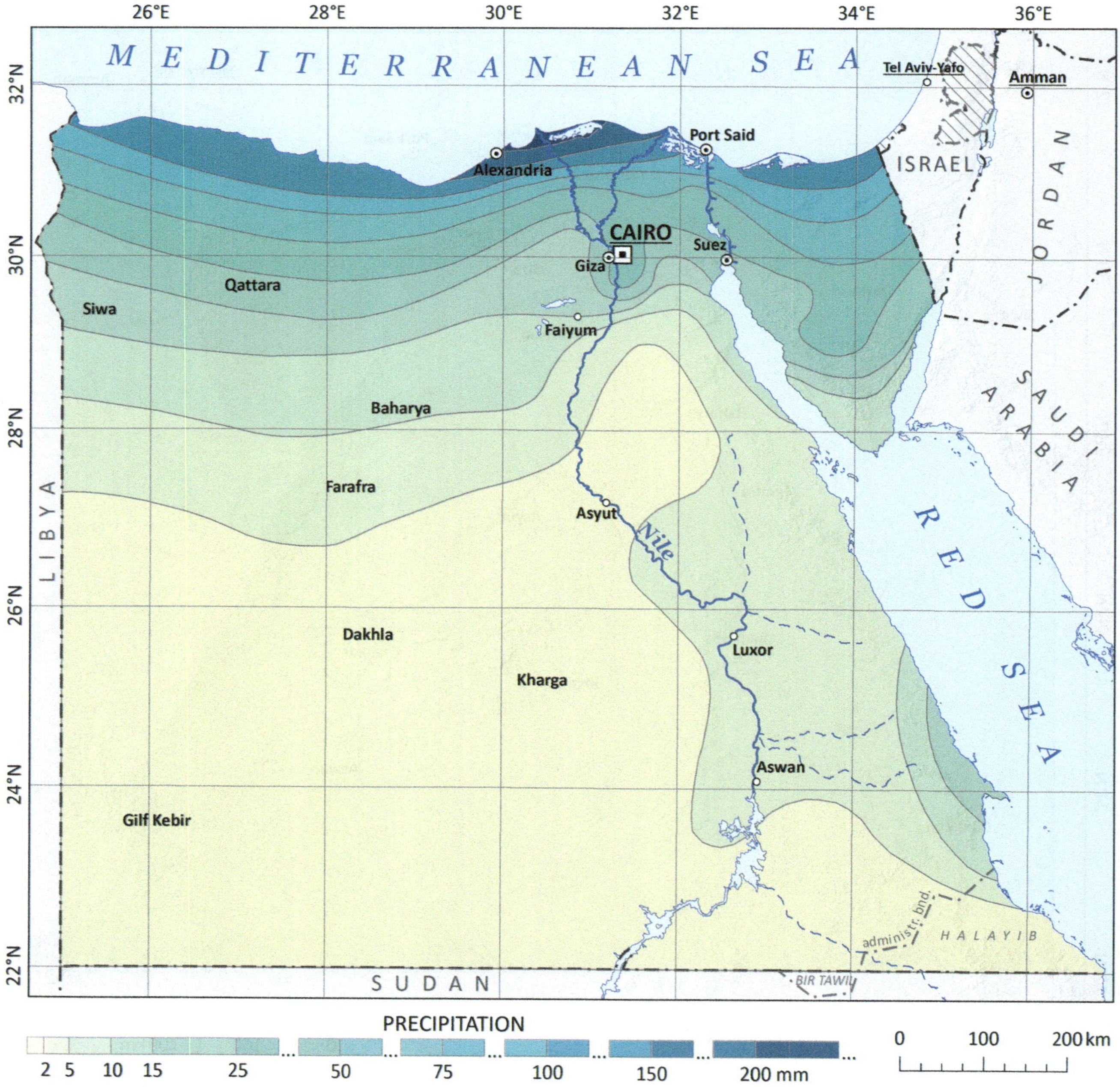

Fig. 3.4 Mean annual rainfall (mm) in Egypt (based on the Egyptian Meteorological Authority data)

rainfall in one day/hour, exceeding in most instances the annual mean of the station. Due to the location of Egypt at the southern margins of the Mediterranean Sea, it is affected by the Mediterranean climatic conditions. The most important of these are the winter cyclones moving mainly from the west to the east. These cyclones, while moving eastward, might shift southward with their cold fronts creating disturbed atmospheric conditions that prevail for a few days, and sometimes a few hours, leading to rainstorms in which rain falls in a relatively intense manner. These extraordinary conditions are greatly localized due to the nature of the intruded cyclones. During these storms, rain falls in large quantities for short periods of time. Examination of climatic records of Egypt revealed that nearly all stations north of latitude 28° N experienced such extraordinary atmospheric conditions. The largest recorded rainfall in Egypt up to 1975 was in Sallum in the extreme west of the country on the Mediterranean coast, where approximately 121 mm fell in one day (November 22, 1947). Lesser rainfall in one day was reported in other stations such as in Sidi Barrani (40–50 mm), in Matruh (55 mm in October, 53 mm in December, and 75.5 mm in November), Dabaa (49 mm in January), Dekhiela (63 mm in December), Alexandria (48 mm in January and 54 mm in December), Cairo (50 mm in December), El Arish (52 in

Table 3.2 Percentage of annual occurrences of wind according to direction

Station	North	NE	East	SE	South	SW	West	NW	Total	Calm
Matruh	17.6	6.7	6.4	5.7	6.1	12.5	17.4	25.2	97.6	2.4
Alexandria	21.1	9.6	6.0	4.9	6.0	7.3	10.6	32.2	97.6	2.4
Port Said	18.5	13.6	5.1	1.9	4.1	12.8	13.9	21.5	91.2	8.8
Al-Arish	16.9	6.8	2.5	5.3	23.1	12.6	12.6	16.4	96.1	3.9
Siwa	11.7	14.6	13.8	9.1	5.3	6.4	15.0	19.2	94.8	5.2
Bahariya	26.5	15.9	3.8	1.6	1.6	2.6	4.4	22.2	78.6	21.4
Kharga	68.9	10.0	1.6	1.4	3.0	0.9	1.4	7.0	92.1	7.9
Cairo	24.1	18.3	7.4	3.2	5.3	6.2	9.5	17.5	91.4	8.6
Minya	59.3	9.7	0.5	3.2	4.3	2.0	3.3	12.8	94.8	5.2
Asyut	30.8	4.0	1.3	2.3	3.5	4.1	10.8	32.2	88.8	11.2
Aswan	57.6	12.7	1.3	1.4	1.9	1.1	4.0	19.2	99.1	0.9
Al-Tor	14.3	4.7	1.2	1.8	1.4	0.9	13.3	52.4	90.1	9.9
Hurghada	24.2	4.7	1.7	2.6	1.5	0.8	15.0	46.8	97.2	2.8
Ras Banas	43.7	6.8	2.6	2.6	1.2	0.2	2.8	27.7	87.7	12.3
East Uwienat	39.3	40.2	2.3	1.4	0.7	1.3	1.0	9.7	97.7	2.3

Source El-Feqy (1999)

October), Suez(50 mm in October), Qena (55 mm in November), and Ras Banas (64 mm in November).

There have recently been two incidences of extraordinary climatic conditions in which large quantities of rainfall occurred during a short period of time over the whole of Egypt. The first one occurred on November 2, 1994, and covered most of Egypt, while the second occurred on October 2010 in Wadi El Arish Basin in North Sinai. In the first case, low pressure prevailed all over Egypt and heavy rain started to fall in the early hours of the day (El-Dessouky 1995). Recorded rainfall in Cairo reached 32 mm, while other stations recorded excessive rainfall but not as much as Cairo. Floods resulting from this excessive rainfall nearly covered the whole of Egypt but in varying degrees. The second case occurred in Wadi El Arish on January 17, 2010. On the evening of that day, 50 mm of rain fell on the northern part of Sinai in only 6 hours. The rain was localized in the central and northern parts of the catchment area of Wadi El Arish. The resulting floods caused severe damage to houses, cultivated land, and industrial establishments. The flowing water washed away trees, cars, trucks, roads, electric transmission towers, and water pipelines (Moawad 2013).

3.1.4 Wind

Wind regime is either unidirectional, bidirectional, or multidirectional. The first two regimes could blow from north, northwest, and northeast. Prevailing wind blows generally from northwest in the northern part (Alexandria) and Red Sea Coasts (Hurghada, Al-Tor), from north in the middle part (Minya, Kharga), and from northeast(NE) in the extreme south. Multidirectional regime prevails only in the NW (Matruh and Siwa) and NE (El Arish) (Table 3.2).

Local wind known as the Khamasin or "fifties," indicating that it lasts for fifty days, blows from the southwest and south during the months of February to May carrying sand and dust from the Western Desert. Wind energy is low to medium, according to the classification of wind energy. Winter and spring are the seasons of strong and gale winds in Lower Egypt, blowing from the west (Soliman 1972). The frequency of strong and gale winds is low and decreases considerably in the southern direction.

3.2 Past Climatic Conditions

"Present is the Key to the Past" (Hutton 1785). This fundamental concept is based upon the fact that "*The same physical processes and laws that operate today operated throughout geologic time, although not necessarily always with the same intensity as now*" (Thornbury 1954). Thus, **past climates** will be recalled depending upon this fundamental concept. Furthermore, the accuracy of the picture of past climates depends upon the data collected from the field, and analysis and interpretation of this data. The reconstruction of past climates is a difficult task and becomes especially so the further we go back in time, since the significant factors affecting and determining the climate are either totally unknown or reconstructed in a very general way. The general understanding of the past climate of Egypt has improved significantly during the last 50 years due to the extensive research carried out, and the advancement of science in general. This has provided new methods and techniques for collating and analyzing the material collected from the field, in addition to the interpretation of space images. Past climates of Egypt have not been the subject of previous work; therefore, the following section depends on relevant studies in which past climates were deduced from the analysis of the collected data. The analysis of past climates of Egypt will be divided into three sections according to geologic periods.

3.2.1 Climate During the Paleogene

The Paleogene is the first period in the Cenozoic era. It is divided into three epochs: Paleocene, Eocene, and Oligocene. All evidence found indicates that during the Paleogene period, Egypt enjoyed a tropical rainy climate. Rohlfs (1875) and Blankenhorn (1900) were probably the first who reported the presence of fluvial and karstic forms in Egypt. These forms indicate that a rainy climate prevailed during their formation. Also, Ball (1939) reported:

> The Oligocene deposits cover… and form a broad band extending from Cairo southwestward for more than 200 km. They overlie the Eocene limestone, as a rule unconformable, and consists of gravels, sands, and sandstones, usually devoid of fossils but containing in some places silicified wood, and at one locality (in the northern part of Fayum Depression) the remains of numerous large-animals (p.23).

Ball (1939) concluded from these characteristics that the deposits are of fluvial/estuarine origin. Moreover that the fossils found in them represent the remains of trees and animals that flourished on land during the Oligocene and that they were swept northward by rivers and became embedded in the deltaic and estuarine sediments formed where the rivers reached the sea. Although it is impossible to reconstruct the courses of Oligocene rivers, evidence points to the presence of a great river that flowed in a more or less northern direction to enter the sea immediately to the north of what is now the Fayum. Other Oligocene rivers flowed into the Red Sea Depression, while numerous others followed more or less in the present-day courses of the principal eastward-draining wadis (Ball 1939).

In 1976, while researching the evolution of Eocene-Miocene sedimentation patterns in parts of northern Egypt, Salem (1976) illustrated that the sediments of this region are deltaic/estuarine forming a huge delta that started to develop in the Late Eocene continuing its development to Lower Miocene, occupying the area to the north of Fayum and Qattara Depression. This indicates that the rainy tropical climate prevailing in Egypt during the Paleogene started at least since Late Eocene and continued without any breaks. Abdel Rahman and El-Baz (1979) and Said (1990a) reached similar conclusions that the Oligocene deposits which contain mixed fluvio-marine fossils, huge silicified tree trunks, forming in some places petrified forests, and remains of various land animals, crocodiles, tortoises and turtles, were laid down by rivers of considerable size. This conclusion conforms to that previously reached by Salem in 1976. Said (1962) found that the southernmost extension of the Oligocene sediments includes pebbles of quartz, chert, flint, and quartzite, and that these sediments were derived from the Nubia Sandstone and Eocene rocks toward the south. Shukri and Ayouti (1954) have noted that sillimanite is abundant in the Oligocene sediments in the east of Cairo, a mineral not present in the west of Cairo. According to Shukri and Ayouti (1954), this mineral originates from the metamorphic rocks to the west of the Gulf of Suez. It can be inferred from this research that the drainage system of the Neogene had tributaries extending southward and westward. With such a large delta and tributaries and the fossil content of the sediments, the conclusion that Egypt enjoyed a rainy tropical climate during the Paleogene is confirmed.

3.2.2 Climate During the Neogene

The climate during the Neogene period (Miocene and Pliocene) does not differ much from that of the Paleogene. This is because the large drainage system which was dominating the scene during the Paleogene extended geologically to Lower Miocene. Said (1962, 1990a) assigned an Early Miocene age for Moghra Formation in northwest Egypt. From the sedimentologic characteristics, Said (1990a) concluded that the climate was certainly tropical to subtropical. Said (1990a) also found that the in situ biota points to a tropical climate and African affinities.

Field work carried out in 2005 in the Sheikh Abdallah area (between Bahariya and Farafra) led to the discovery of evidence indicating that the Western Desert was covered by woodland during Late Miocene (Pickford et al. 2006). Fossiliferous cave breccia at Sheikh Abdallah in the Western Desert of Egypt has yielded 11–10 Ma micro-vertebrate fauna, which contains *Galagidae, Microchiroptera, Macroscelididae, Soricidae, Erinaceidae, and Rodentia*. The locality also yielded remains of frogs, snakes, lizards, and birds. The fauna indicates a mean annual rainfall in excess of 500 mm and perhaps as much as 1200 mm. The development of the Nile Valley in Egypt before the end of the Miocene by about 6-7 Ma indicates that a rainy (probably also tropical) climate prevailed during this period. The Red Sea Mountains, reaching their maximum height, and the fall of the level of the Mediterranean Sea due to the closure of the Strait of Gibraltar, were two very important factors that lead to the increase in rates of vertical erosion of the Nile and the development of the Eonile Canyon and its tributaries in the Red Sea Mountains.

At the beginning of the Pliocene, the sea level of the Mediterranean rose again, allowing enough water to invade the low areas in northern Egypt and the Nile Valley up to Aswan in the far south, converting it into a Marine Gulf (Said 1990a). So far, evidence indicates that it is only in the Late Pliocene that Egypt witnessed the onset of more humid conditions. These conditions converted the Marine Gulf of the Nile into a veritable river channel, named Paleonile by Said (1981). The climate became arid during the early and Middle Pliocene since no evidence was found to prove the contrary.

Buried wadis in the southern plain of the Western Desert are another fluvial forms indicating the prevalence of a nationwide humid climate. This hidden drainage system below the Selima Sand Sheet was discovered by radar imaging in 1981(SIR/A), 1984 (SIR/B), and 1994 (SIR/C) during flights of the US space shuttles. This palaeo-drainage includes wide wadis (10-30 km), a 100-km-long wadis and narrow anastomizing channels that developed at the floor of large wadis (McCauley et al. 1986). Regional reconstruction of these previously unknown palaeo-drainage systems was the subject of several studies. The conclusion reached by most of these studies is that the large wadis are relics of a Cenozoic system that drained the Eastern Sahara and that its narrow channels developed during the Quaternary pluvial periods (McCauley et al. 1982). Also, previous studies postulated that the Red Sea Mountains were the source area of the large wadis long before the development of the Nile Valley (McCauley et al. 1986). It can therefore be said that the buried wadis in southern Egypt reveal new evidence indicating that the climate was rainy during the Cenozoic not only in the northern parts, but also in the southern region, and consequently all over Egypt.

3.2.3 Climate During the Quaternary

The advent of the Pleistocene (from ~2.84 Ma) brought to Egypt a pattern of aridity that sets the tone of the climate prevailing in Egypt, with minor fluctuations throughout the Pleistocene (Said 1990b). During the earliest Pleistocene, the Paleonile stopped flowing in the Nile Valley in Egypt, indicating a long period of aridity (~1 million years) in which Egypt was converted into a desert (Said 1990b). It is expected that during this long arid episode, wind was active and modeled the fluvial sediments that were deposited mainly in basins with thick sections such as the basin of the Great Sand Sea. This long and arid period was followed by a short pluvial period (Idfu Pluvial) during which a highly competent river (the Protonile) flowed in the Nile Valley according to Said (1990b). This suggestion by Said depends on the presence of fluvial gravels in the Nile Valley. This gravel is composed of quartz and quartzite cobbles and gravel that were derived from local sources. At a later time, by the end of Early Pleistocene, a short pluvial occurred, during which the conglomerate of Armant Formation was deposited. This Armant Pluvial is separated from the Protonile Pluvial by a short arid phase (Said 1990b).

The Protonile is succeeded by another two rivers in the Nile Valley: the Prenile and the Neonile (Said 1990b). They are separated from each other by unconformity and a long recession. The mineral composition of the Prenile sediments indicates that the Egyptian Nile was connected for the first time with the Ethiopian highlands (~700,000 years ago) across the Nubian region in northern Sudan and southern Egypt. The Prenile is separated by a relatively long period from the Neonile, which started to flow approximately 200,000 years ago, because the connection with Ethiopian sources was cut off. During this period, the climate changed several times from arid to pluvial during which rainfall was diminishing gradually until it reached the present arid conditions ~5000 years ago.

According to Said (1990b), the total number of major pluvials recognized during the Quaternary was seven. The oldest two pluvials, the Idfuan and the Armatian, are of the early Quaternary period and are not succeeded/preceded by inter-pluvial deposits. The other five more recent pluvials are as follows: Abbassian I, Abassian II, Saharan I, Saharan II, and Nabatian. The Abbassian I Pluvial occurred during Late Middle Pleistocene and is represented by thick locally derived gravel, which overlies the eroded surface of the Prenile and earlier deposits of the Nile Valley. It is concluded from the characteristics of sediments deposited during the Abbassian I Pluvial that this period was characterized by intense rain (Said 1990b). The Abbassian II Pluvial is represented by a gravel bed in the Nile Valley and was laid down after the first Neonile sediments reached Egypt (~400,000 years ago). The gravel of this Pluvial is rich in Acheulian implements.

Acheulian sites were found in the Western Desert in various localities such as the Gilf Kebir (Meyers 1939; McHugh 1982), spring deposits in Kharga (Caton-Thompson 1952), in Bir Sahara-Bir Tarfawi area (Wendorf and Schild 1980), and in Kharga, Kurkur, and Dongul (Hamdan 1987; Wendorf and Schild 1993). After a relatively short period of time, the two short pluvials, named Saharan I and Saharan II, occurred. They were separated from each other in the Nile Valley by a number of thin silt beds and dune sands (Said 1990b). Sediments of Saharan Pluvials contain tools of Mousterian and Aterian cultures. In the Western Desert, previous work reported the presence of sediments of fluvial origin in various localities, such as Bir Sahara-Tarfawi area (Wendorf and Schild 1980), the foot slopes of Abu Tartur plateau (Embabi 1967; Donner et al. 2015), and Kharga (Caton-Thompson 1952), indicating that these two Pluvials occurred in the whole country.

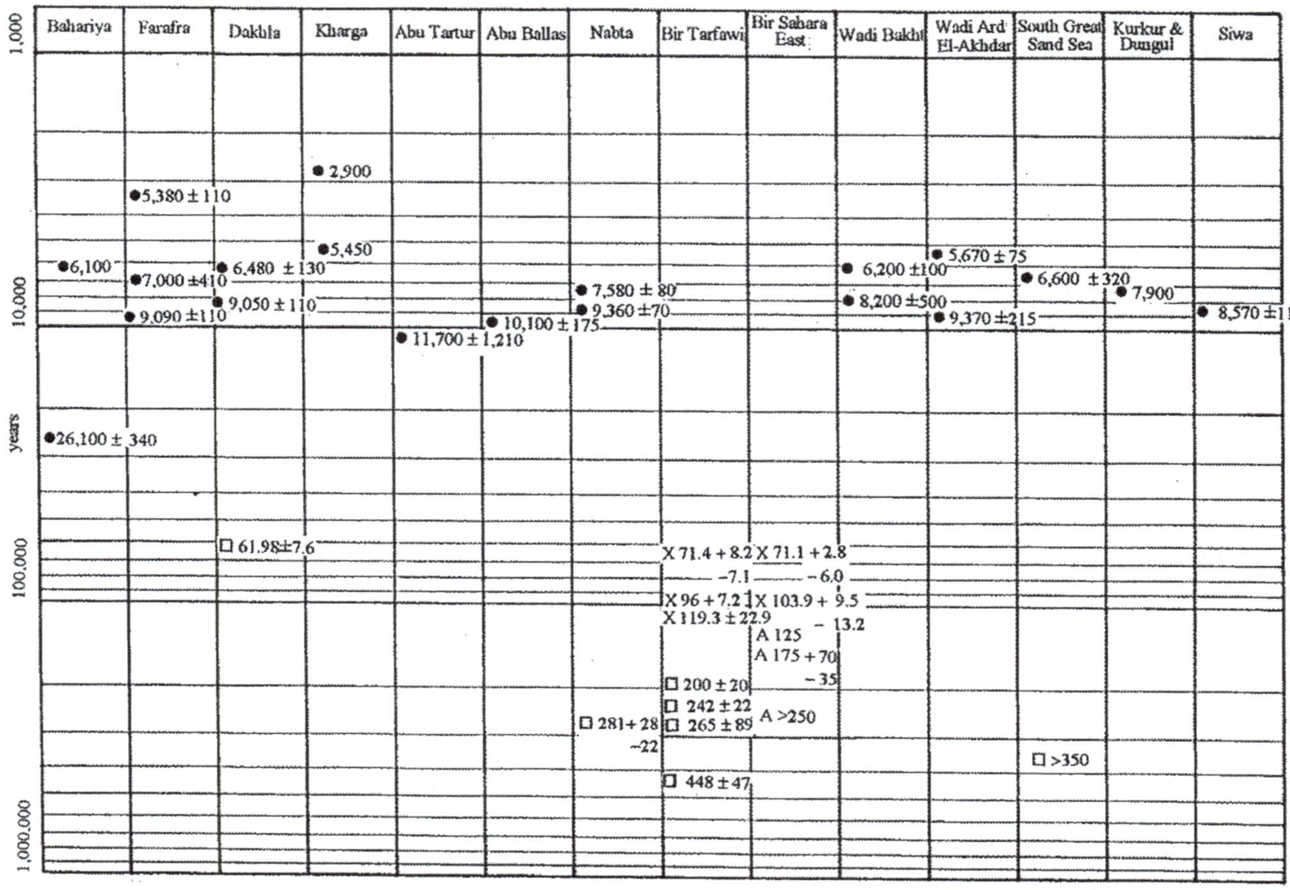

Fig. 3.5 Relative ages of samples from the playas of the Western Desert of Egypt. (*Source 1.* Bahariya: Barich et al. 1991; Pachur and Roper 1984; *2.* Farafra: Barich et al. 1991; Donner et al. 1999; *3.* Dakhla: Brookes 1994; *4.* Kharga: Banks 1984; *5.* Abu Tartur: Pachur and Roper 1984; *6.* Abu Ballas: Pachur and Roper 1984; *7.* Wendorf et al. 1977; Szabo et al. 1995; *8.* Bir Tarfawi: Wendorf et al. 1993; *9.* Bir Sahara East, Wendorf et al. 1993; *10.* Wadi Bakht: Kroplin 1987; *11.* Wadi Ard El-Akhdar: Kroplin 1987; *12.* South of the Great Sand Sea: Haynes 1982; Maxwell 1997; *13.* Kurkur and Dungul: Banks 1984; *14.* Siwa: Hassan 1978); *15.* Abu El Egl (Abu Tartur): Donner et al. 2015. (•) Radiocarbon (BP); (□) U-Series (ka); (X) Optical (ka); (A) Amino-acid racemization

The sediments of the most recent pluvial, the Nabatian, are well recorded in previous works, mainly on the playas of the Western Desert, and are summarized in Fig. 3.5 (Embabi 2004). The chart shows that, apart from those belonging to the Pleistocene, those pluvials fall in the interval 11,000–5000 BP. As can also be seen from the chart, the Nabatian playas were the subject of many studies; therefore, the picture that is drawn of the climate during this period of development is the most accurate among other Quaternary pluvials. One of the most interesting findings is that the climate was semiarid with seasonal rainfall (Embabi 2004).

3.3 Conclusion

Present-day hyper-arid to arid climate prevailed Egypt since Mid-Holocene only. It is characterized by cold to warm rainy winters and hot dry summers. It is also characterized by the occurrence of extraordinary conditions during which temperature rises up to 40–50°C in summer, and decreases in winter to a few degrees above or below zero. These heat or cold waves prevail only for a few days. Normally, rainwater is absorbed by soils or evaporates in very short time. On the contrary, during extraordinary conditions, rain may increase resulting in floods that flow for a short time and distance as well, which might cause damage for some human belongings.

The effect of present arid conditions on the geomorphology of Egypt cannot be ignored. The most significant effect is that of wind action and water action. Wind action may be direct or indirect. Direct effect lies in removal and deposition of loose fine materials, mainly on sand dunes and beaches. Indirect effect is in the initiation of wave action which could lead to coastal erosion in some localities. Water action is effective only during rainstorms which occur in various areas during winter. On the whole, the effect of flash floods is very limited and localized.

Inferences made from inherited landforms and landscapes showed that past wet climates prevailed Egypt during millions of years. This is why past climates were more significant in modeling the landscapes of the Egyptian landmass than the present arid conditions.

The climate of Egypt was humid between Late Eocene and Late Miocene. Since Late Eocene, the whole period of the Oligocene, early, middle, and until Late Miocene (~25 Ma), all available evidences (geomorphological, sedimentological, biological, etc.) indicated that the climate was humid without any arid periods in between. The animal and plant remains indicated that a tropical climate prevailed during this long and humid period. It is the longest pluvial period in climatic history of Egypt, during which major landforms of Egypt, such as the Nile Valley and the main drainage nets of the Red Sea and Sinai Mountains and of the Gilf Kebir, were developed. Furthermore, the first generation of karst forms, especially caves, developed during this period. Most karst forms have degraded by time, and their remains are exposed on the surfaces of the limestone plateaus of the Western and Eastern Deserts.

References

Abdel-Rahman M, El-Baz F (1979) Detection of a probable ancestral delta of the Nile River. In: F El-Baz and D M Warner D (eds) Appolo-Soyuz test project, vol. II, Earth observations and photography, NASA, Washington DC, pp 511–520

Ball J (1939) Contribution to the geography of Egypt. Ministry of Finance Egypt, Government Press, Survey and Mines Department, Cairo

Banks KM (1984) Climates, cultures and cattle: the Holocene Archeology of the Eastern Sahara, Department of Archeology, Southerm Methodist University, Dallas

Barich BE, Hassan F, Mahmoud AA (1991) From settlement to site: formation and transformation of archaeological traces. Scienze Dell' Antichita, stroia Archaeologia Anthropologia 5:33–62

Blankenhorn M (1900) Neues zur Geologie and palaeontologie Aegyptens. II: Das palaeogen (Eozaen und Oligozaen). Z. Deutsch. Geol. Ges. 52:403–479

Brookes IA (1994) Geomorphology and quaternary geolgy of the Dakhla Oasis region, Egypt. Quat. Sci. Rev. 12:529–552

Caton-Thompson G (1952) Kharga Oasis in prehistory. Athelon Press, London

Donner JM, Ashour MM, Embabi NS, Siiriainen (1999) The quatewrnary geology of a playa in Farafra, the Western Desert of Egypt. Ann. Acad. Sci. Fennicae, Geologica Geographica 160:49–112

Donner J, Ashour MM, Brook GA, Embabi NS (2015) The quaternary history of the Western Desert as recorded in Abu El-Egl Playa. Bull. Soc. Géog. d'Égypte 88:1–18

El-Dessouki T (1995) Climatic conditions resulting the floods of November 1994. In Proceeding of the symposium on water resources in the Arab countries (in Arabic) I: 35–60. The Egyptian Geographical Society, Cairo, 26–28 November 1994

El-Feqy A (1999) Wind in Egypt. Dissertation (in Arabic). Ain Shams University, Cairo

Embabi NS (1967) A morphological study of the Kharga Oases Depression, the Western Desert, Egypt. Dissertation, Bristol Unversity, UK

Embabi N (2004) Geomorpholgy of Egypt, vol. I, the Nile Valley and the Western Desert. Special Publication, The Egyptian Geographical Society, Cairo

Hamdan MA (1987) Geomorphology and quaternary geology of Umm El-Dabadib Area, Kharga Oasis, Dissertation Cairo University

Hassan FA (1978) Archeological explorations of the Siwa Oasis Region, Egypt. Current Anthropology 19:146–148

Haynes CV (1982) The great sand sea and selima sand sheet, Eastern Sahara, geochronology of desertification. Sci. 217:629–633

Kehl H and Bornkamm X (1993) Landscape ecology and vegetation units of the Western Desert of Egypt In: Meissner B, Wycisk P. (eds), Geopotential ecology analysis of a Desert Region, Catena Supplement 26: 155–178

Kroplin S (1987) Palaeoclimates evidence from Early to Mid-Holocene playas in the Gilf Kebir, southwest Egypt. Palaeoecology of Africa 18:189–208

McCauley J, Schaber C, Breed M, Grolier C, Haynes V, Issawi B, Elachi C, Blom R (1982) Subsurface Valleys and geoarchaeology of the Eastern Sahara revealed by Shuttle Radar. Science 218:1004–1020

McCauley J, Breed M, Schaber C, Haynes V, Issawi B, El-Kilani A (1986) Palaeodrainages of the Eastern Sahara, The radar rivers revisited (SIR-A/B Implications for a Mid-Tertiary Trans-African Drainage System). IEE Trans. Geo. Sci. Remote Sensing GE 24:624–648

McHugh W (1982) Archaeological investigations in the Gilf Kebir and the Abu Hussein dune field. A basis for comparison with Mars. In: Maxwell T (ed) El-Baz F. Desert Landforms of SW Egypt, Washington DC, pp 301–344

Meteorological Authority (1979) Climatological Normals of the Arab Republic for Egypt up to 1975. Cairo

Meyers OH (1939) The Sir Robert Mond Expedition of the Egypt Exploration Society. Geogr J 93:287–291

Moawad M (2013) Analysis of the flash flood occurred on 18 January 2010 in wadi El Arish, Egypt (a case study), Geomatics. Nat. Hazards Risk. 4(3):254–274

Pachur H-J, Roper HP (1984) The Libyan Western Desert and Northern Sudan during the Late Pleistocene and Holocene. In: E Kelitzsch, R Said, E Shrank (ed) Research in Egypt and Sudan. Rei A/Band 50, Results of the Spcial Reseach Project in Arid Areas, Period 1981–1984, Berlin, pp 249–280

Pickford M, Wanasb H, Soliman H (2006) Indications for a humid climate in the Western Desert of Egypt 11–10 Myr ago: evidence from Galagidae (Primates, Mammalia). Palaeoecology 5:935–943

Rohlfs G (1875) Drei Monate in der libyschen Wüste. Kassel (reprint Köln 1996)

Said R (1962) The geology of Egypt. Elsevier Publishing Co Amsterdam

Said R (1981) Geological evolution of the River Nile. Springer-Verlag, Vew York

Said R (1990a) Cenozoic. In: Said R (ed) The Geology of Egypt. Balkema, Rotterdam, pp 451–486

Said R (1990b) Quaternary. In: Said R (ed) The Geology of Egypt. Balkema, Rotterdam, pp 487–507

Salem R (1976) Evolution of Eocene-Miocene sedimentation patterns in parts of northern Egypt. Amer Assoc Petrol Geol Bull 60:34–64

Shukry N, Ayouti M (1954) The Mineralogy of Eocene and later sediments in the Anqabiya area, Cairo-Suez District. Bull. Faculty Sci Cairo Uni Cairo 32:47–61

Soliman K (1972) The climate of the United Arab Republic. In Griffiths J, Soliman K (eds) The Northern Desert (Sahara), Climates of Africa. Wold Survey of Climatology, Elsevier Publ. Co. Amsterdam, 10:79–92

Szabo BJ, Haynes CV, Maxwell TA (1995) Ages of quaternary pluvials episodes determined by Uranium-Series and radiocarbon dating of lacustrine deposits of Eastern Sahara, Palaeogeogr Palaeoclimatol Palaeoecol 113:227–242

Thornbury W D (1954) Principles of Geomorphology. Wiley, New York

Wendorf F, Schild R (1980) Palaeoenvironment of the Western Desert of Egypt. In: Wendorf F, Schild R (eds) The Prehistory of the Eastern Sahara. Academic Press, New York, pp 224–241

Wendorf F, Schild R, Close AE and Associates (1993) Egypt during the last interglacial: The Middle Palaeolithic of BirTarfawi and Bir Sahara East. Plenium Press, New York

Wendorf F, Close A, Schild R, Said R, Haynes C V, Gautier A, Hadidi N (1977) Late Pleistocene and recent climatic changes in the Egyptian Sahara. Geogr J 143:211–234

UNESCO (1977) Map of the world distribution of arid regions: Explanatory note, M A B Technical Notes 7. Paris

Part II
The Nile

4 Remarkable Events in the Life of the River Nile

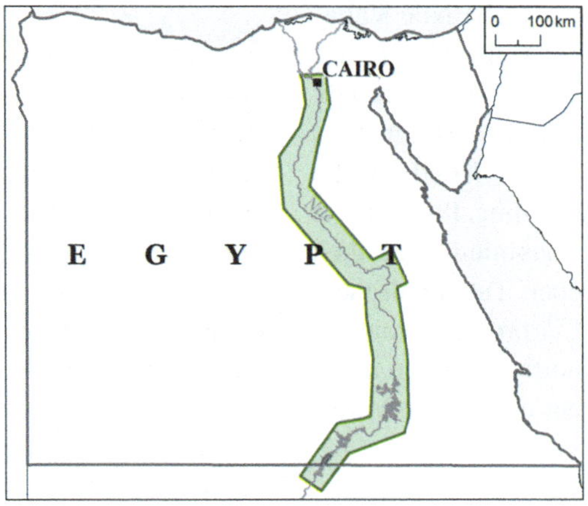

Abstract

The present Nile represents the final stage of a series of rivers that have started millions of years ago. The Nile and its valley originated in Late Miocene times, and since then it has passed through several stages. Some of these stages are significant in determining major changes in its life. Such changes were due to remarkable regional geologic events and can be described as "dramatic events." These events are: the drying up of the Mediterranean Sea, known as the Messinian Crisis in late Miocene; the submergence of the Nile Canyon due to sea level rise during the Pliocene, resulting in an estuary that extended southward to Aswan; and the arrival of African waters in the Middle Pleistocene. The floodplain of the Nile Valley is as recent as the present river, developing only when African waters loaded with silt had arrived in Egypt during late Pleistocene/Early Holocene.

Keywords

The Nile • Floodplain • The Nile Canyon • The Nile Gulf • The Eonile • The Prenile • The Neonile

4.1 Introduction

Previous studies concluded that several drainage systems have developed on the surface of the Egyptian landmass since late Eocene times (Ball 1939; Salem 1976; Abdel Rahman and El-Baz 1979; Bown and Kraus 1988; Said 1990; Issawi and McCauley 1993). Of these former drainage systems, the one debouching in northern Fayum is the most significant because it continued from late Eocene to Middle Miocene ($\sim$25 Ma) according to the age of its deltaic sediments in northern Egypt (Salem 1976). Its huge delta which appears on space images is evidence that this system lasted for such a long period. On the Landsat-TM space image, no comparable form of alluvial origin could have been detected. It also indicates that no significant tectonics have occurred in the Western Desert during this long period.

Sources of this late Eocene–Oligocene–Middle Miocene fluvial system were in Egypt, mostly in the eastern and southern parts the Western Desert. In these source areas, consolidated fluvial sand and gravel could be seen as separate mounds and inverted wadis, whose age was estimated to

N.S. Embabi, *Landscapes and Landforms of Egypt*, World Geomorphological Landscapes,
https://doi.org/10.1007/978-3-319-65661-8_4

be Oligocene (Said 1990). Tectonics, mainly faulting, occurred in Late Miocene resulting in a change in the direction of slopes in various areas, of which those of the Western Desert, and along the Hinge Zone and the path of the Nile Valley are the most significant to the development of the Nile Valley. Geologic sections across the Nile Valley show the presence of several faults on both sides of the Valley. Also geological sections across the Nile Delta indicated that there was a tectonic embayment delimited by faults from the east and west, called "**North Delta Embayment**" by Said (1981). In the meantime, the rising of the Red Sea Mountains reached its maximum and their eastern and western slopes became steeper. Taking all these geologic changes into consideration, it is projected that the water flow has been diverted from the north of Fayum to the east, where it concentrated along the path of the Nile Valley debouching in the Delta Embayment.

4.2 Geologic and Geomorphologic Development of the Nile

The flow of water along a path controlled by tectonics represents the first step in the life of the Nile and its Valley. But the real birth of the Nile occurred when the level of the Mediterranean Sea fell several hundred meters due to the closure of **Gibraltar Strait**, cutting off the recharge of the Mediterranean with the Atlantic water, and drying up the Sea. This event has been called the "**Messinian Crisis**." The drop in sea level, which is in fact the base level not only for the Nile but also for all streams ending in the Mediterranean Sea, resulted in an increase in the energy of water flow and its erosive power. Since the main sources of the Nile water have been in the Red Sea Mountains, and since these mountains have been rising up, the growth of the difference between the source and base level accelerated resulting also in an increase in the energy of the flow and its erosive power. Therefore, the river became very powerful, deepening its channel at a high rate, and created by the end of the Miocene a very deep canyon. This canyon was called "**The Nile Canyon**" and its stream "**The Eonile**" (Said 1981). It can be said that the Messinian Crisis which occurred during Late Miocene (7–6 million years) represents the first among three remarkable events in the development of the Nile.

The second major event is the rise of sea level during Early Pliocene, resulting in the drowning of the Nile Canyon forming a long Marine Gulf that extended to the latitude of Aswan. This Gulf is called the **Pliocene Gulf of the Nile** (Ball 1939).

The third remarkable event is the arrival of the African waters in the Middle Pleistocene, after which African waters became the sole source of water flow in the Nile. The river which carried out the African water is called by Said (1981) the **Prenile**.

4.3 Remarkable Event I: The Messinian Crisis

The Egyptian Nile owes its origin to a major event, which happened due to the closure of Gibraltar Strait during Late Miocene (7–6 million years ago), leading to the discontinuation of water supply from the Atlantic Ocean, and the entailing conversion of the Mediterranean Sea into a great lake(s). This event is known as "The Messinian Crisis" (Said 1981, 1993). It has become known only recently, when the research work of the Deep Sea Drilling Project (an international program for the study of the ocean floor) revealed two geological facts. The first is that there are no marine Late Miocene sediments in Egypt (Said 1981); the second fact is the widespread occurrences of thick evaporites of halite, gypsum, anhydrite, and dolomite beneath the sea floor proved to be of Late Miocene age (Hsu et al. 1973). These evaporites were interpreted to be laid down in a series of shallow salt lakes, representing the relics of a desiccated sea in deep ($\geq$3000 m) dry basins.

Lying in a region of high evaporation rate, Said (1993) speculated that the collected water vapor resulting from the drying of the Mediterranean moved to the land by the northern winds causing heavy rain all the year round along the high and newly elevated mountains of the Eastern Desert. The rivers fed by these rains occupied deeply incised wadis in these mountains, relatively higher than they are nowadays. These streams collected water flowing from the mountains, and supplied the trunk river of the ancestral Nile called by Said (1981, 1993) the "**Eonile**," which occupied a path—tectonically controlled—in the low-lying area between the Eastern Desert and Western Desert. As the level of the Mediterranean Sea fell down, the Nile bed deepened to respond to the new base level, starting from the Hinge Zone which were at present-day Cairo site, resulting in the development of an erosional deep canyon the bottom of which reached depths well below the current sea level (Fig. 4.1). Available data shows that the Eonile had a depth of 170 m at Aswan, 800 m at Asyut, 2500 m north of Cairo, and >4000 m below sea level in the northern fringes of the Nile Delta. From the total load carried out by the Eonile (70,000 km^3 in less than three million years, estimated to be about 20% of the total amount of sediments in the Delta Cone), Said (1981) concluded that the Eonile must have been an extremely vigorous river. One of the most significant consequences of the development of the Nile Canyon is that it divided Egypt into the two distinctive geomorphic regions of the present-day landscape: the Western Desert and the Eastern Desert.

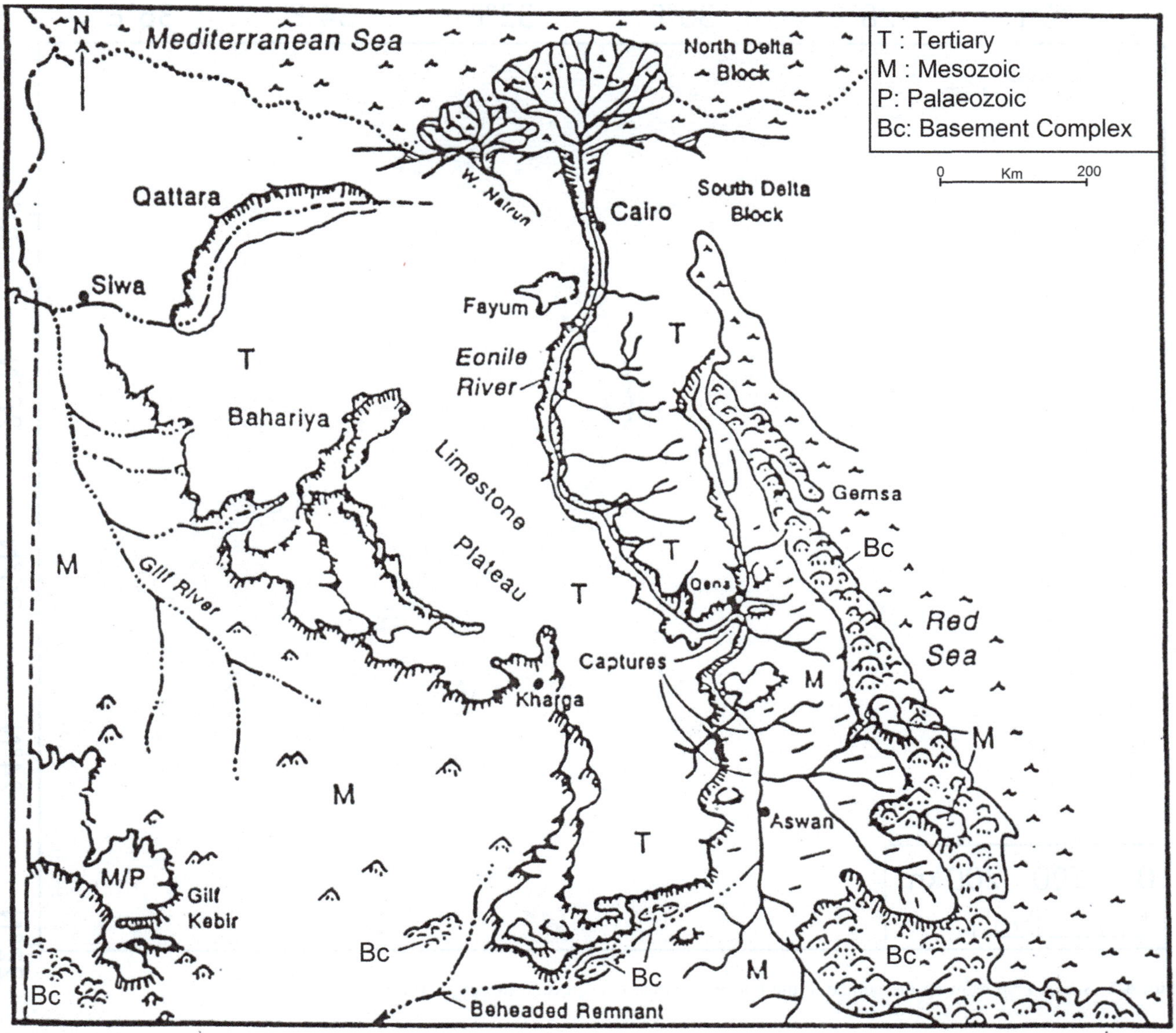

Fig. 4.1 Sketch map showing drainage lines and the Eonile Canyon in Late Miocene. (*Source* Issawi and McCauley 1993)

Comparing the morphological characteristics of the Eonile Canyon with those of the Grand Canyon of Colorado River, Arizona, Said (1993) found that although the Eonile Canyon seems to have been longer and deeper, it had the same width as the Colorado Canyon. Both canyons cross large stretches of desert land but receive sufficient waters from sources beyond to carry their load through successfully. Moreover, both carry equal loads of sediments in the range of 300,000 tons/day.

4.4 Remarkable Event II: The Pliocene Gulf

This event occurred with advent of the Pliocene, some 5.3 million years ago, when the **Eonile Canyon** was transgressed by the advancing waters of the Mediterranean Sea, resulting in the creation of a narrow and long Gulf/estuary that extended inland to the latitude of Aswan (Fig. 4.2) and attaining a level of 180 m a.s.l. by Middle Pliocene (Sandford and Arkel 1929; Ball 1939; Said 1981, 1993). Also, the

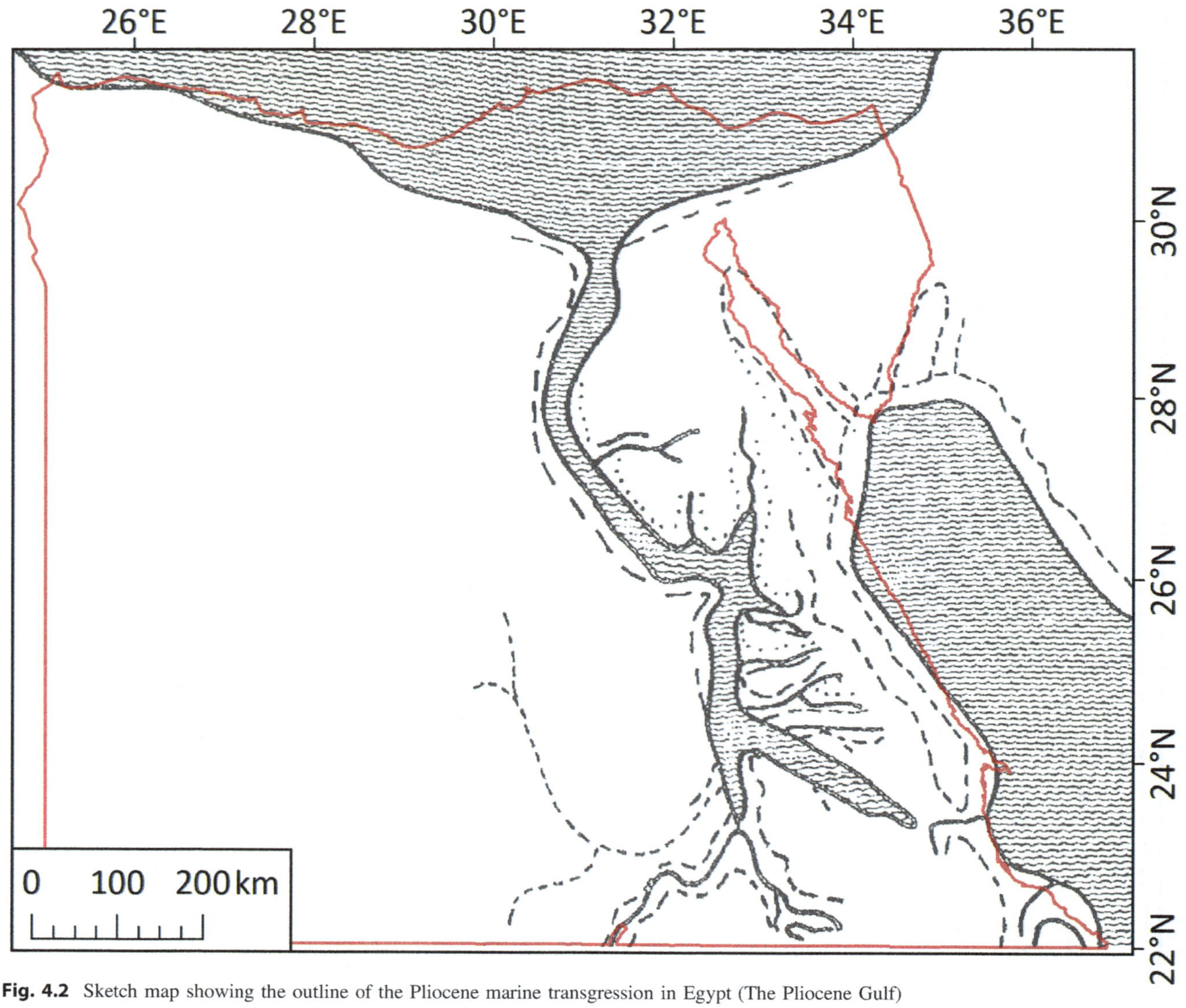

Fig. 4.2 Sketch map showing the outline of the Pliocene marine transgression in Egypt (The Pliocene Gulf)

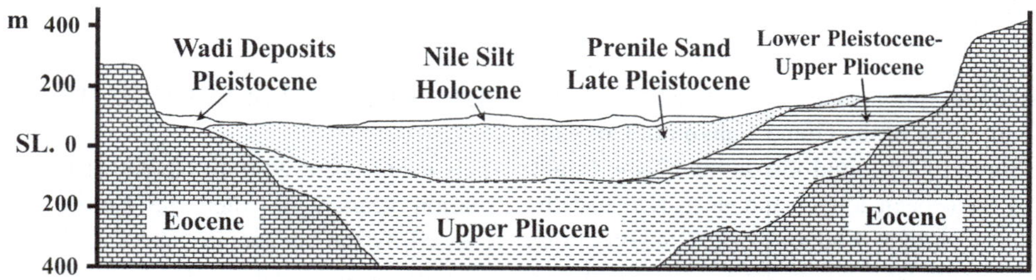

Fig. 4.3 Idealized cross section for the Nile Valley, showing the Eonile Canyon, the Pliocene infilling sediments, and the Nile silt. (*Source* Modified from Fig. 1, Said 1981)

Marine Gulf extended into the lower reaches of some wadis, such as Wadi Qena, in the Eastern Desert, which were captured by the Eonile forming arms, the longest of which was about 30 km (Little 1935–1936). The average width of the **Pliocene Gulf** was about 12 km, except where the Gulf had extended into the mouths of the tributary wadis. The

Gulf was filled by gravels and sands brought into it by lateral tributaries of the Eonile and the downwash from its sides. With this infilling and the rise of the landmass, the sea retreated rapidly from the Gulf by the end of the Pliocene.

The life of the Nile has witnessed three significant changes since the beginning of the Pliocene Marine Gulf. The first is the disappearance of the Eonile River after the Gulf had been filled with sea waters. This is because the Gulf completely submerged the whole Eonile River from Aswan to the Mediterranean and leaving no section saved from drowning.

The second change experienced by the Nile is that huge amounts of marine and continental sediments were laid down and covered the canyon bed and the lower sections of its sides (Fig. 4.3). This event led to two significant changes in the morphology of the Nile Canyon. The first is that the bed of the valley became higher and wider and almost flat due to the deposition of sediments in a marine environment. The other change is that the lower section of the two sides of the Canyon disappeared, and only the upper parts of the sides have been exposed to subaerial changes (Fig. 4.3). Therefore, what is seen nowadays in the field is not the whole wadi sides, but only the upper sections.

The third significant change in the life of the Nile is the development of a new river on the surface of the Pliocene sediments (Ball 1939). It is the second river after the Eonile, which developed in the same valley but in a different environment, and is called "**Paleonile**" by Said (1981, 1993). Although the sources of this river are not known with certainty, Ball (1939) and Said (1993) reached the same conclusion about the source of this new river. Both of them believe that the Red Sea Mountains have been the most probable source of this river. Ball (1939) relying on geomorphologic evidence speculated that the river started to erode its channel in the Pliocene sediments soon with withdrawal of the sea from the Gulf, meanwhile the Red Sea Mountains were thrust up to greater heights, enormous amount of denudation occurred by rain and tributaries cutting the mountains and feeding the "**Paleonile**." Said (1993) depending on paleontological evidences, concluded that the absence of freshwater faunal assemblage belonging to Central Africa in the Paleonile sediments suggests that the river must have had its sources in Egypt, most probably in the highlands of the Eastern Desert.

4.5 Remarkable Event III: The African Prenile

This event occurred in the Middle Pleistocene, some 800–700 ka years ago, when a mighty river, called by Said (1981, 1993) "**the Prenile**" reached Egypt. This river drew its waters from the Ethiopian highlands, when the Atbara and possibly the Blue Nile were able to cut across the Nubian swell, pushing their way to Egypt. It is the first river to develop a connection with African sources. Its sediments are composed of coarse sands, intercalated at the top with dune sands indicating that arid conditions must have prevailed during later phases of the river's history. These sediments can be seen along the banks of the Nile Valley and the Delta margins. In the Nile Delta, the Prenile sands are exposed in many localities as low-lying hills above the modern cultivated silt layer and were first named "**Turtle-backs**" by Fourteau (1915)—since their slopes are convex similar to those of the turtle-backs (sea Chap. 6, Sect. 6.4 for a more detailed description of this microform). The Prenile sediments are also present in the Nile Delta and the Valley, in the subsurface below the modern Nile silt layer. These coarse sands of the Prenile form an underground water reservoir holding the Nile waters, which infiltrate from irrigation canals and the Nile channel itself. Also, the Prenile sediments provide one of the main components used in building and construction nationwide. Sand quarries could be seen dotting the Nile Valley sides and the margins of the Delta.

On average, the thickness of the Prenile sediments ranges from 70 m in the Nile Valley to 300–400 m in the Delta. In some localities, thickness reached 250–1000 m in the Nile Valley and the NE region of the Delta. The greatest thickness of the Prenile sediments has been recorded to the west of the modern channel, especially in the Minya reach.

The lifetime of the Prenile is difficult to assess, because there are no datable fossils in its sediments (Said 1993), but its age is estimated on the basis of available geologic data. Its sediments were found lying below a horizon containing fresh human implements belonging to the Acheulian (lower Paleolithic) age. Artifacts of this age in Egypt are estimated to have an age of about 350–400 ka. Said (1993) concluded that since the artifact horizon has been separated from the Prenile sediments by several intervening events, it is likely that the Prenile River ceased to flow some 400–450 ka ago.

Two significant geomorphologic features developed with the arrival of the Prenile. The first is the formation of a huge delta, and the second is the invasion of the Fayum Depression by Prenile water. According to Said (1993), **the Prenile built up a huge delta** that extended into the Mediterranean Sea and had an area at least three times that of the modern delta. Its sediments were found some 150 km from the modern delta shoreline. The extent and thickness of sediments affirm that the Prenile was a vigorous river, depositing about 100^6 m^3/yr during its life, estimated to be 300–400 ka long (Said 1993). This delta is the largest among three deltas, forming the present-day Nile Delta (see Chap. 6, Sect. 6.2 for more information about the development of the Delta).

The second geomorphologic feature is that **the Prenile water invaded the Fayum Depression**, representing the earliest connection of the Depression with the Nile, forming a huge lake that filled the Depression to an elevation of 43 m a.s.l. (Little 1935–1936; Ball 1939; Said 1993). The connection and disconnection between the Nile and Fayum has been repeated several times until the Ptolemaic time, when water entering the Fayum was regulated (for more information on the relationship between Fayum and the Nile see Chap. 12). The most important geomorphologic result is the transformation of Fayum Depression from a desert to a fluvial Depression, having the similar affinities to the Delta and the Nile Valley. Due to the availability of water and fertile soils, Fayum became the most populated Depression among other Depressions of the Western Desert. Other landforms resulted from the arrival of Nile water to Fayum will be discussed in Chap. 12.

4.6 Other Events in the Life of the River Nile

After the Prenile ceased to flow some 400–450 ka ago, no dramatic changes occurred. The only change was in the flow and the load of the River Nile. This change of flow resulted from climatic changes and from connection and disconnection of African waters. According to Said (1981, 1993), rivers with African connections became incidental and far in between. When they came to Egypt, they were unstable at times, low in most other times and never had the competence of the Prenile. All African rivers had a load of sediments similar to that of the present-day river with regard to texture and composition. Though they differed in regime and quantity of water flow, their sediments were quite similar in aspect; therefore, these rivers were grouped with the modern one/**Neonile**.

Said (1993) grouped the events that occurred during the evolution of the Neonile into four groups. The first set of events started with a wet period, called by Said (1981) the "**Abbassian Pluvial**," whose rain fed several ephemeral streams. This pluvial was interrupted by a short arid interval, during which connection with African sources was resumed, creating a river called by Said (1993) the **Dandara (Alpha Neonile)**. It had a regimen repeated in all the following Neoniles with an African connection. Its sediments were fine-grained and completely different from that of the Prenile. The second set of events occurred during the wet "Saharan" interval, which resulted in a low and erratic stream flow with African connection. This interval coincided with the Middle Paleolithic period in Egypt. The third group of events occurred some 70 ka years ago, when two seasonal rivers from Ethiopia arrived to Egypt. They carried great quantities of silt and spread it on their floodplain in Nubia and southern Egypt. The forth set of events started with the advent of the river, which has been flowing, since 10 ka years ago. Its load of silt, which is brought from Equatorial and Ethiopian Highlands, was laid down on the entire floodplain in southern and northern Egypt. This river developed during the Holocene wet phase (Nabatian Pluvial). Rain of this pluvial increased the flow of the river, allowing it to flow all year round. As Said (1993) concluded the present-day Nile is the child of that phase.

4.7 Conclusion

The Nile originated as a canyon starting from the Hinge Zone at the site of present-day Cairo and extended southward by headward erosion. Its consequences have not been restricted to the Cairo region, but extended to other regions in Egypt. The present Nile represents the last river of several others which flowed in the valley created by the Eonile since 6–7 million years. The development of such a deep canyon kept it from obliteration by erosion as was the case with the previous systems, which flowed to the west of the Nile Valley. One of the most significant consequences of the development of the Nile Canyon is that it divided Egypt into the two distinctive geomorphic regions of present-day landscape: the Western Desert and the Eastern Desert.

The rise of sea level at the beginning of Pliocene has several consequences. It caused the disappearance of the Eonile due to the complete submergence of the Nile Canyon and the complete burial of the Eonile channel by the sediments, which were deposited in the Gulf. Moreover, the lower sections of the sides of the valley canyon disappeared below the sediments of the Pliocene Gulf. As the Pliocene Gulf was filled with sand and gravels, the bed of the Nile Canyon rose gradually resulting in first the decreasing of its gradient and the development of a flat bed on which the Neoniles of Late Pleistocene/Holocene laid down the Nile Silt of the floodplain. After the withdrawal of sea water from the Pliocene Gulf, a second river depending on local rainfall (called Paleonile by Said) developed on the sediments that filled the Gulf, which indicates that the climate at that time was wet and that rainfall was sufficient to produce such a competent river. The sediments of this river developed a large delta, which contributed to the present-day Delta.

Had it not been for African waters, the current Nile would not have come to existence; instead, the Valley and the Delta would have been exposed to the work of wind and flash floods just like any other region in Egypt. The arrival of African waters to Egypt guaranteed water supply not only during its flow but also at later times, since it scoured a channel on the high ground, which separated Egypt and Sudan. The Prenile sand contributed to the evenness of the Valley bed making it easy for the flood water to reach to the farthest areas below the valley sides, depositing it on this flat bed and forming the floodplain. The Prenile sands built up a huge delta in the Mediterranean Sea and contributed significantly to the present-day delta. Furthermore, the Prenile invaded with its waters the Fayum Depression, converting it from a desert to a fluvial Depression. Water entering the Depression created a large lake covering the whole Depression at the beginning, which later turned into smaller lakes. Nile water entering Fayum resulted in firstly the deposition Nile silt on its floor, which provided fertile soils similar to those of the Nile Delta and Valley, and secondly they scoured small channels, used afterward as irrigation canals, and deep wadis (canyons) used as drains for excess irrigation water. Compared to the following rivers fed either by decreasing local rainfall or by the African waters coming from Ethiopian/Equatorial sources, it could be said that the Prenile was the largest and most effective river in outlining the landscape of the modern Valley.

No dramatic events have occurred after the Prenile ceased to flow since about 400–450 ka ago. All events happening during this final stage in the life of the River Nile were concentrated on the Nile floodplain and its channel. The sediment load brought by African rivers and laid down on the Nile floodplain was silt. According to size of the water discharge of the African rivers, size of Nile floods were determined; hence, their contribution to the floodplain and the Nile channel. Until 1970, the changes which have occurred in the Nile channel and the floodplain were the result of water discharge which carried water and silt to the Nile and its floodplain in Egypt. After 1970, however, the Nile discharge has become controlled by the Aswan High Dam and by human interference with erosion and deposition along the Nile sides.

References

Abdel-Rahman M, El-Baz F (1979) Detection of a probable ancestral delta of the Nile River. In: El-Baz F, Warner DM (eds) Appolo-Soyuz test project, vol II. Earth Observations and Photography, NASA, Washington D.C., pp 511–515

Ball J (1939) Contribution to the geography of Egypt. Survey & Mines Department, Government Press, Cairo, Egypt, p 308

Bown TM, Kraus MJ (1988) Geology and palaeoenvironment of the Oligocene Jebel Qatrani Formation and adjacent rocks, Fayum Province, Egypt. US Geol. Survey Prof. Paper No. 1452, 60 p

Fourteau R (1915) Contribution a L'étude des depôt Nilotique. Mem. Institut d'Égypte VIII: 56–94

Hsu KJ, Ryan WBF, Cita MB (1973) Late Miocene desiccation of the Mediterranean. Nature 242:239–243

Isaawi B, McCauley JF (1993) The Cenozoic landscape of Egypt and its river systems. J Geol Surv Egypt 19:357–384

Little OH (1935–1936) Recent geological work in the Fayum and in the adjoining portion of the Nile Valley. Bull Institut d'Égypte 18: 201–240

Said R (1981) Geological evolution of the River Nile. Springer, New York

Said R (1990) Quaternary. In: Said R (ed) The geology of Egypt. Balkema, Rotterdam, pp 487–507

Said R (1993) The River Nile: geology, hydrology and utilization. Pergamon Press, Oxford, p 320

Salem R (1976) Evolution of Eocene-Miocene sedimentation patterns in parts of northern Egypt. Amer Assoc Petrol Geol Bull 60:34–64

Sandford KS, Arkel WJ (1929) Paleolithic man and the Nile-Faiyum divide: a study of the region during Pliocene and Pleistocene time. Orient. Institute, Chicago University X: 77

Lake Nasser Region

5

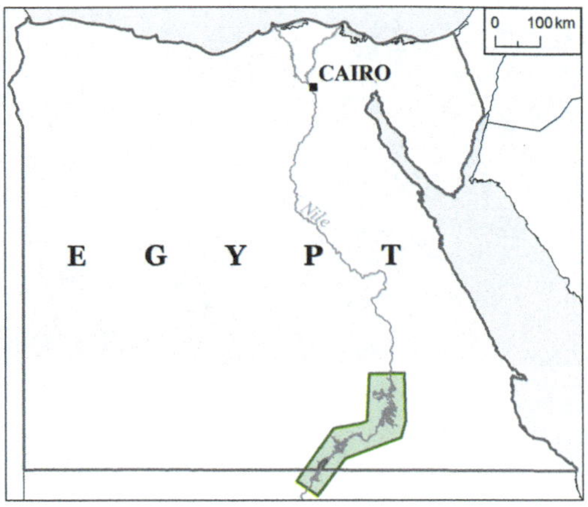

Abstract

Lake Nasser is the most significant man-made form created in Egypt. It came into being after the completion of the High Dam in the 1960s. Its impacts on the geomorphology of Egypt are local as well as national. Its impact on the local geomorphology of the Nubian region cannot be ignored. First, the submergence of the surface area of the Lake resulted in the disappearance of some pre-lake forms, of which the Nubian Nile channel is the most significant lost landform. The second local impact is the development of some submerged forms, such as the many islands in the lake body and the drowned wadis (khors). The third impact is the development of a new local delta in the southern part of the lake basin due the deposition of Nile silt. Nevertheless, national impacts of Lake Nasser have affected the downstream Nile channel, and the Mediterranean coasts of the Nile Delta, due to the deposition of Nile silt in the Lake basin. As a national water storage, Lake Nasser. As a national water storage, Lake Nasser is threatened by several factors. The most dangerous threat comes from Ethiopia, where a new huge dam (El-Nahda/Renaissance) is under construction on the Blue Nile which provides Egypt with more than 50% of its water needs. This dam will deprive Egypt from some of its share in Nile water during filling of the dam and during its operation, resulting in a decrease in the stored water, lowering the lake level, and reduction in electricity production.

Keywords

Egypt • The Nile Valley • Lake Nasser • Submerged forms • Islands • Khors

5.1 General Features of the Lake Region

Lake Nasser is an anthropogenic landform, which was created after the construction of the **Aswan High Dam** at about 6.0 km to the south of the city of Aswan in the late 1960s (Fig. 5.1). It extends from the High Dam in Egypt (23° 58′ N, 32° 58′ E) southwards to Dal Cataract in the Sudan (20° 30′ N, 30° 20′ E) when the lake level reaches 180 m a.s.l. The lake is filled gradually by the excess in floodwater, which comes from the Ethiopian Highlands (Fig. 5.2).The Lake lies in a region characterized by a hyper-arid climate (see Chap. 3, Sect. 3.1.1), and of varied geologic and topographic characteristics. The geological structure is dominated by the basement rocks, which are exposed on the surface in some localities, and are overlain unconformably by the Sabaya, **Lake Nasser** and **Abu Simbel Formations** of the **Nubian Sandstone** (Geological

N.S. Embabi, *Landscapes and Landforms of Egypt*, World Geomorphological Landscapes,
https://doi.org/10.1007/978-3-319-65661-8_5

Fig. 5.1 Lake Nasser:
(**a**) Oblique aerial photograph (taken by the author during a trip from Khartoum to Cairo),
(**b**) Landsat-TM image for the whole Lake

Map of Egypt, scale 1:500,000, 1989). The region is affected by a series of faults and fractures; some of which cut across the lake or run parallel to its shores. Before the lake, the Nubian Nile was a narrow river, which had no flood plain. Controlled by faults, the Nubian Nile changed its trend several times. Korosko bend was and still remains the most prominent change. To the east and west, lies the great sandstone plain, which is best developed on the western side and is called by Butzer (1965) the Lower Nubian Pediplain. Generally, the sandstone plain rises on both sides to about 200–300 m a.s.l. In the low-lying areas, several hills rise up 150–250 m a.s.l., of which **Gabal El-Allaqi** (218 m) and **Gabal El-Mahraqa** (199 m) on the east side are examples. The plain is dissected by several wadis, which are well defined on the eastern side comparatively to those of the western side. **Wadi El-Allaqi**, **Wadi Korosko**, **Wadi Kurkur** and **Wadi Kalabsha** are examples of those wadis. Some wadis open up in semi-closed basins before reaching the Nile.

5.2 Morphometry of Lake Nasser

Previous topography and water level both control the morphometric characteristics of Lake Nasser. Firstly, the general shape and orientation of the lake follow the former Nubian Nile (Fig. 5.2). Secondly, surface area, shoreline length, mean width, mean depth and maximum depth vary greatly with water level changes (Table 5.1). Due to the flattening of the local topography, surface area at 160 m level is less than 50% of that at 180-m level. Shoreline length at 160 m level is approximately 68% of that at 180 m level, increasing rapidly in the rugged areas relative to the flat or undulating areas. Therefore, the shoreline length generally increases with a rise in water level on the more rugged eastern shores compared to the less rugged western shores. The mean width of Lake Nasser at 180 m level is twice that at 160 m level.

5.3 Local Forms of Lake Nasser

Whatever the changes in the water level and their effect on the morphometry of Lake Nasser, the drowning of former topography and the deposition of Nile silt has resulted in the

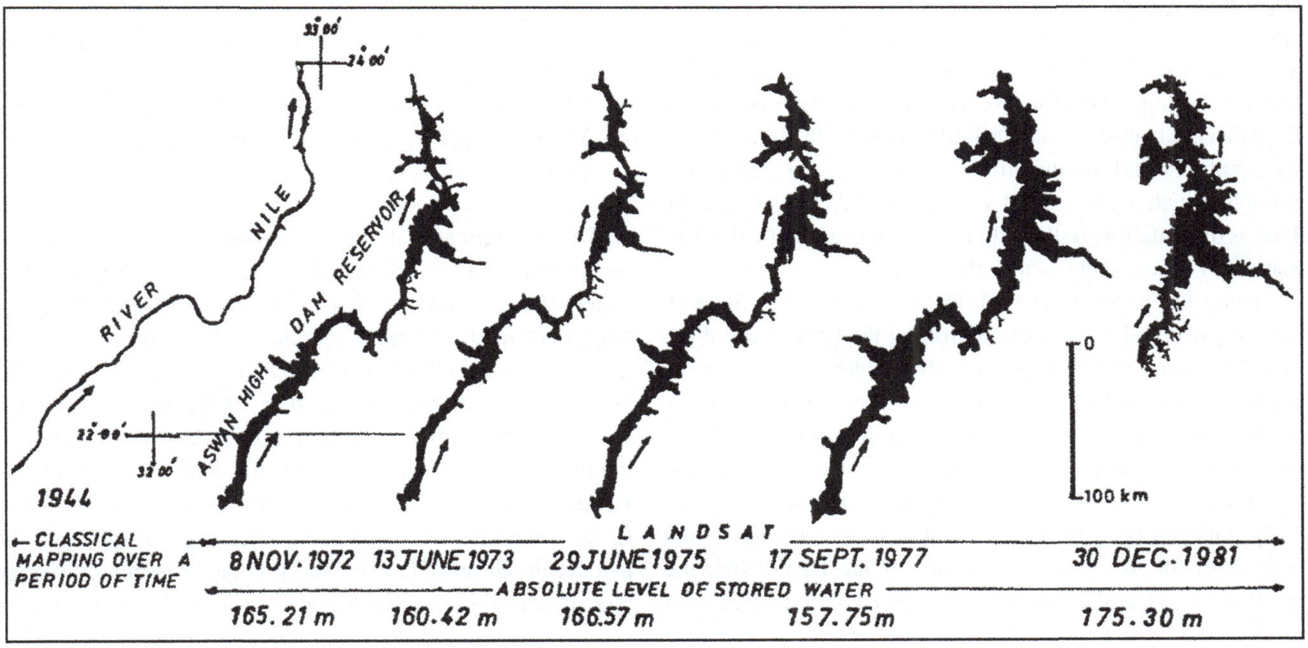

Fig. 5.2 Stages of Lake Nasser development (*Source* Entez 1973)

Table 5.1 Some morphometric characteristics of Lake Nasser

Morphometric characteristics	Egyptian Sector		Total Lake	
	Water level		Water level	
	160 m	180 m	160 m	180 m
Length (km)	291.8	291.8	419.3	481.8
Area (km^2)	2562	5237	3084	6276
Volume (km^3)	53	131	64	158
Shoreline Length (km)	5416	7875	5860	8804
Mean Width (km)	8.85	17.95	7.36	13.03
Mean Depth (m)	20.53	25.01	20.75	25.23
Maximum Depth (m)	110	130	110	130

Source Smith (1982)

development of new forms in the Nubian region of the Nile. These forms comprise the **Nubian Delta**, **khors (embayments)**, headlands, peninsulas and islands.

5.3.1 The Nubian Delta

The most important geomorphologic impact of the development of Lake Nasser is the disappearance of the Nubian channel and all the islands that were present in it beneath the stored water and the deposited silt in the Lake. Since the Lake is acting as a local base level for the upper reaches of the Nile, the narrow Nubian Valley and its channel will be filled gradually by Nile sediments starting from the south and advancing northwards. This filling is happening at present in the Sudan and in the extreme south of Egypt (Fig. 5.3). It is already known that since 1964 when the **High Dam** closed the Nile channel, virtually all the annual alluvium (about 13×10^6 m^3) carried out by the River Nile has been deposited in the lake. Most deposition occurs in a 135 km section extending to the south of Adindan at the Egyptian-Sudanese border (Smith 1982). In time, these sediments extend northward as shown by the successive surveys which were carried out for the lakebed (Fig. 5.3). This sedimentation will lead, in the long term, to the formation of a new delta within the lake.

According to Smith (1982), the development of the new delta is governed by factors controlling the deposition of silt in Lake Nasser. A major factor is the sudden decrease in the velocity of the flow of Nile water as soon as the river reaches the lake in the Sudan. It is expected that deposition should be greatest at the junction area and would decrease progressively downstream. This is confirmed by the measurements in the lakebed (Fig. 5.3). Since Lake Nasser acts as a local base level for the upstream reaches of the River Nile, the pre-flood level is the effective base level for the river as it enters the lake. At this level, the river enters the lake and deposition occurs. The pre-flood level changes annually and therefore base level and location of deposition changes accordingly. The rise of water level during the flood period is another factor affecting deposition in the lake. As a result of this rise, deposition will advance gradually upstream, and the sites of minimum and maximum water level of the lake will control annual deposition each year. Other factors affecting sediment deposition and the development of what may be called the **Nubian Delta** are several. The following are the most significant ones:

(1) Amount of alluvium that enters Lake Nasser during flood period.
(2) Position of the sediment mass deposited during the previous years.
(3) Rate of erosion and removal of the- sediments by the following flood.
(4) Roughness of the lake floor due to buried and unburied islands.
(5) Man-made constructions of the pre-dam period.
(6) Dune sand drifted into the lake along the western side by prevailing winds.

The development of the Nubian Delta is evidenced by the appearance of the silt banks of the Nile during periods of low water. These banks decrease in extension and thickness northward and disappear beyond Abu Simbel at about 280 km to the south of the High Dam (about 40 km from the border). It is expected that the shape of the Nubian Delta will be controlled by the configuration of Lake Nasser, which is longitudinal as a whole. When this delta will appear on Egyptian territory is an issue which is difficult to answer due to the complexity of the factors affecting deposition in the lake basin. Nevertheless, based upon the facts that the annual rate of deposition in the Sudan sector of the lake between

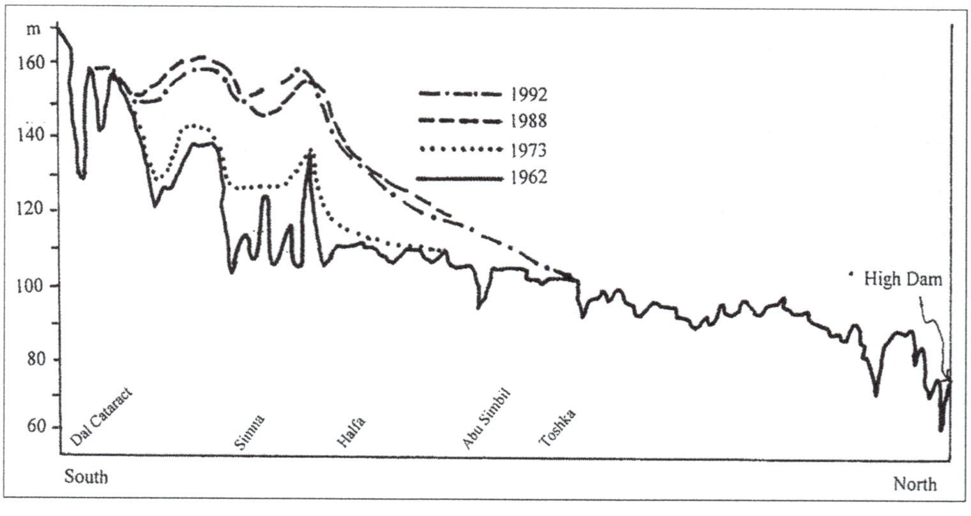

Fig. 5.3 Sedimentation in Lake Nasser bed (*Source* Entez 1973)

Fig. 5.4 Lake Nasser: High Dam, embayments (khors), sand drift, islands and headlands (Google Earth 2014)

1971 and 1979 was about 1.0 m (Smith 1982) and that the level of the lake bed is about 100 m a.s.l. in the southern part of Egypt, in addition to the fact that the maximum water level is 180 m a.s.l., it can be inferred that the emergence of the Nubian Delta requires around 80–100 years from the present time; i.e. by the end of the Twenty-first Century.

5.3.2 Embayments (Khors)

Embayments are the most common landforms along the sides of Lake Nasser and are responsible for - irregularities of the shoreline (Fig. 5.4). Nearly all embayments are drowned wadis, which are known locally as khors. Other embayments are drowned lowlands along the sides of the lake. **Khors** developed long before the construction of the High Dam. In fact, they started to develop after the old **Aswan Dam** had been built and heightened at the beginning of the Twentieth Century. **Khor El-Allaqi**, Khor Dehmit, Khor Sharaf El-Din and **Khor Kalabsha** are examples of pre-High Dam embayments. At the time, the size and number of the khors were small relative to those of Lake Nasser because the maximum level of water storage in the old Aswan Dam was 121 m a.s.l.

Khors are extremely variable in shape, length, width, depth, shoreline length and surface area. Although these characteristics change with water level, embayments can be classified into four types. Simple khors are small, single drowned wadis, which extend inland not more than 1.0 km. Dendritic khors are large and branching drowned wadis of which Khor El-Allaqi is the largest (about 40 km at 150 m a.s.l.). Irregular dendritic embayments are the result of the submergence of some wadis which open at adjacent lowlands. These two dendritic types dominate the shorelines of Lake Nasser since all large embayments belong to these types. The fourth type of embayment has an irregular shape which developed under the effect of the previous shape of lowlands around the lake. Although they are few in number, some of the irregular embayments cover several square kilometers. The combined surface area of the largest 40 khors would occupy about 60% of the total area of Lake Nasser at 180 m level (Entz 1973). High resolution space images (Google Earth 2014) show that the sides of khors all around the lakes are lined by two or three lines of small trees and shrubs (Fig. 5.5a, b). Most probably, these tree lines came into being, and are kept alive, due to the underground water seepage along bed contact of the sandstone. The aquifers of the underground water were charged with water

(a)

(b)

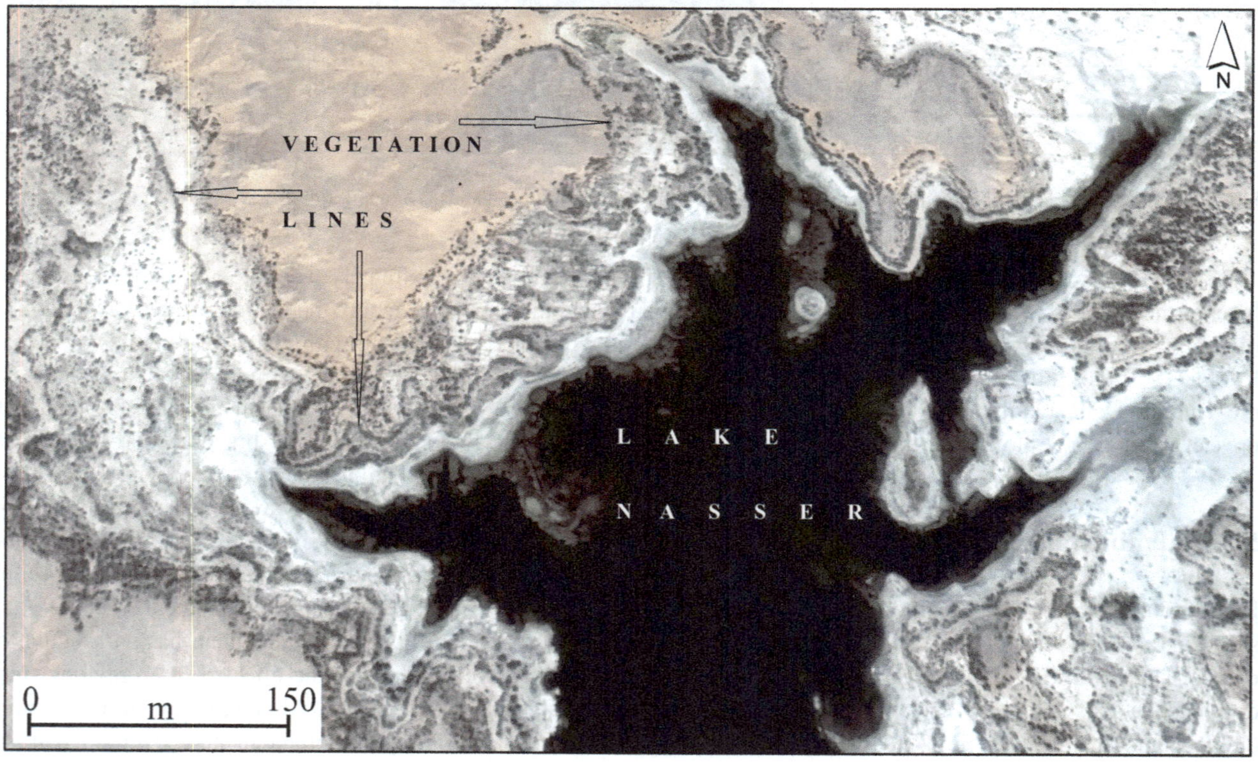

Fig. 5.5 Vegetation along the shores of Lake Nasser: (**a**) Along the shores of an embayment, (**b**) Along a sand beach (*Source* Google Earth 2014)

while the water level of the lake was higher than that of the present day.

5.3.3 Headlands and Peninsulas

The drowning of wadis and lowlands by the stored water in Lake Nasser is accompanied by the development of other landforms, such as headlands and small peninsulas. These forms represent the high ground, hills or interfluves lying between the embayments. Some of these headlands are just small simple land projections several hundred meters long. Some relatively large peninsulas are composed of headlands and khors that are small in size. Their length may reach 7 km and their area 15 km^2. Like embayments, their shape and dimensions change according to seasonal and annual change in the water level of Lake Nasser. Furthermore, whereas some headland or peninsulas might be transformed into islands, others might disappear with the rise of water level.

5.3.4 Islands

Islands represent **submerged hills** or high grounds along the sides of Lake Nasser. Similar to the two previous forms, islands change greatly in size and shape with changes in water level. With the rise in the level of water in Lake Nasser, some islands may disappear; others may become smaller, while others still might come into being as additional land is submerged. According to the topographic maps at scale 1:50,000 (1991) which were compiled from aerial photographs taken in 1988 when the water level was 155 m a.s.l., there were about 2000 islands in the Egyptian sector of Lake Nasser. At that date (1988), there were 65 islands rising more than 180 m a.s.l. Generally, small islands of areas less than 500 m^2 are the dominant, whereas the large ones are very few. The largest island is 9 km long and 3 km wide in the sector of Abu Hur. Large islands are irregular in shape due to the presence of khors and headlands, whereas small ones are circular in general. Only a few islands have names, which are in fact their former names when they were hills during the pre-lake period, such as **Gabal El-Allaqi Island** (218 m) and **Gabal El-Miharrqah Island** (202 m). The number of islands changes according to the level of the lake. All islands whose level is less than 180 m a.s.l. appear and disappear according to changes that occur in the level of the lake due to water discharge and recharge. Islands disappear when submerged in water due to the rise in the level of the lake. Islands also disappear when the lake level decreases and some near-shore islands are reconnected to the mainland.

5.4 Geomorphological Impacts of Lake Nasser

Apart from the local geomorphological **impacts of Lake Nasser** in the Nubian Region, there is the disappearance of the **Nubian Nile** and its islands below the water of the lake and the development of the above-mentioned new forms. There are also other impacts which can be called national impacts, since their effects are nation-wide. These are classified into two groups: the first is concerned with impacts on the Nile Valley, and the second is concerned with those on the coasts of the Nile Delta. The first group comprises two impacts on the Nile Valley. Firstly, no more silt will be spread on the flood plain due to the stopping of Nile floods carrying Nile silt. This indicates that the Nile flood plain will be deprived of the annual additions of Nile silt, and consequently no natural geomorphologic change is expected to occur. The artificial change came from the population of the Nile Valley, especially from those manufacturing building bricks which were made from the annual additions of Nile silt. For several years, this industry was suffering from a shortage of raw material, i.e. silt. Although not legally permitted, the brick manufacturers purchased from poor farmers the upper layers (approximately one or two meters) of silt from their fields. Consequently, there was a human erosion of the surface layer of the Nile flood plain, without any natural compensation, resulting in a temporary deterioration in the fertility of cultivated land. The second impact on the Nile Valley is that the Nile water released from the High Dam - nearly devoid of sediment load, and hence its erosive power increased resulting in some erosion on the Nile bed and sides. Although measurements made after completing the High Dam showed a very low rate of erosion, it represents a permanent problem in the Nile channel.

Geomorphological impacts of Lake Nasser on the Mediterranean coasts of the Nile Delta are discussed in Chap. 6, Sect. 6.5.4.1. These impacts were analyzed in many previous studies, which concluded that the rate of coastal erosion increased after Lake Nasser came into being in the 1970s, due to the trapping of the Nile silt in the Lake basin. The role which the Nile silt played in pre-High Dam times was to compensate the eroded material by waves and coastal current actions. Therefore, there was a natural balance between deposition and erosion along the Delta coast. The trapping of the silt in the lake basin has resulted in an imbalance between rates of erosion and deposition. This is why several sectors of the coast, especially the headlands of the two Nile branches of Damietta and Rosetta, retreated at very high rates. Although, some preventive actions were taken to minimize and even stop this retreat, the goal was not fulfilled as planned. These actions are likely to become obsolete if sea level rises under the effect of global warming.

5.5 Hazards Threatening Lake Nasser

The main purpose of constructing the High Dam on the Nile in Upper Egypt was to control floods and provide water for irrigation, by storing water reaching Egypt from Ethiopian Highlands during times when the demands for economic activities do not need all the water that reaches it, such as during the summer season. At the same time, some of this excess water is needed to compensate the deficiency in water reaching Egypt from the Equatorial sources of the Nile during the winter season. Furthermore, water which is not consumed during the winter may be needed when there is a water deficit due to what is known in Egypt as low floods; or in other words, when the quantity of water is less than the normal quantity reaching Egypt during pre-dam times because of low rainfall on the Ethiopian Highlands.

However, in time, Lake Nasser grew gradually because of the annual excess water. It nearly reached its maximum capacity in the early 1980s. As can be seen from Table 5.1, the maximum capacity of water storage for the whole lake basin is in the range of 150×10^9 m^3, and in the Egyptian section it is 130×10^9 m^3. This water storage is known as century storage, since excess water could remain and be stored in the lake basin for a hundred years or more. Hence any decrease in the storage capacity would affect the quantity of water which could be stored in the lake. Four factors are currently threatening the capacity of water storage in the lake by decreasing the size of its basin, or by decreasing the water reaching Egypt from the Nile sources. First, the deposition of **Nile silt**; second, the deposition of aeolian sand; third, climatic change; and the fourth factor is the construction of dams on Upper Nile reaches.

5.5.1 The Nile Silt

It is well known that the load of Nile water which reaches Egypt from Ethiopian Highlands is in the range of fine sediments (sand, silt, clay), which are collectively called Nile silt-However, grain size analysis of deposited sediments revealed a decrease in size from south to north, where samples are wholly composed of sand in the south whereas clay is predominant in the north (ElDardir 1994). In pre-dam times and during the flood season, the silt was deposited on the Nile flood plain and along the northern coasts of the Nile Delta. By storing the flood water in the lake from the early 1970s, all the silt was deposited in the lake basin, starting in the extreme south and gradually extending northwards (see Fig. 5.3). As mentioned in Sect. 5.3.1, the deposition of the silt resulted in the development of a local delta in the southern part of the Lake, and with the continuous deposition of silt, the delta will be growing so long as the lake is alive. This deposition of silt happens to be filling the lake basin, resulting in a decrease in its water storage capacity. Pre-dam plans estimated that it would take 500 years to fill the lake basin with silt, and the lake would then be obsolete. Due to the intervention of the threatening factors, those five hundred years of the working age of Lake Nasser are subject to change. It has been suggested that in order to increase the planned age, it is possible to dredge the Nile silt from the lake, and spread it on its shores to be used in several activities of which agriculture is a priority. Other researchers suggested extracting minerals of economic value from it such as gold (Hassaan et al. 1995).

5.5.2 Aeolian Sand Drift

The second factor threatening the storage capacity of the Lake is the deposition of aeolian sand in the lake basin. It is a well known fact that there are sand accumulations on the western side of the Nile. It is also a well known fact that some of the sand is drifted into the Nile channel and carried northwards by flowing water in the Nile channel. From the time the lake came into being, all the deposited sand has accumulated in the lakebed, starting along the lake shores and extending into the lake basin. The end result of this accumulation of sand is the same as that of the deposition of Nile silt in the lake basin; i.e. a decrease in the size of the lake basin and consequently a decrease in its water storage capacity. Although the aeolian sand could be dredged from the lakebed like Nile silt, this sand so far has no economic value. Until now, there have been no studies that evaluate the actual threat of aeolian sand, nor how to stop or minimize it.

5.5.3 Present-Day Climatic Change

Present day climatic conditions throughout the world change significantly from one year to the next. Some years, the changes may be slight, causing hardly any problems to economic activities; but in other years, the changes are drastic and cause serious problems to human activities. These year-to-year changes may continue for several consecutive years, resulting in severe problems to national economies. The most significant climatic element relevant to Lake Nasser is rainfall on the Ethiopian Highlands. Rainfall may increase resulting in an exceptional rise in the level of water discharge reaching Egypt. These are considered locally high or very high floods, and the lake basin cannot accommodate this discharge. For those years of exceptional rainfall and water discharge, the Egyptian Irrigation Administration constructed a spillway through which the excess water is discharged into small basins in the Western Desert. Severe problems to the Egyptian economy occur when rainfall decreases to the extent where water discharge is much less than the demands of the economy. This has already occurred several times in Egypt's history from the times of the Ancient Egyptians resulting in famine and a deterioration of the economy (Said 1993).The most recent of these critical conditions happened during 1986 and 1987 (Embabi 1993) when the Nile discharge in Aswan was only 33×10^9 and 35×10^9 m^3 respectively, whereas Egypt's annual demands are 50×10^9 m^3 causing an annual deficit between 15 and 20×10^9 m^3. That time, for the first time in Egyptian history, the country was not affected by economic harshness because water storage in Lake Nasser compensated the deficit in the Nile water discharge.

5.5.4 Construction of New Dams on the Nile Reaches

Without a political agreement between the concerned governments, the construction of new dams on the upper Nile reaches, especially in Ethiopia and Sudan, is the most serious threat to Lake Nasser. The first of these dams is the Grand Ethiopian Renaissance Dam (El-Nahda in Arabic) on the Blue Nile in Ethiopia. Although the precise impact of this dam is not yet known, there are fears of temporary reduction of water availability due to the filling of the dam and a permanent reduction because of evaporation from the reservoir. The filling of the dam will spread over several years, therefore, there will be a loss of $11–29 \times 10^9$ m^3 of water per year, which would cause a large sectors of farmers to lose their income while the reservoir is being filled. It will also affect Egypt's electricity supply by 20–40% while the dam is being built.

The Grand Ethiopian Renaissance Dam could also lead to a permanent lowering of the water level in Lake Nasser if floods are stored instead in Ethiopia. This will reduce the ability of the High Dam to produce hydro-power to the tune of 100 MW loss of generating capacity for a 3 m reduction of water level.

5.6 Conclusion

The creation of Lake Nasser in the early seventies of the 20th Century added to a new regional landform that changed the face of the Earth's surface not only in the southern part of Egypt, but also in the north of the Sudan. Its geomorphological impacts extended also to other regions in Egypt. Therefore, this form has a nation-wide geomorphological impact. Although there are some other anthropogenic forms created or disappeared in Egypt during the 20th Century and the end of the 19th Century such as the Bitter and El-Temsah Lakes along the Suez Canal, the Lakes of **Wadi El-Rayan**, and the temporary **lakes of Toshka** Spillway in the Western Desert, Lake Nasser is the most significant anthropogenic form created in Egypt in modern time. This is because it is not only the largest-, but also owing to its impact on other related forms.

Beside a permanent water body of the Lake, new forms came into being, while other disappeared due to the submergence of pre-Lake surface. The disappearance of the Nubian Nile Channel is the most significant changes in landforms in the southern part of Egypt. The development of new submerged forms: embayments (drowned wadis) and islands (drowned hills), and the development of a local delta at the southern section of the Lake due to deposition of Nile silt are all new—created due to the creation of the Lake.

The deposition of Nile silt in the Lake basin resulted in an increase in the erosive power the Nile water released from the Lake for national economic requirements. This change caused a negative change in the equilibrium state of water flow in the Nile Channel downstream. The other change in Egypt geomorphology is the imbalance which occurred in the state of equilibrium between net erosion and deposition along the northern coasts of the Nile Delta. During pre-Dam times, there was a balance between wave erosion of the Delta coasts and deposition of Nile silt along these coasts. After the development of the Lake, the Nile silt load is deposited in the Lake basin, therefore coasts of the Nile Delta have been retreated in several sectors leading to the destruction of some establishments and the disappearance of some cultivated lands. Although some preventive actions were taken to minimize the rate of coastal retreat, the threat of retreat of the coasts may be heightened if the sea level rises under the effect of global warming.

Nasser Lake as a water reservoir for Egypt is exposed to national and international threats. National threats are due to the deposition of Nile silt and aeolian sand in the Lake basin, resulting in the shrinkage of the storage capacity of water. International threats come from the construction of new dams on the upper reaches of the Nile, especially in Ethiopia from which most of water stored in the Lake comes from there. Of the Ethiopian dams, the Grand Ethiopian Renaissance (El-Nahda) Dam on the Blue Nile is the most dangerous, since it will store a high percentage of water arriving Egypt during the filling year of the dam basin. Also, even after filling the dam Basin, it will store some of the flood water which arrives in Egypt during the high peak of the Blue Nile flow in summer months for the purposes of generating the maximum electricity from the power station of the dam.

References

Butzer K (1965) Desert landforms in the Kurkur Oasis, Egypt. Ann Assoc Amer Geogr 55:578–591

El Dardir M (1994) Sedimentation in High Dam reservoir 1987–1992 and sedimentary futurologic aspects. Sedimentol Egypt 2:23–39

Embabi N (1993) Water resources in Egypt: The Nile Water. (In Arabic), Special Publication, Middle East Research Center, Ain Shams University, Cairo

Entz B (1973) Morphometrics of Lake Nasser and Lake Nubia. Working Paper No. 2, Lake Nasser Development Center, Aswan

Google Earth (2014) Egypt Images Latitudes 22-32 N and Longitudes 22-32 E, http://www.google.com

Hassaan M, ElDardir M, Nassar Y (1995) Alluvial gold placer in main sedimentation area, High Dam Lake, Egypt: The first record. Sedimentol Egypt 3: 125–129

Said R (1993) The River Nile: Geology, hydrology and utilization. Pergamon Press, Oxford

Smith S (1982) Application of remote sensing techniques to the study of the impacts of the Aswan High Dam. Dissertation, University of Michigan

6 The Nile Delta

Abstract

The Nile Delta is one of the large deltas in the world. It occupies at least ~25,000 km^2. It extends below the Mediterranean Sea, covering an area that equals its continental area. The delta is composed of four successive deltas developed by the successive Niles, which occupied the Nile Valley from Late Miocene. Its sediments are composed of sand and gravel overlain by a thin layer of alluvial clay (the Nile silt), with a total thickness of ~4000 m. The delta is characterized by small/micro-forms, especially along the coastal fringes (i.e., coastal lagoons, dunes, marshes, and sabkhas), except for the sandy mounds known as "Turtle-Backs," which spread in several localities in the east. All these forms have been subjected to anthropogenic modifications. The coasts have also been exposed to erosion especially along the promontories of the two Nile branches of Rosetta and Damietta. Due to global warming, it is expected that sea level will rise up inundating the lowlands along the coastal fringes, especially the lagoons and marshy areas, affecting all human activities.

Keywords

The Nile Delta • Egypt • Nile branches • Coastal lagoons • Turtle-backs • Sea level rise and coastal changes

6.1 General Morphologic Characteristics

The **Nile Delta** with its triangular shape is a nearly flat plain (Fig. 6.1). Its surface slopes gently northwards, where the difference in elevation between its apex in the south and the Mediterranean coast is 18 m. The maximum distance from the delta apex, where the Nile is divided into **Damietta and Rosetta Branches**, to the **Borollos promontory** at the coast is about 160 km. Maximum breadth along a straight line from **El-Tinah plain** east of Port Said to Alexandria is ~260 km. Area reported in the literature is ~25,000 km^2. This area represents what is covered by recent Nile silt and the coastal fringe. Sediments of the ancient Niles, called the Paleonile, Prenile, and Neonile, cover wide tracts along the eastern and western margins of the delta (Said 1981). If these tracts are included in the delta area, its surface will increase by ~4000 km^2. It is also mentioned that the ancient sediments (Eonile and Paleonile deltas) extend below the Mediterranean water down to 200 m contour, taking the shape of a cone, representing the submarine continuation of the delta mass. If this submerged area is included, the total area of the delta will almost be twice as that of the present landmass (Schlumberger 1984; Abd El-Motaal 1999).

Exhibiting the fan-like shape, the contour map of the delta (Fig. 6.2) reveals that the slope of the surface of the

N.S. Embabi, *Landscapes and Landforms of Egypt*, World Geomorphological Landscapes,
https://doi.org/10.1007/978-3-319-65661-8_6

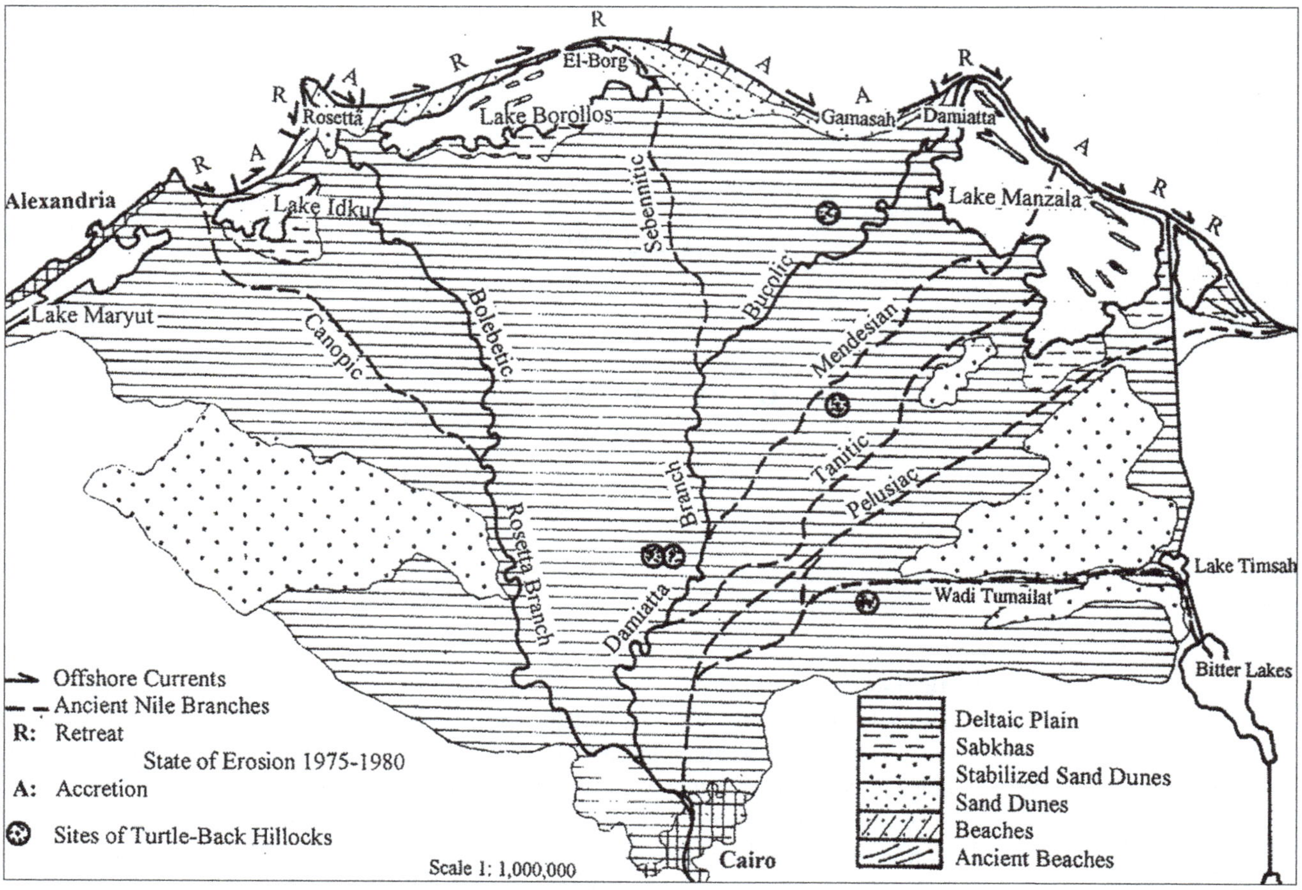

Fig. 6.1 A geomorphologic Map of the Nile Delta (*Source* Compiled from several sources)

ground decreases northwards especially along the coastal fringes, where heights range between two and three meters below sea level. The five-meter contour line divides the delta nearly into two equal halves. With their arcuate pattern, the contour lines show that the middle part between the two Nile branches is higher than the eastern and western parts along the same latitude. It is notable that the slope of the delta surface runs due north in the middle and runs northeast and northwest in the east and west, respectively. In addition, the contour map shows that the surface of the Nile Delta is slightly higher in the east than in the west, resulting in a secondary slight slope toward the west and northwest. Therefore, the lowest parts exist in the northwest around Idku and Maryut lagoons and not around El-Borollos lagoon. As a result of this secondary regional slope, the Damietta Branch is relatively higher and smaller than the Rosetta Branch. Such characteristics of the delta surface are due to the nature of deposition of the Nile sediments in the delta embayment.

6.2 Geology and Evolution of the Nile Delta

Due to its geologic origin as a tectonic embayment, which has been filled with the Nile sediments since the Miocene, the Nile Delta acquired its unique triangular shape. The Nile sediments are composed of sand and gravel overlain by a thin layer of alluvial clay. The thickness of these Nile sediments increases seaward until it reaches ~4000 m at the present coastal fringe. Figure (6.3) shows a composite geologic column of the sub-surface deposits, which represent the sediments of the Nile Delta laid down during the stages of the evolution of the Nile River.

According to Said (1981), the Nile Delta is not a simple delta. On the contrary, it is composed of four overlapping sub-deltas formed by the successive rivers that developed in the Nile Valley. The same names suggested by Said (1981) for successive Niles are used for the sub-deltas as follows: (a) **Eonile Delta**; (b) **Paleonile Delta**; (c) **Prenile Delta**; and (d) **NeoNile Delta**.

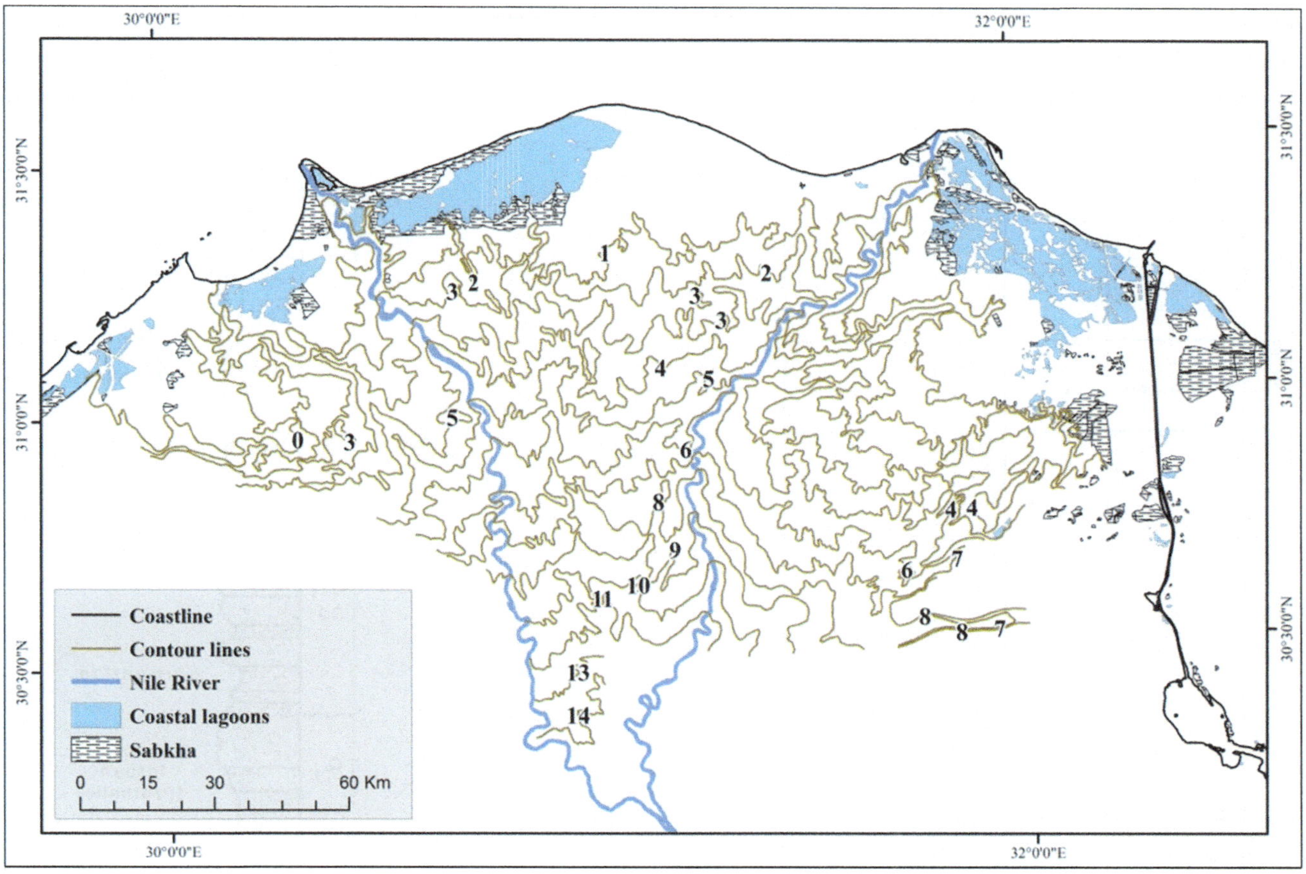

Fig. 6.2 A contour map of the Nile Delta

The earliest of these deltas was the Eonile Delta. It was formed during the Late Miocene as an apron in the North Delta embayment. Its sediments are coarse, derived from the elevated Tertiary rocks of the Eastern Desert. According to Abd El-Motaal (1999), lithological characteristics, thickness, and areal distribution of the Upper Miocene sediments of the delta indicate that they have been accumulated in the form of coalescing fans in front of an active fault scarp called by Said (1981) "**The Hinge Zone**" (Fig. 6.4). These fans are buried underneath the surface and are recognized only in the boreholes of the North Delta Embayment, at depths varying between 3500 and 4000 m.

The second delta developed by the Paleonile, which was debouching in the Pliocene Gulf. This Gulf filled the North Delta Embayment with marine sediments. This is why when the Paleonile arrived, its delta started branching immediately to the north of Cairo. The Paleonile delta is built up of fine-grained sediments. Said (1993) postulates that this delta was of the bird's foot type, because the fine-grained sediments of the Paleonile were not as dense as sea water; therefore, the sediments advanced into the sea along the distributaries of the Paleonile and were deposited along their sides (Said 1993).

The third delta is the Prenile delta. It is composed of an enormous load of coarse sand and gravel, which was brought by this river from Ethiopia. According to Said (1993), this delta was fully formed some 400,000 years ago. At that time, the Prenile delta assumed enormous dimensions. It extended into the Mediterranean Sea with an arcuate shape and an area, at least, four times as that of present-day delta (Said 1993).

Due to the cessation of the flow of the Prenile, aggradation was interrupted and separated by long episodes of erosion, resulting in a net degradation of the Prenile Delta. The Neonile (fourth) delta was built on the relics of that delta, which forms its core, and is responsible for its arcuate forms (Said 1993). Since the sediment load of the Neonile was relatively small (35 million tons/year), the Neonile delta built-up has not resulted in much change in the mass or frontal advance of the Neonile delta during the past 7–8 millennia. These sediments of the Neonile form a relatively thin layer of the Nile silt, overlying the Prenile massive sediments of sand and gravel. Thickness of this top layer of fine material varies between five meters along the delta margins and 20–30 m in mid-delta (Abd El-Motaal 1999). In the meantime, the Neonile delta did not cover completely the

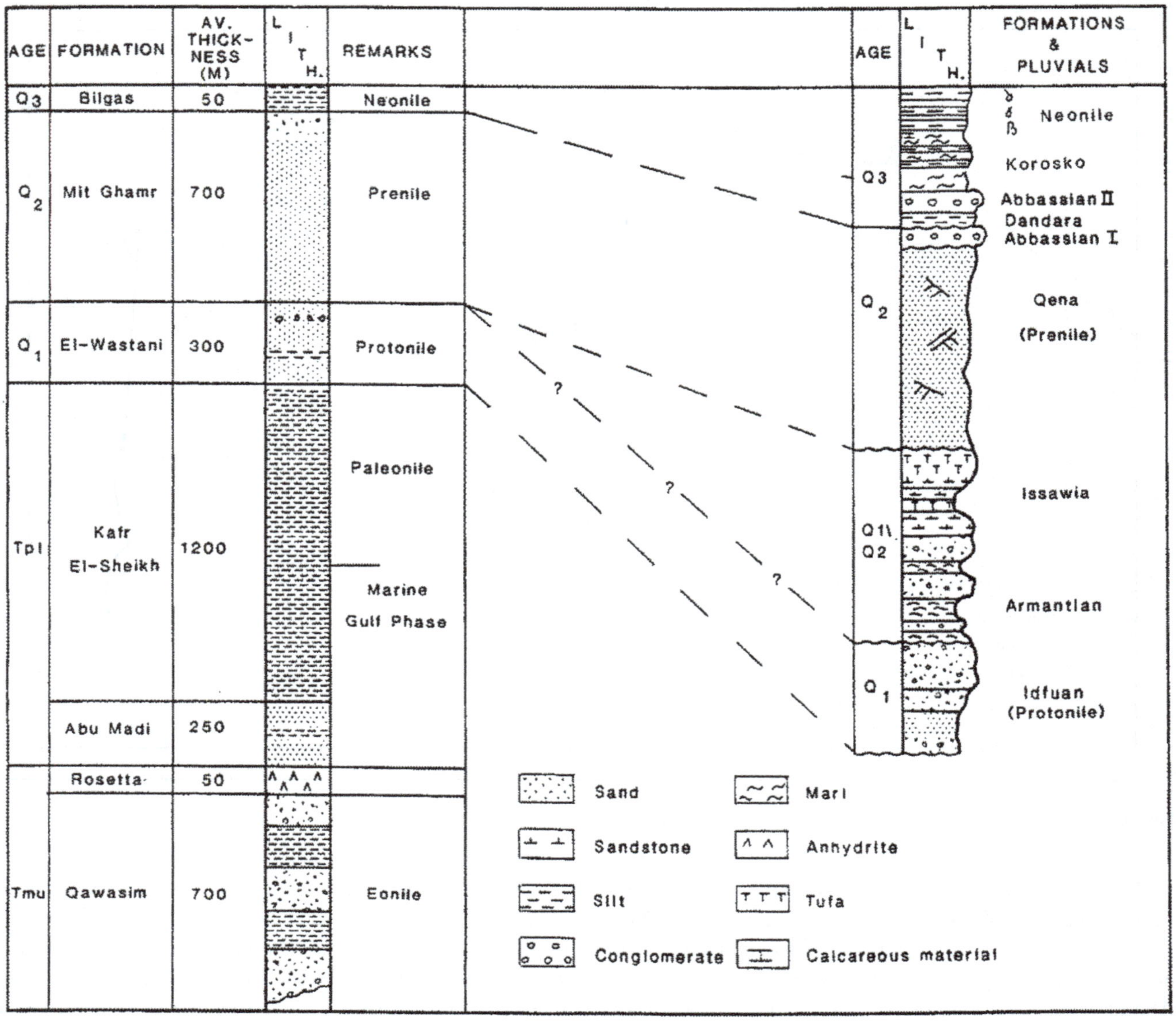

Fig. 6.3 A composite geological column for the Nile Delta sediments (*Source* Said 1990)

Prenile delta, since the sediments of the latter are exposed as islands in the delta plain (Turtle-Backs mounds) and on the eastern and western margins of the Nile Delta according to the Geological Map of Egypt, scale 1:500,000 (Conoco Coral and The Egyptian General Petroleum Corporation 1986–1987). Generally, the increase of sediments thickness was accompanied by subsidence during sedimentation stages on the downthrown side of the hinge zone.

Although it looks uniform and homogeneous, the Nile Delta plain is characterized by several forms. Some of these forms are developed under the present environmental conditions, whereas some others are relict forms, which developed in past environments that were different from those prevailing at present. From the geomorphic point of view, relevant forms are the Nile Delta Branches, the Turtle-Back hills or mounds, the coastal forms, and the Barari (Wasteland) salt marshes flats (Fig. 6.1).

6.3 The Nile Delta Branches

6.3.1 Ancient Branches

At present, the Nile bifurcates into two branches about 20 km north of Cairo. Relicts of ancient branches can be traced at several localities in the delta. Several sectors of present-day irrigation canals and drains occupy those of ancient branches. Some authors (Tousson 1922; Ball 1942) described the Nile Branches in the delta during various episodes. According to these studies, there were seven

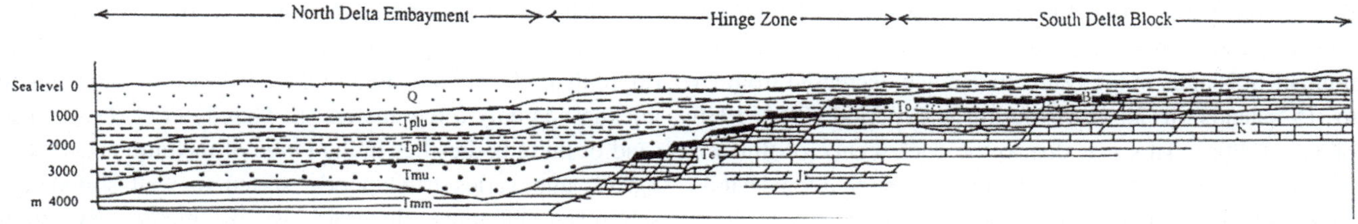

Fig. 6.4 N–S Geological cross-section along the Nile Delta J: Jurassic, K: Cretaceous, Te: Middle Eocene, To: Oligocene, B: Basalt, Tmm and Tmu: Middle and Upper Miocene, Tpll and Tplu: Middle and Upper Pliocene, Q: Quaternary (*Source* Redrawn from Said 1990)

branches during the pre-dynastic period (Fig. 6.1), and then, they were reduced to five branches during the Pharaonic times. Since then, changes in number and course have been governed by the Nile discharge and human activity until they were reduced to two branches fully controlled by humans. The concentration of former distributaries (Four major ones) in the northeastern part of the delta (Fig. 6.1) was partly due to the rapid lowering of this part of the Nile Delta (Stanley 1988). According to Said (1990), the disappearance of the ancient branches occurred during low-flood years, when deposition exceeded erosion. On the other hand, Sneh and Weissbrod (1973) concluded that the degeneration of the lower reaches of the Pelusaic Branch was due to the silting up of the mouth of the distributary by the prevailing W–E long shore current. It was separated from the sea by a series of accretionary coastal sand ridges. According to them, the process of degeneration of the lower reaches of this branch occurred about 2000 years ago (25 AD). Torab (1998) postulated that, beside the effect of low-floods and the silting up of the mouth of this branch, the disappearance of the Pelusaic Branch was due to sea level rise and the submergence of the lower section by seawater. This branch disappeared completely at a faster rate, as the local drainage system was neglected as a result of wars and the region was gradually depopulated. Although the ancient branches disappeared as a feature, they left their imprint on the geomorphological aspects of the delta, such as small sub-deltas in the Mediterranean Sea and islands in Manzala and Borollos lagoons. In the west, the Canopic Branch silted up as result of the re-excavation of the Bolebetic branch, which took place around 300 BC (Sestini 1989). The other branches of the delta started silting up early, but they finally disappeared during the eleventh, thirteenth, and seventeenth centuries when the Nile had exceptionally low discharge (Said 1993).

All branches reached the Mediterranean from the Pelusium Branch in the east to the Canopic Branch in the west, except for one branch, which had its mouth at Lake Timsah or the Gulf of Suez, according to some previous studies (Tousson 1922; El-Shazly et al. 1975; Ramadan 1984; El-Fawal 1992). At first, this latter branch followed the present-day **Wadi Tumaylat**, which crosses—in a W–E direction—the east Delta desert, and then diverted its course to the southeast across the **Bitter Lakes** lowland, after which it was taking a southward direction to the Gulf of Suez. This postulation was based on several pieces of evidence, such as tree trunks and vertebral bones of large fishes in the sands of Wadi Tumaylat, the filling of the wadi with younger sands and clayey sands of fluviatile origin, and the presence of augite in the sediments of the Suez Canal. The augite is probably derived from the Ethiopian highlands through the Blue Nile and Atbara tributaries (Ramadan 1984). The Nilotic origin is further evidenced by the mineral composition of the existing sediments of the Wadi Tumaylat. They are rich in opaque iron oxides, pyroxenes, epidotes, and amphiboles similar to the heavy minerals suite of Nilotic sediments developed after the connection with Ethiopian tributaries (El-Fawal 1992). Trends of paleo-currents of the sediments of the Wadi Tumaylat that indicate persistent eastward flow represent additional evidence that this Wadi was an eastern distributary of the Nile Delta (El-Fawal 1992). The course of this branch was utilized several times in the past to connect the Red Sea with the Nile by a navigable canal. At present, Wadi Tumaylat is occupied by Ismailia irrigation and navigable canal, which branches—upon reaching Ismailia city—into a northern branch heading toward Port Said and a southern one heading toward Suez city. During the Islamic period, the most westerly branch was not a delta branch, but rather the northern extension of Bahr Youssef, which branched from the Nile in the vicinity of Asyut city in the Nile Valley (Embabi 2004).

6.3.2 Present-Day Branches

The two existing branches of Damietta and Rosetta have geomorphological aspects similar to those of the Nile channel in Upper Egypt, such as meanders, islands, and rough bed. Changes of these aspects were recorded in several topographic map series, which have existed since the beginning of the twentieth century. Recent studies (Torab 1990a) on the geomorphological changes in **Damietta Branch** after the construction of the High Dam revealed that its width decreased by approximately 27% (from 275 to

200 m) on average due to decreased water discharge and velocity, and increase in rate of deposition. Also, the total number of islands in the Damietta Branch decreased from 24 to 15 islands in 30 years during the period 1955–1986 (Torab 1990b). This was due to the coalescence of thirteen islands with the banks of the channel, although a few islands were built up during the same period. Decrease in water flow in the late eighties and early nineties resulted in the appearance of new islands, amounting to a total of 23 islands in 1993 (Taha 1997). Due to the increase in meandering, 3.15 km was added to the total length of the channel, in spite of the disappearance of some 800 m at the mouth of the branch by coastal erosion. Similar changes occurred in the **Rosetta Branch** during the same period (EL-Disouki 1997). However, the Rosetta Branch is lower, nearly twice as wide and deeper, carries more sand, and has larger meanders and more point bars compared to the Damietta Branch. Rosetta Branch is more vulnerable to erosion than the Damietta Branch due to the regional slope of the delta surface to the west and northwest. Islands in this branch decreased from 30 islands in 1955 to 23 islands in 1982 and went down to 14 islands in 1993 (Taha 1997).

6.4 Turtle-Back Mounds

First named by Fourteau (1915), the "**Turtle-Backs**" are small low-lying hillocks, with rounded crests having the appearance of turtle-backs as yellow islets amidst the recent black Nile silt. They occur as two groups in the southern and northeastern parts of the delta (Fig. 6.1). They vary greatly in size and relative height. However, those of the southern group are generally higher and wider than those of the northeastern group. Relative height varies between 10 and 11 m in the south, and 2 and 6 m in the north, whereas their area ranges between several square kilometers and a few squares of hundred meters. Some of them carry names, especially the large ones such as "Remal El-Iraqi," lying about 15 km to the east of the city of Shebin El-Kom. Most of them were recently exposed to major changes and modifications due to quarrying their sands to be used as building material.

Due to their peculiar presence amid the Nile silt of the delta, the turtle-backs have been studied by many authors—Fourteau (1915) and Hume (1925)—with a focus on their origin. Structurally, they are composed of non-fossiliferous unconsolidated non-gravelly sand accumulation, with weakly developed parallel lamination and planar cross stratification, with occasional graded lamination (Issa et al. 1990). Mechanical analysis of samples collected from 16 hillocks in south and northeast Delta showed that sand represents between 93 and 99.85% of the sediments, whereas the rest of the sediments are composed of silt and clay (Kholief et al. 1969). This sandy formation extends at the sub-surface, deeper in the southern part of the Delta (more than 60 m) compared to the northeastern parts, where thickness is only a few meters (El-Gamily et al. 1988). The sand of these bodies lies in the range of medium size (1.14–1.6 Ø) and varies from poorly sorted to well sorted. No significant differences in grain size parameters have been found between the southern and northeastern groups. Analysis of grain size parameters revealed that the sediments of the Turtle-Backs were deposited mainly in a riverine environment with beach affiliation for some of the southern group (Issa et al. 1990).

Mineralogical data obtained by Zaghloul et al. (1990a) revealed that sands of the northern group of Turtle-Backs are relatively rich in pyroxene and total feldspars and poorer in opaques with lower quartz/feldspar ratio compared to those of the southern group. These characteristics indicate that the sediments of the northeastern group were deposited in Late Pleistocene to Early Holocene after the waters of the Blue Nile and the Atbara have reached the River Nile in Egypt. On the other hand, the sediments of the southern group are older (Middle Pleistocene) and were derived from the igneous and metamorphic sources of the Red Sea Mountains and the White Nile sediments, which produced the Mit-Ghamr formation (Zaghloul et al. 1990a). In this respect, this interpretation conforms to that by Issa et al. (1990), who found that the characteristics of textural data of the sands of the Turtle-Backs of the northern group seem to be similar to those of Bilqas formation, deposited during the Holocene, and that the sands of the southern group are akin to Mit-Ghamr formation of the Pleistocene age. Said (1981) postulated that the Turtle-Backs are remnants of the Prenile Qena Sands, which were derived from the Ethiopian highland, when the Nile in Egypt was connected for the first time with its African sources during the Middle Pleistocene. These sediments (known as Mit-Ghamr Formation in the delta) extend below the recent Nile silt. No distinction has been made by Said (1981) between the southern and northeastern groups in origin. Said's conclusion agrees with the findings of Kholeif et al. (1969) in that the heavy mineral assemblages of the sands of the Turtle-Backs were derived mainly from the volcanic rocks of the Ethiopian Plateau.

6.5 Coastal Forms

Several forms developed along the coastal fringes of the delta due to land–sea interaction. These fringes do not rise more than 3.0 m a.s.l. and extend about 10 km inland. Along these fringes, several factors acted in the past or are acting at present, such as sea level changes, climatic variations affecting Nile flow and sediment discharge, erosional effect of eastward coastal and offshore currents, human activity since pre-dynastic times, prevailing winds, and neo-tectonics. Due to the interaction of such various factors, numerous coastal forms developed, producing a unique environment differing from the rest of the deltaic plain. Characteristic forms in this environment are the coastal plain, the lagoons, the sand dunes, and the Barari (waste-land) flats (Fig. 6.1).

6.5.1 The Coastal Plain

Although the present coastline is generally convex in outline, it is composed of several concave wide bays and convex promontories. In fact, the concave wide bays are the result of the development of promontories. There are three promontories along the coastline: the Rosetta, Damietta, and Ras El-Borollos (Fig. 6.1). They were built as pro-deltas or delta lobes. The Ras **El-Borollos promontory** was built up as an ancient pro-delta by the defunct central Sebennytic Nile Branch, 2000 years ago (Sestini 1992), whereas the Rosstta and Damietta promontories are young delta lobes formed by the two existing branches. Historical data and other evidence show that these two recent delta lobes continued to grow until early twentieth century, and since then, they have been exposed to erosion, especially as of 1964 onward after the closure of the Nile channel by the Aswan High Dam and the deposition of the Nile silt in **Lake Nasser**. Due to the action of the NW prevailing winds, and the W–E longshore current, the western sides of these two lobes are exposed to erosion more than the eastern sides (Sestini 1992; Tetra Tech 1984). These two promontories record the past advances of the Nile mouths. Upon examining boreholes, it appeared that each advance is reflected as an accretion ridge, which is about a few meters thick overlying first near-shore sandy silt and then marine clay (Sestini 1976). The advance that occurred over the last 2000 years or more has been recorded along some other parts of the coast by ancient beach ridges, such as those of the **El-Tinah Plain** (Sneh and Weissbrod 1973). Some other sectors of the coastline have been unstable or experienced negligible changes. Average high rate of sedimentation of about 5 m/1000 year was estimated from 48 m section of Fadda 1 borehole, near the defunct **Pelusiac Branch** (El-Tinah Plain) deposited in about 10,000 years (Sneh et al. 1986).

The most notable features of the coastline are sand spit beaches and cusps. Spits are periodically growing at Rosetta and Damietta mouths, Borollos, and Manzala lagoon outlets. Beach cusps characterize the shores between Ras Borollos and Gamasa. From the shoreline, flat expanses of beach sand as sandy plains with coastal dunes extend at several localities from Abu Qir bay in the west to El-Tinah plain in the east. The sandy plain has a varying width of 1–12 km in El-Tinah plain and is built up of several layers of marine and Nile sediments, of which the top layer is composed of a series of very low accretionary sand ridges (Sneh et al. 1986). These plains are exposed to flooding during winter storm surges leaving traces of temporary shorelines and deposition of salt and gypsum. At various parts, there are traces of former shorelines, which are marked by accumulations of marine shells and a narrow zone of low dunes, especially from west of Gamasa to **Ras El-Barr** and at El-Tinah plain (Sestini 1976).

6.5.2 Coastal Dunes

Coastal dunes spread along the plains of the northern fringes of the delta at disconnected localities. They appear to the north and south of Lake Idku, on some of the islands of this Lake, at the eastern shores of Abu Qir Bay, around Rosetta Town, and in Mid-Delta between the two Nile branches. Dunes in this sub-region form a dune field, which was named in a former study (Embabi 2000), the "**North Delta Dune Field**." Density of dunes varies greatly in this field, but the largest area covered by dunes with highest density is the one extending from El-Borg to Gamasa. Dune types in this field vary between simple barchans with horns pointing southward, especially in the Mid-Delta sector, to complex and deformed barchans, and small linear dunes. Maximum height is about 20 m in the vicinity of Rosetta and to the east of El-Borg, but the most common height is 2–3 m.

There are two generations of dunes in this field. The first is the older generation composed of low and vegetated dunes with reddish sand. These older dunes exist below the recent coastal dunes in the west and at Kom Mastero east of Rosetta, and in the area of El-Borg, Baltim, and Gamasa. Remnants of these older dunes were found on the islands of Idku and Borollos lagoons, such as Kom El-Akhdar Island (Sestini 1976), and to the south of Idku and Borollos lakes (The Geological Map of Egypt, scale 1:500,000, 1986–1987). Remains of former Islamic settlements are found on these ancient dunes at Kom Mastero and El-Borg. Sand grains of these older dunes are coarser compared to recent ones. This coarse sand is similar to the coarser sand of the Holocene, which was found in boreholes at a depth of 20 m (Sestini 1976).

The second generation is composed of recent dunes, which are higher and sharper than the ancient stabilized ones. Inter-dune areas in these recent dunes appear as innumerable small depressions with flat floors, some of which are occupied by temporary or permanent ponds. These ponds are fed by rainwater stored in dune sands, then percolating and filling the inter-dune areas. Most of the area covered by coastal dunes has undergone reclamation a long time ago, to be used for various purposes. Therefore, the dunes are gradually disappearing due to cultivation and other types of land use. It is expected that these coastal dunes will disappear completely within the next decade or so.

In the North Delta Dune Field, growth of dune belts and linear dunes was measured by two series of aerial photographs, 1955 and 1983, in three areas (Gawish 1988). The rate of annual growth in the downwind direction varied between 0.7 and 1.7 m during the above-mentioned period. Differences were explained by local variations in surface moisture, wind energy, dune height, rainfall, and human activity.

Sands of the Nile Delta Coastal Dune Field mainly originated from the Nile deposits deposited along the coasts of the delta or along the northern margins of the delta (Sestini 1976; Gawish 1988). Due to the almost continuous wetness of the original material, only limited amounts of sand are available for deflation by wind. This explains why this field is limited to a narrow strip along the delta coastal margins.

6.5.3 Coastal Lagoons

6.5.3.1 General Aspects

There are four lagoons along the coastal fringe of the Nile Delta. These lagoons are Maryut and Idku to the west of the Rosetta Branch, El-Borollos in Mid-Delta, and El-Manzala to the east of the Damietta Branch. The largest lagoon is Manzala (1071 km^2), the smallest is Idku (31 km^2). According to archeological remains (Ptolemaic, Roman, and Islamic), some of these lagoons did not exist or were smaller 2000 years ago. From a description by El-Maqrizi (traditional Egyptian historian), Manzala developed only during the sixth century (i.e., 1400 years ago). The development of these lagoons is an outcome of a relatively recent subsidence of the coastal fringes of the deltaee and their flooding by seawater (Ball 1939; Sestini 1976; Stanley 1988; Zaghloul et al. 1990b). From the analysis of field data, it was concluded that Manzala is formed by differential subsidence accompanied by faulting in the neighboring parts. The subsidence may have been caused by some local tectonic activities that prevailed in the area during post-Miocene times, and/or indirectly produced by the structures formed during pre-Miocene time. Rates of subsidence ranged between 0.5 cm/year in Manzala lagoon (Stanley 1988), 1.5 mm/year for the coastal lower delta plain (Sestini 1989), 1.1 mm/year in Kafr El-Sheikh, and 0.7 mm/year at Sherbin, west Manzala (Zaghloul et al. 1990b). These lagoons are exposed to modifications in area and shape as a result of human activity. Their total area decreased from about 10% of the total area of the delta early in the twentieth century to about 7.5% in the 1980s and finally almost 5.0% in the 1990s of that century. Their shores were straightened by land reclamation.

6.5.3.2 Evolution of Coastal Lagoons

Recent studies of some of the delta lagoons (Maryut and Borollos) revealed that these lagoons were exposed to drastic changes during their history. Study of historic evidence and sedimentary history of Lake Maryut by Goodfriend and Stanley (1996) indicated that this lagoon existed as early as 3400–3000 year BC. Since then, the lake has dried up several times and its area has been exposed to expansion and shrinkage according to water supply reaching its basin. During the times of Alexander the Great (300 BC), a much more extensive body of water—known as **Lake Mareotis**—occupied the low-lying areas to the south of the present lake. The lake was supplied by water from the Canopic Nile Branch through a series of canals. By the twelfth century, the lake dried up because the **Canopic distributary** had been silted up and ceased to supply the lake with water.

At the beginning of the nineteenth century (1807), during the Napoleonic Wars, the British filled the basin of Lake Mareotis by seawater by cutting the dyke separating the lake basin from the Abu Qir lagoon to the east. After the closure of the connection to Abu Qir lagoon in 1804, it was reopened once again in 1807, resulting in the flooding of the Mareotis basin with seawater. Due to the separation of the lake from the sea by a dyke in 1808, the lake dried up again until 1892. Since then, the lake basin was isolated from the sea and is fed by Nile water drained from cultivated fields.

From the study of Late Quaternary evolution of the Borollos lagoon region, Arbouille and Stanley (1991) concluded that this lake and its region passed through several stages. A lagoon first appeared in north central Nile Delta plain to the west of the **Sebennitic Nile Branch** during the period 6000–5500 year BP. This lake may be called the "Ancestral Borollos Lake," since it was developed to the north of present-day lagoon, because sea level was lower than it is at present. This lake was bounded by extensive marshes that extended as far south as the Pre-dynastic archaeological site of Buto. This lagoon-marsh complex extended also westward as far as what is now the Rosetta Nile Branch. The ancestral Borollos Lagoon existed also during the following stage (4500–4000 years—BP), but with a smaller area and different configuration. This is explained by Arbouille and Stanley (1991) as the result of the advancement of the delta plain northward or narrowing of the lagoon by the development of a more extensive coastal barrier.

During the third stage (3500–3000 years BP), a lagoon developed at the present location of Borollos lagoon, but with a different shape and wider area. At this time, the coastal barrier east of Rosetta promontory started to develop separating the lagoon from the sea. Moreover, a new Nile Branch, called by Arbouille and Stanley (1991) Pre-Saitic, started to develop and flowed to the south central lagoon margins. In the meantime, due to decrease in Nile floods, the importance of the Sebenitic Branch had diminished. In the fourth stage (2500–2000 BP), the lagoon attained an area and configuration quite similar to those of present day. At the same stage, water flow began in the Bolbitic (Rosetta). The importance of this branch increased during the past two millennia due to the decreased flow in the earlier branch system. Concurrent with these changes in water flow in the Nile branches, the Rosetta promontory developed and the coastal barrier separating Borollos and the Mediterranean Sea extended eastward toward its present position. As the transport of sediments into the lagoon by the active distributaries decreased, and as the subsidence of the delta continued, Borollos lagoon extended southward as recorded by surveys carried out during the nineteenth century. Since then, topographic map series have shown that the southern shore of Borollos lagoon has shifted northward partly by accretion at the mouths of canals and drains and also as a result of extensive land reclamation projects.

6.5.3.3 Geomorphological Characteristics

All coastal lagoons are shallow basins, where water depth varies between 0.5 m and 3.0 m in deep areas, but along their margins, especially the southern ones, depths do not exceed several centimeters. In these very shallow tracts, strong winds might drive off the thin water layer leading to drying up. Except for Maryut Lake, the lagoons are connected to the open sea by one or more openings. These outlets keep the water level in the lagoons similar to or quite near the sea level. Mostly, they are kept open by artificial dredging mainly to improve water circulation for fishing purposes. In the meantime, several agricultural drain channels end up at these lagoons, feeding them with excess irrigation water, drained from the cultivated fields in the delta. This explains why water salinity in the lagoons is lower than that of the sea, especially along the southern margins where the drains enter them. Because it is not connected to the sea, Maryut lagoon is fed mainly by the excess irrigation water drained from the cultivated fields of the western Delta. Also, untreated industrial wastewater is disposed of in the Lake. Because of these conditions, and because it is divided by roads and embankments into several basins, Maryut Lake is more saline than seawater, and its western arm is cut off for the extraction of salt. All lagoons are separated from the open sea by narrow stretches of land of different origins. El-Borollos and El-Manzala are separated by sand barriers or plains, which developed recently, whereas Maryut Lake is separated by a 20-m aeolianite ridge, and Idku Lake is separated by 1–3 m ridge, which follows the course of the former Canopic Branch (Sestini 1992).

Islands are one of the characteristic features of the lagoons. They are small and low, but differ in shape, area, and origin. On the basis of orientation, sediments, and morphological characteristics, many islands of the lagoons appear to be remnants of ancient features such as beach and/or aeolianite ridges, dunes, and riverbanks.

Islands of El-Manzala Lagoon are particularly significant since their number exceeds several hundreds and most of them are linear, and are organized in rows parallel to the present coastal sand barriers, subdividing the lake into several sub-basins. These chains of islands have represented various successive barriers during the growth of the lagoon. In general, the morphology of many islands has been modified by the prevailing N and NW winds and the resulting waves. The N and NE oriented shores are generally sharp with white shelly beaches, **spits** at north and south ends, extending eastward. Protected by the spits, the eastern shores are marshy and irregular, having small lagoons.

6.5.4 The Barari (Wasteland) Salt Marshes Flats

A belt of salt and marshy flats covered with aquatic vegetation (reeds) extends along the southern margins of the lagoons from Maryut Lake in the west to El-Tinah Plain in the east. Most of this belt is a few meters below sea level, where underground water level is high and the soils are saturated with saline water. At present, these areas are subjected to flooding from the lakes during winter storms, whereas in the past they were flooded in summer by the Nile water. These flats are collectively called by locals "*Barari*" (Singular: *Barriya*), which is an Arabic word for unproductive wasteland. In several localities, topographic maps show that local names are indicative of physical conditions, such as "*Mallaha*" (salt pan), "*Birket*" (pond), and "*sayaha*" (flooded flat). At the beginning of the twentieth century, the total area of this belt was about 5000 km^2, i.e., almost twice as that of the lagoons (Willcocks and Craig 1913), and extended about 50 km southward up to the 3.0-m contour line. Due to population pressure, these flats were exposed to land reclamation, and most of them are currently either cultivated or otherwise utilized. To improve land productivity, the salty soils were washed several times and artificially drained into the lagoons by pumping stations. This is why the current total area of the flats is but a fraction of that at the beginning of the twentieth century. The most extensive areas, which still exist, are found at the southern margins of Lake El-Manzala, and its eastern margins to the east of the Suez Canal, where it is known as El-Mallaha and El-Tina plain. These latter areas will disappear in the near future due to land reclamation within the national project of "El-Salam Canal Land Reclamation Project" in north Sinai and west of the Suez Canal. Google Earth images show these areas are under agricultural reclamation since a few years ago, transformed from **salt marshes** and **sabkhas** to cultivated fields.

Only small patches of these marshy, salty sabkha areas are present at the southern margins of El-Borollos and around Idku and Maryut lakes (Fig. 6.1).

6.5.5 Interaction Between the Coastal Fringe and the Mediterranean Sea

Nearly all previous studies on the Nile Delta, such as those of Sestini (1992), Stanley (1990), and Frihy (1988), pointed out that its Mediterranean coasts are exposed to changes due to the action of waves, currents, wind, sea level changes, and human activity. Such changes can be detected in the field and on topographic maps, aerial photographs, and space images. These coastal changes affect not only the landforms which characterize the coastal fringe of the delta, but also the anthropogenic features established in this fringe. The changes which occur/expected to occur in the future are of great concern to the Egyptians due to its devastating consequences on the Egyptian economy, and this is why it is necessary to highlight such changes (see Chap. 24 Geomorphological hazards in Egypt).

6.5.5.1 Future Sea Level Rise

Studies on long-term changes in sea level rise in the Mediterranean region indicate that it is mainly controlled by the interaction of global, regional, and local factors. These factors are 1—Eustatic sea level rise due to global SLR climatic oscillations accompanying greenhouse warming; 2—local tectonism: subsidence/emergence (uplift) of land compaction, neo-tectonism, oil or groundwater pumping); and 3—seasonal oceanographic/meteorological processes (wind, storm surge, wave setup, or standing waves). Of these factors, three are primary for the Nile Delta: glacial melt from Greenland and Antarctica due to global warming, local subsidence of land due to neo-tectonism/gas or groundwater withdrawal.

Among those, glacial melt has been expected to be the dominating factor behind the future rise in sea level. It is predicted that the global sea level would rise by 69 cm by the 2080s and 30–80 cm by 2100 owing to global warming. As the global sea level rises, many low-lying areas of the coasts will be submerged worldwide, and with increasingly severe and frequent storms and wave damage, shoreline retreat will be accelerated. However, impacts of the future sea level rise will not be globally uniform. Low gradient coastal landforms, such as those of the coastal fringe of the Nile Delta, are most susceptible to submergence. Sea level rise along the coasts of the Nile Delta is indicated by tide-gauge records; the year-to-year averages of these records yield measurements of the ongoing rise in sea level. These data comprise the monthly average tide gauge records calculated from the hourly heights of the sea level taken at Alexandria (1944 through 2001; 55 years), Port Said harbors (1926 through 1986; 48 years), and the Borollos (1972 through 2001; 26 years). Results indicate that the mean sea level at Alexandria, Borollos, and Port Said has risen 1.6, 1.0, and 2.3 mm/year, respectively (Frihy 2003). These results when compared with the above-mentioned ones related to global rise; it becomes clear that there is an overestimation in the global prediction models for SLR.

Unfortunately, there are no recent or ongoing measurements with regard to subsidence, but there is ample evidence for continued subsidence in the Nile Delta during the last 2000 years. Evidence includes the Greco-Roman ruins, which lie below sea level along the stretch of Alexandria-Abu Qir (Mahmoud 1995); the occurrence of drowned river banks, beach ridges, and submerged ruins in the lagoons; and the presence of several areas 1–3 m below sea level lying to the south of Lake Borollos, including reclaimed plots where drainage waters have to be pumped upward to the lagoon (Sestini 1992). In addition, the eastern part of Lake Manzala (the northern part of the Suez Canal zone) appears to be subsiding at a rate of 5 mm/year (Stanley 1988). This subsidence could be a contributing factor to the frequent flooding of the region to the east of the Canal.

Subsidence in the Nile Delta could be due to tectonics such as the sector of Alexandria (Mahmoud 1995; Stanley 1988), pressure exerted by the weight of the Nile sediments, or due to extraction of oil and gas, and underground water. This latter factor has not yet been studied, but it is expected to become increasingly significant, since large quantities of natural gas are withdrawn from gas fields discovered in the northern part and offshore the delta. Underground water is increasingly discharged from the delta aquifer to cover the water needs for the newly reclaimed land in the coastal fringe, which is not covered by Nile waters.

6.5.5.2 Sea Level Rise During the Holocene

A number of worldwide investigators have developed chronologies of sea level variations for the last 10,000 years worldwide. They generally agree that about 10,000–8000 years ago, the sea was approximately 30 m lower than at present (Kidson 1982). With the melting of glaciers and thermal expansion of seawater, there was initially a rapid rise in sea level, averaging about 8 mm/year, until approximately 4000 years ago when it slowed to 1–2 mm/year, and then reached its present level 2000 ago. The sea level rise has been estimated to be on the order of 1–2 mm/year, which is similar to that estimated by Stanley and Wane (1993) for west Delta, and quite near to that one estimated by Mahmoud (1995) for SLR in Alexandria region over the last 2000 years. This rate is expected to accelerate during the twenty-first century due to the human-induced global warming.

6.6 Conclusion

The Nile Delta represents a unique region with a landscape of low relief. Apart from the turtle-back mounds, it is a smooth deltaic plain, which slopes gradually in a fan-like shape from its apex toward the Mediterranean Sea. Its development was governed by its origin as a tectonic embayment, which has been filled with the Nile sediments since Late Miocene and this is why the Nile Delta acquired its unique triangular shape. Due to the dramatic events which occurred to the Nile during its development, the delta was built up during multiple stages, each of which represents a sub-delta. Therefore, the delta is a composite one. Since sediments were deposited on the dried floor of the Mediterranean, it took the form of separate fans at the beginning of deposition; then, by continuing deposition, the fans were coalesced. Therefore, it can be concluded that the first stage in the development of the Nile Delta was a coalesced fan stage and not a delta. Local landforms can be classified into two groups: ancient forms which are restricted to turtle-backs mounds, and more recent ones. The mounds are changing only by human action. Their sands are quarried for construction purposes. Therefore, they are expected to disappear in a short time under human erosion. The second group of local landforms is concentrated along the coastal margins. They comprise a wide variety of forms which are exposed to marine and continental processes. Lagoons are the most prominent, since they cover wide areas especially in mid-delta and east delta. These lagoons and other forms (sabkhas, marshes, dunes) are exposed to human action. Therefore, they are shrinking under human pressure. Delta coasts have been exposed to the action of wave and offshore currents since a long time ago, but recently they represent a coastal hazard due to their erosion along some sectors, especially the promontories of Rosetta, Borollos, and Damietta. Another hazard comes from global warming and sea level rise.

References

Abd El-Motaal E (1999) Neotectonics and morphotectonics implications of the Nile Delta basin, Egypt. Z Geomorph N F Suppl-Bd 118:165–181

Arbouille D, Stanley D (1991) Late quaternary evolution of Burullus lagoon region, North Central Delta, Egypt. Marine Geol 99:45–66

Ball J (1939) Contribution to the geography of Egypt. Survey and Mines Department, Ministry of Finance Egypt, Government Press, Cairo

Ball J (1942) Egypt in the classical geographers. Survey of Egypt, Cairo

Conoco Coral and The Egyptian General Petroleum Corporation (1986–1987) The Geological Map of Egypt, 20 sheets, scale 1:500,000, Cairo

El-Disouki S (1997) Some recent geomorphologic changes in the channel of Rosetta Branch (in Arabic). The Arab Geogr J Cairo 29:55–94

El-Fawal F (1992) Sedimentology of the quaternary southwest of Ismailia and its chrono-stratigraphy west of the Suez Canal. In: Proceedings of 3rd conference geology of Sinai and development. Ismailia, pp 141–152

El-Gamily MM, Issa GH, El-Khoriby EM (1988) Geo-electric study on some sand islets (Turtle-Backs) in the Nile delta, Egypt. In: Abstract, 7th symposium on the quaternary & development in Egypt. Mansura University. Mansura, Egypt

El-Shazly E, Abdel Hady M, Salem A, Morsy M, El-Rakaiby M, Aassy I, Kamel A, Meshref W, Ammar A, Meleik M (1975) Geological investigations of the Suez Canal Zone. Remote sensing research project. Acad Science & Research Tech, Cairo

Embabi NS (2000) Sand dunes in Egypt. In: Soliman SM (ed) Sedimentary geology of Egypt I: applications and economics, Special publication. The Sedimentological Society of Egypt. Cairo

Embabi N (2004) Geomorphology of Egypt, vol I. The Nile Valley and the Western Desert, Special Publication. The Egyptian Geographical Society, Cairo

Fourteau R (1915) Contribution a L'étude des depot Nilotique, Mem. Institut d'Égypte, Le Caire T VIII: 56–94

Frihy O (1988) Nile delta shoreline changes: aerial photographic study of a 28-year period. J Coast Res 4:597–606

Frihy O (2003) The Nile Delta-Alexandria Coast: vulnerability to sea-level rise, consequences and adaption. Mitig Adapt Strat Glob Change 8:115–138

Gawish AM (1988) The geomorphology of the coastal zone of the Nile Delta between Damiatta and Rosetta Nile Branches. Dissertation (in Arabic) Cairo University

Goodfriend G, Stanley D (1996) Reworking and discontinuities in Holocene sedimentation in the Nile Delta: documentation from amino acid racemization and stable isotopes in mollusk shells. Mar Geol 129:271–283

Hume W (1925) Distribution of sand deposits in Egypt. In: Geology of Egypt, vol I. The surface features of Egypt, their determining causes & relation to geological structure. Survey of Egypt, Government Press, Ministry of Finance, Cairo, pp 29–81

Issa Gh, El-Sherbini M, Abdel-Daiem A, El-Khoriby E (1990) Field occurrence morphology and sedimentological characteristics of the sand islets (Turtle-Backs) in the Nile Delta, Egypt. Science Bul 17:575–597 (Mansura University)

Kholief M, Helmy E, El-Shahat A (1969) Geological and mineralogical studies of some deposits in the Nile Delta, U.A.R. J Sediment Petrol 59:1520–1529

Kidson C (1982) Sea level changes in the Holocene. Quat Sci Rev 1:121–151

Mahmoud AA (1995) Evidence of the Late Holocene subsidence, Northwest of Alexandria, Mediterranean coast, Egypt. Bulletin de la Societe de Geographie d'Egypte 68:99–112

Ramadan F (1984) Sedimentological studies on the bottom sediments of the Suez Canal. Dissertation, Zagazig University

Said R (1981) The geological evolution of the River Nile. Springer, New York

Said R (1990) Quaternary. In: Said R (ed) The geology of Egypt. Balkema, Rotterdam

Said R (1993) The River Nile: geology, Hydrology and Utilization. Pergamon Press, Oxford

Schlumberger (1984) Well evaluation conference. Egypt, Cairo

Sestini G (1976) Geomorphology of the Nile Delta. In: Proceeding of The Seminar on the Nile Delta Sedimentology. UNESCO/ASRT/UNDP, Alexandria

Sestini G (1989) Nile Delta: a review of depositional environment & geologic history. In: Whateley M, Pickering K (eds) Deltas: sites and traps for fossil fuel. Geological Society of America, Special Publication, vol 41, pp 99–127

Sestini G (1992) Implications of climatic changes for the Nile Delta. In: Jeftic L, Milliman J, Sestini G (eds) Climatic changes and sea level rise in the Mediterranean Region. Edward Arnold, London, pp 536–601

Sneh A, Weissbrod T (1973) Nile Delta: the defunct Pelusiac branch identified. Science 180:59–61

Sneh A, Weissbrod T, Ehrlich E, Horowitz A, Moshkovitz S, Rosenfeld A (1986) Holocene evolution of the northeast corner of the Nile Delta. Quat Res 26:194–206

Stanley DJ (1988) Subsidence in the northeastern Nile Delta: rapid rates, possible causes, and consequences. Science 240:497–500

Stanley DJ (1990) Recent subsidence and northeast tilting of the Nile Delta. Mar Geol 94:147–154

Stanley DJ, Wane AG (1993) Nile Delta: recent geological evolution and human impact. Science 260:635–658

Taha MM (1997) Geomorphology of the Nile depositional islands (in Arabic). Arab Geogr J Egypt Geogr Soc 29:223–229

Tetra Tech Inc. (1984) Shoreline master plan for the Nile Delta Coast, Report to shore protection authority. Ministry of Irrigation, Egypt, Cairo

Torab MM (1990a) Morphology of the Damietta Nile channel (in Arabic). Paper presented in the symposium of geography and cartography in society welfare, 6–8 March 1990. Faculty of Arts, Alexandria University, Alexandria

Torab MM (1990b) Morphology of the Nile islands in Damietta Nile Branch after the construction of the High Dam (in Arabic). Arab Geogr J Egypt Geogr Soc 22:137–165

Torab MM (1998) Causes of disappearance of the Pelusiac Nile Branch, and the environmental hazards facing El-Salam Irrigation Canal in El-Tina Plain (in Arabic). J Humanit 2: 53–100 (Faculty of Arts, Alexandria University (Damanhur Branch))

Tousson O (1922) Mémoire sur les ancienne branches du Nil. Memoire Institut d'Égypte 4:212

Willcocks W, Craig JI (1913) Egyptian irrigation. E & F N Spon, London

Zaghloul ZM, Abdel-Daim AA, Taha AA (1990) Geomorphology, geologic evolution and subsidence of the Nile Delta during the quaternary. Mansura Sci Bull 17:471–495 (Mansura University)

Zaghlooul ZM, Issa G, El-Sherbini M, El-Khoriby E (1990b) Contribution to the origin of Turtle-Backs in the Nile Delta Area, Egypt. Mansura Sci Bull 17: 599–620

Part III

The Plateaus

The Sandstone Plateaus, SW Western Desert

7

Abstract

The sandstone plateaus of Gilf Kebir and Abu Ras in SW Egypt have been known as the Gilf Kebir. Recently, this name became restricted to the eastern plateau, while the western is named Abu Ras. They are composed of horizontal beds of Palaeozoic and Mesozoic sandstones, and are covered by a resistant caprock. Volcanic intrusions of different sizes are scattered on the surface of western central part of the Gilf Kebir Plateau, forming black to dark gray sub-circular constructs on its surface, and on floors and wadi walls. Faults of different sizes and directions cut across the Gilf Kebir and Abu Ras Plateaus, and the southern escarpment of Abu Ras and El-Aqaba. Current landforms are due to the interaction of fluvial, mass wasting, and aeolian processes. Therefore, dominant fluvial forms are wadis dissecting both plateaus, and playas restricted to the upper reaches of some wadis in the Gilf plateau, and Dunes are found in Gilf plateau only, covering the N escarpments and the Depression separating the two plateaus, and the falling and climbing dunes in several E-W wadis dissecting the plateau. Geologically and geomorphologically, it can be said that these are the oldest sedimentary plateaus in Egypt and have been exposed to fluvial processes at least from Late Eocene. When climatic conditions became relatively arid at the beginning of the Pleistocene, sand from the Great Sand Sea encroached on the northern slopes of the Gilf and choking some wadis. Only minor change occurred by fluvial action, such as the Holocene playas at the heads of some wadis.

Keywords

SW Egypt • Gilf Kebir and Abu Ras Sandstone Plateaus • Volcanic Hills • Wadi Canyons • Playa deposit • The Great Sand Seas

7.1 Location and Exploration

The sandstone plateaus of SW Western Desert are characterized by a succession of horizontal sandstone strata of Palaeozoic and Mesozoic age, intruded by volcanic rocks. These platforms are capped by extremely resistant sandstone and exhibit local forms, which indicate that they represent larger sandstone platforms. Discovered and named by Prince Kemal El-Din Hussein in January 1926, these platforms have become well known in the literature since 1928 by their famous name "**The Gilf Kebir**" (Hussein 1928). In early 1933, Penderel discovered—while on a reconnaissance flight along the western escarpments of the Gilf Kebir—the separation between the northern and southern parts of this

N.S. Embabi, *Landscapes and Landforms of Egypt*, World Geomorphological Landscapes,
https://doi.org/10.1007/978-3-319-65661-8_7

extensive plateau (Penderel 1934). Contour lines on topographic maps (scale 1:250,000) and recent radar images indicate that these two parts are separated by a large, wide wadi covered by a sand mantle (El-Baz 1999). The two parts are connected by a steep dissected escarpment called **El-Aqabah** (The Obstacle). Recently, the northern platform has been called "Abu Ras Plateau" on some maps (Geologic Map, scale: 1:500,000, 1987, Sheet No. NF 35 NW), after a promontory carrying the name in the western escarpment of this plateau. Thus, the Gilf Kebir is now restricted to the southern platform. The wadi separating the two plateaus is 20–30 km wide and is floored by numerous linear dunes that migrated southward from the Great Sand Sea. Exploration of this part of Egypt started in the late 1930s and early 1940s (Hussein 1928; Clayton 1933, 1937; Almásy 1936; Shaw 1936; Bagnold et al. 1939). Although the main purpose of these early expeditions was to explore and map this extremely remote region of SW Egypt, they made highly important contributions to the understanding of the geology and geomorphology not only of Gilf Kebir, but also of the southern Western Desert and arid zone geomorphology. Exploration and research were resumed after the World War II, mainly by the Egyptian Geological Survey either independently or in collaboration with foreign expeditions (El-Baz et al. 1980). However, multidisciplinary research attracted more scientists due to the increasing appreciation of the incredible potential of this barren region of SW Egypt. Also, our knowledge increased significantly due to the gradual proliferation of topographic, geologic, and aerial photographs and space imaging coverage.

7.2 Geology

The geology of both Abu Ras and Gilf Kebir Platforms is treated in several studies (Issawi 1973, 1978, 1982; Dardir 1982; Grolier and Schultejann 1982; El-Ghawabi 1987; Hermina et al. 1989; Klitzsch and Schandelmeier 1990). Gilf Kebir is comprised of Late Jurassic to Late Cretaceous clastics, whereas Abu Ras is composed of Silurian, Devonian, and mainly Carboniferous clastics, overlain by a relatively thin cover of Late Jurassic to Middle Cretaceous strata (Klitzsch and Schandelmeier 1990). The Geological Map of Egypt (scale 1: 500,000, Gilf Kebir Plateau Plate, No. NF 35 NW) shows that the Gilf Kebir Plateau is covered by the Sabaya and Magharabi Formations (Upper-middle Cretaceous). These two formations (mainly fluvial sandstone) make up the upper-half to one-third of this plateau, whereas the lower part is built up of the Gilf Kebir Formation (Late Jurassic-lower Cretaceous), which is mainly composed of fluviatile sandstone (Klitzsch and Schandelmeier 1990).

The same Geological Map also shows that Abu Ras Plateau is covered by both the Sabaya and Gilf Kebir Formations. Palaeozoic rocks are exposed not only along the southern and western escarpments of Abu Ras Plateau, including the Aqaba Pass, but also along sides of the wadis dissecting this plateau. The Silurian Um Ras Formation, the oldest Palaeozoic rocks, is exposed at the base of the escarpments and attains a thickness of ~400 m between Um Ras and SW slopes of the plateau (Klitzsch and Schandelmeier 1990). Um Ras Formation is composed of fluvial to deltaic white sandstone intercalated with beds of shallow marine siltstone, shale, and sandstone. The Wadi Malik Carboniferous Formation (marine sandstone, siltstone, shale inter-bedded with fluvial deltaic and tidal sandstone) is best exposed along the sides of Wadi Abd el-Malik and its side wadis (the Geological Map, scale 1:500,000, Plate No NF 35 NW, 1987). These wadis extend from the south and southwest of this plateau to the northern end. At the northern section of Wadi Abd el-Malik, the Wadi Malik formation is overlain by Late Carbonirferous glacial and fluvio-glacial deposits with varves (northern Wadi Malik formation; the Geological Map, scale 1:500,000, Plate No NF 35 NW, 1987). Volcanic intrusions of different sizes are scattered on the surface of the western central part of the Gilf Kebir Plateau. They form black to dark gray sub-circular constructs on the surface, and on floors and wadi walls of some wadis. These intrusions also exist as low sub-circular and arcuate ridges that protrude above the plateau surface (Grolier and Schultejann 1982). The age of these intrusions is estimated to be Late Mesozoic-Early Cenozoic (Meneisy 1990), and recently dated at 38 ± 2 Ma (Kröpelin 1987), whereas other researchers believe that several episodes of volcanic activity were experienced ranging in age from Oligocene or Miocene to Quaternary (Grolier and Schultejann 1982).

The same Geological Map also shows that faults of different sizes and directions cut across Gilf Kebir and Abu Ras Plateaus, and along the southern escarpment of Abu Ras and el-Aqaba. Four main directions can be detected on the maps: E-W, N-S, NW-SE, and NNE-SSW. The bounding faults of NNW-SSE strikes are greater in proportion compared to other trends (El-Ghawabi 1987). This latter study concluded, using the interpretation of Landsat images, that the faults in this region show a system of horst-graben structure similar to that of the Red Sea region. This study postulates that the entire region (including the southern part of the Great Sand Sea) has a topographic expression of major block-faulting structure, where Abu Ras Plateau represents a horst block,

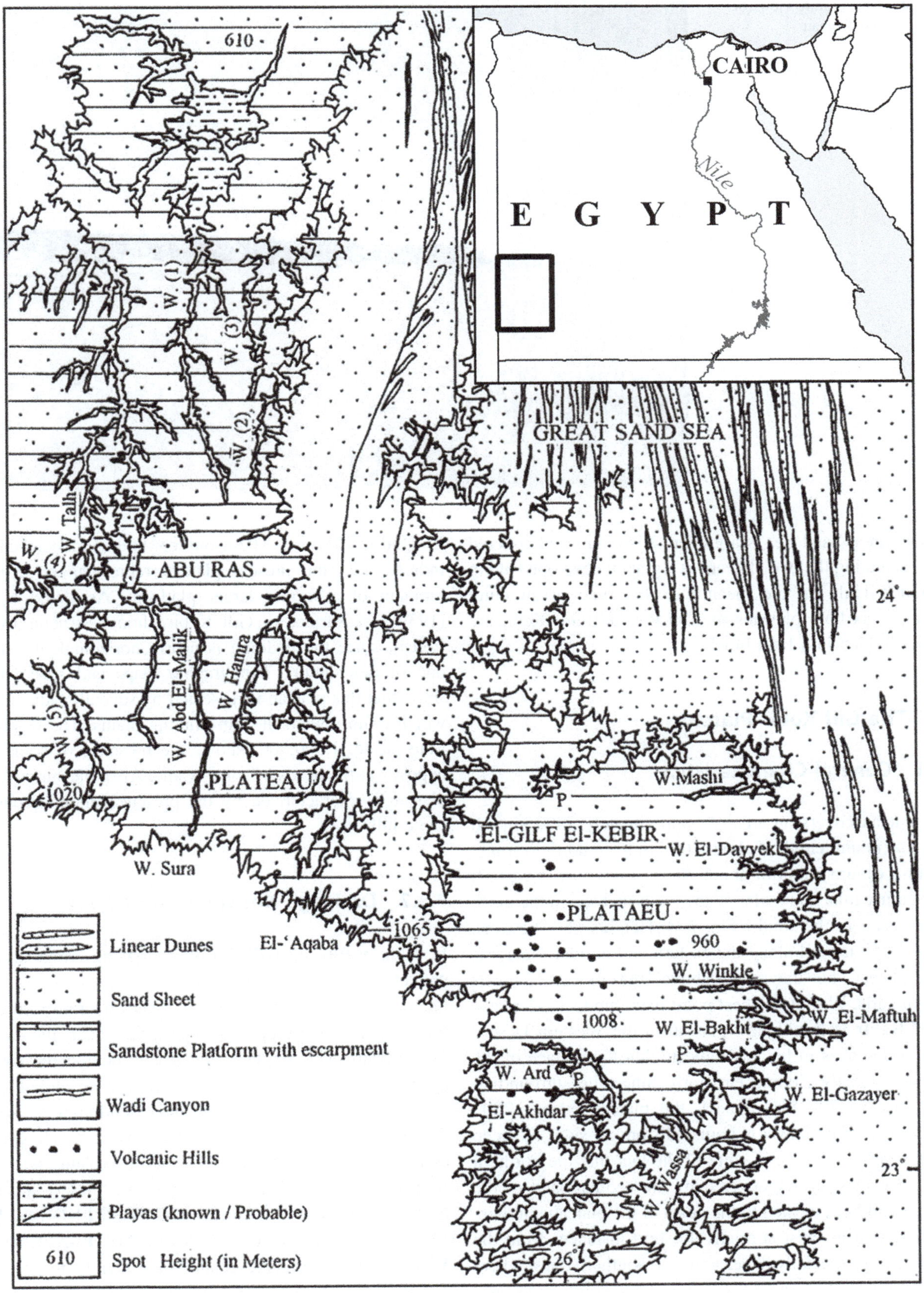

Fig. 7.1 A geomorphologic map of the Gilf Kebir and Abu Ras Plateaus

Fig. 7.2 A panoramic view for the Gilf Kebir Plateau

and the Great Sand Sea is a graben. Also, the sand-covered wadi separating the two plateaus is a graben developed along a wrench fault. In the meantime, this type of structure indicates that both plateaus of Abu Ras and Gilf Kebir are locally divided into fault blocks.

7.3 The Gilf Kebir Plateau

7.3.1 General Characteristics

This plateau covers an area of approximately 7500 km^2. It has a maximum length of about 100 km from north to south, and a maximum width of about 70 km from east to west. It takes an irregular triangular shape, with its base in the north and its head in the south (Fig. 7.1). Heights vary between 850 and 950 m a.s.l along the eastern and northern margins of the plateau, but they reach above 1000 m a.s.l along its western parts. With its highest point rising up to 1091 m a.s.l, **Gilf Kebir** is the highest plateau in the Western Desert. These differences in heights indicate that the surface of the plateau is mostly sloping eastward. Since the surrounding plains rise to 700–800 m a.s.l, the plateau edges—200–300 m high. The plateau is bounded from all sides by steep escarpments (Fig. 7.2), in spite of the fact that the northern escarpments are nearly covered by the aeolian sand of the **Great Sand Sea,** located to the north of the plateau. Outliers of different sizes exist at relatively short distances from the escarpments, especially the eastern and southern ones, where they are relatively large. Of these outliers, some large ones rise to the same level as the nearby escarpments and extend for several kilometres, such as the one named **El-Oqab El Qadim** (950 m). These outliers and semidetached parts indicate that the present plateau represents an erosional remnant of a former larger platform. As stated by Peel (1939), the surface of Gilf Kebir, though structurally controlled, since the bedding is nearly horizontal, is certainly one of denudation. Topographic maps and space images show that this plateau is dissected by several wadis, most of which drain toward the surrounding plains. The Geological Map (scale, 1:500,000) shows the presence of some playa deposits at the heads of some wadis. Also, the surface of the plateau is characterized by the presence of numerous volcanic hills.

7.3.2 Drainage Networks and Basins

In plan, **wadis of Gilf Kebir** follow an irregular radial pattern, with the majority draining eastward. As shown on the topographic maps, the divide, if such a term may be used for a nearly flat-topped plateau, lies near the western escarpments. This explains the reason behind the fact that the most extensive wadi systems are best developed on the eastern and southeastern edges, and extend up to 50 km into the interior, although wadis have developed all over the escarpments of the plateau. This also explains why these wadis carry names. Figure 7.1 shows that these wadis are—from north to south—**Wadi Mashi**, **Wadi El-Diyaq**, **Wadi El-Maftuh**, **Wadi El-Bakht**, **Wadi Ard El-Akhdar**, and **Wadi Wassa**. Drainage basins vary greatly in their geomorphologic characteristics, mainly in area, density, morphometric aspects, and wadi form. Topographic maps (scale 1:100,000) show that the largest wadi basin is that of Wadi

Wassaa, followed by **Wadi Winkle** and Wadi Bakht. It can be also noticed from the maps and from morphometric analysis of drainage networks that the number of the first- and second-order wadis is exceedingly large. The description provided by Peel (1939, 1941) represents the first geomorphological information on these wadis. After 40 years, only two wadis (El-Bakht and Ard El-Akhdar) were investigated in detail (Maxwell 1982). However, it is possible to make some conclusions on the morphology of the wadi canyons in the Gilf Kebir based on the data collected by these previous studies, with the assistance of fieldwork, topographic maps, and space images.

Wadis are composed of straight sections, connected by abrupt bends. These straight sections take four directions: N-S, E-W, NW-SE, and NE-WS. Such a characteristic indicates that wadis are structurally controlled. Previous studies concluded that the straight sections of the wadis are controlled by a joint/fracture system that takes the above-mentioned directions (Maxwell 1982; Hamdan and El-Etr 1999). Wadis can be divided into two types according to shape: canyon-like wadis and shallow short wadis. The canyon-like type extends along the lower reaches of the main wadis and main tributaries. These wadis are narrow with flat beds and steep sides. Their depth varies between 100 and 150 m, and their width between 200 m in their upper sections and 1000 m near their mouths. The shallow and short wadis extend along the upper reaches and represent the first- and second-order wadis. This difference is due to two factors. The first is that the canyon-like sections are the parts most-affected by structural lines, which facilitated vertical rather than lateral erosion. The second factor is the mode of wadi development, which was explained magnificently by Peel (1939, 1941). It was suggested that when rain falls on the plateau surface, the bulk of water does not flow over the surface, but sinks into the permeable sandstone and reappears again at the base of the plateau cliffs in the form of springs. Such springs would lead to spring sapping, attacking the scarps at their base. Therefore, the upper parts of the cliffs would retreat by undermining. As such, wadis would grow and extend headward along the joint lines by undercutting at their heads, starting from the plateau escarpments. Under this mode of development, no extensive widening of these canyons occurs since one proceeds headward as shown on space images and topographic maps (Fig. 7.1). Due to the fact that the bulk of rainwater soaks into the sandstone, little water is left to flow on the surface of the plateau. Therefore, wadis developed under this weak water flow would be shallow and short. This is why first- and second-order wadis, which drain into the **canyon-like wadis**, carry these characteristics. This is also why this type of wadis is imperceptible in the field to the extent that Peel (1941) stated "*there is a surprising lack of evidence of water flowing over the plateau surface to drain into the wadi head.*" Also, this type of wadis does not appear on small-scale maps, such as 1:500,000. Therefore, these wadis did not attract the attention of previous studies, which is a fact that also holds true for the small internal drainage basins developed in the western part of the plateau.

Fig. 7.3 A lee dune developed in the leeward side of Wadi Bakht

7.3.3 Local Landforms

Local landforms developed in the wadis are of different origin. Accordingly, four forms can be distinguished. They are sand dunes, talus cover, **playa flats**, and gravel sheets. Since the E-W orientation of most canyon-like wadis is perpendicular to the N-S direction of sand flow coming from the Great Sand Sea, sand accumulates on the sides of these wadis in the form of falling or climbing dunes (Fig. 7.3). Although these dunes are relatively small forms, they were significant agents in developing lakes in some wadis. Coalescence of some opposite falling and climbing dunes resulted in blockage of wadis, especially in the upper reaches where the width of wadis is small; such dunes acted as dams that trap runoff water flowed in some wadis, forming shallow lakes upstream (Peel 1939). Angular rock fragments, taking the form of talus cover, and talus cones due to weathering and rockfalls exist in the lower sections of wadi sides and on dune surfaces. Such material produced under the present arid conditions represents the latest geomorphic changes in the wadis of Gilf Kebir and indicates that no significant changes occur under current arid conditions. This is due to the fact that Gilf Kebir Plateau lies in the hyper-arid region of the southern Western Desert.

Playa flats are composed of fine materials that were deposited in former lakes. As shown on the Geological Map (scale 1:500,000), these former lakes developed only in small basins at the heads of some wadis. As has been seen in the field in 1978, these lakes developed due to the blocking of some wadis by coalesced falling and climbing dunes (El-Baz et al. 1980). According to previous studies (Kröpelin 1987), these lakes were formed during early and Middle Holocene. These playa flats will be analyzed in detail in Sect. 7.3.4.

Gravel sheets cover wadi beds and the wide mouths of the wadis, which open at the plains surrounding the plateaus especially from the east and south. Astonishingly, the gravel at wadis mouths does not take the form of fans as expected under such conditions. Although this absence of fans was noticed by previous researchers, no explanation was mentioned. In fact, conditions at the mouths of wadis along the margins of Gilf Kebir Plateau are not favorable for fan development, since there is no abrupt change of slope in these topographic settings. All wadi mouths open toward the plains as wide embayments, where slope changes gradually from wadi beds to the plains that surround the plateau (Fig. 7.4). Therefore, gravel was transported by flowing water issuing at wadi mouths and spread as sheets rather than accumulating as fans. **Gravel sheets** at wadi beds are incised by wadi channels, taking the form of **wadi terraces** (Fig. 7.5), which rise several meters (1–2 m) only above the wadi floor in most cases. However, a terrace 10–15 m high, more than 100 m wide, and 3 km long is recorded on the eastern side of Wadi Winkle near its junction with Wadi El-Maftuh. Also, in Wadi Winkle, terraces at different levels were recorded (Kröpelin 1989). Most probably, this incision occurred after a catastrophic draining of lakes due to breaching of dune dams holding the lakes.

7.3.4 The Playas

Playas developed due to the blockage of some wadis of Gilf Kebir by falling and climbing dunes attracted the attention of

Fig. 7.4 Wide mouth of Wadi Bakht at the eastern side of the Gilf Kebir Plateau

Fig. 7.5 A wide terrace (1–2 m) in the lower section of Wadi Bakht

most researchers who have visited this remote corner of SW Egypt starting with the early explorers in 1938 (Bagnold et al. 1939). First described by Peel (1939), playas were studied in detail when scientific research was resumed in this region in the late 1970s and early 1980s (El-Baz et al. 1980; McCauley et al. 1982; Maxwell 1982; Pachur and Röper 1984; Kröpelin 1987, 1989). In the meantime, playa deposits of Gilf Kebir appeared for the first time on Geological Maps (scale 1:500,000, Gilf Kebir Sheet, No. NF 35 NW) (Conoco and The Egyptian General Petroleum Corporation 1987), and therefore, their distribution is now well known. This Geological Map shows that playa deposits are present in the two wadis of El-Bakht and Ard El-Akhdar, which were first recorded in 1938 by Peel (1939) and later studied in detail in 1978 (Maxwell 1980, 1982; McCauley et al. 1982), and in 1982 (Kröpelin 1987). The same map also shows that playa deposits developed in a small basin at the head of an unnamed wadi cutting into the Plateau from the northern escarpment. Due to its location in the far north of Gilf Kebir Plateau, this latter site has not been studied so far. In his study of the playas of the Gilf Kebir, Kröpelin (1989) recorded the presence of playas in a tributary of Wadi

Fig. 7.6 Breached playa deposits at the head of Wadi Bakht

El-Maftuh, lying 10 km to the north of Wadi El-Bakht, which he called Wadi Winkle. Pachur and Röper (1984) recorded similar playa deposits in the upper courses of Wadi Mashi, and in another wadi, they called "Contrast Wadi."

7.3.4.1 Wadi Bakht Playa

In this wadi, more than 10 m of inter-layered lake and aeolian deposits were deposited upstream from a dune dam (Maxwell 1980). This dam was breached, and fluvial erosion has exposed a cliff in the playa sediments (Fig. 7.6). The wadi floor above the dune is graded to the level of the playa surface. A longitudinal W-E profile across the playa surface showed a gradient of 0.0028 over a distance of about 1775 m (Maxwell 1980), and a gradient of 0.0044 in a distance of about 950 m (Kröpelin 1987), indicating that the surface of the playa is almost horizontal above the dune barrier. This also evidences the cumulative effect of fluvial erosion and deposition long before the past 7000 years since man has lived on the latest lake bed (Maxwell 1982).

As described by Peel (1939), deposits of the playa consist of thin alternating layers of mud and red sand, overlain by a six-inch mud cover. The stratigraphic column of the playa sediments measured by Kröpelin (1987) shows that it is composed of a sequence of homogeneous clastics with no indication of erosional disconformities. The section also shows that the mud cover is thick (6 feet = $\sim$180 cm) and not as thin as Peel (1966) mentioned and that the underlying bed consists of cross-bedded medium sand of aeolian origin, representing the former fossil blocking dune. Among the other findings by Kröpelin (1987) is that between 8.1 and 7.9 m below the surface, the underlying dune sands inter-finger with the overlying playa sands, indicating that deposition of playa sediments kept pace with the growth of the falling dune. Although Kröpelin (1987) stressed that the playa sands were laid down in a playa area (humid environment), and were not wind-blown sand deposited after the desiccation of the playa lake, no mention is made about their origin. However, later studies show that this playa sand is similar to that found in the Sherika playa of Kharga Depression, and was interpreted as wind-blown sand deposited in the playa lake (Embabi 1968–1969). The playa sands of Wadi Bakht might have originated in a similar way or have been washed by runoff from aeolian sands laid down in the surroundings. The age of these playa sediments varies between 8200 ± 500 bp at a depth of 7.2 m below the top of the column, and 6080 ± 420 BP at a depth of 2.4 m below the top of the column (Kröpelin 1987).

7.3.4.2 Wadi Ard El-Akhdar Playa

This wadi is a tributary of **Wadi Wasaa** and lies about 35 km up valley from its junction. In this wadi, a playa is developed in its upper reaches in a circular, amphitheatre-like Depression, which is 2200 m long and 750–1500 m wide, with a total area of about 3 km^2 (Kröpelin 1987). This Depression is surrounded by numerous basaltic hills; one of which forms the northern side of the gorge at the mouth of the amphitheatre. The eastern part of the playa surface is dissected by a network of small channels, which run from the edges of the Depression into a 50–80 m wide main channel that cuts up to eight meters into the playa deposits. The narrow mouth of the Depression, which is 600 m long and 50 m wide, is situated at the eastern end of a main channel below a basaltic hill. This channel of the playa deposits occurred as a direct consequence of the abrupt breaching of the dune blocking the wadi (Kröpelin 1987, 1993). Within this gorge, a sharp-crested N-S falling dune extends from the basaltic hill approximately two-thirds of the way across the wadi (Maxwell 1982). On the southern side of the gorge, a 3.4 m section of sediments is exposed. This section is composed of structureless beds of poorly sorted pebbles and cobbles set in a matrix of silt and clay, alternated with well-sorted red sand lenses with low-angles, cross-stratification dipping westward (Maxwell 1982). These characteristics indicate that during the accumulation of the gravel and other coarse materials under gravity, a falling dune has experienced several phases of extension across the constriction. Therefore, it was concluded that although the blockage of the wadi seems to have been initiated by rockfalls from the southern slope, a falling dune formed in the lee of the basaltic hill was responsible for the complete closure of the constriction (Kröpelin 1987).

The playa deposits of Wadi Ard El-Akhdar differ from that of Wadi Bakht in several aspects. Here, the deposits cover the entire floor of the basin and have been extremely eroded during later periods of fluvial erosion (Maxwell 1982). Examination of a section located 640 m up wadi from the fossil dune dam revealed alternating layers of coarse sand, silt, clay and fine, angular gravel, some of which contain root structures and bioturbation features (Kröpelin 1987, 1989). The sediments take the form of seven single fanglomerates at the peripheries of the playa (Kröpelin 1993). It was concluded from these characteristics that the sediments were washed from the surroundings by sheet floods over a short distance, after exceptionally heavy rainfall (Kröpelin 1987, 1993). Radiometric ages show that sedimentation occurred between about 9400 and 5000 BP (Kröpelin 1987).

7.3.4.3 Other Wadi Playas

Remnants of playa sediments were found in two localities in the upper reaches of Wadi Winkle—(Kröpelin 1989). These deposits were laid down in palaeolakes dammed by falling dunes, which developed at the leeward side of the wadi. Similar deposits were also found—as mentioned earlier—in another two wadis, Wadi Mashi and Contrast Wadi, to the north of Wadi Winkle.

7.3.4.4 Environmental Conditions of Playa Development

The presence of playa deposits in some of the wadis of the Gilf Kebir indicates that certain environmental conditions prevailed in these wadis. The most important aspects of these conditions are as follows:

(1) The development of narrow wadis with a general W-E trend;
(2) Deposition of falling and sometimes climbing dunes at the upper sections of the wadis, blocking the wadis, and damming any runoff that might occur;
(3) The accumulation of runoff or underground water in the basins developed due to the blockage of wadis, resulting in the development of lakes; and
(4) The deposition of fluvial sediments and aeolian material in lakes developed in the basins dammed by dunes.

7.3.5 The Outlier Remnants

As mentioned earlier, steep-sided conical or flat-topped outliers of different sizes lie at relatively short distances from the escarpments of the Gilf Kebir, especially the northern, eastern, and southern ones where they are relatively large. These outlier remnants differ in shape and size. Some large flat-topped ones rise to the same level of the nearby escarpments and extend for several kilometres. The largest flat-topped outlier lies to the north of the plateau and extends about 20 km from east to west, and about 12 km from north to south (Fig. 7.1). Other examples of these large outliers are the one named **El-Oqab El-Qadim** (920 m), lying to the south of Wadi El-Dayyiq, and the one which lies 30 km to the north of El-Oqab El- Qadim (916 m), with a maximum length of 5 km and maximum width of 3.5 km. Other flat-topped outliers are smaller in size, but still rise to heights similar to those of the margins of the plateau. Only, when taking the conical shape, they become lower than the margins of the plateau. These conical hills were called "**inselbergs**" in some previous studies on the Gilf Kebir (Peel 1941; McCauley et al. 1982). Therefore, these outlier remnants can be classified into three categories. The first category comprises the large flat-topped remnants, which rise to the same level of the adjacent plateau and extend several kilometres in length and width, and will be called "mesas." The second category includes the small flat-topped remnants, which extend less than a kilometre with heights similar to those of the adjacent plateau, and will be called "flat-topped outlier hills." This group of hills lies at greater distances from the edges of the plateau than the first group. The third group covers the small conical remnants with rounded crests, which rise to heights less than those of the plateau and lie at a longer distance from the edges of the plateau than the first or second group of remnants. This group will be called "conical hills" or "inselbergs" since they stand isolated on the pediplain and are separated by relatively long distances. Another feature, which can be correlated with the outlier remnants, is the semidetached segments or spurs, extending from the edges of the plateau (Fig. 7.1). Although outlier remnants are classified into three groups, every kind of transition from one group to another or between the outliers and the semidetached segments can be observed in the field or on maps.

These characteristics suggest that the outlier remnants and the semidetached parts represent an evolutionary system. When the semidetached segments are completely separated from the plateau edges, they become mesas or flat-topped hills as long as they still keep the horizontal caprock. When the caprock disappears, the mesas and hills gradually become conical hills (inselbergs) standing in groups, chains, or in isolation. Nearly all previous studies believe that dissection and separation of outlier remnants and the change from one stage to another are carried out by running water and mass wasting processes. On the other hand, some studies postulate that flat-topped hills are being reduced—under the present hyper-arid conditions—to conical inselbergs mostly by mass wasting and wind erosion, with only rare and sporadic spring sapping and slope wash (McCauley et al. 1982). Since wind is now the dominant agent in this region, these studies believe that wind can remove and transport the comminuted slope debris. But, a question still remains unanswered, namely how is the slope debris comminuted? In the meantime, it was stated in this study that the conical shape of the outlier remnants is a form of dynamic stability that reached an equilibrium state under variable wind conditions (McCauley et al. 1982). Although winds in SW Egypt are highly directional, it was postulated in this latter study that wind is topographically deflected, based on some field examples of local micro-features such as a falling dune, which takes a star-like form in **Wadi Bakht**. This interpretation of the development of the conical shape by wind action (abrasion and deflation of weathered rockfall debris) cannot be accepted. This is because the conical form is the least-resistant shape to wind force; therefore, wind will only remove fine loose particles, if there are any, since disintegration rates are very low in this hyper-arid region. Even wind abrasion rates, if occurring when wind is laden with sand, are also very low due to the highly resistant sandstone bedrock. In conclusion, it can be said that the conical hills acquired this form under wetter climatic conditions different from the present-day hyper-arid climate. Under such past climatic conditions, spring sapping, which does not take place currently, was the principal agent in removing the flat caprock through undermining and rockfall. This conclusion is indirectly confirmed by the vast number of slumped huge limestone blocks, which can be seen along the escarpments of the Eocene plateau overlooking Kharga Depression and the Abu Tartur Plateau. As mentioned in Chap. 11, along these escarpments, several lines of the slumped blocks follow each other from the mid-slope up to the edges of the plateau. Recently, their age has been estimated to be Early Quaternary or Late Tertiary (Busche 2001).

7.3.6 The Basaltic Hills

One of the characteristic forms of the surface of Gilf Kebir Plateau is the **basaltic hills**. Unfortunately, very limited information is available on these landforms. A preliminary Geological Map of Gilf Kebir produced from the analysis of Landsat image (1131-08,141; acquired in December 1, 1972) (Grolier and Schultejann 1982) shows that there are 33 volcanic exposures, some of which extend for more than 4 km and in an irregular form. The map also shows that these volcanic hills are concentrated in the SW part of the plateau and extend northward from the upper reaches of **Wadi Ard El-Akhdar** for about 55 km.

Measurements made on height found that there are two basaltic hills at the head of **Wadi Winkle** that rise 40 m above the surrounding plateau on top, and another hill about 90 m high rises on the eastern side of the Depression at the head of Wadi Ard El-Akhdar (Kröpelin 1989). This latter basaltic hill displays typical columnar structure on parts of the slope, which are not covered by rock debris. In Wadi Winkle, it was found that on both sides of a basaltic hill, there is a continuous gradient from the plateau surface to the wadi floor. This characteristic furnishes evidence on the age of the valley formation, since the basaltic debris (lava) seems to have accumulated in the already existing wadi (Kröpelin 1989). In Wadi Ard El-Akhdar, basaltic talus scree intercalates with playa sediments south of the main channel, indicating that it must have been originated from a basaltic hill (Kröpelin 1989). This means that the basaltic debris must have been transported by a sudden torrential slope wash, or moved down the slope under a seismic accident during the development of the playa. This is why some studies believe that volcanic activity occurred during several episodes, ranging in age from Oligocene or Miocene to Quaternary (Grolier and Schultejann 1982). According to Meneisy (1990), the analysis of some basalt samples from the Gilf Kebir gave an age of 75 ± 3 Ma.

7.4 The Abu Ras Plateau

7.4.1 General Characteristics

Although topographic maps show that it has irregular margins, **Abu Ras Plateau** takes a rectangular shape, having a maximum N-S length of 165 km, and an average width of 50 km. This means that it covers an area of 8250 km^2. As mentioned earlier in the previous section, this plateau is separated from Gilf Kebir by a sand-filled gap. It is bounded from the south by a relatively straight, high (more than 1000 m a.s.l), and steep escarpment, which is locally called **El-Aqabah**. Although extending a few kilometres into the Libyan Territories in its northern parts, the western escarpment of Abu Ras Plateau runs in an irregular N-S line. In the north, the plateau is merged imperceptibly into a subdued small sandstone plateau, which could be considered its northern extension since it rises to 550–650 m a.s.l. The northern edges of this extension represent a wall that checked the linear dunes of the Great Sand Sea. Topographic maps (scale 1:100,000 and 1:500,000) show that heights on the surface of **Abu Ras Plateau** are less than 1000 m a.s.l, except along the southern escarpment, where the surface atop the edge rises between 1000 and 1065 m a.s.l. Height decreases mainly northward, until it becomes about 600 m a.s.l in the extreme north. It also decreases eastward (850–800 m), and westward (750–650 m). This is why the plateau surface is sloping northward on the whole.

Abu Ras Plateau differs from the Gilf Kebir Plateau in some geomorphic aspects. Firstly, it is highly dissected by well-developed drainage networks (Fig. 7.1). Therefore, only small parts of its original flat surface are left. This difference in dissection was explained by the silicification of the top layer due to volcanic activity in Gilf Kebir, resulting in the resistance of the surface to erosion (Hamdan 1999); nevertheless, it could be due to the high density of faulting in Abu Ras. Secondly, most of its surface drains northward. Thirdly, no volcanic intrusions were found on its surface, or in the vicinity. Fourthly, no playas similar to those of the Gilf Kebir are reported in the heads of its wadis. Fifthly, only three wadis carrying names (**Wadi Abd El-Melik** and its tributary **Wadi Talh**, and **Wadi Hamra**) exist. This is probably due to the remoteness of the plateau. Due to this characteristic and relative inaccessibility, studies that have been carried out on Abu Ras are of the type of reconnaissance and exploration (Clayton 1933; Penderel 1934; Maxwell 1982; Hamdan and El-Etr 1999). Therefore, the treatment of the geomorphology of this plateau is dependent on the interpretation of the topographic and geologic maps, aerial photographs, and space images. Results of previous studies using space images will be mentioned whenever relevant.

7.4.2 The Escarpments of the Plateau and the Outlier Hills

The escarpments of Abu Ras Plateau can be easily delineated in the south, west, and east. On the northern side, the plateau merges gradually into the rocky ground lying to the west of the Great Sand Sea. Relative height of the escarpments is 250–300 m in the west, 300 in the south, and 150–200 m in the east, indicating that the least relative height is that one in the east. This is due to the filling of the gap separating this Plateau from the Gilf Kebir Plateau by drift sand and linear dunes of the Great Sand Sea. Large sections of the eastern, western, and southern escarpments—though dissected by wadis—exhibit a linear trend (Fig. 7.1), which coincides with some of the fault lines appearing on the Geologic Map (scale 1: 500,000; sheets No NG 35 SW and NF 35 NW). This characteristic indicates that these linear sections can be classified as fault-line escarpments.

Apart from the wadis dissecting the escarpments, the most prominent landforms that can be detected on maps and space images are outlier hills and mesas standing at a near distance from the escarpments. These outliers, which are similar to those of the Gilf Kebir Plateau, represent remnants of former parts of the main plateau that have been separated by the wadis, indicating that the present plateau covered a wider area compared to the present one.

7.4.3 The Wadi Morphology

The most notable feature of Abu Ras Plateau is the prevalence of the steep-sided wadis that deeply incise the plateau surface. The topographic maps (scale 1:100,000; Sheets No. NG 35 A6, NG 35 A6) show that the plateau is highly dissected by an integrated drainage system, which differs significantly from that of Gilf Kebir. In this system, nearly all the large basins are elongated in form with S-N trend, since the main wadis are much longer than their main tributaries, and take the same trend. As mentioned earlier, only few wadis carry names: **Wadi Abd El-Melik** and its main tributary **Wadi Talh**, **Wadi El-Qubba**, **Wadi Soura**, and **Wadi Hamra**, which are the main tributaries of the Wadi separating Abu Ras Plateau from Gilf Kebir Plateau (Fig. 7.1). In Sect. 7.4.4, the morphological characteristics recognized from space images, aerial photographs, and topographic maps will be analyzed.

7.4.4 Characteristics of the Wadis

Although wadis of Abu Ras have diverse morphological characteristics, the following common aspects can be easily recognized on maps, aerial photographs, and space images. Along several sections, the wadis show a linear trend, mainly S-N, SE-NW, and E-W. Apart from the wide basins, which developed along some sections, the width of the main wadis and their main tributaries does not change significantly along their longitudinal profiles. All these common characteristics can be interpreted in terms of control exerted by the fault system, which characterizes this plateau. Along fault lines, running water concentrated eroding the wadis, directing water energy vertically rather than laterally. This was aided by the relatively steep gradient of the plateau surface, where height falls from about 1000 m a.s.l in the south to about 700 m a.s.l in the north along a distance of about 150 km (about 1:500). All wadis are filled with aeolian sand, which drifted from the Great Sand Sea lying to the north of the plateau. Therefore, their beds and the lower parts of their sides are obscured by this sand. In spite of this, the upper parts of wadi sides appear to be steep. Another aspect of the wadis is the development of irregular or circular semi-closed basins along their courses or at their mouths. These basins vary in size from small ones (2–3 km wide), such as those of Wadi Hamra, to large basins, such as those of **Wadi Abd El-Melik** (15 km wide), which lies to the south of the junction of **Wadi Talh** (Fig. 7.1). Beds of these basins lie 150 m below the surrounding surface of the plateau. Another internal basin is the one in which Wadi No.1 ends (Fig. 7.1). It seems that these basins developed in areas where wadi erosion was intense due to structural control, and lithological weakness. It is expected that some lakes were developed in these basins or at least in some of them. Lake sediments deposited in these expected lakes cannot be recognized on aerial photographs or space images; in addition, they are hidden by aeolian sand filling the basins. Therefore, field investigation is indispensable to discover such an aspect.

It is not possible to identify all morphometric characteristics of the wadis given the lack of required tools. Only length and width of the main wadis were accurately measured on photomosaics, scale 1:100,000. These measurements showed that Wadi Abd El-Melik is the longest (154 km). It is also the only one that cuts across the plateau from the extreme south to nearly its extreme north (Fig. 7.1). It is the only wadi that has eleven tributaries that are longer than 10 km. It is also the only wadi with a maximum width of 2.5 km. All these characteristics indicate that this wadi has the largest basin area. Wadi No. 1 is the second longest wadi (72 km). It lies to the east of Wadi Abd El-Melik, starting at a point in the center of the plateau, and debouching at an internal basin between the northern edges of Abu Ras and the degraded plateau in the north. Its width varies between 1.0 and 0.5 km, and has two tributaries, which are more than 10 km long. The depositional basin where this wadi ends is semi-closed, since it opens toward a similar one through a narrow gap in a ridge separating the two basins. These basins were—most probably—former lakes that have been drained when they were filled with sediments. The other large three wadis (**Wadi Hamra**, Wadi No.2, and Wadi No.3) are all parallel tributaries of the hidden wadi developed in the gap between **Abu Ras and Gilf Kebir Plateaus** (Fig. 7.1). Except for the wide basins, the maximum and the minimum width of the three wadis is almost similar (Max. width 1.5–2.5 km, Min. width 0.4–0.5 km). Wadi Talh is one of the largest tributaries of **Wadi Abd El-Melik**, and it has been counted as an independent wadi since it is one of the famous wadis mentioned in previous studies and on maps.

There are two other relatively long wadis cutting across Abu Ras Plateau. The first (Wadi No.4, Fig. 7.2) is in the northeast corner to the south of the lower section of Wadi Abd El-Melik, taking a general E-W trend, with a total length of about 32 km. However, the lower half of this wadi lies in the Libyan Territories. The second wadi lies in the southwest (Wadi No.5, Fig. 7.1), with a total length of about 34 km. It takes a general S-N trend, starting from the southern escarpment and ending in the western side, separating the most SW part of the plateau as a large mesa. Other wadis in Abu Ras Plateau with significant forms are a small wadi in the western part of the southern escarpment (Fig. 7.1), which is called **Wadi Soura** (the Valley of Pictures). The sides of this wadi are characterized by two features. The first is that they are decorated with many drawings, and this is why it is called Wadi Soura. One of the most interesting drawings could be called the swimmers' plate, where some ancient locals are swimming or diving. The second feature is the development of caves at the base of wadi sides. As mentioned earlier in this Chapter (Sect. 7.2), the base of the wadi sides where caves were recorded is composed of white sandstone, which belongs to the Silurian rocks of Umm Ras Formation (Geological Map, scale 1:500,000). Unfortunately, no detailed studies were conducted on these caves, and no investigations were made on the possible location of the pools or lakes in which the locals were swimming or diving.

7.5 The Gap Between Abu Ras and Gilf Kebir Plateaus

Since some wadis cutting across the eastern edge of Abu Ras end in the longitudinal Depression (the gap) separating the two plateaus, it might be appropriate to investigate its morphological characteristics and its relationship to the two plateaus. This gap measures 15–50 km in width and about 150 km in length. The bed of this sand-filled gap lies at 850–900 m a.s.l (i.e.,100–150 m below the surrounding plateaus). As indicated earlier, this gap was interpreted by previous studies as a wide wadi filled with the oncoming sand from the Great Sand Sea. Since it is filled with sand, the original form of the wadi and its morphological characteristics have been hidden below the sand cover, and its original bed must have been lower than the actual level of the gap appearing on maps. Fortunately, recent radar images revealed some of the hidden features of this gap (El-Baz 1999). The images confirmed that this gap is occupied by a large straight wadi having the same S-N trend, parallel to the trend of the main wadis of Abu Ras. Its straightness confirms the previous conclusion that this wadi developed in a graben separating the two plateaus along wrench faults (El-Ghawabi 1987). The **radar images** also reveal that the original width of the wadi bed varies between 5 and 10 km, and is as wide as the gap appears on maps or space images. Wadi sides appear irregular due to the dissection by numerous tributaries, which meet the main wadi almost at right angles—a common characteristic in the wadis cutting across Abu Ras Plateau. This means that the wadis cutting the eastern escarpments of the plateau represent tributaries for this main wadi occupying the graben.

7.6 Evolution of Gilf Kebir

This analysis of the geomorphological aspects of the sandstone plateaus of the **Gilf Kebir and Abu Ras** shows that they are remnants of wider ones and that they were exposed to dissection for a long span of time. It also indicates that the region enjoyed wetter climates in the past in contrast to hyper-arid climate nowadays, as shown in Chap. 3. According to McCauley et al. (1982), much of the degraded drainage basins, now invaded by wind-blown sand, are relics of an ancient (Mid-Tertiary?) landscape that had probably reached a mature stage of fluvial dissection. This conclusion is similar to that reached in Chap. 3, Sect. 3.2.

Streams of the mid-late Tertiary age drained the highlands of the Gilf Kebir and Abu Ras to the north, east, west, and south. This complies with the postulation of Issawi and McCauley (1993) that there was a drainage system draining the Gilf Kebir northward toward the basin of the Great Sand Sea during Late Eocene–Oligocene or Oligocene only, and they called it the Gilf Drainage System. McCauley et al. (1982) made an attempt to restore the mid-Tertiary drainage patterns, and concluded that the Gilf Kebir was formerly a major drainage divide in SW Egypt. By the onset of present hyper-arid climatic conditions, the trunk streams became defunct, and their valley slopes were reduced by mass wasting and wind erosion (McCauley et al. 1982).

7.7 Conclusion

The sandstone plateaus in the southwest of the Western Desert are characterized by a landscape developed by erosion and tectonics during a long period of time. Age of their rock formations dates back the Paleozoic and Mesozoic. These formations include the only formation of glacial origin of Middle/Late Carboniferous age in Egypt. The development of this type of formation was interpreted by the fact that Egypt at the time existed in the southern hemisphere near the Arctic Circle (Klitzsch 1989). However, the characteristics of forms and landscape of this region suggest that they are the product of the interplay of several processes, of which fluvial, aeolian, and weathering and mass wasting are the most significant. The location of these plateaus to the south of sand flow of the Great Sand Sea resulted in not only the burial of the northern slopes of the two plateaus and filling the wadis dissecting them, but also in the development of sand dams across the W-E drainage lines dissecting the eastern part of the Gilf plateau. This happens when the plateaus are checking the flow of sand coming from north, but at the western and eastern sides, the sand flow is unchecked by any obstacle and continues southward.

Fluvial processes were responsible for the development of the drainage systems on the Gilf Kebir and Abu Ras Plateaus. Among these drainage systems is the one trending northward to the Great Sand sea basin. Toward this huge basin, a drainage system carried and deposited large amounts of sediments during Middle Miocene and has been called "**The Gilf System**" in a previous study (Issawi and McCauley 1993). In another study, it was postulated that the major stream (the Gilf) is responsible for the development of Qattara Depression (Albritton et al. 1990). Another mega-Paleo-drainage system draining the eastern slopes of the Gilf Kebir is revealed by Radar Topographic Data and is called "Toshka Drainage System" (Ghoneim and El-Baz 2007). It is draining internally in the Toshka Depression where it formed the largest paleo-lake in the Western Desert known as Bir Sahara-Bir Tarfawi playa (Embabi 1999). Other forms are playas developed in the upper sections of some wadis, terraces, and fans at wadi beds and at their mouths.

It can concluded that, fluvial action is by far is the most effective process in shaping the landscape of these two plateaus since its effect extends beyond the fluvial forms to the sediments deposited in the Great Sand Sea. Such sediments were exposed to wind action after the onset of arid conditions, and the development of the largest sand sea in Egypt.

References

Albritton CC, Brookes JE, Issawi B, Swedan A (1990) Origin of the Qattara Depression, Egypt. Bull Geol Soc Am 102:952–960

Almásy LE (1936) Récent explorations dans le Désert Libyque (1932–1936). Publ Spéc Soc Géogr d'Égypte, Cairo

Bagnold R, Peel R, Myers O, Winkler H (1939) An expedition to the Gilf Kebir and Uwienat, 1938. Geogr J XCIII:281–313

Busche D (2001) Early Quaternary landslides of the Sahara and their significance for geomorphic and climatic history. J Arid Environ 49:429–448

Clayton PA (1933) The western side of the Gilf Kebir. Geogr J LXXXI:254–259

Clayton P (1937) The southwestern desert survey expedition 1930–1931. Bull Soc Royale Géogr d'Égypte 19:241–265

Dardir A (1982) Basement rocks of the Gilf-Uweinat area. In: El-Baz F, Maxwell T (eds) Desert landforms of Southwest Egypt: a basis for comparison with Mars. NASA, Washington, DC, pp 67–78

El-Baz F (1999) Groundwater Concentration beneath sand fields in the Western Desert of Egypt: indications by radar images from space. Egypt J Remote Sens Space Sci 1:1-21

El-Baz F, Boulos L, Breed C, Dardir A, Dowidar H, El-Etr H, Embabi N, Grolier M, Haynes V, Ibrahim M, Issawi B, Maxwell T, McCauley J, Moustafa M, Yousif M (1980) Journey to the Gilf Kebir and Uwienat, Southwest Egypt, 1978. Geogr J 146:51–93

El-Ghawabi M (1987) Use of Landsat-images for interpretation of the geologic evolution of the Gilf Kebir area, SW Egypt. Proceedings of international workshop on remote sensing and resource exploration. ICTP, Triesta, Italy

Embabi NS (1968–1969) The semi-playa deposits in the Kharga Oases Depression. Bull Soc Geog d'Egypte 41:73–87

Embabi NS (1999) Playas of the Western Desert, Egypt. Ann Acad Sci Fenncae, Geol Geogr 160:5–47

Ghoneim E, El-Baz F (2007) The application of radar topographic data to mapping of a mega—paleodrainage in the Eastern Sahara. J Arid Environ 69:658–675

Grolier M, Schultejann P (1982) Geology of the southern Gilf Kebir plateau and vicinity, Western Desert, Egypt. In: El-Baz F, Maxwell T (eds) Desert landforms of Southwest Egypt: a basis for comparison with Mars. NASA, Washington, DC, pp 189–206

Hamdan A, El-Etr H (1999) Quantitative photolineation analysis and some geomorphic and structural aspects of the Gilf Kebir and Abu Ras Plateaus, south Western Desert, Egypt. Egypt J Remote Sens Space Sci 1:101–144

Hermina M, Klitzsch E, List F (eds) (1989) Stratigraphic Lexcon and explanatory notes to the geological map of Egypt, Scale 1:500,000. Conoco Inc, Cairo

Hussein P (1928) L'exploration du DésertLibyque. La Géographie, 50:171–183 and 520–336

Issawi B (1973) Contribution to the geology of Uwienat-Gilf Kebir area, Western Desert, Egypt. Internal Report, Geol. Survey of Egypt, Cairo

Issawi B (1978) New findings in the Gilf-Uwienat area. Ann Geol Surv Egypt 8:275–293

Issawi B (1982) The geology of the south-western desert of Egypt. In: El-Baz F, Maxwell T (eds) Desert landforms of Southwest Egypt: a basis for comparison with Mars. NASA, Washington, DC, pp 57–66
Issawi B, McCauley J (1993) The Cenozoic landscape of Egypt and its river systems. Ann Geol Surv Egypt 19:357–384
Klitzsch E (1989) The Paleozoic chapter 4 in stratigraphic lexicon and explanatory notes to the geological map of Egypt, scale 1:500,000. Hermina M, Klitzsch E, List FK (eds) Conoco Inc., Cairo:57–76
Klitzsch E, Schandelmeier H (1990) South Western desert. In: Said R (ed) The geology of Egypt. Balkema, Rotterdam, pp 249–257
Kröpelin S (1987) Palaeoclimatic evidence from Early to Mid-Holocene playas in the Gilf - Kebir, southwest Egypt. Palaeoecol Afr 18:189–208
Kröpelin S (1989) Untersuchungen zum sedimentationsmilieu von Playas im Gilf Kebir (Südwest Ägypten). In: Kuper HR (ed) Forschungen zur Umweltgeschichte der Ostsahara. Heinrich-Barth-Institut, Köln, pp 184–305
Kröpelin S (1993) Geomorphology, landscape evolution and paleoclimates of Southwest Egypt. Catena Suppl 26:31–65
Maxwell T (1980) Geomorphology of the Gilf Kebir. In: El-Baz et al (eds) Journey to the Gilf Kebir and Uwienat, Southwest Egypt. 1978, Geogr J 146:76–83
Maxwell T (1982) Erosional patterns of the Gilf Kebir plateau and implications for the origin of Martian Canyon lands. In: El-Baz F, Maxwell T (eds) Desert landforms of Southwest Egypt: a basis for comparison with Mars. NASA, Washington, DC, pp 281–300
McCauley J, Breed C, Grolier M (1982) The interplay of fluvial, mass-wasting and eolian processes in the eastern Gilf Kebir region. In: El-Baz F, Maxwell T (eds) Desert landforms of Southwest Egypt: a basis for comparison with Mars. NASA, Washington, DC, pp 207–239
Meneisy M (1990) Volcanicity. In: Said R (ed) The geology of Egypt. Balkema, Rotterdam, pp 157–172
Pachur H, Röper H (1984) The Libyan "Western Desert" and Northern Sudan during the Late Pleistocene and Holocene. In: Klitzsch E, Said R, Schrank E (eds) Research in Egypt and Sudan, Band 50, Results of the Special Research Project in Arid Areas, Period 1981–1984, Berliner geowissenschaftlicheAbhandlungen, Reihe A, Band 50:249–284, Berlin
Peel R (1939) The Gilf Kebir. Part 4, in: An expedition to the Gilf Kebir and Uweinat, 1938. Geogr J XCIII:295–307
Peel R (1941) Denudational landforms of the Central Libyan Desert. J Geomorph 4:3–23
Peel R (1966) Landscape in Aridity. Inst Br Geogr Trans 38:1–23
Penderel H (1934) The Gilf Kebir. Geogr J LXXXII:449–456
Shaw WBK (1936) An expedition to the southern Libyan Desert. Geogr J LXXXVII:193–221

The Karstified Carbonate Platforms in the Western Desert

8

Abstract

The Karstified Carbonate Platforms occupy vast areas in the Western Desert (WD). They are divided into two groups according to age: the Palaeocene-Eocene platforms occupying the eastern part of the WD, and the Miocene ones covering the northern part. This chapter deals with the first group because it accommodates a variety of forms representing the carbonate platforms compared to the Miocene plateaus. The Palaeocene-Eocene platforms are composed of a succession of beds of carbonates and clastics, where limestone/chalk covers their surfaces. Height ranges between +500 m a.s.l on Abu Tartur Plateau and 47 m a.s.l in El-Bahr Depression. Karst forms are dominant, due to the dominance of carbonate rocks on the surface of the ground and the past pluvials that have prevailed in Egypt since the Late Eocene. Other forms include those of eolian, structural, and fluvial origin. Of these forms, bedrock flutes of various sizes from minute (a few mm deep) to giant flutes (several meters deep and several kilometers long) cover wide areas in various localities. Although wind erosion played a major role in the development of most of them, karstification initiated the development of the large and mega types. Remnants of an ancient drainage system can be observed in all the platforms; inverted wadis (Oligocene?) characterize the central parts of the platforms, to the west of Ghard Abu Moharik.

Keywords

Carbonate karstified platforms • Cone/Karst Hills • Terra rossa soils • Silica gravel sheets and concretions • Kharafish • Rock yardangs • Giant flutes

8.1 Introduction

The carbonate platforms in the Western Desert are characterized by a succession of alternating hard and soft beds of limestone/chalk and shale/clay, varying in age from Upper Cretaceous/Palaeocene to Middle Miocene. The bounding escarpments, which represent—at the same time—the walls of the adjacent low-lying areas, were considered in some previous studies as cuestas such as those of **El-Tih** and **El-Egma** in central Sinai (Awad 1952). In general, these plateaus have been classified in previous studies into two wide platforms: the Eocene platforms, in which the top layers are limestone and chalk that belong to the Palaeocene and Eocene periods; and the Miocene platform, which is topped by Middle Miocene carbonates. Names are given to some of these platforms in part or in full. **El-Diffa** or

N.S. Embabi, *Landscapes and Landforms of Egypt*, World Geomorphological Landscapes,
https://doi.org/10.1007/978-3-319-65661-8_8

Marmarica is the name of the northern Middle Miocene plateau, **El-Quss Abu Said** Plateau bounds the **Farafra Depression** from the west, and **Abu Tartur Plateau** is the part of the Eocene plateau lying between the **Kharga and Dakhla Depressions**. Other names are given to some sections of the Eocene plateau, such as **Gebel El-Gir** for the section of the plateau, bound by the Qena Bend of the Nile Valley, and **Sinn El-Kaddab** for the extreme southern part of the Eocene plateau. Due to differences in age, lithological aspects, and consequently in the forms exhibited, these vast platforms can be divided into two groups: the Palaeocene-Eocene platforms and the Miocene platforms. This chapter will deal with the former, since they are the most extensive platforms in the Western Desert and are home to a variety of forms representing key characteristics of carbonate platforms.

8.2 General Characteristics

Previous studies addressed the whole area between the Nile Valley in the east, the line of Depressions in the west, and **the Fayum Depression** in the north—which is covered by Palaeocene-Eocene carbonates—as one platform called the Eocene limestone plateau. This is probably due to two facts, the first is that most of this vast area is covered by Eocene carbonates, and the second is that only small detached (**El-Quss Abu Said**) or semidetached parts (**Abu Tartur Plateau**) are named. Also, available data sources, dating back to two decades, showed no much difference between the various parts of these vast carbonate platforms. Although several forms are repeated in different localities such as Depressions, wadis, and yardangs, the pattern of distribution of these forms and other forms makes it impossible to divide this vast region into several characteristic platforms.

Topographic maps (scale 1:250,000) show that maximum heights are not much over 500 m a.s.l, such as in **Abu Tartur** (560 m), **Gabal El-Gir** (560 m), and the plateau to the west of Aswan (590 m). Minimum heights are generally less than 200 m a.s.l, such as in the vicinity of Bir Murr (175 m), to the east of Ghard Abu Moharik (150 m), and in **El-Bahr Depression** (47 m). Local relief is generally low (10–20 m), but in some localities might reach more than 200 m.

Although the cap rock is composed of Palaeocene-Eocene carbonates, the geologic map (scale 1:500,000) shows wide local lithological variations, and this is why there are more than 15 formations of carbonates exposed on the surfaces of the plateaus. Examples of these variations are the white neritic chalk and chalky limestone (Palaeocene Tarawan Formation), the dense thickly bedded limestone with characteristic concretions and local flint bands of Drunka Formation (Lower Eocene), and the shallow marine limestone interbedded with shale and sandy shale (Middle Eocene Wadi El-Rayan formation). However, reference will be made to relevant formations as they are discussed. Although the main carbonate outcrops end along the eastern extension of the edge of the plateau known as the Sinn El-Kaddab (latitude 22° 40′N), the last remaining Eocene limestone outlier lies approximately 60 km to the south of the edge of the Eocene plateau (**Barqat El-Shabb** 370 m). This means that this plateau retreated northward 60 km since the exposure of the Lower Eocene formations, about 50 million years ago.

Local structural lines developed in several localities on the Palaeocene-Eocene platforms, but with varying densities and extensions. These structural lines comprise faults and small domes, taking mainly an E-W trend to the south of latitude 24°N, and N-S trend between 24° and 26°N. Fractures and joints—with various orientations—can be seen on the Geological Map spreading all over the platforms. As will be seen in the following analysis, these structural features were responsible for the development of some forms or were a controlling factor in the development of local relief features characterizing the surface of some parts of the platforms.

Upon closer inspection, it can be said that the forms of the carbonate platforms are the product of the interaction of karstification, wind action, water action, and local structures. Apart from the eolian landforms that developed by wind action and structural forms, all other forms are inherited from past climatic conditions. Topographic maps, air photo mosaics (scale 1:100,000), and space images show that there is no pattern of distribution for these forms. However, except for features originated by local structures, other forms spread all over the surfaces of these Palaeocene-Eocene platforms. Therefore, it was found that it is not possible to divide this group of carbonate platforms into separate ones with distinguishable landforms, and it is more convenient to discuss the surface forms according to their origin. Accordingly, landforms can be classified into four groups: karst, structural, eolian, and fluvial forms. Of these forms, karst ones are the most widespread, due to two factors: The first is that the carbonate rocks on the surface of the platforms are nearly 100% exposed, and the second is that these carbonate rocks have been subjected to wet climates since the end of the Eocene as mentioned in Chap. 3 (Sect. 3.2.2). It could be even said that nearly every centimeter is affected by karst processes. This is why these platforms are described as karstified platforms in this chapter.

8.3 Karst Forms

The interaction between the carbonates and the karstification processes produced a wide variety of **karst forms**. Moreover, due to the fact that these carbonate platforms represent the largest oldest extension of carbonates in Egypt, it can be said also that all types of karst forms in Egypt are represented on these platforms. In spite of this fact, there are some forms that are more widespread than others. The most widespread forms are **Solution Depressions**, residual hills, residues (undissolved fine materials), and cave systems.

8.3.1 Solution Depressions

These Depressions vary in size from a few meters to several kilometers. They spread all over the surfaces of the Palaeocene-Eocene platforms. This is probably due to the fact that the carbonate formations are highly fractured. Therefore, solution rates increase at points where surface water percolates at fracture crossings. Some of the Depressions initially were sinkholes, absorbing all surface water reaching them by drainage lines. Examples of such small Depressions exist on the surface of the plateau just to the north of Farafra Depressions, and to the east of Ghard Abu Moharik. In the latter, the terrain is characterized by numerous small low-lying holes. Density, shape, and size of these Depressions are a function of local structure and stage of development. However, local variations in these

Fig. 8.1 Minute solution hollows on the surface of the plateau between Dakhla and Farafra

characteristics cannot be treated at present due to lack of detailed topographic maps or aerial photography coverage. Small swallow holes, tabular, and/or funnel-shaped, 2–15 m wide and up to 30 m deep, are commonly aligned along fracture and joint lines and were seen in the field on the surface of the plateau between Dakhla and Farafra (Fig. 8.1).

There are no local names for these closed basins. Only two of the relatively large Depressions are called **El-Bahr**, the Arabic word for the sea. One of these two Depressions lies to the NE of Bahariya Depression, to the east of the northern part of Ghard Abu Moharik, and the second lies atop of **El-Quss Abu Said Plateau** to the west of Farafra Depression. The first one is probably the largest Depression on the surface of the Eocene plateau, since its maximum N-S length is about 40 km, and its maximum E-W breadth is about 23 km, as measured on the topographic sheets (Ghorabi, El-Qarah El-Hamrah, and Darb El-Bahnasawi scale 1:100,000). The same maps show that the lowest point in this Depression is 47 m a.s.l (28° 29′N, 29° 27′E), which is most probably the lowest point on the Eocene plateau. The rim of the Depression is delimited by the 150 m contour line. This means that this Depression is about 100 m deep. The second Depression is deeper (145 m), since the lowest point at its floor is 180 m a.s.l and the highest contour line at the edge of the Depression is 325 m a.s.l as indicated on the topographic sheet Qasr El-Farafra (scale 1:100,000). Most probably such a name (El-Bahr) was given to the flatness of their floors and the appearance of mirage from a distance, making floors look like a sea. It seems that this name is a common local name, since it applies to other flat areas.

8.3.2 Residual Karst Forms and Residues

8.3.2.1 Cone/Karst Hills

The most widespread residual forms and features include **cone/karst hills**, **chert gravel sheets**, and **silica concretions**. **Residues** are the fine calcareous materials that did not dissolve and remained on the surface of the ground where the **terra rossa** and **iron and manganese oxides** are widespread. Cone hills separated by small inter-depressions are distributed all over the surfaces of the Palaeocene-Eocene carbonate plateaus. There are several areas where some concentrations of these hills are separated by cockpits. The largest and most well-known concentration lies in the area to the northeast and east of Bahariya Depression (Fig. 8.2). It was found that concentrations of the **cockpit-conical hill landscape** in some areas arise from the presence of high-density fracture systems. As indicated in previous studies, this form with inter-depressions represents an evolution stage in the karst system known as cone/cockpits karst in tropical karst lands (Sweeting 1958). Although these combined forms (cone hills and inter-depressions) were known to those who studied the geomorphology of Bahariya Depression (Ball and Beadnell 1903; El-Kayali 1980), their karst origin has been recently revealed by El-Aref et al. (1987). These hills vary greatly in size even in the same field (Fig. 8.2) and density from one field to another. Sides of most hills are dissected by small drainage lines arranged in a radial pattern (Fig. 8.2). All these differences represent various stages of development, and the effect of local structural lines controlling their dissection. Karst areas of advanced stage of degradation are mainly characterized by the presence of widely spaced and low cone hills in a nearly peneplained terrain. The hilly landscape disappears, and the surface of the plateau becomes mostly flat with some shallow basins with playas/fluvial deposits in the final stage of karst planation.

8.3.2.2 Chert Gravel Sheets

Angular gravel sheets cover wide areas on the Eocene plateau between the Nile Valley and Kharga, Farafra, and Bahariya Depressions. The main component of this gravel is silica. Its angularity indicates that the gravel has not been

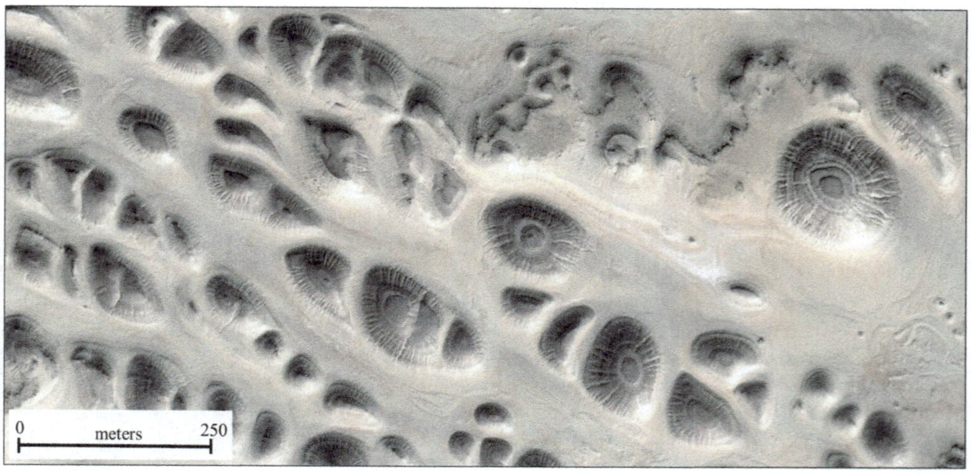

Fig. 8.2 Space image of a field of cone karst hills, east of Bahariya Depression (*Source* Google Earth 2012)

transported by any water flow, and that it was transported for a very short distance. The source of this gravel is the silica bands that developed in some of the Palaeocene-Eocene carbonate beds, such as the siliceous limestone of the Garra, Kurkur, Drunka, Dungul, and Naqb formations. These silica intercalations take the form of thin layers (5–10 cm) or nodules in the limestone beds. After the dissolution of the upper limestone strata, these silica layers and nodules are left on the surface of the ground, exposed to weathering processes, and disintegrated into small fragments. During the process of dissolution of carbonates, oxide residues were leached downward toward the horizon lying below the gravel sheets. Therefore, the gravel sheets represent a residual form—similar to melon concretions—that developed after the karstification processes removed the host limestone rocks (Fig. 8.3)

8.3.2.3 Silica Concretions

One of the prominent and well-known features of the limestone plateau between the Nile Valley and the mega-Depressions of Kharga, Farafra, and Bahariya is the clusters of well-rounded silica concretions, resembling "**melon fields**" (Beadnell 1909; Caton-Thompson and Gardner 1932). Most of these concretions are spherical in shape, but some of them acquire an elliptical form. Their diameter varies greatly between 20 cm and more than 100 cm. Their distribution on the surface of the plateau does not show any specific pattern. Fields of melon concretions may cover areas reaching hundreds of square meters, which may also comprise hundreds of melons (Fig. 8.3). However, the density of melons may vary from one field to another, depending on the original number of concretions in the host rock.

Geologically, the concretions show various stages of exposure from the host rock. The host rock is limestone, which belongs to the Palaeocene/Eocene period. Field study showed that the broken concretions have concentric structure. This means that the concretions attained their structure and their shape before their exposure on the surface of the ground, and that the concretions had been formed concurrently with the deposition of the limestone (Embabi 1967). After the exposure of the host (limestone) rock to karstification processes, the concretions were progressively exposed on the surface of the ground because they are more resistant to the chemical action of water and other erosion processes than the host limestone. As stated by Caton-Thompson and Gardner (1932) "*the flinty concretions [are] left free on the surface by the removal of the less resistant surrounding rock.*" How much limestone has been removed by solution is a matter which cannot be estimated. This is because the maximum thickness of the limestone bed, which hosts the concretions, is unknown. Also, the thickness of the remaining part of this bed is unknown.

8.3.2.4 Calcareous-Terra Rossa Soils

The **terra rossa soils** cover wide areas on the surface of the Palaeocene-Eocene carbonate platforms. Unfortunately, they cannot be detected from topographic maps or aerial photographs, because these tracts are covered with lag gravel or coarse sand and do not have any specific topographic expression. Field studies revealed that these soils are present on flat surfaces or low-lying grounds. These soils represent the insoluble fine residues left on the surface of the ground after the dissolution of the carbonate rocks. Due to differences in mineral composition, the color of the soils varies between dark and light brown to gray and white.

Red loamy clay flooring most of the numerous shallow Depressions in the plateau between the Nile Valley and Kharga Depression was recorded a long time ago (Caton-Thompson and Gardner 1932). Its origin was also known as an insoluble residue of the limestone, washed into the low-lying areas by occasional runoff, and stained red by the iron content in the limestone itself. The brilliant red hue

Fig. 8.3 Large field of melon concretions at the floor of one of the Giant flutes, SW Dairut Town in The Nile Valley

is due to the presence of anhydrous iron (hematite and geothite). Red soils and terra rossa residues are found to develop under three different conditions.

(1) Floors of shallow Depressions, where soil sediments are washed in by surface water. This type was first reported by Caton-Thompson and Gardner (1932) on the Eocene plateau during their trip from Abydos to Kharga in 1930. Red colluvial material is also found in topographic hollows on different tufa surfaces in Kurkur (Butzer 1964);
(2) Below the silica gravel sheets, where insoluble residues are left in situ after the dissolution of the limestone and are protected by the gravel. This type has been seen to spread in wide areas also on the Eocene plateau between Kharga Depression and the Nile Valley (Fig. 8.4a);
(3) Along joint walls in the limestone that caps the plateaus, where red wash-in cases with a variable addition of detritus material are preserved in the cracks and joints (Fig. 8.4b). This accumulation occurs when surface water infiltrates into these openings. This type is seen on the surface along road cuts or in limestone quarries, e.g., in the vertical cuts of the limestone quarry of Assyut Cement Factory, and the escarpment of the Palaeocene-Eocene plateau between Kharga Depression and the Nile Valley.

Ancient consolidated thick red soils were observed in the field to the southeast from the Bahariya Depression on high grounds spread in a disorganised manner. This indicates that these red soils were developed on higher surfaces and then transported from their original surfaces and deposited under such conditions. It also indicates a previous cycle of karstification at higher levels.

All previous studies concluded that these sediments of red soils and terra rossa reflect more intensive weathering under seasonally wet, subtropical, or tropical climate (Butzer 1964; El Aref and Lotfy 1985; El Aref et al. 1991).

8.3.2.5 Iron Oxides

The economic **iron ore deposits** of Bahariya Depression in the Western Desert are confined to karstified features and represent karst residues. The main iron ore deposits of Bahariya Depression occur in three localities within the cone/karst terrain along the main north–eastern escarpment of the Depression. It was suggested that the iron deposits were developed by lateritization during the post-Eocene karstification of the limestone cap rock (–El Aref and Lotfy 1985). Iron oxides derived from the limestone rich in iron carbonates accumulated in karst Depressions and other low-lying areas that acted as traps for the leached (lateritized) materials. The essential minerals constituting the iron ores are siderite, geothite, hematite, manganese oxides, and barite. These minerals exhibit various megascopic and microscopic fabrics, such as crustified structure, cockade structure, large-scale concentrations and algal pisolites, and preserved micro-cellular structures and tissues of plant remains. Such fabrics indicate depositional and lithification processes from weathering solutions, with direct or indirect organic influences (El Aref and Lotfy 1985).

8.3.2.6 Cave Systems

Caves are recorded in several localities on the surface and subsurface of the Eocene carbonate plateau within the mega-depressions of Bahariya, Farafra, Dakhla and Kharga, and the Nile Valley. All recorded cave systems were found quarried or degraded, except for the **Djara Cave**, which was found intact below the surface. Although our knowledge on the surface features of this desolate part of Egypt increased significantly in the last ten years due to the increased numbers of scientific expeditions (Brook et al. 2002), not much information was collected on karst forms in general and on caves specifically. Therefore, it is recommended that further geomorphic research should be oriented toward karst features in this part of Egypt. However, available information on caves and degraded caves in this carbonate plateau indicates that there were several cycles of karstification (Embabi 2004). The oldest cycle goes back to Early Eocene–Oligocene times, when Egypt as a whole was enjoying a tropical wet climate. Caves developed during this early cycle are degraded, and only remnants are found exposed on the surface, such as those found the area known as the **Crystal Mountain** between Bahariya and Farafra Depressions. In this area, there are several remnants of **degraded caves**, such as stalactites, calcite crystals, and cemented breccia.

8.4 Structural Forms

Although the available geologic maps show that local structures spread all over the Paleocene-Eocene platforms of the Western Desert, they also show a concentration of these local structures in several areas. The first one is characterized by the presence of hills of volcanic origin. They are scattered in the northern part of the Eocene plateau mainly along the fault line, which extends in a NW direction from El-Bahnsa Town to the north of Minya city in the Nile Valley. However, the largest volcanic expression lies to the south of this fault line and appears on the topographic map of Manfalut (sheet No. 36 M2, scale 1:100,000) as a dome called **Gabal El-Gibeil**, which means the little mountain (290 m), rising about 100 m above the surrounding surfaces. Although this hill measures 5 km from north to south, and 4 km east to west, the volcanic material covers a small area only at the top of the hill—according to the Geological Map (scale 1:500,000). This might indicate that some of the molten

Fig. 8.4 Terra Rosa Soils: **a** on the surface of the ground protected by chert gravel. **b** In a road cut, where the red Terra Rosa covers the surface and fills the gaps below the surface of the ground

material intruded into the sedimentary cover, resulting in the dome-shaped hill. Examples of other volcanic hills that carry names are **Qaret El-Soda** (the Black Hill), El-**Qarrat El-Zurq** (the Blue Hills), and **Qaret El-Dabaa**.

A second area where surface features are related to local tectonics concentrate lies in the northern part between the latitude of Luxor and the latitude of Asyut in the Nile Valley, and a third one in the southern part, to the south of latitude 25°N. Examination of the Geological Maps reveals that faults represent the main structure in both areas. In the northern area, fault lines occur along several sections of the escarpments of the plateau, which overlook the Nile Valley. These sections are relatively straight, and the plateau attains higher levels compared to adjacent sections. Similar sections are present at the northern escarpments of Farafra Depression and the eastern escarpments of El-Quss Abu Said Plateau. Therefore, it could be said that the development of the escarpments of the Eocene plateau at the above-mentioned localities can be considered as fault line scarps. In this northern area, faults are also present on the plateau surface, but no surface expression could be detected so far. In the southern area, there are two sets of tectonic lines: a dominant east–west set particularly west of longitude 32° 30′E and a secondary N-S set mainly along longitude 32° 30′E. Previous studies (Embabi 1967; El-Etr et al. 1982) and aerial photographs showed that small closures (centroclines, plunging anticlines and synclines, basins and domes) are frequently observed along fault lines. Therefore, one can argue that local structures in this part of the Palaeocene-Eocene plateau are composed of fault lines and small folds. In addition to these compound structural lines, there are some relatively large circular structures, considered to be unrelated to any of the faults and of different origin than the small closures. It was found that some specific forms owe their existence to the development of these local structures. Three types of structural forms could be recognized: fault line escarpments, anticlinal and synclinal hills, and anticlinal Depressions. In addition to major fault line escarpments, such as the E-W sections of the Eocene plateau looking over the southern part of Kharga Depression and the Sinn El-Kaddab escarpments, secondary and minor fault line escarpments also extend along the plateau surface (Fig. 8.5). This map shows that there are, at least, 10 major and secondary fault line escarpments with a general E-W orientation from latitude 25° to 23°N. Therefore, this suggests that fault line escarpments represent one of the most conspicuous structural features that characterize the Sinn El-Kaddab Plateau.

Folded hills can be classified according to their structures into two sets: anticlinal hills and synclinal hills. The anticlinal hills can be divided into two groups: concentric hills and noncircular hills. In the first group, the hills take a circular form under the control of the original structure (domes). These hills are composed of a core, surrounded by several circular parallel low ridges (cuestas) that rise gradually from the peripheries to hill center in some cases or stand higher than the core. Although they are generally small and low, size and height of these circular hills are controlled by the original structure size. Topographic sheet (scale 1:100,000) of Paris village shows that the axis of the largest circular hill measures about 30 km, and that the core of the hill rises to 400 a.s.l. The map also shows that some of the circular ridges of this hill rise 431–436 m a.s.l, and that the 350 m contour line delineates the base of the hill. As Fig. (8.5) shows, this hill lies on the plateau surface just to the east of Gabal Abu Bayan El-Bahari in the southern part of Kharga Depression (24°30′N, 31° 00′E). One of the smallest circular hills has a length of about 4 km, a width of about 3.5 km, and a height of 20–30 m above the surrounding grounds. These circular ridges of such structures take the form of an onionskin. Noncircular anticlinal hills developed over small anticlines plunging in one or both directions, with an axis normally parallel to the fault line along which they developed. Some of these anticlinal hills are asymmetrical because they are truncated by the accompanied fault lines. The Geological Map (scale 1:500,000, Sheet No NF 36 NW, El-Saad El-Ali) (Conoco and The Egyptian General Petroleum Corporation 1987) shows that in some cases, older rocks (Cretaceous or basement) are exposed in the core of these structures. In general, these hills vary in size between several hundred of meters to a few kilometers, with a relatively high core and fringing circular ridges. An example of these noncircular anticlinal hills is the one developed to south of the fault line escarpment of Sinn El-Kaddab in the vicinity of **Dungul** (Fig. 8.6). This figure shows that the highest point in the core of this hill rises up to 287.5 m a.s.l, i.e., about 100 m above the surrounding surfaces. The Geological Map shows that the core of this hill is composed of basement rocks.

In some other cases, the closure structures are inverted topographically, so that synclinal basins are expressed as hills and anticlinal highs are inverted into basins (Fig. 8.5). Due to the absence of detailed geologic mapping, it is impossible to identify all cases of inverted forms. However, available maps allow the identification of some examples of such forms. The first example is the two anticlinal basins, which developed on the plateau surface lying to the east of Paris village in the southern part of Kharga Depression (Fig. 8.5). The northern basin is composed of a single anticline, and the southern basin is composed of two anticlines that are separated by a fault line. This is why the northern basin is smaller (6.25 km × 2.5 km) than the southern (7 km × 5 km). The second example is of an inverted synclinal structure, which forms **Gabal Marawa** that lies at about 30 km to the east of the escarpment of **Sinn El-Kaddab Plateau** (23° 30′N, 32° 30′E).

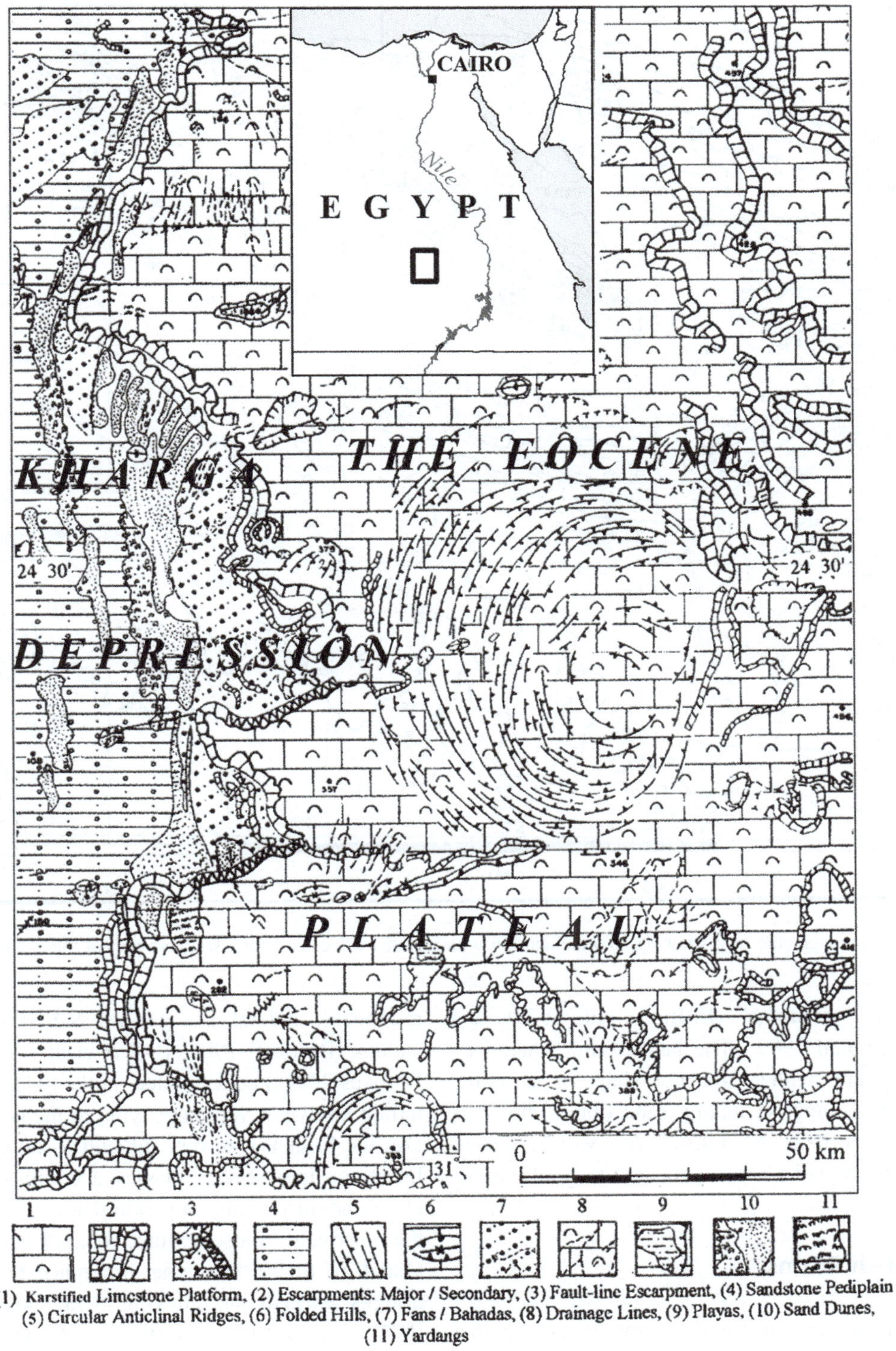

Fig. 8.5 Geomorphological map of the karstified carbonate plateau to the east of Kharga Depression

8.5 Eolian Forms

This group of forms can be divided into erosional forms, developed by wind abrasion, and depositional forms, developed due to the deposition of wind laden sand. Normally, erosional eolian forms take linear shapes as parallel ridge-and-trough system, with individual features ranging in size from minute ones that can be detected only in the field to relatively large features, which can be recognized on relatively large-scale air photos or high-resolution space images. This pattern of forms was given various terms in previous studies, such as "**Eolian Erosional Lineations**"

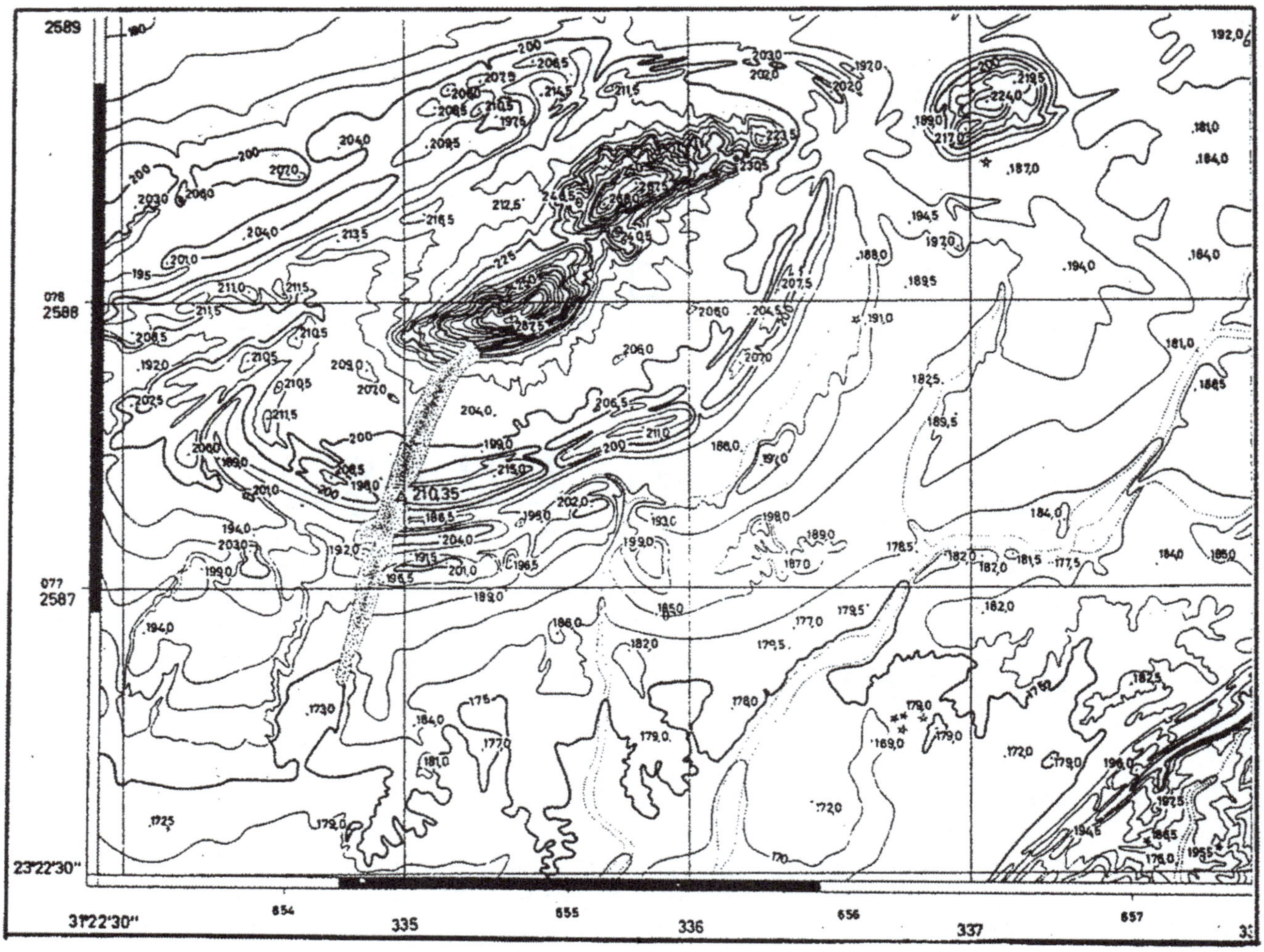

Fig. 8.6 Extract of a topographic map showing an anticlinal hill with its outer ridges and core, southern part of the carbonate plateau

(Brookes 2001a), "**yardangs**" (El-Baz et al. 1979), and "**Kharafish**" (Beadnell 1909). The latter term "Kharafish" will be used, where neither size nor shape is at issue. Depositional forms comprise mainly sand dunes, either as sand seas or as dune belts. Kharafish forms will be treated first since they cover wider areas compared to sand dunes.

8.5.1 Kharafish Forms

Kharafish is a local name in Egypt for rough surfaces that are characterized by ridges and troughs/grooves/corridors. Beadnell (1909) was the first to use this local term to describe the hummocky country, which is made up of innumerable elongated hillocks (yardangs) lying on the limestone plateau between Kharga Depression and the Nile Valley. Examination of topographic maps (scale 1:25,000), air photo mosaics (scale 1: 100,000), and field investigations revealed that this type of topography spreads to the south of 26° 30′N, usually occurs in fields, varies greatly in roughness, and takes a general N-S orientation. Previous studies (Beadnell 1909; El-Baz et al. 1979; Haynes 1980; Brookes 2001a) concentrated on the fields of the very rough terrain known as yardangs, lying to the north of Kharga and Dakhla. In fact, it is very difficult to divide these kharafish lineations into several classes according to size. Furthermore, the only clear distinction that can be made is between the relatively coarse-textured kharafish, and fine-textured kharafish. In the first type, difference between ridges and troughs exceeds 50 cm (Fig. 8.7a), known as yardangs (El-Baz et al. 1979), and the fine kharafish terrain, where difference between crests and separating floors does not exceed several centimeters (Fig. 8.7b, c), known as primary flutes (Breed et al. 1997). It seems that these two types evolve independently from each other. No evidence was found in the field indicated that the two forms represent two evolutionary stages of one form. In the meantime, no mention is made in any previous study that any one of them represents a stage of development for the other one. Therefore, the kharafish forms will be treated in the following

Fig. 8.7 Kharafish of different sizes: **a** yardangs in the field to the north of Kharga, **b** small yardangs to the east of Kharga, facing Doush Village in the southern part of Kharga, **c** polished limestone with micro-flutes on the plateau to the east of Kharga

subsection as two separate groups: (a) yardangs and (b) primary flutes.

8.5.1.1 Yardangs

As mentioned earlier, extensive fields of countless **yardangs** developed on the surfaces of the Palaeocene-Eocene carbonate platforms of the Western Desert. Seen on maps and air photos, these fields are concentrated in the southern part (nearly to the south of latitude 28°N) and can be divided into three groups. The first group spreads on the platforms to the north of Kharga-Dakhla Depression. The second group lies on the plateau to the east of Kharga Depression, whereas the third group developed on the surface of the southern part of the Eocene plateau, named in this analysis as "Sinn El-Kaddab Plateau" after the well-known edge given this name.

The first group of **yardang fields** extends about 250 km from east to west and several tens of kilometers from north to south, as revealed by space images. This group was partly mapped from a Landsat image 190/042 (Fig. 8.8) in a previous study (Brookes 2001a). This map shows three of the four fields of yardangs, and that the three fields are interconnected and extending westward from the Kharga-Asyut Highway atop the escarpments of Kharga-Abu Tartur-Dakhla (Brookes et al. 1993). The fourth field lies to the east of the highway and to the northeast of the other three fields. This latter field extends approximately 65 km from north to south and about 35 km from east to west, with a general orientation from NNW to SSE. The first of the other three fields lies to the north of Kharga Depression (the Kharga field), and the second field covers the area lying to the north of Abu Tartur Plateau with smaller fields on the surface of this plateau (the Abu Tartur field). The third field extends along a strip (10–15 km wide) on the plateau surface just to the north of Dakhla Depression (the Dakhla field). In the Kharga field, countless large yardangs, which occur isolated and in clusters, are hundreds of meters long and tens of meters high (El-Baz et al. 1979). The maximum extension of this field is about 80 km from north to south and 90 km from east to west. A reconnaissance visit in September 1978—made by the author with some members of EL-Baz Expedition to the Gilf Kebir and Uwienat (El-Baz et al. 1980)— showed that some of the yardangs in this field have irregular tops, with a mantle of grayish-pink residual material. In the meantime, some other yardangs are flat-topped and retain some of the original weathered surface. Small ridges are intensely fluted by wind, and the surface glistens when this fluting occurs on crystalline limestone (Fig. 8.7). This field lies in the path of sand flow of Ghard Abu Moharik (Fig. 8.8). The Abu Tartur field extends from **Ain Amur embayment** to **Wadi El-Batikh** Plateau (about 50 km), with an average width of about 15 km. On the **Abu Tartur Plateau** itself, there are small clusters of yardangs in the east, the center, and the west.

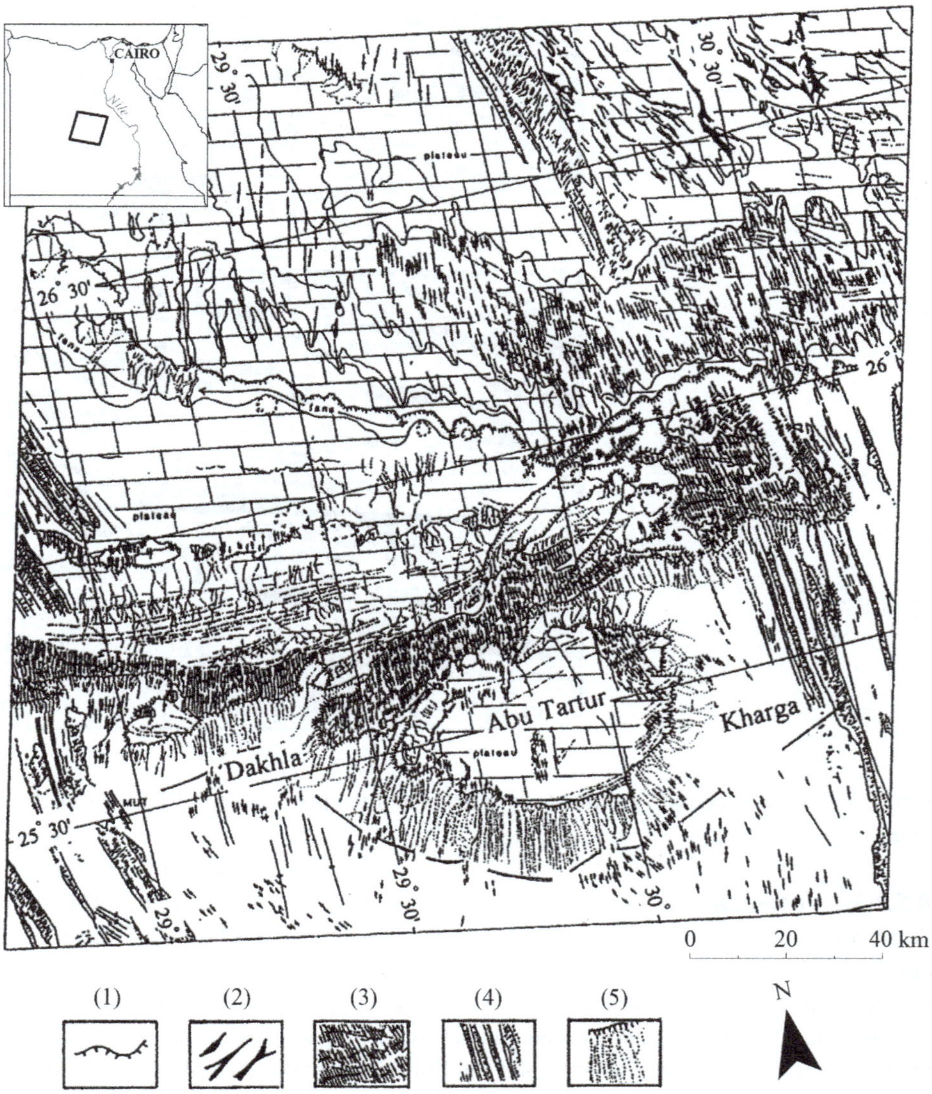

Fig. 8.8 Geomorphological map of the plateau to the north of Abu Tartur, Dakhla, and Kharga (*Source* Modified from Brookes 2001a)

(1) Escarpment, (2) Giant Flutes, (3) Fields of Yardangs, (4) Linear Dunes, (5)Wadis / Drainage Lines

A previous study (Ashour 1974) examined some of the yardangs on the surface of the eastern part of Abu Tartur and found that their length ranges between 50 and 350 m, width between a few meters and 50 m, and height between a few meters and 15 m. These ridges are separated by troughs that are covered by fine brown material, with a thickness up to 5 m. This study also found that the surface of some yardangs is covered by a calcareous crust (caliche), whereas some other yardangs are completely composed of this calich. Along the sides of the yardangs, some solution hollows are present. **The Dakhla Yardang field** extends—unbroken—westward from Darb El-Tawil to West Mawhoob Plateau, with an average width 5–15 km.

The second group of yardang fields is distributed on the surface of the plateau to the east of Kharga Depression. This group comprises two major fields. The first field covers the area that lies to the east of the Kharga escarpment from longitude 30° 50′E to 31° 30′E and extends southward from latitude 26° 00′N to 25° 30′N (about 55 km). It seems that the innumerable elongated hillocks reported by Beadnell (1909) are only the smaller ones in this extensive field of yardangs and not a part of the field to the north of Kharga as reported in some previous studies (El-Baz et al. 1979). This is because Beadnell described the surface forms while crossing the Eocene plateau from the Nile Valley to Kharga Depression. In this field, Beadnell (1909) found that every portion of the exposed rock surfaces is deeply scoured, and that the furrows are separated by upstanding ridges—often sharp enough to injure the feet of a person or a beast. Troughs in between are half buried in drift sand. The second major field lies near the Nile Valley on the promontory of the plateau known as **Gabal El-Gir** (the Limestone Mountain). In this relatively high promontory, the surface is cut into linear ridges that take a NNW-SSE trend. These ridges cross the wadi sides, which lie at right angles to the direction of the ridges.

The third group of yardang fields is distributed on the **Sinn-Kaddab Plateau**. Yardang fields of this group are concentrated in two areas: one in the western part of the plateau and the other in the eastern part. In the western area, yardangs developed mainly on the relatively high ridges of the dome-like features and edges of Depressions or on structural surfaces. Ridges and escarpments in this area lie in the path of the sand flow coming ahead from the northern direction. Therefore, the local projecting features acquired two patterns: a circular pattern that developed under structural control, and a N-S linear pattern that developed by wind abrasion along the weak lines in the circular ridges (Fig. 8.9). Some of these ridges look like a jigsaw, and this is why one of them is called "**Gabal Sharshar**", which indicates that this "Gabal" is dissected as a small ridge-trough system. In some cases, the intervening surfaces are also dissected by wind abrasion, but also into wide and long ridges compared to those developed on structural circular edges. In other cases, however, yardang density increases on low-lying grounds compared to higher surrounding surfaces.

The second area where yardangs are concentrated on the Sin El-Kaddab Plateau lies near its eastern escarpment facing the outlier named **Gabal El-Barqa**, between latitudes 24° 35′N and 24° 00′N, and centered around longitude 32° 15′E. Although the width of this field is variable, it is generally wider in the north (45 km) than in the south (5 km). In this field, air photo mosaics (scale 1:100,000) show that there are sand flows in this area, and that sand accumulated in inter-yardangs corridors but without developing any type of dunes. In the north–eastern cluster of yardangs, the sand filled the corridors, leaving only the tops of the yardangs exposed. This eolian sand coverage of yardang corridors indicates a cycle of sand deposition and not one of sand erosion. Yardangs in this area exhibit different sizes and shapes, such as long and slim ones adjacent to short and wide ridges. Although the general trend of yardangs in this area is 355°, there is a small cluster of yardangs with an orientation of 290° (24° 10′N and 32° 10′E).

8.5.1.2 Primary Flutes

This type of flutes forms the **fine kharafish terrain**, where difference between crests and separating floors does not exceed several centimeters. Flutes are found on all types of limestone and chalk. They have rounded floors and are symmetrical in cross section. Flutes may end with what is called "vortex pits." Flutes and vortex pits grade into each other. Many **primary flutes** are crossed by minute secondary flutes, transverse or oblique to primary wind direction (Whitney 1983). In the field, flutes were found to develop on flat surfaces, on ventifacts, and on yardang prows and flanks, but not on the leeward side of yardangs (Fig. 8.7). Flutes—though representing fine textured kharafish—can be classified into micro-flutes and macro-flutes. Micro-flutes are a few millimeters only in width and depth, whereas macro-flutes reach several centimeters in depth and width. Micro-flutes (grooves) were found to develop on homogeneous (Tarawan) chalk; therefore, abrasion by wind is also uniform. On the contrary, macro-flutes develop on inhomogeneous silicified or crystalline limestone, where rates of abrasion are unequal. Therefore, softer areas become deeper and wider and develop into macro-flutes. It was also found that micro-flutes never grade into macro-ones. Furthermore, it seems that some of the micro-flutes developed during an arid phase in the past. This is because some flutes are stained and not glistening, and that their sides are dissected by minute rills, which develop on carbonate rocks by the action of water from dew.

8.5.1.3 The Development of Kharafish

All previous studies agreed that flutes and yardangs in the Western Desert are developed by the erosional action of wind. Near Kharga, Beadnell (1909) observed that the

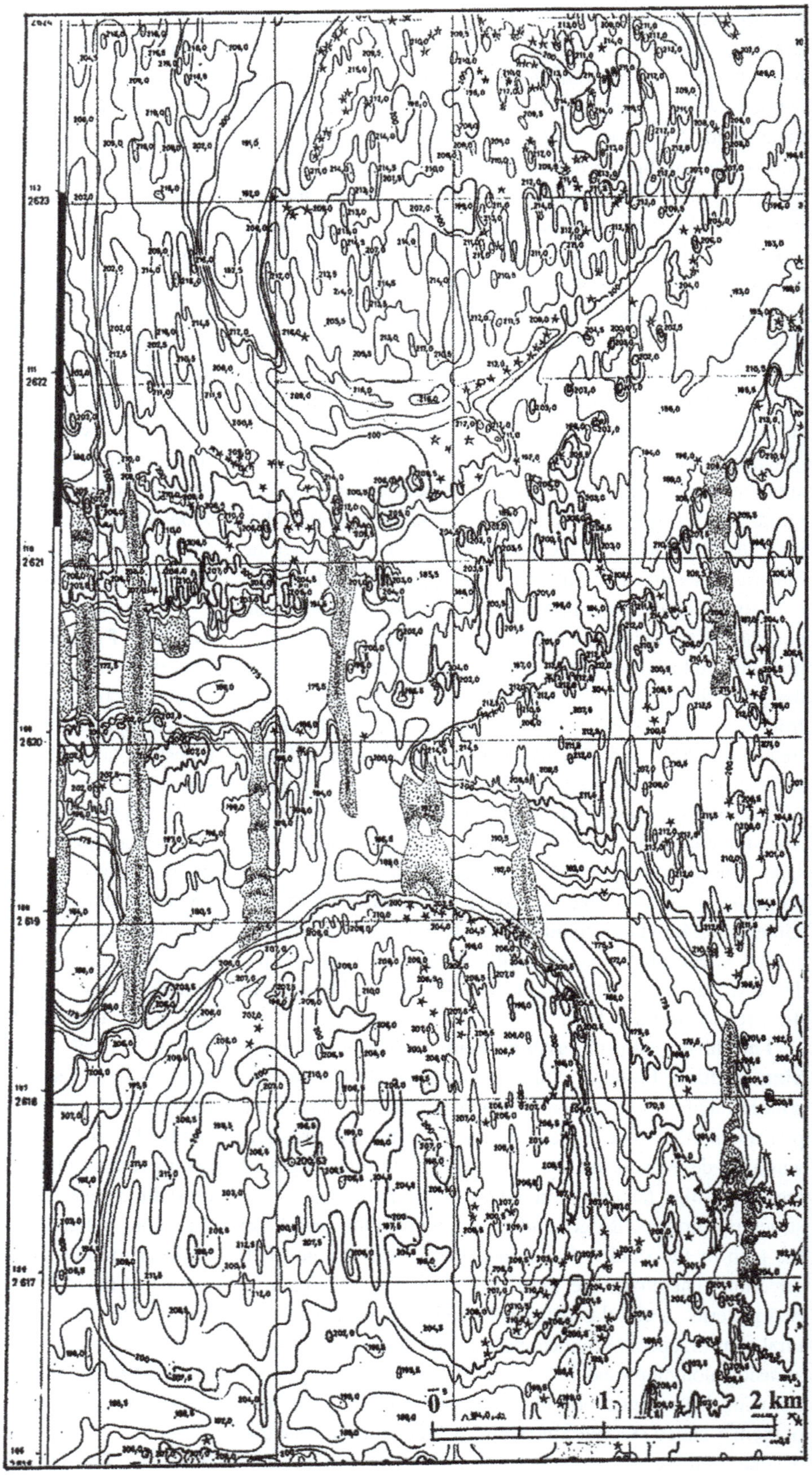

Fig. 8.9 Extract of topographic map (scale 1:25,000) showing circular structural ridges cut into yardangs

hillocks of the kharafish are separated by deep troughs half buried in drift sand, all lying parallel in the direction of the prevailing winds. El-Baz et al. (1979) concluded from their study on the yardangs in the field of North Kharga that they are eroded by a combination of abrasion and deflation. The most obvious prerequisites are strong, unidirectional or reversing winds, and great fetch over barren continuous rock exposures. In addition to deflation and abrasion processes, Breed et al. (1997) postulate that running water acts to influence yardang field development. Yardangs may be initiated along former stream courses that are enlarged and modified by wind. Wind is unquestionably the main yardang-forming agent (Cooke et al. 1993). Brookes (2001a) concluded that the development of AELs (Eolian Erosional Lineations) to the north of Kharga and Dakhla Depressions requires initial selective incision of the substrata by wind erosion in order to isolate positive forms between erosional grooves (which become troughs, then corridors as they deepen and widen). Incisions progressively focus airflow, which exploits lithological/structural weaknesses, commonly joints, to erode transversely and attack an upwind-facing prow. The presence of micro/macro-flutes on yardangs prow and flanks is taken as clear evidence that wind erosion is the main agent in the development of yardang fields. However, no previous attempt was made to evaluate the role of other factors in the initiation of yardang development. Also, no attempt, except that of Brookes (2001a), was made to explain the presence of yardang fields in one locality and their absence in another.

Data collected from field observations, topographic, and geologic maps provides an answer to the above queries. One of the most interesting findings is the relationship between the size of flutes and homogeneity of bedrock material on flat surfaces. As mentioned earlier, micro-flutes developed on homogeneous limestone and chalk, where no differential wind abrasion occurs. On the other hand, macro-flutes developed where rock composition is not homogeneous. By analogy, yardangs, which are separated at present by corridors/wide flutes, do not develop on homogeneous rocks, unless there is another factor assisting in the dissection of the bedrock by wind abrasion, such as structural weaknesses (e.g., joints) as mentioned by Brookes (2001a). In fact, no previous study reported that the initial stage of yardang development was seen in the field, a stage where joints are widened by wind abrasion but without isolation of positive forms. This means that if joints are a necessity in the incision of substrata by wind action, wind cannot widen these joints to develop yardangs unless these joints are widened by another agent. Along the joints, rain water or surface water may infiltrate, dissolving C_aCO_3 and widening these openings. This type of karstification was seen along road cuts, where joints are filled with brown residues that resulted from the infiltrated surface water. Therefore, it can be said that karstification of jointed limestone is a significant factor in widening joints and in yardang development. Karstification plays another role in yardang development also through the dissolution of C_aCO_3 from surface layers of siliceous limestone containing silica bands. With time, silica bands appear on the surface of the ground and are left as a lag gravel armor protecting the fine residues resulting from dissolution of carbonates. Areas covered with such chert gravel were found to be devoid of yardangs (Brookes 2001a).

Although wind flutes represent a direct form of wind abrasion, they provide evidence at the same time that wind action is but one of the current main agents in modeling yardangs. Therefore, yardang fields are expected to be present in areas where there is a sand flow, on condition that other initiating factors are fulfilled.

8.5.2 Sand Dunes

Sand dunes spread on the Paleocene-Eocene plateaus in various localities as sand seas, dune fields, minor dune belts, and sand accumulations. Although characteristics of some sand seas in the Western Desert are treated in Chap. 14, the present analysis stresses those aspects relevant to the geomorphic characteristics of these platforms. Distribution of sand seas and dune fields (Fig. 13.1) shows that there are two sand seas (Ghard Abu Moharik and Farafra) and one dune field (South Rayan) on the Paleocene-Eocene plateaus. Only the northern section of **Ghard Abu Moharik Sand Sea** (about 300 km) extends on the Eocene plateau in a linear form from NNE of Bahariya Depression to the northern escarpment of Kharga. Similarly, only a small section of Farafra Sand Sea crosses the Palaeocene plateau from Farafra Depression to Dakhla Depression. As stated in Chap. 14, the northern section of Ghard Abu Moharik is composed of two separate bodies. The first represents the main body of Ghard Abu Moharik, whereas the second lies just to the east of this first part and extends for about 70 km only from latitude 28° 30′N to 27° 50′N. The secondary body of Ghard Abu Moharik reaches its maximum width in a basin-like area, where its lowest point lies at 65 m a.s.l. As indicated in Chap. 14, it is clear that local inherited forms and pre-dune topography of the surface of the plateau controlled the development of this sand sea. At the upwind section of this sand sea, linear dunes developed in the lee side of several isolated conical (most probably of karst origin) hills. Dune density and width of the sand sea increased in low-lying areas, whereas dune density decreases and barchans develop on relatively high ground and margins of the sand sea. Since it covers other surface forms, and taking its size into consideration, this sand sea represents the most recent major form developed on the Palaeocene-Eocene plateau. Sands of Farafra Sand Sea contributed to the development of yardangs lying on this part of the plateau crossed by sand dunes. The **South Rayan dune field** extends for about 185 km from the southern part of Wadi El-Rayan to the latitude of the city of Dairut in the Nile floodplain. Here also, pre-dune topography of the southern part of Wadi El-Rayan controlled the development of this dune field into linear dunes in the northern part, then into barchans on the plateau surface in the southern part.

In addition to the above-mentioned sand seas and dune fields, air photo mosaics (scale 1:100,000) and topographic maps (scale 1:250,000) show that small clusters of sand dunes and sand accumulations spread in several localities due to local factors. Examples of dunes are those which extend on the southern part of the Eocene plateau. Dune belts in this area developed due to the climbing up of dune sand from Kharga dune belts, eroding the circular edges of local the anticlinal structures (Fig. 8.9) in the southern part of the plateau as mentioned earlier in the section of yardangs (Sect. 8.5.1.1). Falling dunes of various sizes developed at the escarpment of the Eocene plateau overlooking the Nile Valley at several localities from Qena Bend to Lake Nasser. Some of these dunes extend in the downwind direction and terminate at the shores of Lake Nasser. Other examples include two small barchan belts developed on the surface of the Eocene plateau. These two belts lie between latitudes 26° 00′N–25° 50′N and longitudes 31° 10′–31° 20′E, and between 25° 50′–25° 40′N and 32° 00′–31° 50′E, respectively.

8.6 Fluvial Forms

Fluvial forms developed on the surfaces of the Palaeocene-Eocene platforms can be classified into erosional and depositional forms. Erosional forms include drainage lines, wadis, and giant flutes.

8.6.1 Drainage Lines

Drainage lines are dried up shallow watercourses, where water flowed during past pluvial periods, or where water may flow at present after heavy rain showers. Short drainage lines, only a few kilometers in length, spread all over the surface of these platforms. They drain either in small closed basins or in relatively low-lying flats, which prove that they represent previous internal watercourses. This also indicates that most of the former surface water was diverted to subsurface underground water, which was either stored in underground aquifers or appeared along the plateau escarpments in the form of springs. To what extent do these drainage lines represent previous interconnected drainage systems or several drainage systems, or represent recent independent small, shallow drainage systems? This question does not have a straightforward answer due to the little available evidence. However, available evidence suggests that these drainage lines do not represent relicts of a previous larger system or systems. Detailed topographic maps show that there is no general regional slope in any direction, instead slope changes locally along very short distances. In addition, these previous watercourses do not form integrated drainage nets, forming in very few cases poorly organized networks. This characteristic indicates that—with the exception of the plateau margins overlooking the Nile Valley—these carbonate plateaus are internally drained. Also, Geological Maps show that the surface limestone and chalk beds are highly fractured, which means that most of the surface water will be transformed into underground water after a short time and a short distance of flow. Although fluvial forms spread on most of the surfaces of the Palaeocene-Eocene plateaus, there are wide areas that appear on the topographic maps (scale 1:100.000) devoid of any trace of such forms. One example is the area lying to the north of Kharga Depression covered by the topographic sheet of Aqabat El-Ramliya (Sheet No 36. I 2), between latitudes 26° 00′N and 26° 30′N and longitudes 30° 30′E and 31° 00′E. This absence of fluvial forms could be explained by the same assumption that surface water has never had the chance to flow even for a short distance. Instead, surface water percolated through the limestone cap rock, dissolving the surface layer, carrying carbonates to underground aquifers, and leaving the residues (terra rossa and chert gravel) on the surface of the plateaus.

8.6.2 Wadis

Wadis are restricted to the margins of the Palaeocene-Eocene carbonate plateaus, which slope toward the Nile Valley, Kharga-Dakhla Depression, and the southern peneplain. Although they vary greatly in size, some of the wadis extend for several tens of kilometers inside the plateau. These relatively large wadis are characterized by well-developed drainage nets, such as Wadi Rimidin and Wadi Abu Salam at the latitude of the city of Idfu in the Nile Valley. Most of these wadis carried huge quantities of fluvial materials, which were deposited as fans and bahadas at the base of the escarpments of the plateau, such as those which developed along the margins of the Nile floodplain. Some of these fluviatile materials appear on the Geological Map (scale 1:500,000) as belonging to the Pliocene period. Therefore, it can be concluded that some of these wadis have developed at least since the Pliocene period.

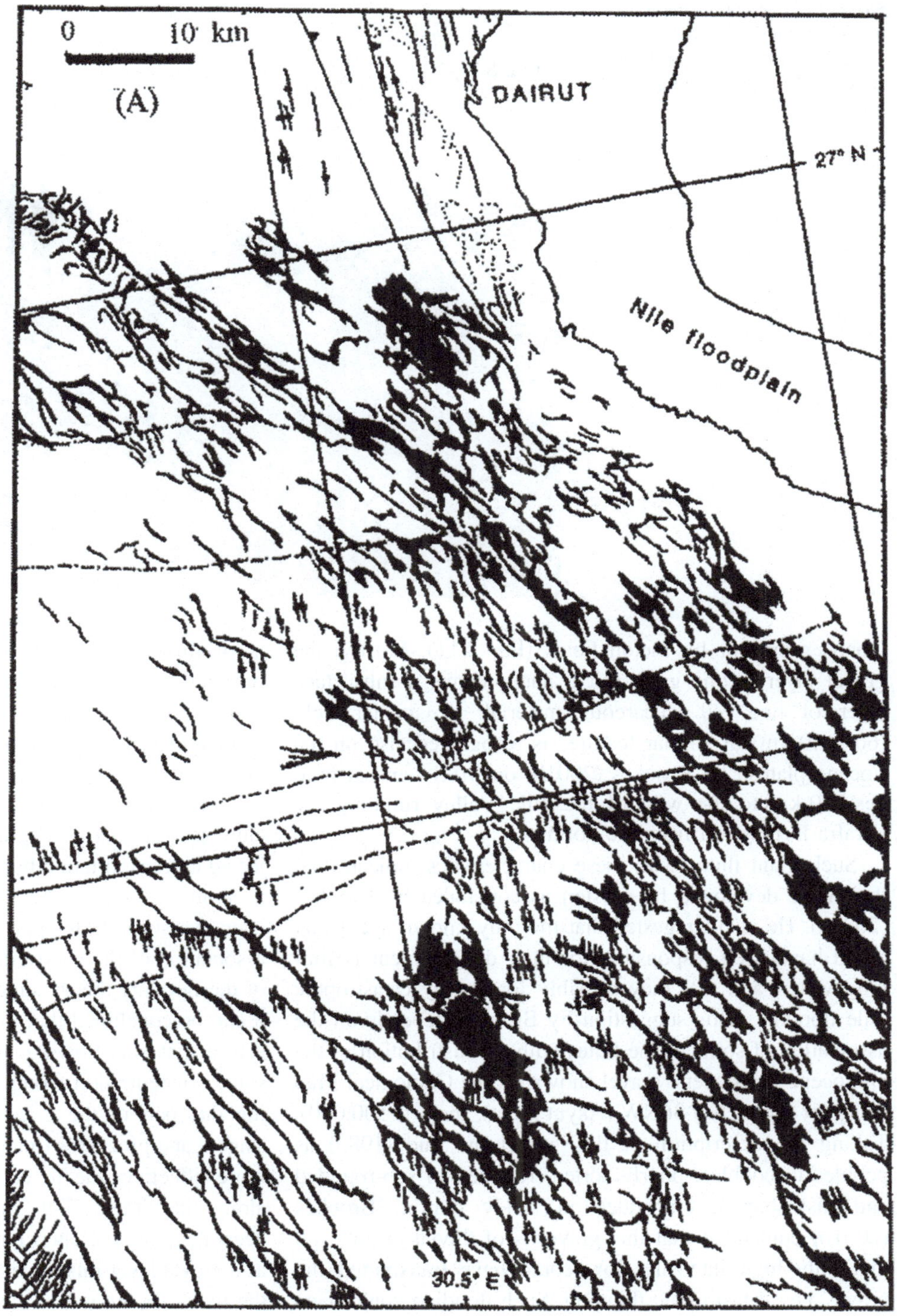

Fig. 8.10 Map of giant flutes to the west of the Nile Valley (*Source* Brookes 2001b)

8.6.3 Giant Flutes

"**Giant flutes**" are a term used by Brookes (2001b) to denote darker markings that appear on MSS Landsat images of the dissected Eocene limestone plateau west of the Nile, north and south of latitude 27°N. The length of these flutes is 1–30 km, and the width is between 100 m and 10 km. All flutes are curved or sinuous, branching, taper at both ends, and form a large field whose NW-SE length is 225 km and NE-SW width is 125 km (Fig. 8.10). Although Brookes (2001b) was not able to examine these marks in the field, field examination by the author confirmed Brook's interpretation of space images that these marks represent negative relief. In the field, they appear as corridors separating parallel low ridges (less than 10 m). Their floors are of irregular width and covered either by angular to sub-angular lag

Fig. 8.11 Field photograph of two giant flutes with melon concretions strewn on their floors, separated by a giant yardang (ridge)

(chert) gravel or by melon fields (Fig. 8.11). Mostly, the gravel layer is one grain thick, and protects a subsurface layer of fine red calcareous material. A smaller field (60 $\times$ 40 km) of similar features is recognized also on the Eocene plateau by Brookes (2001b) on MSS Landsat images 250 km to the west of the Nile Valley rimming the Farafra Depression from the north.

Such giant flutes with these characteristics indicate that they were developed by erosion as concluded by Brookes (2001b). He also suggested that the only known agent that could have been responsible for their development is running water, and that they possibly formed by catastrophic Nile flooding. This suggestion by Brookes depends on the association of some of the flutes with the distribution of the Oligocene-Pleistocene gravel shown on the Geological Map of Egypt (Sheet NG 36NW (Asyut sheet), scale 1: 500,000). Taking the description of this gravel by Said (1983) as pebbles to cobbles, poorly sorted, rounded to sub-rounded and disk-shaped, embedded in sandy matrix, Brookes (2001b) concluded that this gravel is of fluvial origin, and that only high intensity flow could have transported the gravel, i.e., catastrophic flooding. Such flooding could have been caused locally by erosional and/or tectonic breaching of a dam across the Nile Valley drainage, raised by tectonics or developed by a dyke.

Alternative origins, bedrock structure, karst or rapid marine regression were excluded by Brookes, since they are unsatisfactory. However, although Brookes (2001b) was correct in his interpretation that these giant flutes are eroded by fluvial action, the catastrophic flooding theory is unacceptable. This is because this theory depends mainly on flute/gravel association, and on the inference that this gravel is of fluvial origin. In fact, distinction should be made between consolidated fluvial gravel covering nearly flat expanses as dissected patches, and the angular and sub-angular chert gravel found at the floors of the so-called giant flutes, which developed from the silica bands as a residual lag after the dissolution of the host limestone. The angularity of this chert gravel indicates that it was not subjected to transportation by water flow—or if so, then only for a short distance. Therefore, although the morphologic characteristics of the giant flutes indicate a fluvial origin, the association of the chert gravel with the flutes is not in favor of the flooding theory suggested by Brookes (2001b). Here, it can be concluded that these flutes were initiated by water action that was flowing in the form of a sheet flow, with some concentration along weak structural lines characterizing this part of the Eocene plateau. Following this initial stage, karst processes (mainly dissolution of limestone) were more effective due to concentration of any surface water along the flutes. With time, the flutes are widened and deepened, and the chert bands disintegrate into small pieces as a gravel layer mixed with the fine residues on the floors of the flutes. In the meantime, winds deflated the fine material from the surface layer, leaving the gravel and the melon concretion as an armored surface protecting the fine sub-layer from erosion (Fig. 8.11).

8.6.4 Playas

In addition to the above-mentioned depositional fluvial forms developed as fans and bahadas sourced by the wadis dissecting the margins of the plateaus, the only form, which can be detected on topographic maps, air photos, and Google Earth, is the playas. At the floors of the closed basins, former watercourses deposited all clastics, carried out from the surroundings. It is expected that the load of these watercourses was mostly fine sized (silt and sand), and that the water flowing to the basins formed temporary lakes. A playa of this type can be seen at both sides of the Cairo-Bahariya highway near the entrance of the Bahariya Depression.

8.6.5 Consolidated Fluvial Gravel

Consolidated Fluvial Gravel was recorded during a trip to the Djara Cave from Bahariya as a cap layer on small low-lying hills of similar height on the surface of the Eocene plateau between Bahariya and Ghard Abu Moharik. A section of one of the hills was measured at the site (27° 55′N, 29° 33′E) showing that the total thickness of this gravel cap layer is about 1.5 m and is divided into four thin horizons. The top one is a 50 cm conglomerate horizon and is composed of well-rounded gravel 2–3 cm, underlain by a 30 cm conglomerate horizon of large boulders. The lowest sub-layer is composed of a 60 cm sub-angular conglomerate, overtopped by a very thin layer of fine materials. On stratigraphic grounds, the consolidated gravel is deposited by a post-Eocene drainage system, and it can be correlated with the Oligocene Nekheil Formation in the Wadi Nekheil-Quseir area, Red Sea region, which carries similar sedimentological characteristics (Hermina and Lindenberg 1989; Said 1990). This conglomerate layer is considered by Said (1990) as inverted wadi deposits, but this is not possible since it spreads over a wide area in sheet form rather than in a wadi bed form. The fluvial consolidated gravel layer represents a remnant of a wider gravel sheet that has been deposited by an Oligocene drainage system. This drainage system flowed westward with its headwaters in the eastern side of the newly elevated Red Sea Mountains, as suggested by some previous studies (Said 1990; Issawi and McCauley 1993). The gravel of the conglomerate seems to have been derived from the Eocene and Cretaceous rocks, which covered the Red Sea Mountains. Therefore, this gravel was transported over some distance due to the expanse of land from the Red Sea Mountains in the east to the middle of the Eocene plateau in the west. In the meantime, space images show inverted wadis to the west of Ghard Abu Moharik, taking the form of sinuous linear ridges, with a total length extending approximately 50 km from SW to NE (Fig. 23.10a).

8.7 Conclusion

Due to the interaction between the carbonate rocks exposed on the surfaces of the Palaeocene-Eocene platforms, and the wet climates that have prevailed in Egypt since Late Eocene, karst forms dominate their surfaces, and karstification played the major role in modeling them. This is why they are called in this study "**Karstified Carbonate Platforms**". The presence of degraded karst forms on their surfaces indicates that karstification played another role which is the reduction of the surfaces of such platforms. The presence of the degraded karst forms also indicates that there were at least two cycles of karstification. The first was an early cycle during which the degraded forms were developed. This cycle is mostly pre-Miocene, during which karstification was very intense because Egypt had a hot tropical climate, as indicated by evidence from the degraded cave system exposed on the surface of the platform in the area between Bahariya and Farafra (Pickford et al. 2006). During this cycle all the degraded forms (caves, terra rossa soils, cone/karst) developed and degraded at the end of this long period, and became exposed on the surface of the ground, resulting in the reduction of the surfaces of the platforms. The second cycle covered the Plio-Pleistocene period, during which all non-degraded karst forms were developed.

References

Ashour MM (1974) The geomorphology of Abu Tartur Plateau. Dissertation (in Arabic), Ain Shams University, Cairo

Awad H (1952) A new type of desert cuesta in Central Sinai. In: Proceedings of the XVII International Geographical Congress, Washington

Ball J, Beadnell H (1903) Bahariya Oasis: Its topography and geology Egypt. Survey Department, Cairo

Beadnell H (1909) An Egyptian Oasis (Kharga). John Murray, London

Breed C, McCauley J, Whitney M, Tchakarian V, Laity J (1997) Wind erosion in drylands. In: Thomas D (ed) Arid zone geomorphology: process, form and change in drylands, 2nd edn. John, Chichester

Brook G, Embabi N, Ashour M, Edwards R, Cheng H, Cowart J, Dabous A (2002) Djara Cave in the Western Desert of Egypt: morphology and evidence of quaternary climatic change. Cave Karst Sci 29:57–66

Brookes I (1993) Geomorphology and quaternary geology of the Dakhla Oasis Region, Egypt. Quat Sci Rev 12:529–552

Brookes I (2001a) Aeolian erosional lineations in the Libyan Desert, Dakhla Region, Egypt. Geomorphology 39:189–209

Brookes I (2001b) Possible Miocene catastrophic flooding in Egypt's Western Desert. J Afr Earth Sci 22:325–333

Butzer K (1964) Pleistocene palaeoclimates of the Kurkur Oases, Egypt. Can Geogr 8:125–141

Caton-Thompson G, Gardner E (1932) The prehistoric geography of Kharga Oasis. Geogr J LXXX:369–409

Cooke R, Warren A, Goudie A (1993) Wind erosion. In: Cooke R, Warren A, Goudie A (eds) Desert Geomorphology. UCL Press, London, pp 290–306

El Aref M, Lotfy Z (1985) Genetic karst significance of the iron ore deposits of El Bahariya Oasis, western Desert, Egypt. Ann Geol Surv Egypt 15:1–30

El Aref M, Abou Khadrah A, Lotfi Z (1987) Karst topography and karstification processes in the Eocene limestone plateau of El Bahariya Oasis, Western Desert, Egypt. Z Geomorph N F 31:45–64

El Aref M, El-Dougdoug A, Mesaed A (1991) Landform evolution of ferricrete duricrusts, El-Heiz Area, El-Bahariya Depression, Western Desert, Egypt. Egypt J Geol 1–2:1–39

El-Baz F, Breed C, Grolier M, McCauley J (1979) Eolian features in the Western Desert and some implications to Mars. J Geophys Res 84:8205–8221

El-Baz F, Boulos L, Breed C, Dardir A, Dowidar H, El-Etr H, Embabi N, Grolier M, Haynes V, Ibrahim M, Issawi B, Maxwell T, Moustafa A, and Yousef M (1980) Journey to the Gilf Kebir and Uwienat, Southwest Egypt, 1978. Geogr. Journal 146: 51–93

El-Etr HA, Yehia MA, Dowidar H (1982) Fault pattern in the southWestern Desert. Middle East Research Centre, Ain Shams University. Sci Res Ser 2:123–151

El-Kayali M (1980) Bahariya Depression: a geomorphological study (in Arabic). Dissertation, El-Minya University, Egypt

Embabi NS (1967) A geomorphologic study of Kharga Oases Depression, the Western Desert, Egypt. Dissertation, Bristol University

Embabi NS (2004) The geomorphology of Egypt: landforms and evolution I: the Nile Valley and the Western Desert. The Egyptian Geographical Society, Cairo

Haynes C (1980) Geological evidence of pluvial climates in the Nabta area of the Western Desert, Egypt. In: Wendorf F, Schild R (eds) Prehistory of the Eastern Sahara. Academic Press, New York, pp 353–371

Hermina M, Lindenberg HG (1989) The Tertiary. In: Hermina M, Klitzsch E, List F (eds) Stratigraphic lexicon and explanatory notes to the geological map of Egypt, scale 1: 500,000, Conoco Inc Cairo. pp 141–218

Issawi B, McCauley J (1993) The Cenozoic landscape of Egypt and its river systems. Ann Geol Surv Egypt 19:357–384

Pickford M, Wanas H, Soliman H (2006) Indications for a humid climate in the Western Desert of Egypt 11–10 Myr ago: evidence from Galagidae (Primates, Mammalia). Palaeoecology 5:935–994

Said R (ed) (1990) The geology of Egypt. Balkema, Rotterdam

Sweeting M M (1958) The karstland of Jamica. The Geog. Jour. 124: 184–199

Whitney MJ (1983) Eolian features shaped by aerodynamic and velocity processes. In: Brookfield ME, Ahlbrandt (eds) Eolian sediments and processes, Elsevier, Amesterdam, pp 223–245

9 The Region of El-Galalah Plateaus NE the Eastern Desert

Abstract

The plateaus of El-Galalah El-Bahariyah (Northern), El-Galalah El-Qibliyah (Southern), and Wadi Araba represent a distinguished geomorphologic region, overlooking the Gulf of Suez, and occupying the NE corner of the Eastern Desert. Both Galalahs rise several hundred meters asl, reaching in some localities more than a thousand meters, while the floor of Wadi Araba rises only 120–250 m asl. The region is structurally dissected by faults of different orientations and is composed of a broad swell, where Wadi Araba represents its axis, and the two Galalahs represent the two flanks of this anticline. Accordingly, this Wadi can be considered an inverted topography. The regional geomorphology is simple, since the Region is composed of three units: the two cuestas of the two Galalahs and the anticlinal Wadi Araba. On the surfaces of these regional morpho-tectonic units, several erosional forms developed, of which fluvial and karst forms are the most widespread due to the presence of carbonate rocks and past humid climate. Caves and cone karst are the most widespread karst forms. Apart from Wadi Araba, wadis, as fluvial forms dissect all surfaces of the two plateaus. These wadis are characterized by meandering, developed due to the low gradient of the plateau surfaces, and entrenching due to tectonic uplifting of the two plateaus. Analysis of Landsat-TM images revealed that these wadis developed multi-cyclic fans in Wadi Araba. The volcanic cone, discovered in Wadi Araba, is used to estimate the age of this Wadi.

Keywords

El-Galalah El-Bahariyah Plateau • El-Galalah El-Qibliyah Plateau • Wadi Araba • The Eastern Desert • Incised Wadis • cone karst • Caves

9.1 Introduction

This region includes the two Galalah plateaus of El-Bahariyah (the northern), El-Qibliyah (the southern), and Wadi Araba which separates the two plateaus. This region lies in the NE corner of the Eastern Desert, where the plateaus overlook the Gulf of Suez by steep scarps, and Wadi Araba debouches in the Gulf (Fig. 9.1). The two Galalahs and Wadi Araba represent a distinctive geologic and geomorphologic region; therefore, they are treated jointly in this chapter.

The Northern Galalah is bounded by steep cliffs from north, east, and south. The northern cliff is also the southern

N.S. Embabi, *Landscapes and Landforms of Egypt*, World Geomorphological Landscapes,
https://doi.org/10.1007/978-3-319-65661-8_9

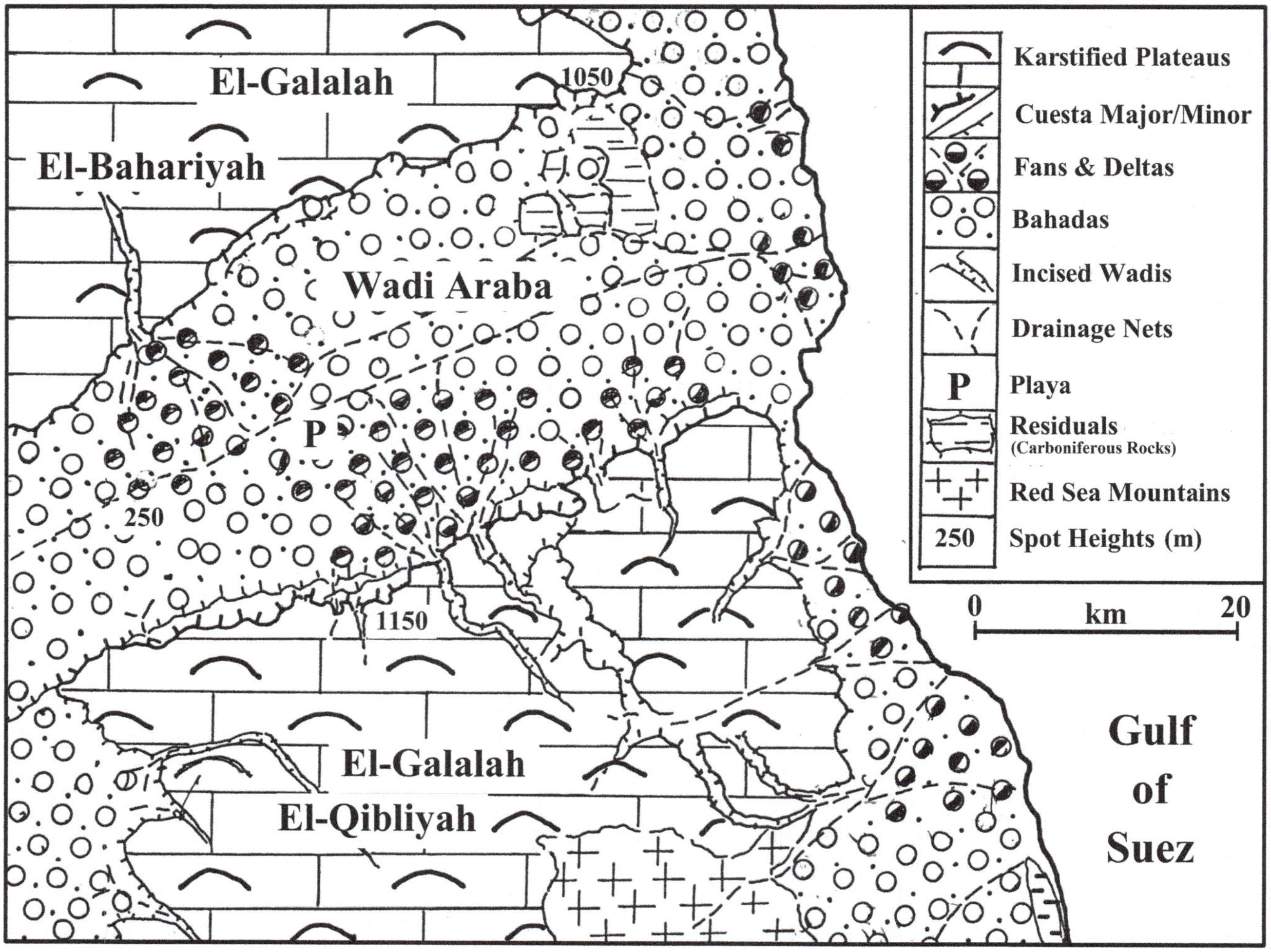

Fig. 9.1 Geomorphologic map of the Region of El-Galalah El-Bahariyah, El-Galalah El-Qibliyah, and Wadi Araba (Compiled from Google Earth 2014, topographic maps, scale 1:50,000)

side of Wadi Akheider, and the same applies to the southern side of the plateau where the steep cliffs represent the northern side of Wadi Araba. On the eastern side, the cliffs are overlooking the Gulf of Suez, without nearly any coastal plain. The highest point is 1249 m asl, whereas the average height is approximately 1000 m a.s.l, but altitude decreases westward and reaches 400–450 m, and the plateau merges with El-Maaza plateau. Although the landscape takes the form of a tableland, it is dissected by deeply incised wadis especially along the margins characterized by cliffs. Most of the wadis extend into the heart of the plateau, transforming parts of its surface into a highly dissected landscape.

The Wadi Araba is very wide, with an average width of 25 km. It extends from SW to NE for approximately 80 km. In this wide wadi, only one main channel runs from west to east and was fed by wadis from the two Galalahs. However, topographic maps show that most of its tributaries come from the Southern Galalah. Due to change in slope angles, wadis issuing from the two Galalahs developed large fans on Wadi Araba floor before debouching in the main channel. Along the southern side, some of the fans coalesce forming compound fans.

The Southern Galalah is bounded by a steep cliff from north and east. The highest point is 1349 m asl, with an average height of approximately 1000 m a.s.l. The height of the plateau surface decreases westward until its surface merges with that of El-Maaza plateau. This plateau is dissected by numerous wadis outnumbering those of the northern plateau, and this is why its surface and cliffs are more dissected.

Due to the major role of the geological background in shaping this region, a brief geologic account will be given before tackling the geomorphologic characteristics of the two Galalahs and Wadi Araba.

9.2 Geological Background

This section deals with the geological aspects affecting the landscape of this region and its characteristic forms. The analysis makes use of geologic maps that cover the whole region at a scale of 1:500,000 (Conoco Coral and The Egyptian General Petroleum Corporation 1987), and those that cover some parts of it at a scale of 1:100,000 (Conoco Coral, Inc., Egypt 1982–1983). The analysis depends also on data provided in previous studies.

9.2.1 Stratigraphy

Previous studies and the geologic maps of this region show that the general stratigraphic succession is composed of various types of rocks, ranging in age from the Carboniferous to Quaternary (Sadek 1926; Kostandi 1959; Said 1962; Abdallah and El-Adindani 1963; Omran 1977). The Geological Map shows that the two Galalahs are mostly underlain by Middle Eocene limestone, since areas underlain by Lower/Upper Eocene sediments are relatively very small. This limestone also appears in the cliffs of the two plateaus. This Middle Eocene limestone is the most widespread rock exposure in this region. However, Wadi Araba is covered by Quaternary fluvial materials mainly in the form of fanglomerates, and some small rock exposures, of which those belonging to the Carboniferous period are the widest. This Carboniferous exposure in Wadi Araba consists of shale, sandstones, marl and is known as Rod El-Hamal Formation. It is one of three small outcrops belonging to this period in Egypt, the first of which is in western Sinai, and the second is in SW corner on the Western Desert. Other sedimentary rocks, whose age is post-Carboniferous and pre-Quaternary, consist mainly of clastics such as shale, marl, and sandstone, with some thin limestone beds. They appear along the escarpments of the two plateaus and the small exposures on the floor of Wadi Araba.

In addition to the sedimentary cover, volcanic rocks, mainly basalts, exist in some localities such as small exposures on the surfaces of **El-Galalah El-Bahariyah** and **Wadi Araba** only (Abdallah et al. 1973). No interpretation is given in any previous study for the absence of volcanic rocks on the surface of **El-Galalah El-Qibliyah**. A close relationship was found between the occurrences of the volcanic basalt and the faults in Wadi Araba (Abdallah et al. 1973). These volcanic materials were brought along the main fault and fracture planes in positions where these planes intersect with each other.

9.2.2 Structure and Tectonics

While analyzing structure, regional and local structures are differentiated. Regionally, nearly all previous studies concluded that a broad, gentle, plunging anticline accommodates the three units of this region, where Wadi Araba lies along the axis of the anticline, and the flanks of Northern Galalah and Southern Galalah Plateaus dip in opposite directions: NW and SE, respectively (Nakkady 1955; Said 1962; Abdallah and El-Adindani 1963; Omran 1977). Nonetheless, Abdallah and El-Adindani (1963) suggest that Wadi Araba can be considered as a large faulted upthrown block, but in their conclusion they stated that the Carboniferous rocks exposed in the Wadi floor suffered some folding. Also, Ramdan and Khalil (1995) and Höntzsch et al. (2010) concluded from their studies that only the northern escarpment of El-Galalah El-Qibliyah (the southern side of Wadi Araba) is a monocline and its beds are dipping southward, whereas the southern escarpment of El-Galalah El-Bahariyah (the northern side of Wadi Araba) is bounded by a fault line. Although faulting is the dominant tectonic feature, it is more likely that it was not the main structural element contributing to the regional configuration of this region. Regional faulting trends are NE-SW and NW-SE. This type of faults bounds the escarpments of the two plateaus and two sides of Wadi Araba. Some faults cross the whole region from north to south; they also cross each other, dividing the plateaus into several blocks. Analysis of the Apollo-Soyuz space images by Abdel-Rahman and El-Etr (1979) revealed that the north-northeast faults divide the northern part of the Eastern Desert into five blocks. Three of these blocks (the two Galalahs and Gabal Ataqa) are structural highs, while the other two blocks (Wadi Araba and Wadi Akheidir) are two structural lows (grabens). It was concluded from this study that Wadi Araba represents an unsolved problem regarding the relationship among its geologic aspects and its topography.

Local structures are composed of faults and folds, as can be seen on the Geological Maps (scale: 1:100,000). Geological Maps also show that local faults are the dominant tectonic feature. They extend for short distances compared to regional ones and spread all over the region, bounding small local cliffs. These local faults and joints played a major role in the development of some sections of the wadis in both El-Galalah El-Bahariyah/El-Qibliyah and the development of karst forms on the surfaces of the plateaus. Folding occurs on a small scale in a few localities. As such, a limited role of these local folds is expected in the development of landforms.

9.3 Cuesta Topography

As mentioned in Sect. 9.1, this region is composed of three geomorphic units. They are the two carbonate platforms of North and South Galalahs separated by Wadi Araba. Taking into consideration the geologic aspects of this region, Wadi Araba can be described as an inverted anticlinal wadi. The two Galalahs can be considered as two opposite facing huge cuestas, since their beds dip at low angles in opposite directions. According to some relatively recent geological studies (Ramdan and Khalil (1995); Höntzsch et al. (2010)), only the Southern Galalah escarpment which looks over Wadi Araba can be considered as a huge compound cuesta, whereas that of the northern Galalah is a fault line escarpment.

On the surfaces of these three geomorphic units, several groups of forms have developed. The first group comprises fluvial forms, and the second consists of karst forms. They have not been formed under present arid conditions. They are inherited from past periods of humid climates, especially the one which extended from Late Eocene to Upper Miocene/Early Pliocene. In addition to these major groups of forms, there are other forms of different origin but with limited distribution, mostly on a small scale. Therefore, forms of this region are grouped into four groups: the cuesta topography, the fluvial forms, the karst forms, and other minor forms. Due to shortage of geomorphic studies on this region, analysis of the characteristics of landforms will depend on the interpretation of Google Earth images and topographic maps. Relevant previous studies will be taken into consideration where necessary.

Cuesta as a geomorphic feature is composed of two slopes: the dip slope and the back slope/cuesta escarpment. In fact, there are only two cuesta escarpments representing the two sides of the Galalahs and currently the sides of Wadi Araba. Similar cuestas were recorded in the northern escarpment of Dakhla Depression in the Western Desert (Brookes 1993) and in the escarpments of El-Egma and El-Tih plateaus in central Sinai (Awad 1952). The dip slope represents the surfaces of the plateaus, either in this region or in any other region with low-dipping beds. There are some other escarpments bounding the Northern Galalah from the north and east and the Southern Galalah from the east and south, but they are not cuestas, because the dip of beds is not broadly concordant with surface topography.

Lithological variations resulted in several small cuestas at the lower section of El-Galalah El-Qibliyah escarpment (Fig. 9.2) and along the upper section of the escarpment of

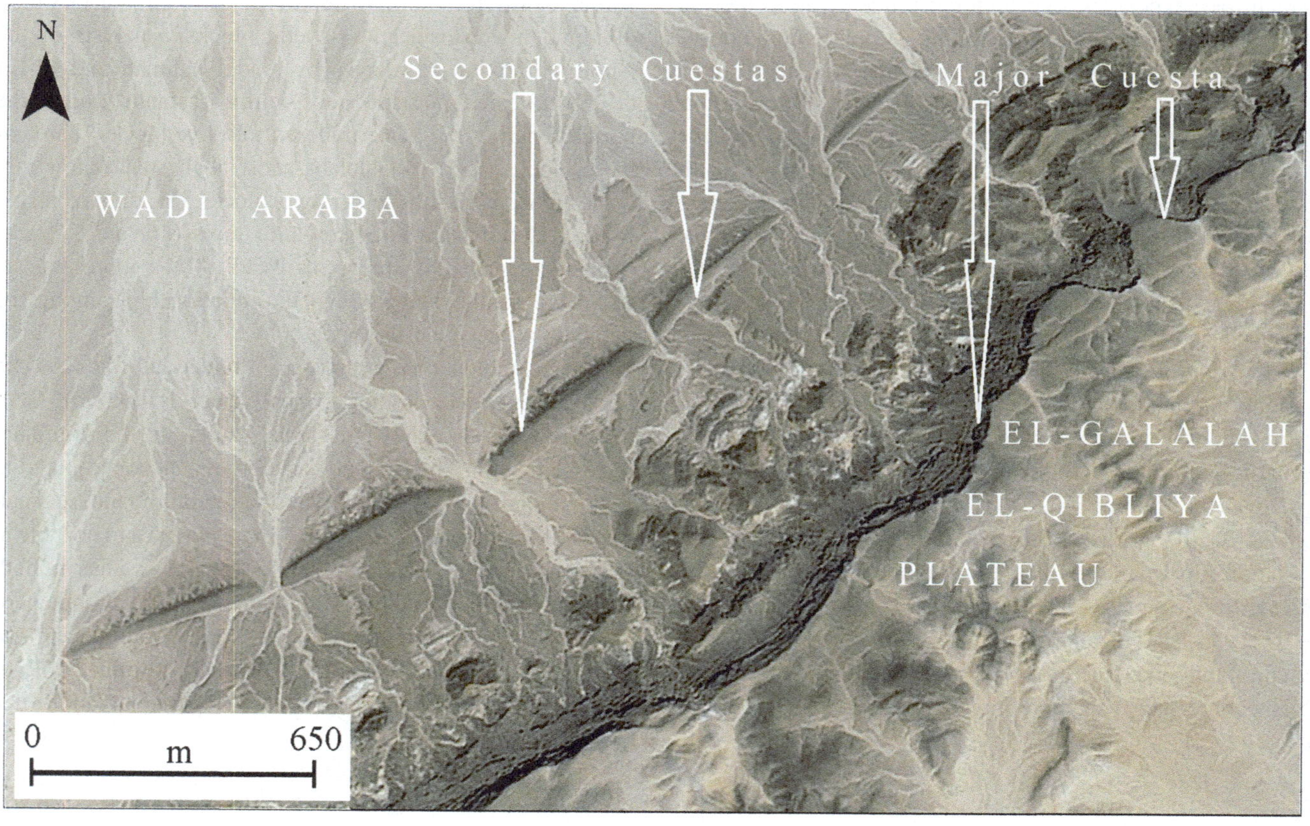

Fig. 9.2 Major and secondary cuestas of El-Galalah El-Qibliyah (*Source* Google Earth 2014)

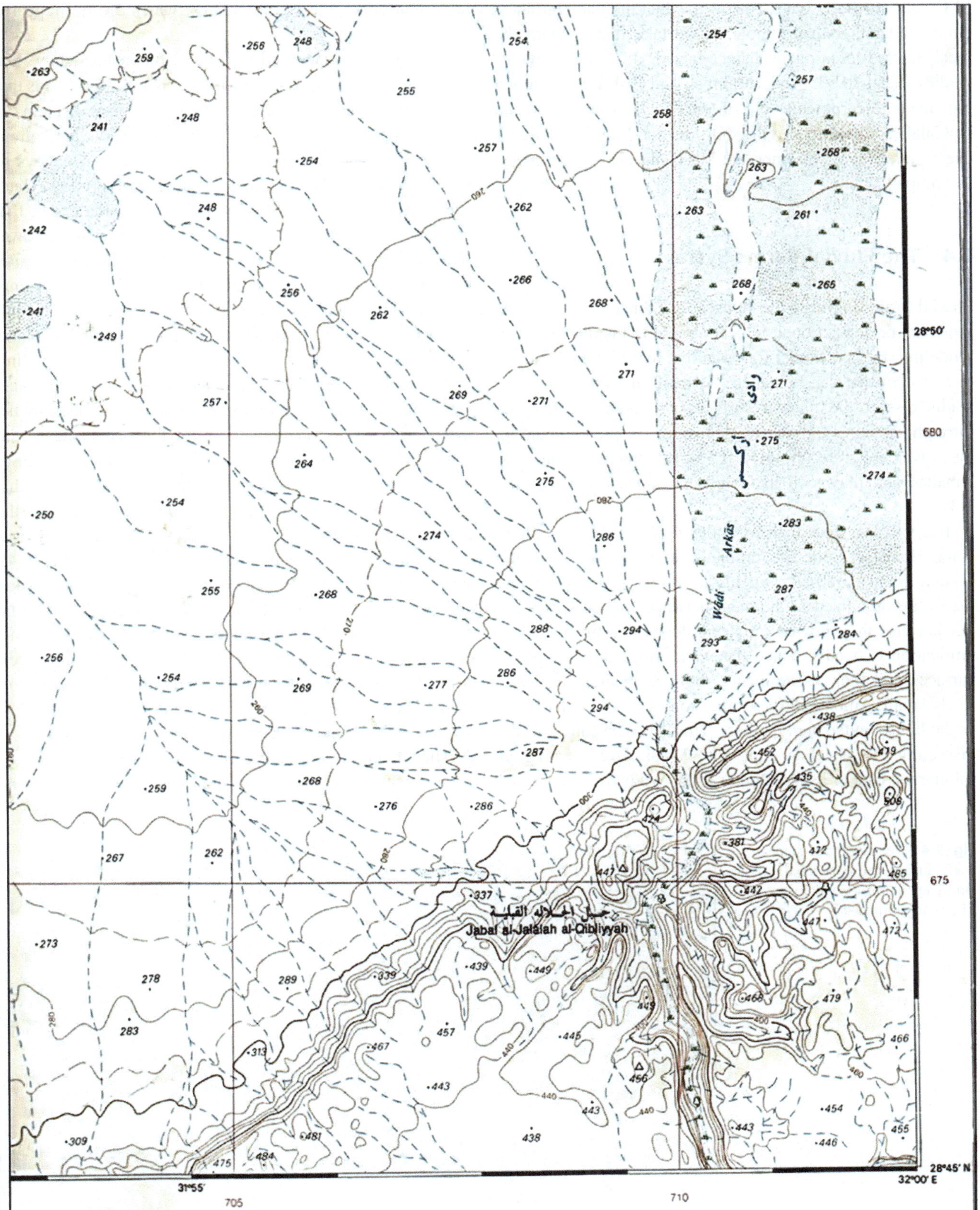

Fig. 9.3 Extract of topographic map (scale 1:50,000 showing the Wadi-Fan System of Wadi Arkās in El-Galalah El-Qibliyah plateau and Wadi Araba

El-Galalah El-Bahariyah. This means that the main cuesta is compound rather than simple. If we take into account the local faulted escarpments, recognized in some previous studies (Saleh 1991; Abdel Rahman and El-Etr 1979) along the main escarpments of El-Galalah El-Bahariyah and El-Galalah El-Qibliyah, it could be said that the major cuestas are not only compound but could also be described as complex.

9.4 The Fluvial Forms System

Fluvial forms dominate the landscape of this region, either as erosional or depositional, indicating a long history of fluvial modeling. In fact, fluvial forms in this region are developed in a system that can be called "**Wadi-Fan System**", where wadis dissect the plateau surfaces and their escarpments through erosion, and where deposition is predominant in the lower areas, resulting in the development of fans and bahadas and playas in the low-lying areas (Figs. 9.2 and 9.3).

Except for a few small localities in Wadi Araba—where some drainage lines are drained internally—the regional system drains externally, mainly to the Gulf of Suez. Some wadis are organized in drainage networks, while others are not because they flowed on steep slopes that do not allow unification of wadis. Since the wadis originate on plateau surfaces and their escarpments, their basin areas are relatively small. Among all the wadis, comes Wadi Araba Basin as the largest outsizing all others. This is because most wadis dissecting the plateaus drain into Wadi Araba, acting as tributaries. Due to its structural origin, Wadi Araba acquired unique morphologic characteristics, which are not repeated in other wadis.

Topographic maps and Google Earth images show that wadis are characterized by two main features. The first is that wadis for the most part or at least large sections of them are meandering, and the second is that many wadis are entrenched or incised. Meandering occurs normally in valleys of humid climates; when gradient becomes low/very low, water speed becomes slow/very slow and weak to the extent that erosion reaches its minimum power and its erosive power becomes ineffective. On the contrary, lateral erosion becomes more pronounced due to channel anastomosing, resulting in gradual meandering. Meandering of streams has traditionally been ascribed to alluvial/flood/delta plains (Thornbury 1958). It was also said that meandering is limited to large rivers which are mature/old in the Davisian cycle of erosion. None of these conditions applies to the meandering wadis on the surface of the two Galalah Plateaus. Meandering within the two Galalahs could have developed under one of two conditions: The first is that development of meanders occurred on the surfaces of the plateaus, characterized by low gradient that allowed lateral erosion to be more effective than vertical erosion (Fig. 9.4). With time, meanders developed due to variations in erosion and deposition on both sides of the channels. The second condition under which meanders developed is that channel erosion was governed by joints and small faults characterizing the surfaces of the two Plateaus. Fluvial erosion followed these structural weak lines, and by time meanders formed with the aid of lateral erosion.

The second aspect characterizing the wadis is that they are incised and entrenched looking like gorges (Figs. 9.5 and

Fig. 9.4 Meandering of some wadis in between cone/karst landscape atop El-Galalah El-Bahariyah plateau

Fig. 9.5 Incised deep wadis at the edge of El-Galalah El-Qibliyah Plateau in the vicinity of Wadi El-Adaid (Photo by Wafaa Khalaf)

Fig. 9.6 Gorge in Wadi El-Nikat with vertical walls on the surface of El-Galalah El-Qibliyah (Photo by Wafaa Khalaf)

9.6), given the similarity between their width and depth (almost 400 m). Incision can be seen also in some meanders on both Galalahs (Fig. 9 7). Entrenching means that erosion rates increased due to change in base level. Was such a change in base level due to the elevation of the two Plateaus? Or was it due to a decrease in sea level? The more probable answer is that the two Plateaus were tectonically raised rather than postulating that the sea level decreased. This is because no feature was found in Wadi Araba, such as raised terraces or incised channel potentially indicative of a

Fig. 9.7 Entrenched meanders atop El-Galalah El-Qibliyah plateau (*Source* Google Earth 2014)

Fig. 9.8 Cut in a wadi fill in El-Galalah El-Qibliyah plateau (Photo by Wafaa Khalaf)

decrease in sea level. On the contrary, sections of wadis on the edges of the Plateaus were incised deeper compared to upper sections of the same wadis. This indicates that incision took place first at the edges of the Plateaus, where water velocity increased first, then with headward erosion, incision extended inside the Plateaus. Another feature which can be taken as evidence for the increased erosive power of flowing water due to tectonic rise is the dissection of the oldest sediments which were deposited in some wadi beds (Fig. 9.8), such as those of Wadi Arkas in El-Galalah El-Qibliyah. In favor of this postulation, some previous studies showed that two different episodes (Late Cretaceous and Early Miocene) of tectonic activity occurred in the northern part of the Suez Rift and adjacent areas (Mostafa and Khalil 1995), potentially indicative of a probable effect of tectonic rise in entrenching wadi channels.

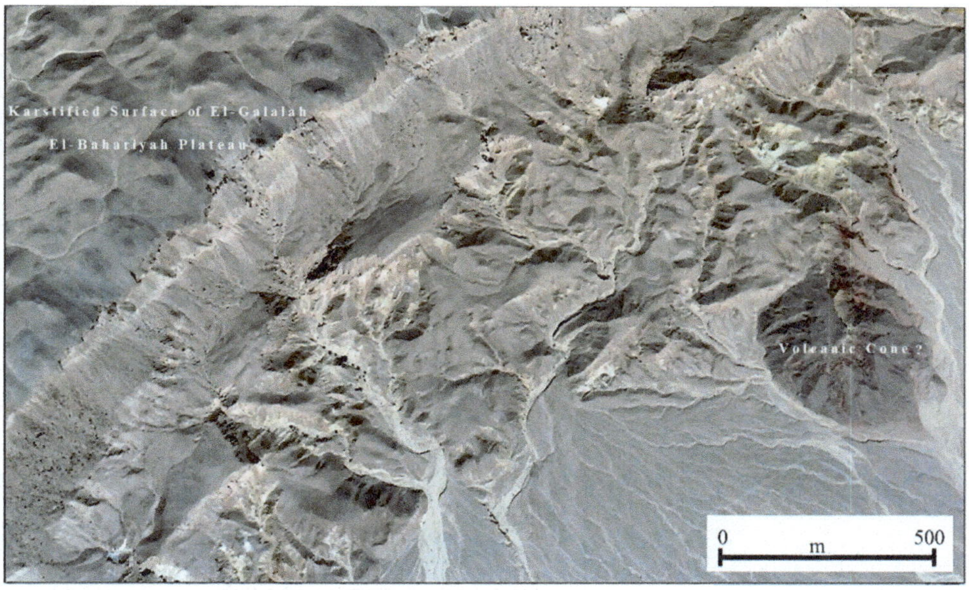

Fig. 9.9 Composite cuesta of El-Galalah el-Bahariyah plateau, and a volcanic cone (?) at its foot slopes (*Source* Google Earth 2014)

9.5 Wadi Araba

Wadi Araba extends ~80 km from west to east, with a width varying between 20 km in the western upper parts and 30 km in the central section (Fig. 9.1).The lowest parts of its floor vary between 120 and 130 m a.s.l, while its sides rise up to 700–800 m asl, indicating a depth reaching approximately 600–700 m. As such, this Wadi did not originate as a normal fluvial valley and acquired its characteristics due to tectonic processes. Another significant feature which looks like a volcanic cone was discovered on the northern side of the Wadi (Fig. 9.9). If this feature is a real volcanic cone, it can be taken as evidence to determine the age of Wadi Araba. This aspect will be considered later. At its floor and at the foot slopes of its sides, several fluvial forms gather, representing the second group of the **Wadi-Fan System**.

Fans and bahadas dominate the landscape of Wadi Araba. Fans vary in size from relatively large to tiny. The largest fan is that belonging to **Wadi Arkas** of El-Galalah El-Qibliyah. Its maximum length and breadth are ~10 km. In fact, this fan is composed of two fans, since the fan of Wadi Arkas overlaps and unites with the eastern one forming a compound fan. These compound fans are dissected by shallow/deep channels (Fig. 9.2), which might continue after leaving the fan for short distances carrying a load of fine sediments, ending in small basins forming **playas** (Fig. 9.3). These maps also show that the playas occupy the lowest points in Wadi Araba (241–242 m asl). Length and width of playa basins vary between 300 and 1500 m. Their small areas and the short drainage lines ending in the basins indicate that these playas are the most recent fluvial forms in this region. They developed at a time of limited seasonal flow during which rainfall was low. This time was most probably during the Holocene wet period.

Landsat and Google Earth space images show that the surface of the fans has different tones, indicating that their development took place during several cycles. To verify this inference, the Landsat images of the largest two fans of Wadi Askhar and another wadi in Wadi Araba were analyzed and classified to bands 3, 2, 1 according to true color. This analysis revealed four cycles of development (Fig. 9.10). This figure illustrates that fans from the oldest phase occupy the smallest area. In addition, the area of fans from the oldest phase is highly dissected. These two characteristics are due to the fact that the oldest sediments were exposed to burial by more recent deposits, but the higher parts were not buried; therefore, they were exposed to erosion for a longer period compared to younger phases. The sediments of the recent phase occupy the largest area within the fans due to a fact that they were exposed to a shorter period of erosion and due to the fact that they cover topographic surfaces inherited from the other cycles.

No information is available concerning the age of the phases of fan development. This aspect needs further field studies and age dating for samples representing each stage of development. However, due to the spread of the sediments of the last phase in Wadi Araba floor and the floors of the tributaries, it can be argued that this phase occurred during the Quaternary wet periods. Most probably, other phases took place during earlier periods; the oldest was during pre-middle Miocene, whereas the second and third phases occurred mainly during post-Middle Miocene. In this connection, the age of Wadi Araba comes up, when did this Wadi start to develop? No answer was given in previous

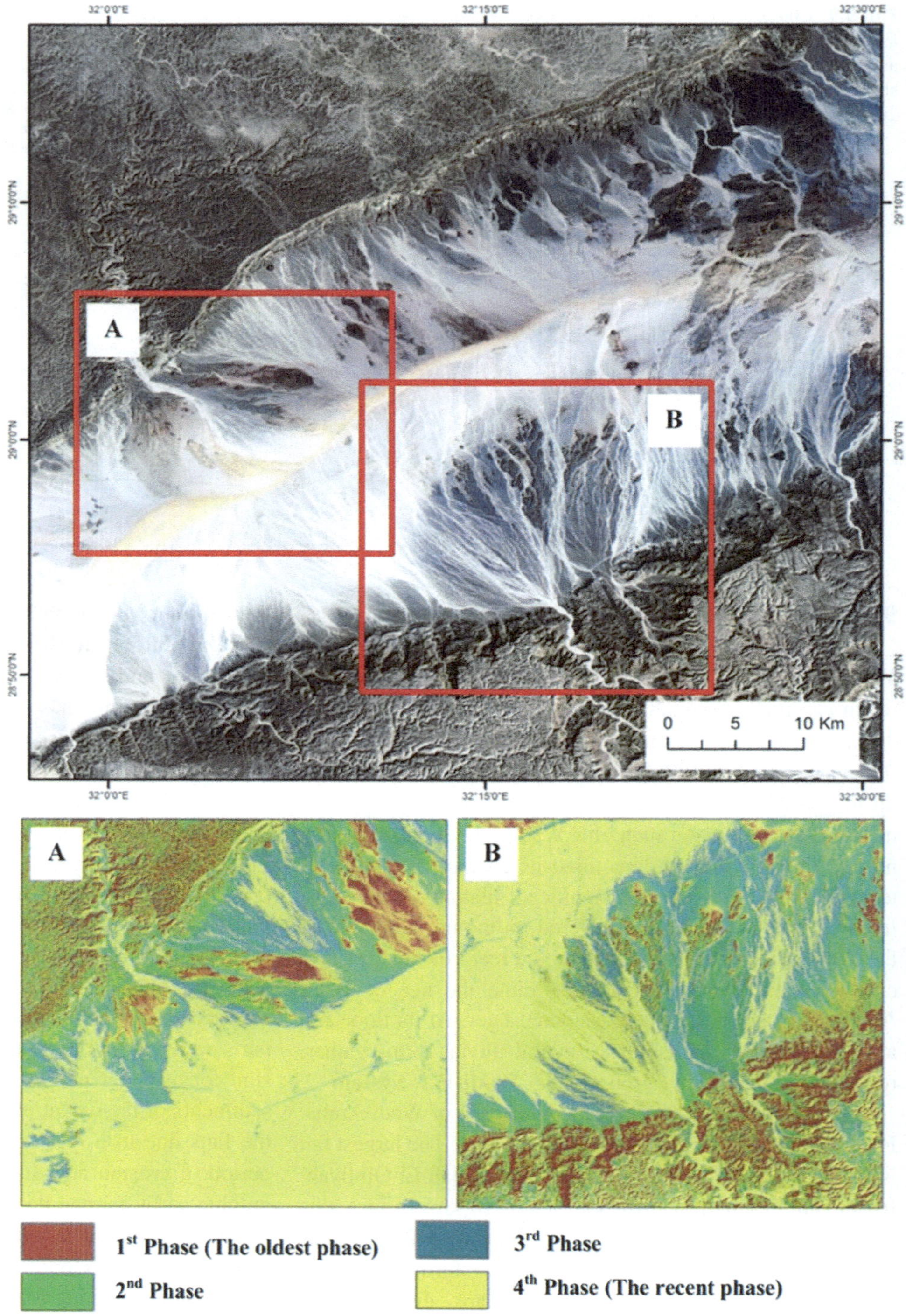

Fig. 9.10 Classification of phases of development of two large fans in Wadi Araba (*Source* Landsat-TM images, (Processed by Moawad B Moawad)

studies. The presence of the volcanic material in Wadi Araba indicates that this Wadi anti-dates the volcanic activity that led to the formation of the cone. According to Meneisy (1990), the youngest Tertiary volcanicity episode, so far dated in Egypt, is that of the Middle Miocene (15 Ma). Therefore, if the volcanic activity in Wadi Araba occurred during the episode of the Middle Miocene, it can be said that the age of Wadi Araba is pre-Middle Miocene.

9.6 Karst Forms

Regionally, the Galalahs could be considered as two karstified tablelands, with varying degrees of karstification, since they are underlain by carbonates and enjoyed humid climates in the past. In spite of this fact, karst studies on this region are still scarce. Only caves as karst forms have been known for almost two decades, and some of them have been

Fig. 9.11 Small cave discovered by Miss/Wafaa Khalaf in El-Galalah El-Qibliyah Plateau

Fig. 9.12 Small solution hollows on a limestone block (Photo by Wafaa Khalaf)

Fig. 9.13 Cockpits in between cone karst landscape of the surface of El-Galalah El-Qibliyah plateau (Photo by Wafaa Khalaf)

studied. The caves studied are: **Nimir Cave** (Goodman et al. 1992), **Wadi Adayd Cave** (Hobbs and Goodman 1995), and St. Antony Cave (Haliday 2004). More caves are expected to exist in this region, as well as other karst forms. Recent exploration in this region discovered some new small caves in El-Galalah El-Qibliyah plateau (Fig. 9.11). In addition to caves, solution hollows of varying size are present everywhere in the Galalahs, of which the tiny ones are the most widespread (Fig. 9.12). Examination of Google Earth images revealed the presence of other karst forms on the surfaces of both Galalahs, of which cone karst is the most prominent and predominant. Vast tracts of the surface of the plateau are dissected into small low-lying rounded/semi-rounded hills (Fig. 9.4). Figure (9.13) shows the presence of cockpits lying in between the cones atop El-Galalah El-Qiblìyah. According to El-Aref et al. (1987), these intra-pits widen by time at the expense of cones due to erosion until the cones vanish or become low/very small mounds, separated by shallow Depressions so that the surface becomes undulating, and the landscape can be described as a "Degraded Karst Landscape." If this landscape represents the final stage in the evolution of karst cycle, the present landscape represents the mature stage according to El-Aref et al. (1987). It is possible that other karst forms and sediments developed during the cycle of karstification, but they become degraded and disappeared over time.

This cone karst landscape of the Galalahs is very similar to the one developed on the Eocene plateau in the area lying to the east of Bahariya Depression and studied by El-Aref et al. (1987). Taking into consideration that the cone karst in the two regions developed on the same type of formation which is the Eocene limestone, it could be suggested that both have been formed contemporaneously. El-Aref et al. (1987) concluded that the karstification processes in the east Bahariya Depression seem to have developed during the

Fig. 9.14 Climbing dunes on the slopes of a steep minor escarpment in El-Galalah El-Qibiliyah plateau (*Source* Photo taken by Wafaa Khalaf)

Fig. 9.15 Polished limestone blocks by wind action on a wadi bed in El-Galalah El-Qibliyah plateau (*Source* Photo taken by Wafaa Khalaf)s

Oligocene times. Therefore, this age can be applied to the Galalah karst landscape. This conclusion is consistent with previous views on the past climates in Egypt.

9.7 Other Minor Forms

Beside these major landforms of fluvial and karst origin, there are some others of different origin, which can be described as minor forms/features due to their small size and limited distribution. In addition, nearly all of these forms developed on the surfaces and slopes of major ones. The most widespread feature is the talus rock fragments which cover nearly all slopes except for the steep segments (Fig. 9.5). This talus cover developed by weathering and mass movements. Are there several generations of such cover? Most probable, there are some generations due to the long geomorphic history of this region but needs field investigations.

The second minor form is of aeolian origin. Sand accumulations and sand polished surfaces are recorded in a few localities in El-Galalah El-Qibliyah plateau (Figs. 9.14 and 9.15). Sand accumulated on north-facing slopes as climbing dunes, indicating that the sand is coming from the north. Most probable, sand is originated from the sediments of the floor of Wadi Araba where large quantities of fluvial sediments were laid down. Sand carried out by wind acted as an erosive power on some projected limestone blocks which lie in their path of flow (Fig. 9.15).

9.8 Conclusion

The Galalah Region represents an independent morpho-tectonic unit in the Eastern Desert. Folds and faults controlled the development of regional and local forms. The two Galalahs and Wadi Araba form together a regional swell, with an axis running in a W–E direction along Wadi Araba; the Northern and the Southern Galalah Plateaus represent the two flanks of the broad swell. Since the beds in these two plateaus dip at low angles in opposing directions, their escarpments take the form of cuestas, similar to those of El-Egma and El-Tih plateaus in Sinai and those of Dakhla Depression in the Western Desert. Faults are present all over the region. They delimit the escarpments of these two plateaus overlooking the Gulf of Suez and large sectors of the escarpments overlooking Wadi Araba. Surfaces of the plateaus are dissected by several small faults, which trend almost in all directions. Of all rock exposures in this region, two are of special geomorphologic and geologic importance. The Carboniferous rocks exposed in Wadi Araba floor represent one of three exposures nationwide. Although they are among the oldest formations, their exposure is indicating that the Wadi occupies the heart of the regional anticline which dominates the structure of this region. The second type of rocks of geomorphic significance is the exposed Eocene limestone atop the two plateaus. During past pluvial periods, these carbonate rocks were subjected to karstification and produced several karst forms, of which four caves were reported in previous studies, and cone karst, which was discovered in the present study through the examination of Google Earth images. Since knowledge on karstification of this region is scarce, its karst forms need further studies.

This region is dominated by Karst and fluvial forms. This is a result of a fact that Egypt enjoyed past humid climates over a long time span in the Tertiary, during which fluvial action and karstification of carbonates modeled the landscape of this country. These two types of forms developed side by side on the surfaces of the two Plateaus. Since most of karst forms are tiny/very small, only cone karst forms have been recognized on the high resolution images of Google Earth. This is why further field studies are recommended. Fluvial forms are fitted together in a system called "Wadi-Fan System." In this system, wadis as erosional forms developed on the surfaces of the plateaus, while fans as depositional forms were formed at the floor of Wadi Araba and other foot slopes of the escarpments.

The wadis on the plateau surfaces have two geomorphic characteristics: (a) Nearly all wadis are meandering, especially along the lower and middle reaches, and (b) Wadis are incised, and incision becomes more pronounced near the edges of the escarpments. Meandering was acquired in the early stages of wadi development due to the low gradient of the plateau surfaces, whereas incision took place due to tectonic uplifting, rather than due to sea level lowering. Similar characteristics were observed at the upper reaches of Wadi El Arish on the surface of El-Egma Plateau in the Sinai Peninsula and the upper reaches of Wadi El-Asyuti on the Eocene limestone to the west of Wadi Qena in the Eastern Desert.

Fans are the most widespread depositional form at the foot slopes of the escarpments. They are best exemplified in Wadi Araba, where wadis dissecting the plateaus and the cuesta escarpments laid down huge quantities of sediments in the form of fans of different sizes matching the size of the wadi. Many are ideal cones of fluvial deposition, while others are modified due to the deposition of sediments in various stages, overlapping the neighboring fans. All fans are dissected by shallow/deep braided channels which unite with the main channel of Wadi Araba. In some instances, some of these channels drain into small closed basins, forming playas. Analysis of Landsat-TM images of two large fans in Wadi Araba revealed a four-phase deposition. The sediments of the fourth phase spread also on the Wadi Araba floor and the beds of the two wadis, indicating that it is the most recent one. The surfaces of the oldest phase occupy very limited areas, because most of them are hidden under more recent sediments of phases three and four, and of highly dissected. No history dating has been carried out, but geologic maps of Wadi Araba (Conoco Coral 1982–1983) show that wadi deposits, which cover the floor of wadis, belong to the Quaternary. Since it was estimated that the age of Wadi Araba is pre-Middle Miocene, it can be argued that the oldest phase of fan deposition belongs to this early stage of development of Wadi Araba, while phases two and three are post-Middle Miocene.

Although the minor forms do not change the landscape of the Galalahs, their role is of local significance.

References

Abdallah A, El-Adindani A (1963) Note on Cenomanian-Turanian contact in the Galalah plateaus, Eastern Desert, Egypt. Egypt J Geol 7:171–172

Abdallah A, Meneisy M, Shaaban M (1973) A geological studies on the volcanic rocks, Wadi Araba and Abu Darag Area, Gulf of Suez Region. Egypt J Geol 17:111–123

Abdel-Rahman M, El-Etr H (1979) Structural pattern of the Northern part of the Eastern Desert of Egypt, Apollo-Soyuz Test Project, Summary Science Report. In: El-Baz F, Warner D (eds) Vol. II: Earth observations and photography. NASA, Washington DC, pp 87–96

Awad H (1952) A new type of desert cuesta in central Sinai. In: Proceedings of the XVII International Geographical Congress, Washington

Brookes I (1993) Geomorphology and quaternary geology of the Dakhla Oasis Region, Egypt. Quat Sci Rev 12:529–552

El Aref M, Abou Khadrah A, Lotfi Z (1987) Karst topography and karstification processes in the Eocene limestone plateau of El Bahariya Oasis, Western Desert, Egypt. Z Geomorph N F 31:45–64

Goodman S, Hobbs J, Brewer D (1992) Nimir Cave: morphology and fauna of a cave in the Egyptian Eastern Desert. Palaeoecol Afr 23:73–90

Google Earth (2014) Egypt Images Latitudes 22-32 N and Longitudes 22-32 E, http://www.google.com

Haliday W (2004) Caves and karsts of Northeast Africa. Inter J Speleol 32(1/4):19–32

Hobbs J, Goodman S (1995) Leopard-hunting scenes in dated rock paintings, from the Northern Eastern Desert of Egypt. Sahara 7:7–16

Höntzsch S, Scheilbner Ch, Kus J, Marzouk A M, Rasser MK (2010) Tectonically driven carbonate ramp evolution at the southern Tethyan shelf: the Lower Eocene succession of the Galala Mountains, Egypt. Facies: 22p

Kostandi A (1959) Facies maps for the study of Paleozoic and Mesozoic sedimentary basins of Egyptian region. In: Proceedings of 1st Arab Petroleum Congress, vol. 2, Cairo

Meneisy MY (1990) Volcanicity. In: Said R (ed) Geology of Egypt. Balkema, Rotterdam, pp 157–172

Mostafa AR, Khalil MH (1995) Superposed deformation in the northern Suez rift, Egypt: Relevance to hydrocarbons exploration. J Petrol Geol 18(3):245–266

Nakkady SE (1955) The stratigraphy and geology of the district between the northern and southern Galala plateau. Bull Inst Égypte 36:254–268

Omran A (1977) Geophysical studies on Wadi Araba and Ras Gharib areas, Gulf of Suez, Egypt. Dissertation, Assiut University

Sadek H (1926) Geography and geology of the district between Gebel Ataqa and El-Galalah El-Bahariya, Gulf of Suez. Geol Survey of Egypt. Paper No.40: 120p

Said R (1962) The geology of Egypt. Elsevier Publishing Co Amsterdam

Saleh K (1991) The geomorphology of the southern escarpment of El-Galalah El-Bahariyah (in Arabic). Dissertation, Ain Shams University

Thornbury WD (1958) Principles of Geomorphology. Wiley, New York

Part IV

The Mega Depressions

The Qattara Depression

10

Abstract

Lying in the northern part of the Western Desert, the Qattara Depression is the largest ($\sim$45,000 km^2) and deepest (−145 m) Depression in Egypt. It is also the nearest Depression to the Mediterranean Sea, with only 56 km separating them. The Depression is divided into two geomorphologic units: the escarpments and the floor. The escarpments bound the Depression Floor from the north and west, but the floor rises gradually in the east and south up to 200 m asl. Rising to $\sim$200 m a.s.l, the escarpments overlook the floor by steep slopes, and are dissected by small, short wadis. The escarpments descend to the Depression Floor in a step-like form due to the effect of lithologic variations. Sabkhas, sand dunes of South Qattara Sand Sea and East Qattara Dune Field, salt crusts, and wadi fans are the dominant landforms of the floor. With the exception of two little oases, the Depression Floor is uninhabited due to the high salinity of discharged underground water. Geological and geomorphologic characteristics suggest that Qattara can be classified as a poly-genetic Depression, which started to develop after the retreat of the Middle Miocene Sea.

Keywords

Qattara Depression • Sabkhas • Salt crusts • Sand dunes

10.1 Introduction

Depressions are one of the most characteristic forms in the Western Desert. They distinguish this desert from other geomorphologic regions of Egypt. There is no other region with such a vast area in which Depressions spread in every part, not only in terms of numbers, but also in terms of surface area. In this desert, there are countless Depressions of various sizes, of which only nine are huge with an area exceeding 1000 km^2, and existing 200–400 m below the surrounding plateaus. Five of these huge Depressions have their floors below sea level. In addition to these physical characteristics, underground water is tapped at the floors of these huge Depressions, providing a livelihood for a relatively small number of the population mainly living by agriculture in separate oases. Utilizing this underground water, several development schemes have been adopted since the beginning of the twentieth century.

Examination of topographic maps of various scales shows that there are a countless number of Depressions in the Western Desert representing a continuous series of various sizes. These maps also show that this series of Depressions forms a hierarchical system, which has a wide base of very small (minute) units and a narrow peak of huge basins. In the meantime, it was observed that most of the relatively large

N.S. Embabi, *Landscapes and Landforms of Egypt*, World Geomorphological Landscapes,
https://doi.org/10.1007/978-3-319-65661-8_10

Table 10.1 Basic morphometric characteristics of the various orders of the Depressions of the Western Desert of Egypt

Order	Description	Length (km)	Breadth (km)	Depth (m)	Area (km^2)
1st	Minute	0.001–0.01	0.001–0.005	Few cm	10–50 m^2
2nd	Micro	0.1–1.0	0.05–1.0	0.1–1.0 cm	0.05–1.0
3rd	Small	2.0–10	1.0–5.0	1.0–15.0	2.0–50
4th	Large	10–50	5–25	5.0–100	100–1000
5th	Mega	50–350	25–250	200–500	>1000

Source Prepared from topographic maps, aerial photographs, and field observation

Depressions are composed of two or more small ones. This characteristic suggests that—in the hierarchical system—the larger Depressions developed by the coalescence of smaller ones rather than by the retreat of their sides. Therefore, the Depressions of the Western Desert may be classified, according to size, into orders in which larger Depressions are formed by the unification of smaller ones. This ordering system of Depressions is similar, to a great extent, to the system of stream ordering in drainage nets, where a first order of streams forms a second order and so on, when unified. Table 10.1 illustrates a suggested ordering of the Depressions of the Western Desert, resulting from the study of available topographic maps, aerial photographs, and field observations.

10.2 General Characteristics

First discovered in 1926 (Ball 1927), then surveyed in 1927 (Ball 1933), the location, size and general characteristics of the Qattara Depression have become well known. A few separate studies were conducted on this depression, but the most comprehensive are those dealing with the Qattara hydro-solar power project.

Table 10.2 shows that the **Qattara Depression**, which is the closest to the Mediterranean Sea (56 km only), is the largest and deepest among the mega-Depressions of the Western Desert. However, the area (19,500 km^2), which was stated by Ball (1933) and which was cited in all the following literature, is smaller than the actual one, because as Ball reported this examined area represents only the area that lies below sea level. Thus, when all other areas sloping toward the floor from the surrounding high grounds, in the east (up to 200 m contour line) and in the south (to South Qattara Sand Sea), are added to the Depression, its area becomes nearly 45,000 km^2 as measured on topographic maps (scale 1: 250,000). This means that less than half of the area of the Depression lies below sea level, a characteristic that distinguishes the Qattara from other mega-Depressions. In the southwest lies the lowest point in Qattara, which is calculated by Ball (1933) to be −134 m as; however, a recent survey (Institut Géographique National and Société Française Stréotopographié 1977) calculated the point to be −145 m. Except for the two very small oases of Moghra at the north-eastern tip and Garra at the western margin, the Qattara Depression could be considered uninhabited, which is another feature that distinguishes Qattara from other mega-Depressions. This is mostly due to the fact that its

Table 10.2 Morphometric characteristics of the mega-Depressions of the Western Desert

Depression	Max length (km)	Max breadth (km)	Area (km^2)	Lowest point (m)	Highest Point (m)	Depth (m)	Orientation
Qattara	340	245	45,000	−145.0	200	345	E-W
Siwa	95	50	2950	−18.5	200	217	E-W
W. El-Natrun	65	24	1100	−24	200	204	NW-SE
El-Fayum	100	80	4370	−45	350	395	E-W
W.El-Rayan	96	78	5040	−64	350	414	NW-SE
El-Bahariya	95	42	2800	118	300	182	NE-SW
El-Farafra	150	150	15,600	25	330	305	NE-SW
El-Dakhla	175	50	6250	109	450	341	E-W
El-Kharga	200	85	12,250	2	471	469	N-S

Compiled from: Topographic maps (scales 1:250,000 and 1:100,000)

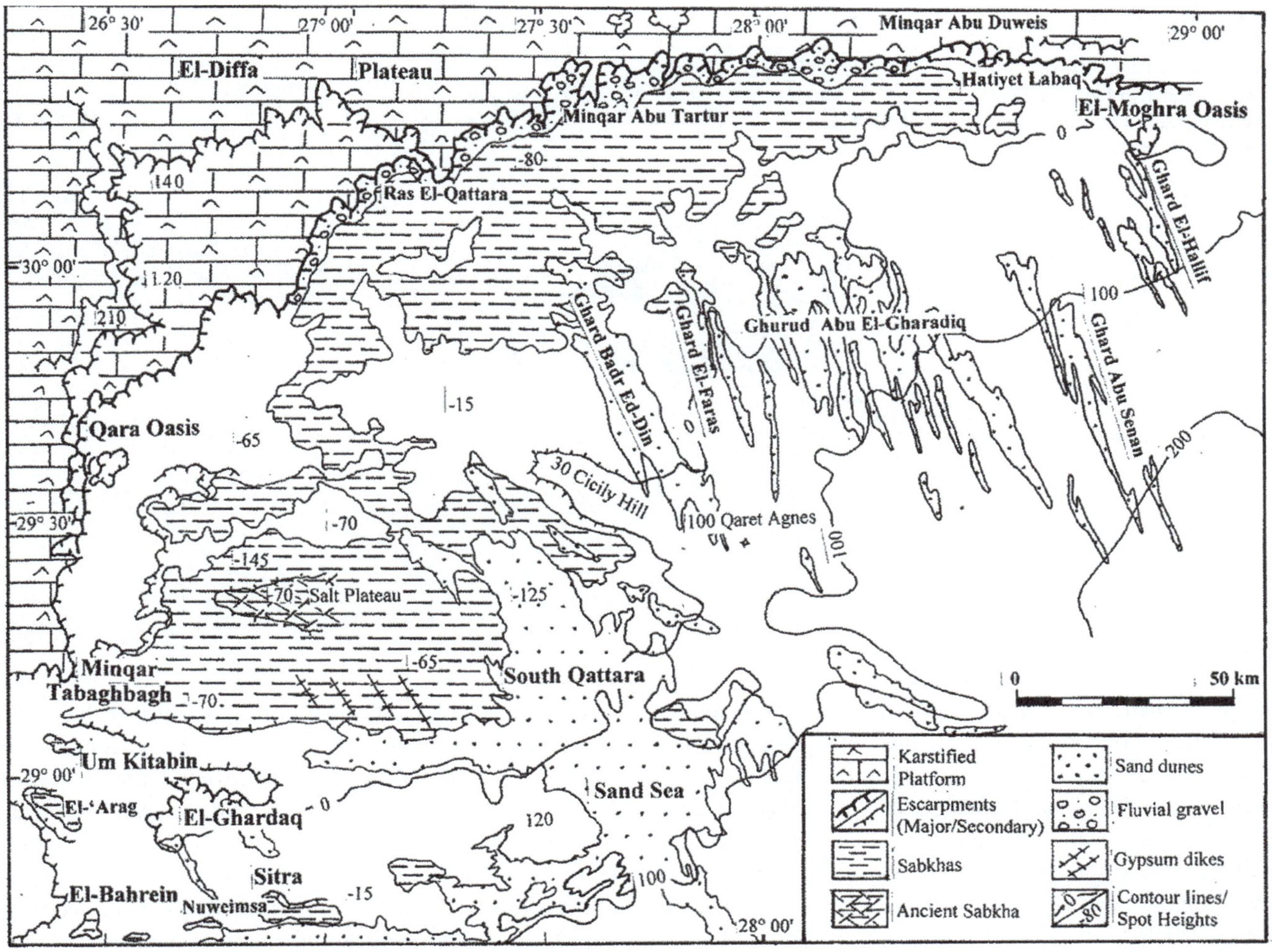

Fig. 10.1 Geomorphological map of Qattara Depression (Compiled from: Topographic Map, Scale 1:500,000; Geological Map, Scale 1:500,000; and Landsat TM images)

floor is covered by sabkhas and that the discharged underground water is highly saline.

Although it is the largest Depression, only few detailed studies were conducted on some of the features of the Qattara Depression. Therefore, its geomorphologic characteristics are not well known. The following analysis is based on the few studies and on the interpretation of available topographic and geologic maps, aerial photographs, and space images. When seen on a geomorphologic map (Fig. 10.1), the Qattara Depression can be divided into two major geomorphologic units: the escarpments and the floor.

10.3 The Qattara Escarpments

Steep escarpments bound the Depression from the north and west without any break (Fig. 10.2). The relative height of the escarpments increases from about 100 m to the north of Moghra to about 275 m at the center of the northern section, then decreases to about 100 m along the western section. Due to lithological variations, the escarpments take a step-like form, being composed of two or more steps. This is clear along the western and SW parts, in the surroundings of Qarah Oasis. The Qattara escarpments change their orientation gradually in curved form from E-W to N-S, but can be divided into three sections according to their orientation. The first section takes an E-W orientation from Moghra Depression extending to approximately 140 km, the second section has a NE-SW trend for another 120 km, and the third has a N-S trend for about 70 km. This means that the total length of the escarpment is about 330 km. In spite of these general trends, the escarpments are characterized by many irregularities, which appear as promontories (locally known as Ras or Minqar) separated by wide wadis (in some instances are called Naqb). Examples of the promontories are Minqar Abu Dwais in the east, Ras El-Qattara about 120 km westward, and Ras El-Munkhafad and Minqar Tibaghbagh in the west (Fig. 10.1). Examples of wadis are

Fig. 10.2 Qattara steep escarpment as it appears along the highway cut from Alamain to Cairo—Bahariya highway (Note the Depression Floor is in the background of the photo)

Naqb El-Hadi, Wadi Shark El-Masrab, Naqb El-Shotayrat along the northern escarpment, and Wadi Hamid, Naqb El-Abyad, and Wadi Ghuweidit along the western escarpment. It was found that these irregularities developed under the control of local joints and faults. In general, wadis dissecting the escarpments are short and occupy small basins (Mostafa 1993). They end at the feet of the escarpments forming fans in most cases, and extend for short distances only into the plateau surface.

10.4 The Depression Floor

The Depression Floor rises gradually from the lowest point (−145 m), which lies in a small basin in the south-western part, to 200 m in the east and northeast, and to about 150 m in the south. No attempt has been made to explain this characteristic, except for a recent study on the effect of salt weathering in developing the Qattara Depression (Aref et al. 2002). This study postulates that the high salinity and the sodium chloride content in the groundwater table and the high rate of evaporation lead to crystallization of halite at or near the surface of outcropping rocks, generating increased pressures that cause the bedrock to disintegrate. Dissolution of halite during wet periods leaves fine-grained material, which can be easily blown away by northerly winds. In contrast, salt weathering is ineffective in the eastern part of the Depression due to the low salinity of the groundwater. El-Bassyoni (1995) found that the salinity of groundwater in sabkhas increases from NE at the **Moghra Oasis** (43,596 ppm) to SW at Ghazalat area (421,105 ppm). Wind deflation is also less effective in the eastern part due to the cohesion of wet sediments, the permanent vegetation cover, and precipitation of halite cement between sandy sediments. In spite of the overall decrease in surface level toward the southwest, local variations occur in various parts of the floor in the form of low-lying hills and mesas with relative heights of 10–20 m, or in the form of small basins with depths of several tens of meters below the surrounding grounds. The most widespread forms are created by seepage of underground water and include sabkhas, salt crusts, gypsum dykes, and lakes. The second group of forms is of aeolian origin, and comprises sand dunes and sand sheets. The third group comprises forms of fluvial origin, resulting from deposition (wadi fans). Finally, there are landforms developed from karstic processes and salt weathering, comprising local basins and other microforms.

10.4.1 Sabkhas and Associated Forms

10.4.1.1 Sabkhas

The most outstanding form in the Depression Floor is the sabkhas. It covers the lowest parts that lie below sea level. First mentioned by Ball (1933), the sabkhas of the Qattara were studied in detail after 50 years by El-Bassyoni (1983). Both Ball (1933) and El-Bassyoni (1995) estimated the area of sabkhas to be 5800 km^2 (i.e., about 8% of the total area of the Depression). Based on aerial photo interpretation and field work carried out along with several profiles, El-Bassyoni (1995) distinguished two broad types of sabkhas in the Qattara: barren sabkhas and vegetated sabkhas. Both categories exhibit remarkable variations in their morphologic and lithic structures, extent, and vegetation cover due to several factors among which relative height, water table, chemical properties of ground water, drainage characteristics, bedrock types, seasonal climatic variations, and prevailing wind are the most important (El-Bassyoni 1995). Subtypes of the barren sabkhas are flat floes, hard rough salt crust, and hard salt crust; while vegetated sabkhas are divided into salt marshes, sabkha with bushes, and sabkha strewn by scattered bushes, and covered with coarse sand.

These types of sabkhas do not form a single continuous tract, do not lie at a constant level, and did not develop at the same time. The largest continuous sheet of sabkha extends from near the northeast tip of the Depression, past the feet of the northern escarpment, and then splits into several patches near the middle part of the Depression. Further southward, the sabkhas of the western part occupy several separate areas of the Depression Floor (Fig. 10.1). The different types and sometimes the same type of sabkha lie at different levels (El-Bassyoni 1995). Topographic maps (scale 1:100,000 and 1:250,000) describe the surface of the sabkhas according to morphology and composition as hard sabkhas, soft-wet sabkhas, clayey surface mixed with gypsum, and plain covered by a hard clayey layer.

A characteristic feature of the Qattara sabkhas is **gypsum dykes**. They appear on topographic maps (scale 1:100,000) (Sheets: Minqar Hamed NH 35 F3 and East Minqar Hamed NH 35 G1) as parallel lines with N-S orientation. These gypsum dykes belong to the type of dykes, which are known as fissure-filling dykes, where secondary materials are precipitated in joints or fractures in the host rocks. Therefore, it is expected that the gypsum dykes of Qattara were formed at a later stage after the drying up of sabkhas.

10.4.1.2 Salt/Evaporite Crusts

These crusts can be divided into two different types: crusts that cover the surfaces of present-day sabkhas, and **old Quaternary crusts**. Salt crusts were found to differ in thickness and surface features. Thin and fragile salt crust predominates at the feet of the northern escarpment, whereas thick and solid crust covers small patches in other parts. Hard salt crusts with polygonal forms and rough structures occupy wide areas in the NE part of the floor (Joint-Venture Qattara 1981). Remnants of old salt crusts are represented by what has been known in previous studies as "**Salt Plateau**". This term was coined by Ball (1933) for a tabular mesa capped by a thick bed of massive translucent rock-salt, very different from the sabkha in the center of the western part, which is only 25–30 m above the surrounding areas existing at approximately −100 m. This "Salt Plateau" extends for about 37.5 km from east to west and has a breadth of 6–9 km from north to south; therefore, it is the largest exposure of thick salt crust (El-Bassyoni 1995). The study of El-Bassyoni (1995) revealed that the salt crust which caps this plateau is not a massive translucent salt bed as maintained by Ball (1933), but is rather composed of thin successive layers of salt, with gypsum pockets and nodules, (5–10 cm). According to the findings of a recent study (Aref and Hamdan 2003) the surface of this "Salt Plateau" is composed of aggregated white to gray, hard, and large (up to 30 cm in size) anhydrite nodules. Because the anhydrite aggregates are more resistant and durable compared to the surrounding Moghra clastics, the "Salt Plateau" stands higher than the surrounding areas. Due to its relative height and hardness, this crust is considered by El-Bassyoni (1995) as the oldest hard rough salt crust in the Depression (Late Pleistocene). Other old Quaternary evaporite sequences that form buttes and mesas are found in other areas at a higher level (−25 m at Talh Fawakhir and Ain Abdel Nabi, and −75 m at Bir Hussain) than that of the "Salt Plateau" (Aref and Hamdan 2003). Since it has no relationship to the present-day ground water aquifer, this relatively high evaporitic crust indicates the presence of a previous lake or sabkha at a higher level than the lower sabkhas existing nowadays. It also affords as Ball (1933) pointed out, evidence of a past dry climate that has prevailed in this part of the Western Desert for a long span of time.

Most of the groundwater that evaporates in the Depression comes from the Moghra Aquifer system, which is recharged from four sources: the Nubian sandstone aquifer, Nile water, rain water, and seawater (Aref et al. 2002). Groundwater seepage in the western Qattara is recharged from the relatively free zones of the Nubian sandstone aquifers, and is normally present directly below the salt crust of the sabkhas (El-Bassyoni 1995). This fact evidences that the sabkhas of the Qattara Depression were formed due to continental seepage of underground water into the Depression. This indicates that the Depression forms a sump for the great subterranean water of the Western Desert. Extensive evaporation of the groundwater exposed to or slightly near the surface in the lower parts of the Depression results in the formation of salt crusts. The groundwater in the sabkhas ranges in salinity from brackish at the peripheries to highly saline toward the inner zones of the Depression (El-Bassyoni 1995). This salinity of groundwater is acquired during its seepage through the younger rocks before reaching the Qattara Depression.

10.4.1.3 Lakes

Locally, seepage of groundwater in relatively small basins in the NE tip and the southern part of the Qattara Depression leads to the development of small permanent lakes. Examples of these permanent water bodies are the twin lakes of **El-Bahrain Sitra Lake, and Nuweimsa Lake** along the southern margins of the Depression and **Moghra Lake** at the NE corner of the Depression (Fig. 10.3). It seems that water

Fig. 10.3 Moghra Lake in the extreme eastern tip of Qattara Depression, and the sabkha that surrounds the lake from the south (Note the appearance of Qattara escarpment in the background of the photo)

Fig. 10.4 Newly developed shallow lake in the southern part of Qattara (Note the grasses which grew in the middle of water, which is indicative of the small depth of the lake)

discharge occurs along some deep fracture lines. Water seepage due to road works was observed in the Nuweimsa area during a field trip in February 1996 (Fig. 10.4). The presence of these permanent lakes indicates that natural underground water discharge exceeds the rate of evaporation. A ring of sabkhas surrounds all these lakes (Fig. 10.3). Although the lakes have appeared on topographic maps since the 1927 survey of Ball (1933), no detailed studies have been made yet on changes of water level. In addition to these natural lakes, there are some small ones resulting from the discharge of artesian wells, which were dug in the seventies of the twentieth century in the western part of the Depression in the region of El-Qarah Oasis (El-Bassoyni 1995).

10.4.2 Sand Dunes

Sand dunes cover wide areas in the southern and eastern parts of the floor of the Qattara Depression (Fig. 10.1). These dunes belong to the **South Qattara Sand Sea** and the westerly dunes (Badr Ed-Din, El-Faras, Abu Ghardik, Abu Senanm, Abu Rakham) of the **East Qattara Dune Field**. Dune coverage in the South Qattara Sand Sea only equals 6800 km^2. However, the location of these dunes in the eastern and southern parts is evidence that their sand is deflated from the disintegrated Moghra clastics exposed at the Depression Floor during its development. This conclusion is confirmed by the development of **El-Ghorood El-Sood (the black dunes)** at the western margins of South Qattara Sand Sea. It was found that the deflated Dabaa black shale (Upper Eocene–Oligocene) represents the second component in these black dunes (El-Bassyoni 1995). The discovery of these black dunes ascertains the role of wind deflation as an agent in the development of the Depression or other Depressions in the Western Desert of Egypt. Other aeolian forms comprise sand sheets and trapped barchanoids in wet sabkhas.

10.4.2.1 South Qattara Sand Sea

It extends for ~200 km in E-W direction from **El-Bahrain Lake** to **Al-Ghard El-Kabir** (Fig. 10.5). Due to its irregular shape, width varies between 10 and 150 km. This irregular shape is due to the roughness of the surface of the ground at the southern peripheries of the Qattara Depression. The irregular shape and the development of various forms and patterns with changing trends make the South Qattara Sand Sea one of the most complex sand seas in Egypt. However, for the purposes of study, the **South Qattara Sand Sea** can be divided into two parts according to dune trend and relative location. The first one is in the east with a NW-SE trend, and the second lies to the west of the first and takes a general E-W trend (Fig. 10.5). Both divisions are connected by a saddle of dunes.

The Eastern Division of South Qattara Sand Sea is characterized by four aspects. The first aspect is the prevalence of linear dunes, with local changes in pattern and trends. The second aspect is the development what has been called in previous studies "**El-Ghorood el Sood**" (El-Bassouny 1995; Aref et al. 2002). These dunes acquired

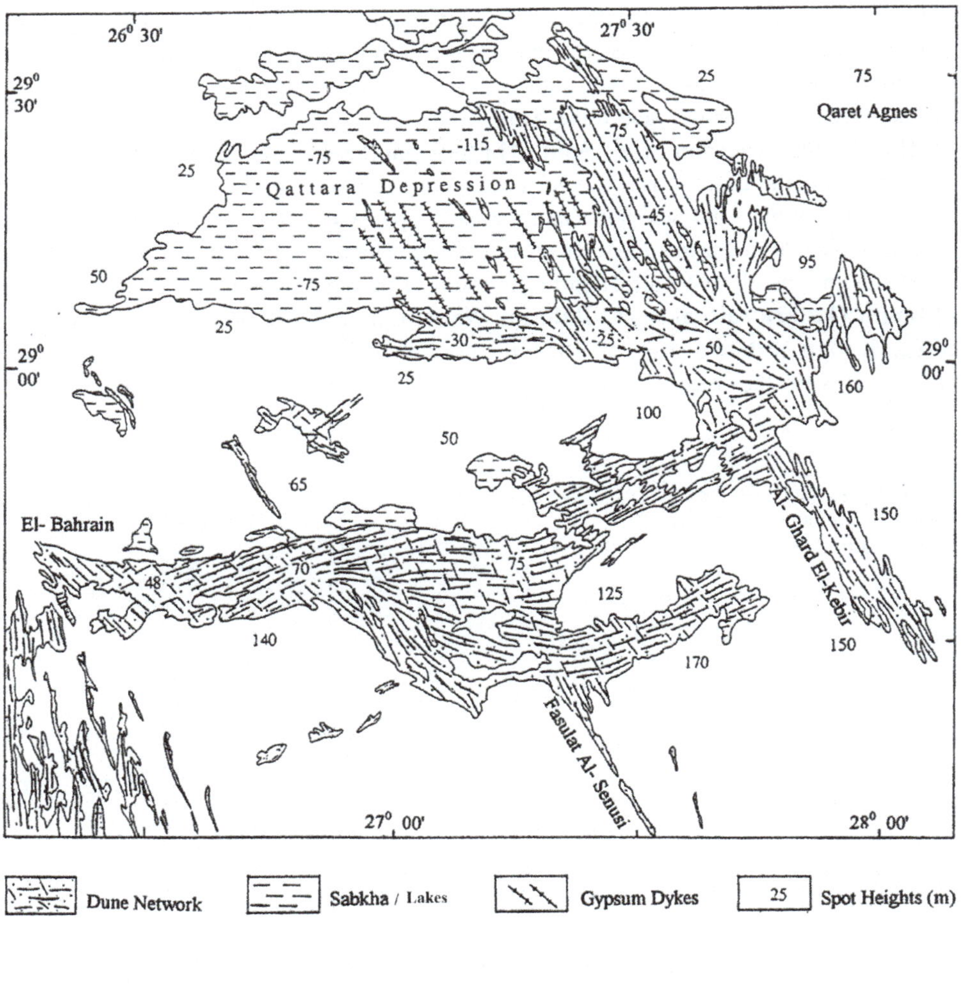

Fig. 10.5 Dune pattern in the South Qattara Sand Sea. *Source* Aerial Photos, Scale 1:60,000, Topographic Maps, Scale 1:250,000 and 1:100,000

their name from the high content of black shale fragments derived from the Dabaa formation. They developed in a large basin at the northern part of this Eastern Division of this Sand Sea. The third aspect is the development of **star dunes** (Fig. 10.6), a rare form of dunes in Egypt since it develops due to a multi-directional wind regime. Unfortunately, no wind data is available in this region, but the presence of this form of dunes is an evidence of such a wind regime. Here, the star form has a larger number of arms compared to the one identified in previous studies in other deserts in North Africa or in Arabian Peninsula. This is the first time star dunes are recorded = in Egypt. The fourth aspect is that the dune pattern changes according to the wind regime and surface topography. Dunes spread all-over the area in a parallel pattern with low density, and two different orientations: NW-SE and E-W trend (Fig. 10.5). Branching, parallel, and network patterns can be recognized readily in various localities in this part of the South Qattara Sand Sea. This is probably due to local wind system, in which direction changes locally in such a small area. At their southern extension, similar dunes developed in a linear shape known as **Al-Ghard El-Kebir**. This Ghard is ~50 km long and 3–10 km wide (Fig. 10.5). It is not a simple linear dune, but rather a complex of dunes of different forms predominantly organized in a linear form. In general, dune density is high in this Ghard, but it decreases at the southern margins with its dismantling into separate linear dunes (Fig. 10.5).

The Western Division of the **South Qattara Sand Sea** is smaller than the Eastern one. It covers nearly one third of the total area of the Sand Sea. It has an irregular shape, since sand is deposited in various local basins of different shapes. It extends in a nearly E-W direction to the south of the line of **South Qattara salt lakes** (**El-Bahrain**, **Neweimsa**, and **Sitra**). It starts with two arms, of which the northern one is coalesced with the Eastern Division of this Sand Sea (Fig. 10.5). Although linear form is predominant, pattern and direction change from one locality to another according to local conditions. It starts in the east with two wide arms separated by a high ground, then after ~25–35 km they unite forming the widest patch in this division (40 × 50 km). From here, the dunes extend westward as a wide (~15 km) belt directly to the south of the lakes of South Qattara until they coalesce with the dunes of the **Great Sand Sea**. It seems that this belt lies in a low-lying area which slopes toward the line of the South Qattara Lakes and sabkhas. Although the southern boundary is straight, several dunes come off with NW trend, of which **Fasulat El-Senosi** is the largest (30 km long).

The linear form is dominant, but with different trends and patterns. NW-SE and NE-SW directions prevail in this Western Division, where the network pattern is predominant (Fig. 10.2), as is the case along the northern peripheries of the Great Sand Sea, indicating bidirectional wind regime in this region. The analysis of climatic data of Siwa Meteorological Station, the nearest station ~200 km to the west, confirms this inference (Embabi 1998).

10.4.2.2 East-Qattara Dune Field

This field is the largest dune field in Egypt (Table 10.1), but sand coverage is one of the lowest. It was studied by Ali (1993). It extends 250 km from E to W, from the NE tip of Qattara Depression to ~30 km west of Wadi El-Natrun. It also extends 150 km from N to S where some dunes in the east approach the Cairo–Bahariya highway. The surface of the ground on which this field developed rises from 15 to 20 m below sea level in the west to more than 200 m a.s.l in the east. This indicates that most of this field lies in Qattara Depression, which is delimited by 200 a.s.l contour line.

This field is composed mainly of compound parallel linear dunes extending in the SSE direction (Fig. 10.7).

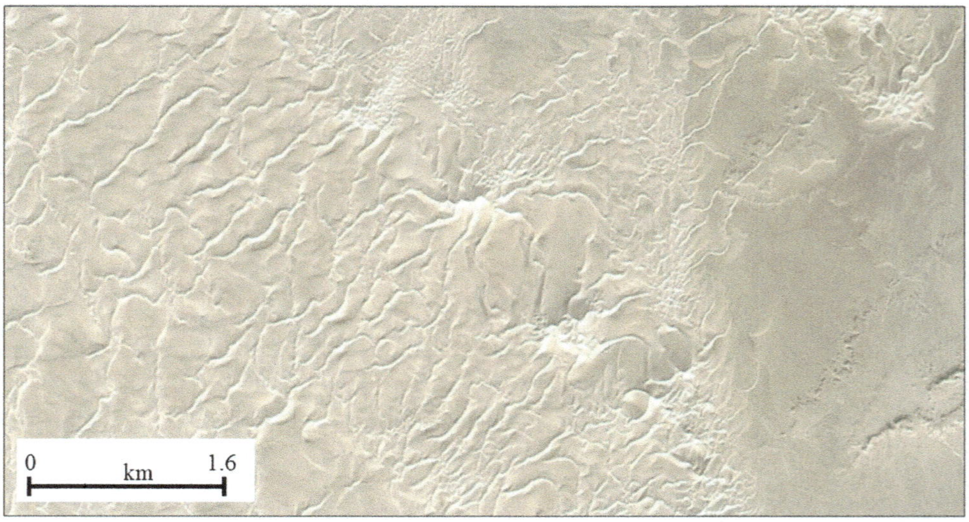

Fig. 10.6 Star dunes in the South Qattara Sand Sea. *Source* Google Earth (2014)

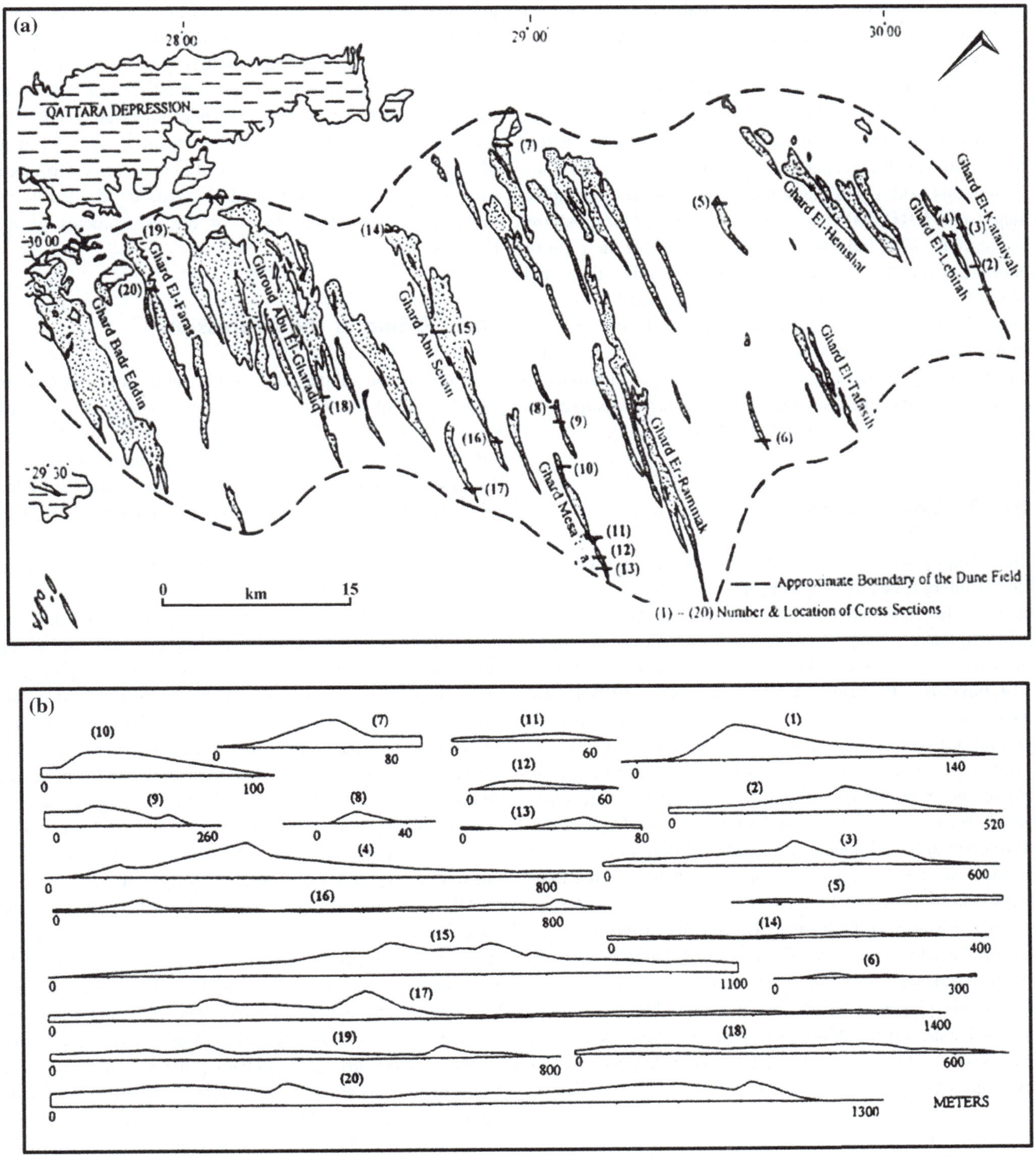

Fig. 10.7 East Qattara Dune field. **a** Distribution of dunes in the field. **b** Cross sections in various parts of dunes. *Source* Ali (1993), with some modifications

Topographic maps show that all dunes carry names, such as **Ghard Abu Sinan**, **Ghard ar-Rammak**, **Ghard Misaada**, **Ghard Abu El-Gharadiq**, and **Ghard El-Henaishat**. This characteristic distinguishes this field from other dune fields in Egypt. Their length varies between 6.0 and 120 km, width between 0.4 and 17.4 km, and height between 18 and 120 m (Ali 1993). In general, dunes become wider and more complex westward. The complex and wide ones are composed of several ridges, which coalesce in an upwind Y-junction, forming in most cases one ridge at the downwind section of the dunes. This explains why the width of many dunes decreases in the downwind direction.

However, nearly all dunes in this field are exceptionally wide at their upwind section. At these upwind sections, sand is shaped into small barchanoids, representing the second dune form in this field. Upon examining these areas on topographic maps (scale 1:100, 000), it was found that they are local shallow Depressions. This indicates that the initiation of these linear dunes started in these Depressions as sand accumulations, and as sand supply continued, dunes grew southward by time. It seems that their growth still continues, since field measurement showed that they grow at their southern tips at different rates nowadays (Ali 1993). In cross section, sand ridges such as those of Farafra and the Great Sand Sea are composed of sharp crests overriding wide plinths or **whale-backs** (Fig. 10.7).

10.4.3 Closed Basins

Mega-Depressions are characterized by the presence of closed internal basins. These basins in the Qattara Depression can be classified into two groups. The first one comprises relatively large basins developed along the outer margins of the floor such as **El-Moghra**, **El-Qarah**, **El-Ghardaq**, **El-Arag**, **Umm Ketabin**, **El-Bahrain**, **Nuweimsa**, and **Sitra**. Their length and breadth exceed several kilometers, and depth is several tens of meters. Topographic maps show that the Depressions of El-Bahrain, Nuweimsa, and Sitra lie along an E-W line and are elongated. Except for Moghra and Qarah—these Depressions developed in the Mokattam limestone (Middle Eocene), and are all bound by major or secondary fault lines. This indicates that this group of basins developed by karstic processes under the control of fault lines. The second group of basins lies on the sabkha surfaces. They are irregular and relatively smaller and shallower compared to the first group. Since they are developed in Moghra clastics, these Depressions are not of karstic origin. Most probably, they have developed due to salt weathering and the deflation of weathered product by wind action.

10.4.4 The Wadi-Fan System

Google Earth images show the presence of two interrelated fluvial forms. The first are small wadis dissecting the escarpments of **Qattara Depression**. They are restricted to the escarpment slopes, since they extend short distances into the plateau surface, and terminate at the foot slopes of the escarpment (Fig. 10.1). The second form is alluvial fans developed at the feet of the northern escarpments, separating the steep slopes of escarpments and the flat surfaces of the sabkhas (Fig. 10.1). In some instances, fans extend as broad loops into sabkhas and stand at higher levels compared to sabkha floors. Some of these fans exhibit several features of weathering and wind deflation, while others are incised by recent channels (El-Bassyoni 1995). Due to the development of several beside each other, alluvial fans coalesce forming a narrow depositional belt (bahadas) separating the low-lying sabkhas and the steep slopes of the escarpments of the Depression.

10.5 Origin of the Depression

The discussion about the origin and development of Depressions of the Western Desert has continued for many years. Different and varying views offered by previous studies on geologic and geomorphologic characteristics were discussed in a previous research by the author (Embabi 2004). In this previous study, Depressions of the Western Desert were classified according to origin into three groups: (1) Tectonic Depressions; (2) Morpho-climatic Depressions; and (3) Polygenetic Depressions.

Embabi (2004) classified Qattara Depression as belonging to the third group of polygenetic Depressions. This group comprises Depressions, developed by exogenic processes under the control of geologic structures. As indicated in this Chap. (Sects. 10.3 and 10.4), the Qattara Depression developed where Marmarica limestone (Middle Miocene) overlies the Moghra clastics (Lower Miocene). Local landforms are of diversified origin: fluvial (wadis-fans system), karstic (micro-small closed Depressions, and gypsum dykes), related to underground water seepage (Sabkhas, salt pans "Salinas"), salt weathering, and aeolian processes (sand dunes).

All previous studies indicated that each of the above-mentioned agents and processes contributed to the development of Qattara Depression, and none of them could be said to have been the sole agent. Ball (1927), Knetsch and Yallouze (1955) and Said (1960) concluded that wind erosion contributes to slow deepening of the micro-Depressions on the plateau surface by deflating the residuals after the decay of the limestone at their floors. The role of wind in modeling the Qattara Depression is not limited to erosion. As mentioned earlier, sand dunes represent a characteristic feature in this Depression. Sand derived from Moghra beds was transported southward and was deposited in the east and south part of the Qattara Depression.

Fluvial action working on facial boundaries of limestone of Middle Miocene and the underlying clastic beds of Moghra Formation (competent and incompetent) have created the initial stage of Qattara Depression. By back-wearing, competent beds retreated northward and westward forming escarpments and cuestas that currently surround the Depressions from northern and western sides. The presence of a cuesta along the northern side of Qattara Depression is taken as basis for a hypothesis suggesting that the Qattara began as a stream valley incised nearly along the regional strike of the Miocene Formations (Albritton et al. 1990). The transformation of this valley with external drainage to an internally drained basin may have occurred due to one of two possibilities as suggested by Albritton et al. (1990). The first is by wind during an interval of arid climate, whereas the second is by karst processes through the infiltration of groundwater in the carbonate rocks when the stream was active. Furthermore, Said (1960) postulated that the process of parallel retreat of the Qattara escarpment has been accomplished by wind erosion after the removal of the limestone cap rock due to the chemical action of water. This cannot be accepted, since the work of wind erosion is governed by the physical properties of wind that do not produce equal rates of erosion along the slopes of high escarpments.

Karstification processes working on the surfaces or sub-surfaces of the limestone El-Diffa (Marmarica) Plateau through the chemical action of percolating water resulted in down wearing of the surfaces. Karstification processes originated the micro-Depressions of the Miocene plateau by solution, collapse or subsidence. Very minute solution holes/basins (10–100 cm) were recorded on El-Diffa Plateau. As postulated by Said (1960), these minute and micro-Depressions represent an embryonic stage in the development of greater Depressions. At a later stage, they grew laterally to coalesce forming larger ones. Many of the coalescing Depressions, which developed on the limestone plateaus, could be detected on large scale topographic maps.

Recent field studies and petrographic examinations of core samples of the bedrock of the floor of the Qattara Depression indicated that **salt weathering** is prevalent in the western part in marked contrast to the eastern part (Aref et al. 2002). Therefore, salt weathering provided fine-grained debris that is easily deflated by wind, and this explains why the western part of the floor is topographically lower (−145 m). These circumstances occurred during certain periods when the piezometric pressure was high enough to raise the underground water to the surface. Such changes in the level of underground water in the past were correlated with sea level changes (Gad 1992). This is because there is a slow migration of underground water toward the sea, which acts—in general—as a base level for this migrating water. The rise in sea level is reflected in the rise of the water table, and on the contrary, a lowering of sea level is reflected in a lowering of the water table. This relationship was more obvious not only in Qattara Depression but also in the northern Depressions of the Western Desert compared to southern ones. This might explain why the floors of the northern Depressions are all below sea level, and why permanent lakes and sabkhas occupy their floors.

Probably the most important group of weathering processes is those responsible for **cliff sapping**, which is the dominant factor in scarp retreat (Dumanowaski 1960; Embabi 1967; Ashour 1974; Albritton et al. 1990; Brookes 1993; Busche 2001). Along the escarpments of Qattara and other Depressions undermined steep limestone and chalk cliffs with hanging sections and slumped blocks lying below the cliffs were recorded in several localities, and can be detected on some large-scale aerial photographs. The presence of several generations of slumped blocks indicates that undermining and slumping are repeated several times and that this process played a major role in slope retreat. During wetter phases, sufficient water must have been reached the contact zone between the limestone cap rock and the underlying clastics. Water sapping resulted in slope recession and the collapse of the hanged blocks. Naturally, weathered material and collapsed blocks were washed by water action.

It is clear from the preceding discussion that several agents contributed to the development of Qattara Depression. Therefore, **it is a polygenetic Depression**. But when did it start to develop? Since Qattara developed in Lower and Middle Miocene sediments, it was born by the end of Middle Miocene times. At these times, tropical wet climate prevailed in Egypt as indicated by the fossils found on the Eocene Plateau between Bahariya and Farafra (Pickford et al. 2006). This means that the Qattara Depression started to develop by fluvial action.

10.6 Utilization of the Depression for Power Production

As an uninhabited huge Depression (total area 45,000 km^2, with 19,500 km^2 below sea level), deep to the point −134 m according to Ball 1933, and being the nearest Depression to the Mediterranean, Ball (1933) suggested connecting Qattara with the sea by an open channel and a tunnel, and establishing a hydro-electric power station to produce electricity. In the late seventies and early eighties of the twentieth century, the Egyptian Government conducted additional detailed feasibility studies (Joint-Venture Qattara 1981; Institut Géographique National and Société Française Stréotopographié 1977). All these studies and other scientific research agreed with Ball's views (1927) that the Qattara Hydro-Solar Project is feasible, and that the project when executed will help develop new communities and new industries around **the Qattara Lake**. It was argued that it would be attractive, for instance to the salt extraction industry, and it would enable the establishment of desalination plants to provide the new communities with their needs of potable water. However, under the pressure of some opinions claiming that water from the Qattara Lake will leak into the Nile Delta fertile soils, which could be spoiled by salt precipitation, a decision was taken by the Egyptian Government to postpone any further action. Another concern raised is that under the weight of the huge quantity of water that will be stored in the Lake, strong earthquakes could occur, resulting in a serious damage to the Nile Delta populated villages, towns, and infrastructures.

10.7 Conclusion

Qattara Depression is discovered relatively recently, in 1926 (Ball 1927), surveyed in 1927 and described for the first time by Ball (1933). It is a unique Depression among other mega-Depressions in the Western Desert. It is the largest and deepest erosional Depression not only in Egypt, but also in Africa. It is also the only uninhabited mega-Depression except for the two small oases in the extreme eastern (El-Garah) and western (El-Moghra) tips of the Depression. It is the nearest Depression to the Mediterranean Sea, and this is why Ball (1933) suggested to connect by a open canal and a tunnel—the Depression with the sea and install a hydro-electric power plant at the northern escarpment benefiting from the water fall resulting from the entrance of sea water into the Depression.

The Qattara landscape is dominated by sabkhas and salt crusts, and sand dunes. Underground water discharged freely played a major role in developing this landscape. The presence of old salt crusts composed of several layers of different composition indicates that there were several phases of wetting and drying by underground water. Salt weathering was a significant process in disintegrating the sandstone at the Depression Floor, and consequently explains the deepening of the Depression. Wind deflated the disintegrated sand, forming the dunes in the southern and eastern fringes of the Depression. Multi-directional wind regime at the southern fringes resulted in the development of star dunes. This form of dunes is recorded for the first time in the present study. Water action as surface flow and karstification played a secondary role in dissecting the northern and eastern escarpment, and the dissolution the limestone formations, resulting in the development of small Depressions, some of which are filled with discharged underground water along the southern fringes. It can be concluded from this analysis that the Qattara Depression is a polygenetic Depression initiated by fluvial and karst processes in the Late Middle Miocene times.

References

Albritton C, Brookes J, Issawi B, Swedan A (1990) Origin of the Qattara Depression Egypt. Geol Soc Amer Bul 102:952–960

Ali A (1993) The geomorphology of sand dunes of south and southeast Qattara Depression. Dissertation, Ain Shams University

Aref M, El-Khoriby E, Hamdan M (2002) The role of salt weathering in the origin of Qattara Depression, Western Desert, Egypt. Geomorphology 45:181–195

Aref M, Hamdan M (2003) Sedimentology and environmental interpretation of the evaporite deposits in the Qattara Depression, Western Desert, Egypt. Sedimentol Egypt 10:193–213

Ashour M (1974) The geomorphology of Abu Tartur Plateau. Dissertation (in Arabic), Ain Shams University

Ball J (1927) Problems of the Libyan Desert. Geog Jour LXX:21–38, 105–128 and 209–224

Ball J (1933) The Qattara Depression of the Libyan Desert and the possibility of its utilization for power production. Geogr Jour LXXXII:289–314

Brookes I (1993) Geomorphology and Quaternary geology of the Dakhla Oasis Region. Egypt. Quaternary Science Reviews 12:529–552

Busche D (2001) Early Quaternary landslides of the Sahara and their significance for geomorphic and climatic history. J Arid Environment 49:429–448

Dumanowaski B (1960) Notes on the evolution of slopes in arid climate. Z GeomorphSuppl 1:178

El-Bassyoni A (1983) Field study of sabkha area in Qattara Depression. Qattara Hydro-Solar Project Report, Qattara Project Authority

El-Bassyoni A (1995) Sabkhas of Qattara Depression, Western Desert, Egypt: A survey. Sedimentology of Egypt 3:13–26

Embabi N (1967) A geomorphologic study of Kharga Oases Depression, the Western Desert, Egypt. Dissertation, Bristol University

Embabi N S (1998) Sand seas of the Western Desert of Egypt. In: Alsharhan AS, Glennie KW, Whittle GL, St. C. Kendall CG (eds) Quaternary Deserts and climatic change. Proceeding International Conference on Quaternary Deserts an Climatic Change, Al-Ain, United Arab Emirates, 9–11 Dec. 1995:495–509. Balkema. Rotterdam

Embabi N (2004) Geomorpholgy of Egypt. Vol. I, the Nile Valley and the Western Desert. Special Publication, The Egyptian Geographical Society, Cairo

Gad T (1992) Variables influencing the development of the depressions of the Western Desert in Egypt, with special reference to Dakhla and Kharga. Bull Soc Géogr d'Égypte 65:55–86

Google Earth (2014) Egypt Images Latitudes 22-32 N and Longitudes 22-32 E, http://www.google.com

Institut Géographique National and Société Française Stréotopographié (1977) Survey of the Qattara Depression, surface and volume. Qattara Project Authority, Cairo

Joint-Venture Qattara (1981) Study of Qattara Depression: Feasibility Report. Vol. III, Part I: Topography, Regional Geology, and Hydrogeology. Qattara Project Authority, Ministry of Electricity and Energy, Cairo

Knetsch G, Yallouze M (1955) Remarks on the origin of the Egyptian oasis depressions. Bull Soc Géogr d'Égypte 28:21–33

Mostafa M (1993) El-Dffa Plateau: a geomorphologic study. Dissertation, Ain Shams University

Pickford M, Wanas H, Soliman H (2006) Indications for a humid climate in the Western Desert of Egypt 11–10 Myr ago: evidence from Galagidae (Primates, Mammalia). Palaeoecology 5:935–994

Said R (1960) New light on the origin of the Qattara Depression. Bull Soc Géogr d'Égypte 33:37–44

The Kharga-Dakhla Depression

11

Abstract

The Kharga and Dakhla Depressions are one Depression, since they are encircled by steep escarpments from the east and north without imperceptible breaks and changes. This mega-depression extends for about 350 km east–west, but has a varying north–south extension of 200 km along its eastern boundary in Kharga and about 50 km in Dakhla. Surrounding plateaus rise 300 m a.s.l east of Kharga, 500 m on Abu Tartur, and 300–350 m asl on the northern plateau of Dakhla. Nearly, all the escarpment alignments are controlled by tectonics. Slope form along escarpments is also controlled by lithological differences and marked by several forms such as small wadi systems. The slopes of the escarpments are covered with various types of superficial deposits of different origins. Undermining of the upper segment of the slopes and the presence of several generations of collapsed limestone and chalk blocks indicate that slope retreat played a major role in the development of the Kharga-Dakhla Depression. However, the Depression Floor is an undulating plain since there are numerous landforms, both positive and negative that spread over its surface.

Keywords

Kharga-Dakhla Depression • Scarp retreat • Playas • Tufa • Yardangs • Spring mounds

11.1 Introduction

The Kharga and Dakhla Depressions could be considered as one very large Depression, since high escarpments encircle them from the east and north without any break. Although there is relatively high ground separating them to the south of the Abu Tartur Plateau, the difference between Kharga and Dakhla is imperceptible. Similar forms have developed in both of them, and human occupation history is almost similar. Some previous studies considered both Kharga and Dakhla as one unit (Hamdan 1960) and both of them accommodate the oldest human tools and remains of prehistoric cultures in Egypt. The following analysis will consider them as one large Depression. Reference will be made to it as the **Kharga-Dakhla Depression** and at the same time, local characteristic forms will not be ignored. There are numerous studies on the geomorphology and archaeology of Kharga and Dakhla from the early years of the twentieth century.

11.2 General Characteristics

Lying about 150 km to the west of the Nile Valley, the Kharga-Dakhla Depression extends approximately 350 km from the Kharga eastern escarpment to west Mawhoob Basin

N.S. Embabi, *Landscapes and Landforms of Egypt*, World Geomorphological Landscapes,
https://doi.org/10.1007/978-3-319-65661-8_11

in the west of Dakhla. The north–south (NS) extension is irregular. It measures 200 km in Kharga and about 50 km in Dakhla. The Depression is encircled by steep escarpments from the east and north, while it is open in the south and west where the surface merges imperceptibly with the surrounding areas. The level of the ground rises from 100 m to 200 m a.s.l on the floor of the Depression to 300 m to 500 m a.s.l on the surrounding plateaus. In central Kharga, an area as low as two meters above sea level is recognized on the topographic maps (scale 1: 25,000) lying to the east of **Gabal El-Qarn**. The area of Dakhla is 6250 km^2, while that of Kharga is 12,250 km^2, resulting in a total area of 18,500 km^2. If the area lying to the south of **Abu Tartur** is added to that of Kharga and Dakhla, the area will increase to more than 20,000 km^2.

The Geologic Map of Egypt (Fig. 2.1, Chap. 2) shows that this region is composed of limestone (Lower Eocene)/ chalk (Palaeocene) as a cap rock for the plateaus, and clastics of shale, clay, sandstone (Cretaceous) forming the floor and the escarpments below the cap rock. Quaternary fluvial, karst, and aeolian sediments unconformably overlie the bedrocks. The natural discharge of underground water from the Nubian sandstone on the floors of Kharga and Dakhla made possible for prehistoric people to gather, live, and cultivate the famous oases in the Depression. Ancient Egyptian and Greco-Roman heritage has been found there. Moreover, the attempts to reclaim new land on the artesian water of the Nubian aquifer since the beginning of the twentieth century indicate that the oases have remained a home for human settlers and new societies in the desert outside the Nile Valley. The geomorphologic maps of Kharga and Dakhla (Figs. 11.1 and 11.2) show that they can be divided into three geomorphic units: the fringing plateaus, the escarpments, and the floors.

11.3 The Surrounding Plateaus

Although the plateaus lying to the east and north are not an integral part of the Kharga-Dakhla Depression, their forms are highly correlated with those developed in the Depression itself. Space images, aerial photographs, and topographic maps show that these plateaus comprise four groups of landforms. The first group includes structurally controlled forms, which appear as parallel east–west cuesta ridges with smooth slopes and relatively steep scarps to the north of Dakhla. On Kharga Plateaus, there are some circular forms just to the west of the southern half of the escarpment developed along fault lines that extend to the Kharga escarpments. The second group is of aeolian origin and can be divided into two types: extensive fields of rock yardangs ending at the edge of the plateaus lying to the north of both Kharga and Dakhla, and also sand dunes of both the Farafra Sand Sea and Ghard Abu Moharik before entering the Depression. The third group of landforms is of fluvial origin and also comprises two types: joint-controlled wadi systems that dissect the cuesta ridges on the Dakhla Plateau and the heads of some wadis extending into the Depression; playas in some of the closed basins, such as Miramar and Pharoas, to the north of Dakhla. The fourth group is of karstic origin, where small solutional features and precipitated materials (calcite) were observed in the field by the author and recorded by Kleindienst et al. (1999). Solution cavities were also recorded in the boreholes on the Abu Tartur Plateau (Said 1990). Their age is estimated to be Late Tertiary (Kleindienst et al. 1999).

11.4 The Escarpments

Escarpments encircle the Kharga-Dakhla Depression from the east and north. Although they are characterized by a steep face in the top section conformable with the calcareous (limestone/chalk) cap rock and a concave slope composed mainly of clastic beds of shale, clay, and sandstone, there are some significant variations not only in their local orientation, but also along their slopes. Accordingly, they will be divided into three sections: the Kharga escarpments, the Dakhla escarpments, and the Abu Tartur escarpments.

11.4.1 Alignment and Slope Form

In Kharga, the alignment of escarpments is straight along most of their sectors, irrespective of orientation (Fig. 11.1). This is due to the effect of faults controlling the scarp retreat (Embabi 1967). Another characteristic is that it takes a step-like form along several sectors, due to the presence of a sequence of limestone/chalk and shale/clay beds, ranging in age from Upper Cretaceous to Lower Eocene, and forming three slope sequences. Of special importance is the middle slope sequence since its surface takes the form of a platform along the central part of the eastern escarpment (Fig. 11.1). This platform is composed of Tarawan Chalk and Dakhla shale. On the surface of this platform, an erosion surface is recorded (Embabi 1967). The first slope sequence is characterized by the presence of several lines of fallen rock blocks due to undermining of the top limestone layer. The characteristics of these fallen blocks (weathered, dark color, broken into smaller pieces, and their presence amid aeolian sand) indicate that they developed in the past and not under the present arid conditions (Fig. 11.3a), indicating the significance of this process in scarp retreat and the development of the Depression. The third slope sequence along the escarpments developed on the Cretaceous Duwi formation (phosphate, shale, and oyster beds) as dissected cuesta.

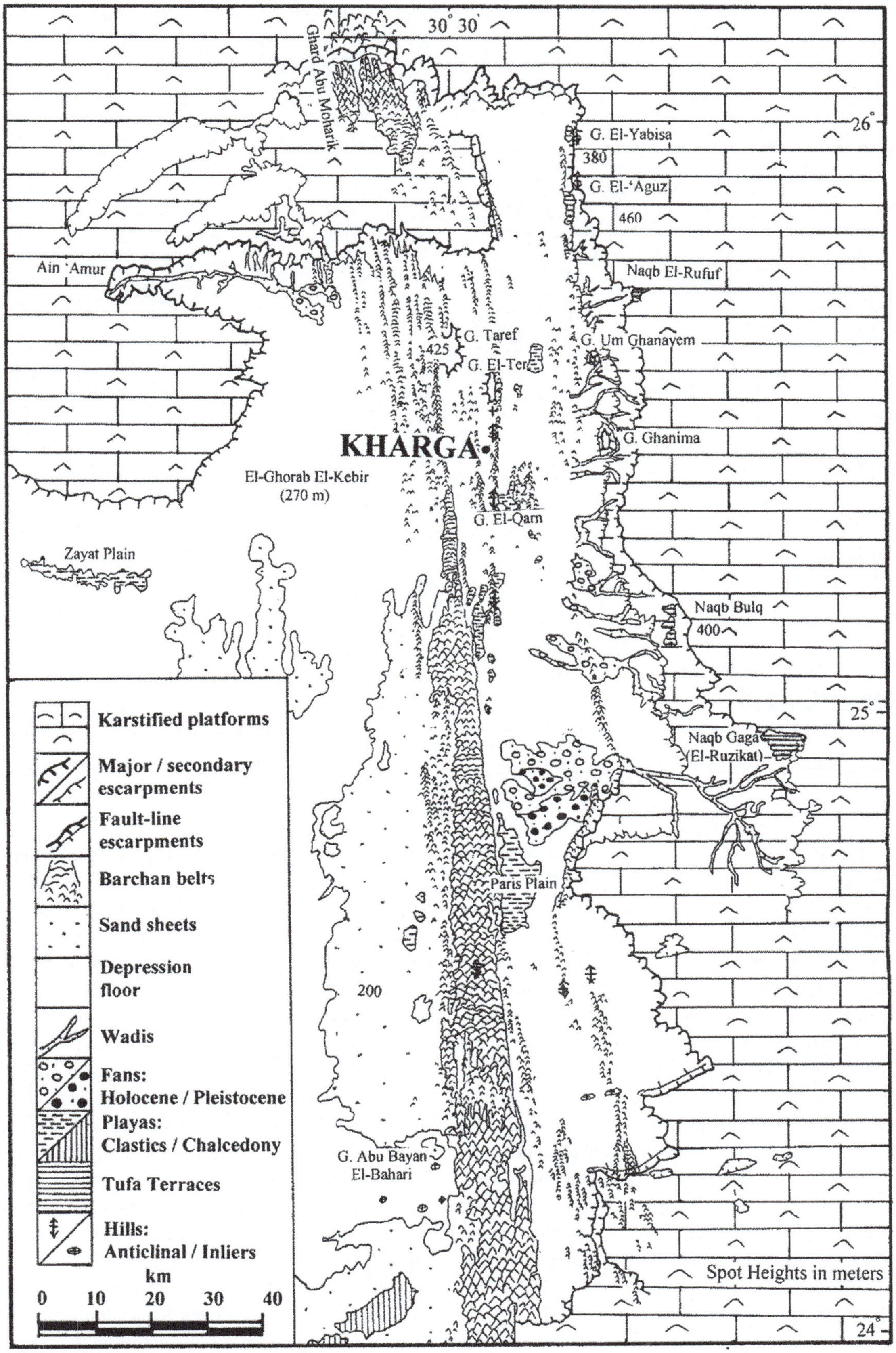

Fig. 11.1 Geomorphologic map of Kharga (Compiled from: Geologic Map, scale 1: 500,000; Landsat-TM Images; Topographic Maps, scale 1: 250,000; Field Work)

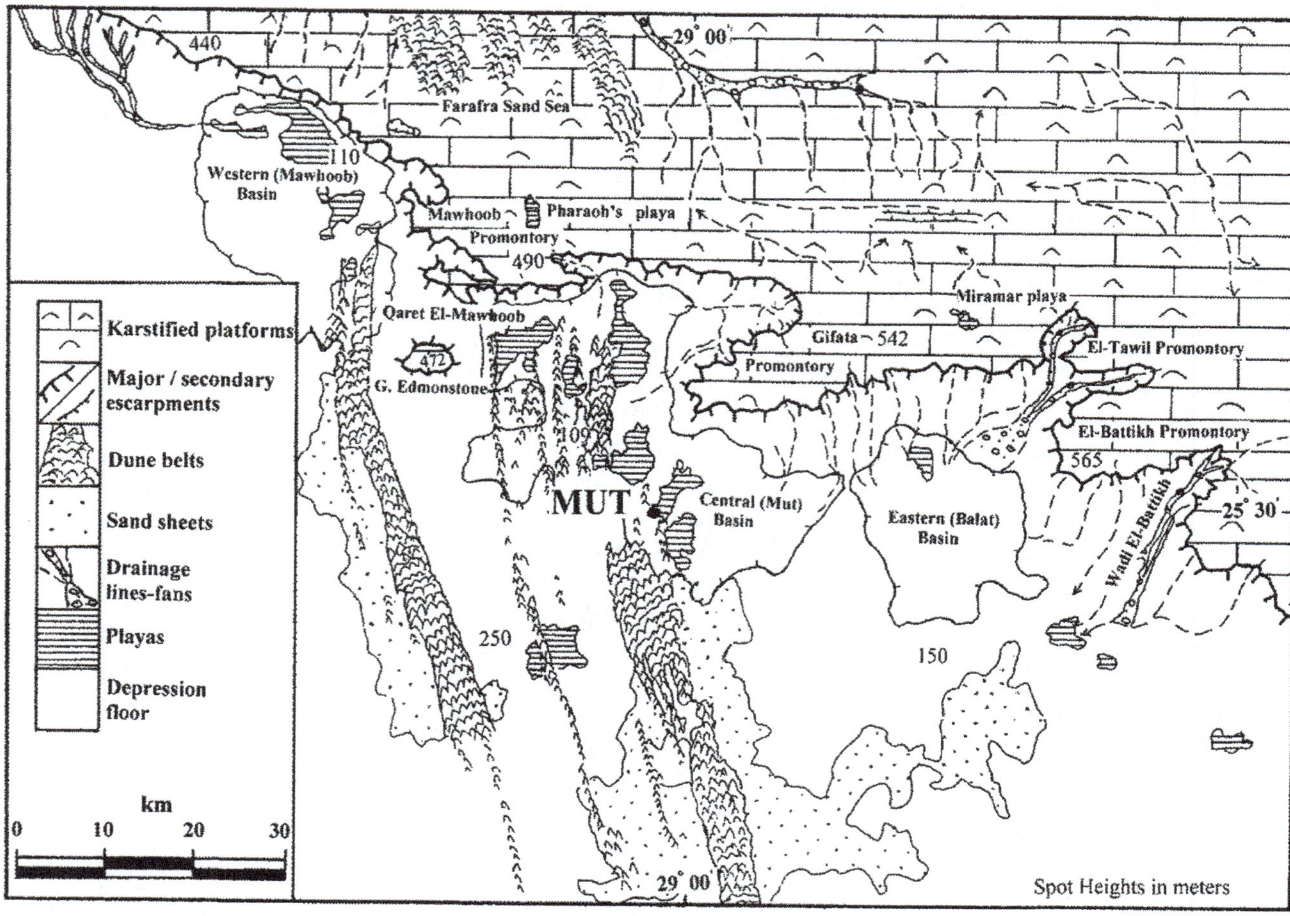

Fig. 11.2 Geomorphologic map of Dakhla (Compiled from: Landsat-TM Images; Topographic Maps, Scale 250,000; Brookes 1993; Field Observation)

This cuesta appears only along the eastern escarpment due to the low dip of beds toward the east (Embabi 1967). In this respect, this cuesta is equivalent to that one developed on the Duwi formation in Dakhla (Brookes 1993).

The **Kharga escarpments** are dissected by small incised wadis, which become shallow and braided at the piedmont zone. Due to this change, water flow, during wet phases, deposits its load that is transported from the escarpment slopes on the Depression Floor. The coarse material (boulders and gravel) is deposited first in the piedmont zone as fans and sheets, then the fine sediments (sand, silt, and clay) settle in the basins where water flow terminates and forms seasonal or temporary water bodies (playas). In general, drainage basins are small, except for the basins of **Wadi El-Midauwara**, in the central part of the eastern escarpment with its tributaries extending at the Eocene escarpment, and **Wadi Ain Amur** on the NW escarpment (Fig. 11.1). The length of Wadi El-Midauwara is approximately 65 km from its head in the east to the peripheries of its fan in the west; whereas that of Wadi Ain Amur is about 40 km. As shown in the figure, Wadi El-Midauwara has deposited a large compound fan, with an east–west length of approximately 25 km. This fan is composed of two types of deposits: fanglomerate and wadi deposits. The fanglomerate sediments were most likely laid down at a phase later than the gravel sheets of the Dakhla Piedmont (Brooks 1993). Similar gravel was recorded in the vicinity of Gabal El-Ter and Gabal Taref, 15 m above the surrounding surfaces (Embabi 1967) and in several localities along the eastern escarpment (Caton-Thompson 1952). Wadi deposits shown on the Geological Map represent a third phase of fluvial deposition as wadi fans and along wadi beds. It is the most recent phase in the development of the piedmont of Kharga escarpment. Cycles of deposition and erosion along the escarpments of the Kharga Depression require further research, since most available knowledge depends on field observations rather than detailed studies.

Although the **Dakhla escarpment** has a general E-W orientation, it is characterized by several acute changes that take the form of embayments and promontories (Fig. 11.2). Embayment sectors represent wide wadis developed on anticlinal swells, whereas promontories are developed on

Fig. 11.3 Field photographs for (**a**) Kharga and (**b**) Dakhla escarpments, showing slope form and superficial material of sand and collapsed blocks

synclines as reversed topography (Brookes 1993; Hermina 1990).The upper segment of the escarpment, like similar ones in other mega-depressions, is a free face of calcareous formations (Paleocene chalk and limestone), 20–40-m thick with widened vertical joints and tilted blocks (Fig. 11.3b). The lower segment of the escarpment is underlain by the Paleocene Dakhla shale and is covered by aeolian sand and a mantle of fallen limestone blocks and weathered shale. It is preserved on ridges separating the steep narrow wadis that dissect the slopes of the escarpment (Fig. 11.3). These wadis have widened due to slope retreat. This process occurs by cliff sapping, which leads to the undermining of cliffs and the collapse of hanging blocks, forming a scree cover. Those fallen materials are exposed to further weathering and removal by slope wash. Therefore, it could safely be mentioned that slope retreat occurs as in Kharga along the slopes of escarpments through several cycles of cliff sapping, undermining, falling of hanging blocks, and the removal of scree mantle by slope wash.

The Abu Tartur escarpment differs greatly in its alignment and in slope form from those of Dakhla and Kharga since it takes a circular shape (Fig. 8.8, Chap. 8). The foot slopes of the plateau extend southward as straight linear sandstone ridges separated by straight wadis. Those two characteristics are mostly due to a structural effect, since previous studies postulated—taking its relative height into consideration—that the **Abu Tartur Plateau** and its extension southward represent a wide swell separating Kharga and Dakhla from each other (Shata 1953, 1959). The straightness of the wadis dissecting the foot slopes of the Abu Tartur escarpments suggests that these wadis developed under the control of a system of parallel joints formed during the same time as the uplift of the Abu Tartur Plateau.

11.4.2 Superficial Sediments and Associated Landforms

On the slopes of the escarpments of the Kharga-Dakhla Depression, fluvial, karst, aeolian sediments, and weathering materials have been recorded according to previous studies. Fluvial gravels spread as sheets or as wadi fill. In Kharga, a gravel sheet has been recorded on the flat surface which developed on the Tarawan Chalk of the middle platform of the eastern escarpment (Fig. 11.4). This gravel sheet is dated as Upper Acheulian by Caton-Thompson (1952). Similar gravel has also been recorded on the sandstone ridges of the northern escarpment in Umm El-Dabadib area and Naqb Bulaq in Kharga (Caton-Thompson 1952). In Dakhla, gravel sheets have also been found on the surfaces of Duwi Cuesta (Brookes 1993). Many wadis dissecting the escarpments are filled with fluvial gravel; some of which are dissected by fluvial action forming small terraces (Fig. 11.5).

Tufa is a chemical calcareous sediment ($\geq 90\%$ C_aCo_3) developed in several localities on the eastern and northwestern escarpments of Kharga, which were first studied by Caton-Thompson and Gardner (1932). Tufa deposits were recorded in eight separate localities (Fig. 11.1). They are **Ain Amur**, **Naqb Asyut**, **Gabal El-Aguz**, **Naqb El-Rufuf**, **Naqb Bulaq**, **Naqb El-Matana**, **Naqb Gaga**, which is also called **El-Ruzaiqat** or **El-Midauwara** (Caton-Thompson and Gardner 1932; Ashour 1974; Kleindienst et al. 1996; Sultan et al. 1997; Hamdan et al. 1999), and **Naqb Doush** (Brook et al. 2003). Such a large number of tufa localities have not been recorded in any other Depression or region in Egypt. This is mostly due to the fact that local hydrogeologic conditions were favorable for the formation of tufa in a number of localities in Kharga, but not in others. Along the eastern escarpment of Kharga, there is a structural line of faults and folds, facilitating the discharge of underground water containing calcium carbonates (Embabi 1967). In Kharga, tufa is classified into two categories: the Plateau Tufa occupying the upper slopes of the escarpments, and Wadi Tufa laid down at the wadis and lower sections of the escarpments

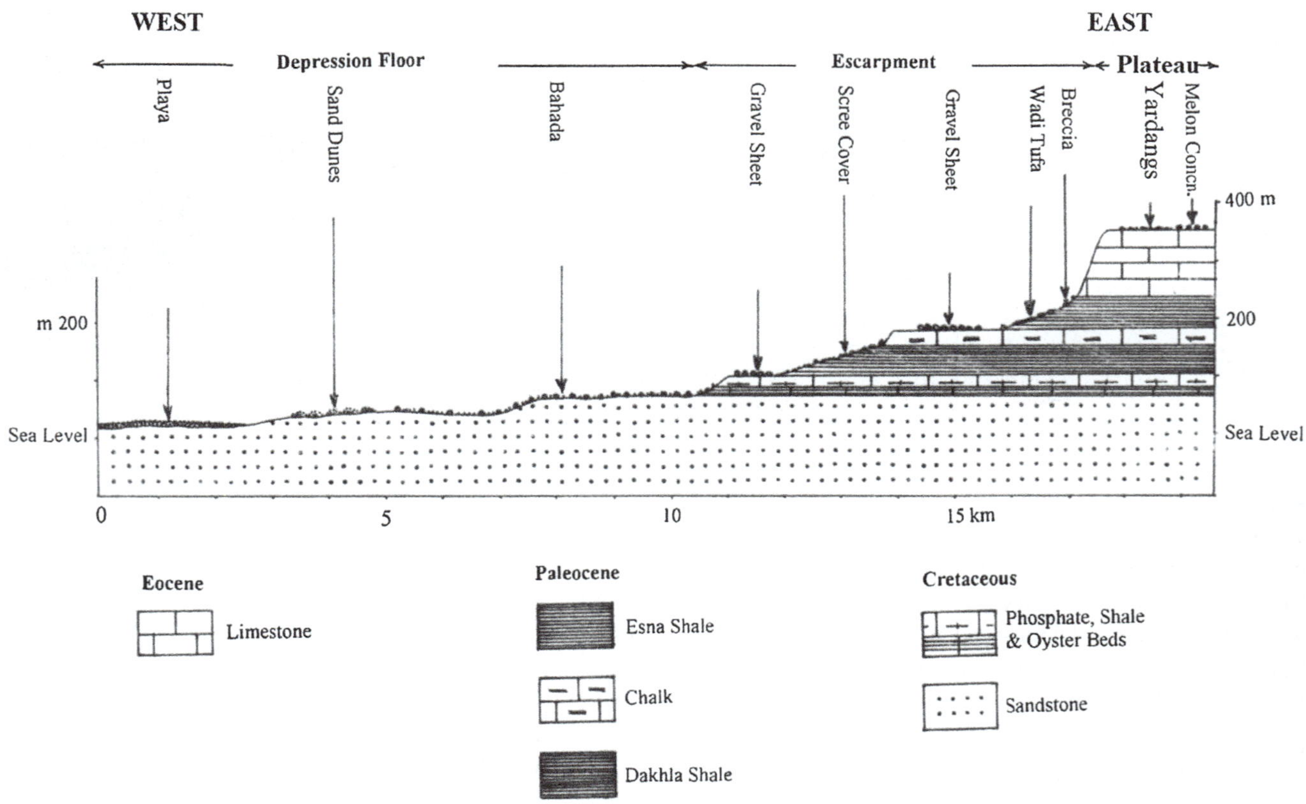

Fig. 11.4 Schematic E-W geologic cross section in Kharga, showing the distribution of superficial deposits

Fig. 11.5 Field photograph of a gravel terrace in Wadi Naqb El-Rufuf, the eastern escarpment of Kharga

(Caton-Thompson and Gardner 1932). The deposits are composed of thin laminae of calcareous material which developed by slow flowing water (surface and underground) moving on the slopes of the escarpments (Fig. 11.6). Tufa takes three depositional forms: sheets, terraces, and bulbous flowstone (Embabi 2004). Some of the tufa terraces which developed on both sides of the upper sectors of Naqb (wadi) El-Rufuf at the eastern escarpment resulted in a step-like slope form that gives this wadi its local name, Rufuf (Fig. 11.6). The age of tufa deposits has been estimated in previous works to be Plio-Pliestocene for Plateau Tufa (Caton-Thompson and Gardner 1932), or from **Acheulean** to **Middle Palaeolithic** for **Wadi Tufa** (Caton-Thompson and Gardner 1932). Also, ages of 338–450 ka are given for the oldest tufa, whereby ages of 225–286 ka and 157–190 ka are given for more recent tufa (Hamdan et al. 1999; Sultan et al. 1997).

An interesting aspect in Dakhla and Abu Tartur is that tufa deposits have been found in situ in a few localities

Fig. 11.6 Tufa deposits at Naqb El-Rufuf at the eastern escarpment of Kharga (**a**) tufa terraces and (**b**) curved tufa lamina at the edge of one the terraces

(**Teineda**, **Balat**, **Mut**) along their escarpments (Mahmoud et al. 2002). In other areas, only some tufa blocks were found in Dakhla (Brookes 1993; Kleindienst et al. 1999) and on the escarpment slopes of **Abu Tartur** (Ashour 1974). The interpretation is that tufa deposits must have been formed at an earlier stage and were mostly removed during the retreat of the escarpment. On the other hand, the absence of tufa in place/few tufa occurrences could be due to the northerly bedrock dip, which diverted most percolating water northward. Although cliff sapping did occur and had led to rock slides and collapses, it seems that other conditions necessary for tufa development (such as water volume and temperature, calcium carbonate content in water) were not fulfilled. This situation is repeated in Kharga, where no tufa was recorded on the south-facing escarpments. Preliminary U/Th dating on a large tufa block found below the escarpment to the north of Balat village indicates that it was deposited more than 350 ka (Kleindienst et al. 1999). The U/Th determination of another tufa boulder collected from the surface of the escarpment north of **Mawhoob village** indicates that water was sapping from the face of the escarpment at 176 $\pm$ 14 ka to 170 $\pm$ 12 ka (Brookes 1993). Another tufa block found further east of Balat locality was also given a U/Th dating between 122 $\pm$ 9 ka and 145 $\pm$ 6 ka (Kleindienst et al. 1999). Although the finding of these tufa blocks in different localities indicates that spring tufas developed in Dakhla and were eroded during subsequent erosional episodes, they developed in fewer localities than in Kharga. On the other hand, their age indicates that tufa deposition occurred during Middle to Late Pleistocene, which is similar to that of Kharga.

Weathering products range in size from very fine particles to huge blocks. They cover the lower section of escarpments and rarely mantle the whole slope. They can be classified into two types: **Quaternary Breccia** and recent scree. The Quaternary Breccia is composed of a consolidated unsorted angular mixture of limestone blocks, boulders, and fine material. In Kharga, several hillcocks known as flatirons are capped by residual consolidated breccia at the foot slopes of the first slope sequence of the main escarpment at **Naqb El-Rufuf** (Fig. 11.7) (Embabi 1967). Similar breccia was found on the NE escarpment at **Naqb Assyut** (Embabi 1967; Busche 2001) and the western escarpment facing Naqb Asyut covering most of the slope (Embabi 1967). It has also been recorded in **Naqb Bulaq** overlain by wadi tufa (Caton-Thompson 1952). Lithological characteristics suggest that this breccia is composed of collapsed blocks formed due to cliff sapping which undermines the steep limestone and chalk cliffs leading to the development of hanging sections along the cliffs, and finally the collapsing of rock blocks resulting in scarp retreat (Fig. 11.3a). This Quaternary Breccia is recorded on the escarpments of Dakhla (Brookes 1993) and Abu Tartur (Ashour 1974). One of its main characteristics is that the blocks are weathered in different degrees, so the lower, or older, ones are more weathered than the higher, or younger, ones. The presence of several generations of slumped blocks indicates that undermining and slumping were repeated several times and that this process played a major role in slope retreat (Embabi 1967; Busche 2001). Most probably, the Quaternary Breccia developed during an arid phase. This arid phase was placed between the two wet phases of the Plateau Tufa and Wadi Tufa (Caton-Thompson 1952). It was also detected as a surface of unconformity in the playa sediments of the yardang field in the Sherika plain in northern Kharga (Embabi 1968–1969).

Due to the present hyperarid climate which is characterized by rare rainfall, only slight weathering occurs and weathering products remain on slopes for long periods of time. Recent talus cover is loose and shows no cohesion. It is characterized by angularity of individual rock fragments. It covers the escarpment slopes from the base of the steep hanging cliffs to the base of the lower slope taking two forms: Scree Mantle and Talus Cones. The hanging cliffs are evidence that undermining occurs under present arid conditions, most probably at a low rate. This suggests that undermining and collapse blocks too occurred during an arid to semiarid period in the past.

Aeolian Sand only covers wide areas on the slopes of the escarpments of Kharga-Dakhla. In fact, these sand accumulations represent a portion of the sand flow entering the Kharga-Dakhla Depression from the two sand seas of Farafra and Ghard Abu Moharik lying to the north of the Depression. In Dakhla, the sand of the **Farafra Sand Sea** descends from the escarpment, mainly covering the lower section of the slope (Fig. 11.3b). In Kharga, the sand of **Ghard Abu Moharik** covers the slopes of the northern and the eastern escarpments. Sand on the northern escarpment like that of Dakhla forms descending dunes, whereas in Kharga it climbs and falls on the slopes of local landforms (such as wadis) since the sand flow is perpendicular to the slopes of the escarpment under the effect of prevailing northern winds.

11.5 The Depression Floor

The floor of the **Kharga-Dakhla Depression** represents a pediplain characterized by an undulating surface, and the presence of basins and hills of different sizes, developed at the expense of retreating **cuesta escarpments** that encircle it from the north and the east. It is lower than the surrounding plateaus by approximately 200–300 m and this is why it has also been termed lowland (Brookes 1993). The Dakhla floor is composed of three large basins (Fig. 11.2). On the contrary, Brookes (1993) thought that there is two basins only on the floor of Dakhla. This is because he excluded the basin of West Mawhoob which lies to the west of **Gabal**

Fig. 11.7 Quaternary Breccias in Wadi Naqb El-Rufuf **a** breccias hillocks at the foot slopes of the northern side of the wadi, **b**, **c** detailed photographs for the angular constituents of the breccias

Edminston. In these large basins, there are a number of small basins. The same holds for the Kharga and Abu Tartur lowland. In these basins, various Quaternary sediments and landforms developed, among which playas, sand dunes, and spring mounds are the most widespread. Hills of various origins represent another prominent landform category in the lowlands of the Kharga-Dakhla Depression.

11.5.1 Playas

11.5.1.1 Distribution

Playas are the most widespread landforms in the lowland of the three regions of Kharga, Dakhla, and Abu Tartur. Although sand dunes cover wider areas than playas, they are not present in the three sub-regions, since they are absent in Abu Tartur. An accurate distribution and indication of the areas of different playas cannot be made due to two factors: The first is that most playas are cultivated because their sediments represent good soils for land reclamation and cultivation. The second factor is that playas cannot be detected easily on aerial photographs or space images because their reflectance is similar to that of surrounding surfaces in many localities.

However, Figs. 11.1 and 11.2 show that playas are distributed along the central lines of both Dakhla and Kharga lowlands, representing the local internal basins that received flowing water and fluvial sediments through drainage lines coming down from the escarpments. Some playas lie just below the Dakhla Depression escarpments in the piedmont zone (see Fig. 11.3b). Although no age is given, the sediments of these latter playas were considered as Pleistocene Lacustrine sediments Facies A (Brookes 1993). Samples taken from the basin, which is called **Bir El-Gabal playa**, near El-Qasr were found to be of the Holocene age, ranging from 5 ± 0.4 ka to 6.4 ± 0.5 ka (Brook et al. 2003).

The Geological Map of Egypt (scale 1: 500,000) shows that there are more than 100 playas whose area exceeds two square kilometers in the Western Desert. The map also shows that these playas are concentrated in the southern part of the desert, namely in the mega-depressions, and lie mostly at the foot slopes of limestone plateaus or sandstone escarpments. Of these 100 playas, there are 25 in Dakhla and 21 in Kharga. The largest playas are present in two separate areas in the southern part of the Western Desert. The first area lies around Bir Tarfawi and Bir Sahara east and extends about 35–40 km from east to west or from north to south. The second area lies to the south of the Kharga Depression along **Darb El-Arbain**, extending about 80 km from north to south and 2–10 km from east to west (Fig. 11.1). These two groups of playas are different from others in that they are composed of calcium carbonates in the **Bir Tarfawi-Bir Sahara** area and **chalcedony** covered playa along the Darb El-Arbain area (The Geological Map, scale 1: 500,000; McKenzie 1993; Issawi 1971). They are also called **Armoured playas**, as in the case of **Darb El-Arbain playas** (Said 1980, 1990) or **Kankur playas**, as in the case of the Bir Tarfawi—Bir Sahara East playas (McCauley et al.1986; Wendorf et al. 1993) since they are covered by a layer of white carbonate or chalcedony nodules up to 15 cm in diameter. According to the Geological Map of Egypt (scale 1: 500,000), they appear to be of Quaternary age.

Two playas are recorded at the foot slopes of the **Abu Tartur escarpment**. The first lies in **Abu El-Egl** area (25° 18′ 45″N, 29° 32′ 25″E) at about 38 km east of Teneida village (Ashour et al. 2005; Donner et al. 2015). The second playa lies at the middle part of **El-Zayat plain** (25° 12′ 55″N, 29° 47′ 13″ E) at about 65 km to the east of Teneida village and extending for about 50 km in the eastern direction with an average width of approximately 2.5 km (Embabi 1999). The length of the Abu El-Egl sediments is approximately 3 km from north to south and its maximum width is 1 km from east to west. Age dating of the sediments of Abu El-Egl playa showed the presence of two phases of development: Late Pleistocene and Holocene. The topographic maps show that the Abu El-Egl wadi ends in the lowland lying to the south of the escarpment of Abu Tartur; therefore, a dam must have been constructed to develop a closed basin where a lake could be formed. So far, no evidence was found to indicate how that could have happened. The deposits of the **Zayat plain** were laid down in an elongated closed basin at the foot slopes of Abu Tartur Plateau.

11.5.1.2 Types of Playa Sediments

Playa sediments can be classified, according to their composition, into three categories: clastic sediments, chemical sediments, and organic sediments.

Clastic Sediments: Clastic sediments represent the predominant materials of playas not only in the Kharga-Dakhla Depression, but also in the Western Desert (Embabi 1999). They are generally composed of fine-grained clastic sediments of clay, silt, and fine sand, which have been deposited horizontally by fluvial action as very thin layers, 1–10-cm thick. In several localities, separate sand layers are present between the clay-silt layers. These sandy layers have been interpreted in the Sherika plain playa 10–15 km to the north of Kharga town as representing arid cycles since they were deposited by the wind; consequently, the sediments are called semi-playa deposits (Embabi 1968–1969). In Umm El-Dabadib area at the foot of the northern scarp of the Kharga Depression, very thin layers of sand intercalated with silt layers (9–28 cm) are present at the base of some of the stratigraphic sections of the playa deposits. This has been interpreted as indicating drier conditions than those prevailing during the deposition of units in the upper parts of the sections (Hamdan 1987). In Dakhla, repetition of quartz-rich beds in the Pleistocene lacustrine sediments near El-Qasr is taken to imply a continual renewal of sand supply between rain storms and the prevalence of a strongly seasonal climate with intense rains alternating with hyperarid windy seasons (Brookes 1993). In the Bir El-Gabal playa, which lies at the base of the Escarpment of Dakhla Depression at about 5 km NNE of El-Qasr village (25° 44′ 27″N, 28° 55′ 19″E), several sand layers up to 20-cm thick have been observed.

Some of the playas exhibit gravel bands derived from the surrounding areas. In the Gabal El-Ghorab playa at the foot of the Kiseiba escarpment in the southeastern part of the Western Desert, the gravels are embedded in silty matrix, representing beach and shore facies and grading down slope into laminated silts and fine sands (Schild and Wendorf 1984). Flattened gravels in some playas represent gravel reworked by wave action along the peripheries of the playa lakes in the Kiseiba region (Issawi and Hinnawi 1984). In the Miramar playa, the fine sediments (silt–clay) are intercalated by several limestone gravel beds with angular bedrock fragments near the eastern margin of the basin, which were derived from tributary wadis and basin walls by episodic rain-fed runoff (Brookes 1993). It was inferred that, at least eight times, runoff was intense enough to transport gravel into the lake. The sediments of this same playa are topped by 25 cm of crudely bedded fluvial gravel, consisting of unsorted, platy fragments of friable limestone, brought by torrential flows from the southern wadis (Brookes 1989,1993; Kleindienst et al. 1999). In the Zayat playa, rounded gravel carried out by waters in wadis issuing from the escarpment of the Abu Tartur Plateau was deposited along the northern peripheries (Embabi 1999). In the Bir El-Gabal playa (Dakhla), gravel is embedded in silt layers indicating an increase in water flow energy. Erosion of fine sediments exposed the embedded gravel partially or wholly on the slopes of yardangs (Fig. 11.8).

Chemical Sediments: These sediments precipitate from water, which is highly charged with various types of dissolved minerals. Previous work and field investigations revealed four types of chemical sediments: These are calcareous, chalcedony, gypsum, and salt sediments. In calcareous material, calcium carbonates constitute 50–100% in many samples taken from several playas in the Western Desert such as Miramar playa of Dakhla (Brookes 1989). The high carbonate percentages indicate that there must have been a sufficient supply of calcium ions to produce the calcareous sediments.

As shown on the Geological Map of Egypt, scale 1: 500,000 chalcedony deposits developed only along Darb El-Arbain to the south of Kharga Depression (Fig. 11.1). Issawi (1971) was the first to record and map them as huge chalcedony sheets, with a milky to earthy white color, forming a hummocky surface, which rises up to 30 m above the surrounding surfaces. The chalcedony was found in the form of nodules, aggregates, and concretions ranging from few millimeters to 10 cm in diameter. The top surface of the sheets is covered with a thin film of hard chalcedony nodules, which covers a very soft powder that formed of fine granular aggregate chalcedony and clay. Issawi (1971) also found that the chalcedony sheets are aligned in a north–south direction perpendicular to the main east–west structural lines, which characterize this part of the Western Desert. Issawi (1971), therefore, concluded that the occurrence of chalcedony sheets is mainly related to structural features (faulting, up-arching movements, or basaltic eruptions). Along the structural lines, silica solutions rose up and spread over the surrounding surfaces to form chalcedony sheets by the dehydration crystallization of silica gels. Chalcedony patches were also recorded in several basins in Nabta region as nodules ranging from a few millimeters to as big as 20 cm in diameter in extreme cases, and they occur in a matrix of maroon kaolinite mixed with clay which is rich in soluble non-carbonate calcium and sodium salts (Haynes 1980). According to Haynes (1980), chalcedony seems to have been formed in standing bodies of water rich in sodium carbonates and silica having a pH of 9.5 or higher, and that salt and silica formed as evaporites in basins.

Salt and gypsum sediments are less frequent than the abovementioned sediments that are chemically precipitated. Salt has been recorded as surface crusts in several localities associated with playa clastics. Areas mentioned in previous studies are Miramar playa atop the plateau to the north of Dakhla (Brookes 1989, 1993), Abu El-Egl playa, and El-Zayat playa (Ashour et al. 2005). Salt crusts were found to consist of white to yellow salts with a thickness which does not exceed a few centimeters. All studies concluded that the salt crust developed through capillary action. Gypsum has been reported from both Abu El-Egl playa and

Fig. 11.8 White gravel embedded in fine sediments exposed by wind action in Bir El-Gabal playa, Dakhla

El-**Zayat playa** (Brookes 1993; Ashour et al. 2005). Brookes (1993) referred to the sediments of these playas as Pleistocene Lacustrine Sediments, since its age determination by Th/U yielded 61.98 ± 7.6 ka, and the sediments were deposited in saline lakes. OSL age dating of a sample from **Abu El-Egl playa** was carried out in the dating laboratory at Helsinki University (Ashour et al. 2005). This sample, composed of quartz-rich sand from the base of the sediments, was found to be of the Pleistocene age, which is 110 ± 60 ka. Radiocarbon dating of some pieces of ostrich egg shells found on the top of the sediments gave an age of 20,580 ± 280 years, indicating and confirming that these sediments are over 20,000 years old. Furthermore, nearly all artifacts found around this playa are from the older **Middle Paleolithic period** (Ashour et al. 2005). The section is mainly composed of gypsiferous clay, marl, dolomitic bands, gypsum, siltstone, and anhydrite, with a thickness varying between 8 and 75 cm. It has been concluded that these sediments developed in a lagoon environment.

Organic Sediments: These sediments are very rare since the environmental conditions necessary for their development occur only in a few localities. In the Kharga-Dakhla Depression, they have been recorded in only one locality, as a thin (1–2 cm) black organic bed in Abu El-Egl playa (Ashour et al. 2005). It seems that organic beds of Abu El-Egl represent the algal mats which develop in shallow lagoonal waters characterizing lagoon margins.

11.5.1.3 Cycles of Sediment Deposition

Seasonal and Short-Term Cycles: Playa sediments exhibit several features, which indicate several cycles of deposition. In some playas, there are desiccation cracks and clay curls on the upper surface of silt layers indicating seasonal drying up. These phenomena have been observed in the playa of Bir El-Gabal in Dakhla and have also been recorded in the **Umm El-Dabadib playa** in northern Kharga (Hamdan 1987). Cut and fill structures which have been observed in the playa sediments of the El-Zayat plain and Abu El-Egl (Embabi 2004) and in Umm El-Dabadib (Hamdan 1987) may indicate short-term cycles of erosion and deposition during the formation of playas.

Long-Term Cycles: The stratigraphy of some playas also reveals features indicating that there were several long-term cycles of playa formation during episodes of increased rainfall separated by periods of wind erosion during arid phases. Features include unconformity surfaces and layers of aeolian sands and dunes, which normally separate fluvial sediments of silt and clay. In the **North Kharga playa** (Sherika playa), two surfaces of unconformity have been recorded at the base of yardang, separating the upper layer from the lower ones which indicate two periods of wind erosion during two arid phases (Embabi 1968–1969). In **Umm El-Dabadib** at the foot of the northern escarpment of the Kharga Depression, Hamdan (1987) recognized two phases of playa development: An old playa (Late Pleistocene or Early Holocene), and a younger (Holocene) playa, but without mentioning any basis of classification or describing features separating the two phases. Nevertheless, the detailed micro-stratigraphy of playa deposits of several sections shows the presence of some aeolian sand layers with abrupt boundaries, indicating phases of fluvial interruption.

In the Dakhla Oasis region, Brookes (1989, 1993) recognized two major types of lacustrine and playa sediments. The first type is named Pleistocene Lacustrine Sediments. These are deposited in basins but have various facies due to changing environment of deposition. These sediments consist of massive pale-brown muddy sands, sometimes with couplets of small-crystal gypsum, clastics, and large-crystal gypsum that include 10–20-cm beds of biogenic lacustrine marl, and sometimes with isolated low-amplitude sandy bars and dunes with reverse beds. As mentioned earlier, a sample of the lacustrine marl from Abu El-Egl playa yielded a Th/U isochron age of 61.98 ± 7.6 ka (Brookes 1993), whereas a thermoluminescence (TL) dating of a quartz-rich sand sample from the base of the playa sediments gave an age of 160 ± 60 ka. Another type of Middle Pleistocene playa sediments termed Calcareous Silty Sediments (CSS) is also found in some basins in southeast Dakhla (Churcher et al. 1999). CSS are relatively thick-bedded and massive, varying from silts and clays, or marls, to sandy silts and contain large diameter (>10 mm) stem casts of "reeds" suggestive of lacustrine deposition rather than a seasonal pan. The age of the CSS faunal assemblage is estimated to be at 200–300 ka and has been found consistent with the conclusion that they are from the Middle Pleistocene age based on elevations and field relationships (Churcher et al. 1999). The second type of sediments is called Early Holocene playa. The playa sediments consist of horizontally stratified pale-brown silty clay, with variable and usually minor sand layers intercalated by several limestone gravel beds. A sample of charcoal from Early Neolithic occupation site yielded a radiocarbon age of 8170 ± 110 BP.

11.5.1.4 Yardangs

In many areas, the playas have been extensively eroded and lowered with numerous **yardangs** as evidence of the once higher playa surface. These streamlined erosional ridges have long been known, not only in the Western Desert of Egypt, but also in most of the major deserts of the world (Embabi 1999; McCauley et al. 1977) and they typically occur in large fields (Fig. 11.9). Yardangs of the Kharga Depression were first mentioned by Ball (1900), Beadnell (1909), and Bagnold (1939) and were described as "mud lions" or "sitting sphinxes" or simply as hummocks. Yardangs are less frequent in Dakhla than in Kharga and are absent in the **Zayat plain**. There has been no interpretation

Fig. 11.9 Field of yardangs in Bir El-Gabal playa in Dakhla

of this phenomenon in previous works, but it is possible that the high percentage of gravel in sediment composition in Zayat playa made local conditions not in favor of yardang development. Also, the sediments are compact and devoid of a system of cracks or joints conformable with the wind system, resulting in unfavorable conditions for the development of yardangs. A third environmental aspect in Abu Tartur and some areas in Dakhla is the absence of sand (the eroding tool) in winds occurring in the playa areas which are devoid of yardangs.

Morphometric Characteristics: Detailed measurements have been made on yardang dimensions in several localities in the Western Desert. These localities in the **Kharga-Dakhla lowland** are as follows: the Umm El-Dabadib playa at the foot-lopes of the northern escarpment of Kharga (Hamdan 1987), in a playa 3 km to the north of Kharga town (Goudie et al. 1999), **Sherika playa** 30 km to the north of Kharga Town, and **Bir El-Gabal playa** about 30 km to the west of Mut. The length varies between 3 and 30 m in the North Kharga playa and 60–65 m in the Umm El-Dabadib and Bir El-Gabal playas, respectively. The width varies between 0.5 and 2.5 m in North Kharga playa, and the height between 0.5 and 5.0 m in North Kharga playa and 11 m in Bir El-Gabal in Dakhla. This size of yardangs has been classified as meso-yardangs in the recent literature on arid geomorphology (Cook et al. 1993). In general, it was found (Embabi 2004) that as the length of yardangs increases, they get wider and higher. Although correlation coefficients show that there is a positive relationship between those three parameters, the relationship is relatively weak between the height and the other two parameters (length and width) in all playas. When height is plotted against length and width, and width against length, it can be seen that the relationship is nonlinear between height and the other two parameters and is linear between length and width. This type of relationship was found in the sample (50 yardangs) of Goudie et al. (1999) 10 km to the north of the town of Kharga.

These positive relationships indicate that yardangs pass through a cycle of development, whereby their size becomes smaller by time. In a previous study in the Rogers Lake in California, the mean ratio between the width and length of yardangs was taken to indicate their stage of development (Ward and Greeley 1984). Most of the isolated (mature) yardangs in Rogers Lake have a width/length ratio of about 1:4 as do some yardangs in Arizona, Peru, Iran, and South Africa. It has also been found that this ratio is similar to that of the equilibrium form of a body immersed in moving fluid (Ward and Greeley 1984). When a comparison was made with the mean ratios of the fields of yardangs of the Kharga-Dakhla lowland, it was found that only two fields at Kharga achieved mean ratios more or less similar to the equilibrium (mature stage) form, while that of **Bir El-Gabal playa** indicated that the yardangs were still in an early stage of development.

Fig. 11.10 Examples of shapes of yardangs: (**a**) Dome-like shape in Farafra, (**b**) Streamlined shape, north Kharga playa, (**c**) Irregular shape, Bir El-Gabal playa, (**d**) Sphinx-like shape, Bir El-Gabal playa, Dakhla

Shape

Previous studies found that the most recurring shape of playa yardangs is the asymmetric streamline or the linear with a steep windward face and a gentle lee slope. This is why yardangs are described in the geomorphic literature as streamlined erosional ridges such as those in Fig. 11.10. Irregular or dome-like shapes are common in some fields of playa yardangs (Fig. 11.10). There are only three fields where yardangs tend to take a streamline shape (**Bir El-Gabal**, North Kharga, **Umm El-Dabadib**), whereas those of the other fields tend to deviate from the streamline form taking dome-like or irregular shapes. Linear shapes are found in environments with a unidirectional wind regime as in Kharga and Dakhla (Embabi 1968–1969; Hamdan 1987; Goudie et al. 1999). Very rarely, some linear yardangs might acquire certain shapes like the sphinx (Fig. 11.10). The windward face of many large yardangs is very steep (Fig. 11.10). The leeward ends of many yardangs are tapered and are lower in general than the windward side (Fig. 11.10).

This data shows that the predominant azimuth of the long axis of the yardang in the Umm El-Dabadib playa (Kharga) is 350°–360°, and 345°–360° in the North Kharga and Bir El-Gabal (Dakhla) playas. The azimuth of other yardangs in Kharga-Dakhla lowland is conformable with the wide unimodal northern wind regime.

Normally, yardangs exhibit undercutting by wind abrasion on the windward face and at the base of the sides, but less pronounced than on the face. Only in Farafra, where removal of the playa sediment exposed the original floor of the basin, undercutting of the windward face exposed the soft chalk bedrock. Occasionally, yardangs might exhibit wind grooves on the upper parts of the yardang sides and on their tops. In some fields, a thin surface layer of silt and clay covers the sloping sides of the yardangs. This surface layer is mostly formed due to the alternation of wetting and drying by dew or drizzle which occurs occasionally. In some fields, the surface layer looks like a crust and is dissected by rills that seem to be active during heavy rain storms which occur in winter or spring every 5–10 years. Collapsed blocks can be seen along the undermined sections, especially at the base of the windward face. As undermining proceeds by time, collapsing of the overhang is repeated until the yardangs look like pedestals, and in a later stage, the whole yardang collapses and can be seen in the field exposed to wind abrasion.

11.5.2 Sand Dunes

Sand dunes are present in Dakhla and Kharga only due to their location at the end of the sand flow of both the **Farafra Sand Sea** and the **Ghard Abu-Moharik Sand Sea**. In Kharga lowland, the sand dunes represent the Back Erg Section of Abu Moharik Sand Sea as will be mentioned in Chap. 14. Therefore, their characteristics will not be discussed here, since they will be treated in Chap. 14.

In the Dakhla lowland, three groups of barchan belts, transverse, and linear dunes have developed in the western part of the Depression (west of Mut longitude), representing the southern extension of the Farafra Sand Sea. As in Kharga, the western belt is the largest due to its location along the flow of the main sand stream. In general, dune belts are mostly oriented north–south, and then gradually change to SSE at their southern extension as wind flows downwind. Here also, wind direction, sand supply, and local relief are the main controlling factors of dune form, density, and orientation of its belts. In addition to dune belts, sand sheets cover various parts of the Depression Floor; the most extensive sand sheet is that in the southwest.

11.5.3 Spring Mounds

Spring activity is relevant to playa development, since many lakes owe their existence totally to spring discharge. Evidence of old spring activity is known in many localities in the southern part of the Western Desert from geomorphologic evidence and in areas that have no source of water today. Most important feature, the result of this activity, is **spring mounds**. These appear in the form of small hillocks, mostly conical in shape. They range in height from a few to 20 m. Each mound has a vent that is cylindrical in shape and is surrounded by layers of clay, ocher, sand, or carbonates, depending on the composition of the oozing waters of the spring (Said 1980).

Spring mounds are micro features associated with artesian ground water discharge. Their height varies between 2 and 25 m, with modal values of 15–25 and 6–2 m (Brooks 1993; Kleindienst et al. 1999). Since no active springs are present in Kharga-Dakhla today (Kleindienst et al. 1999), all spring mounds were built by defunct artesian springs, representing remnant erosional forms around fossil spring vents. No attempt has been made to know their number, but there are hundreds of ancient spring mounds in Kharga and Dakhla, and most of them are generally related to sandstone–mud–stone contacts (Brookes 1993; Kleindienst et al. 1999).

In the meantime, they are clustered; a result of continual discharge through localized fractures. Their lithology and structure appear to be uniform, characterized by internal structure of loose, unstratified, greyish-white fine sand, and silt with rare or common sandstone blocks, with horizontal beds near the summit that was probably deposited in impounded pools (Brookes 1993). Artifacts found in spring mounds suggest several generations of fossil spring activity, since the oldest spring material contains Upper Acheulian tools in western Dakhla, and there was artesian spring activity in Mid-Holocene times (Kleindienst et al. 1999).

Examples of **fossil spring mounds** are recorded in Kharga, Dakhla, and the extreme southern part of the Western Desert. In the northern part of the Kharga lowland, about 10 km NE of Kharga village, Caton-Thompson and Gardner (1932) recorded a group of crater-like spring mounds with concave slopes on the top of low ridges. One of these mounds, called Gabal Kharran, is 0.5-km long, 0.3-km wide, and 5.0-m high. The sediments are composed of an alternating series of loam, clay, and sand in all mounds, except one where gravel was found beside the fine sediments. Caton-Thompson and Gardner (1932) presumed that the origin of this gravel is in the eastern scarp of the Depression and was washed down 18 km on the Depression Floor. The thickness of sediments in most of the mounds is more than 9 m. These sediments do not show any kind of stratification, but are contorted, flung up chaotically from the underlying **Nubian sandstone**, indicating that they represent times of great activity when the underground water burst through the rock floor along the lines of structural weakness. Caton-Thompson and Gardner (1932) attributed to these active episodes the accumulation of sandy silt deposits (probably in pools or small lakes) on the low ground at the base of the mounds. These deposits were subsequently shaped into yardangs by wind action. They found that the deposits of the different mounds are not of the same age, implying that spring activity happened at different times. The collected tools from the area cover a long span of Paleolithic and Neolithic times.

11.5.4 Hills and Mesas

Hills in the Kharga-Dakhla Depression have three different origins. The first hills are **anticlinal/synclinal hills**, the second group are **outlier hills** detached from the adjoining plateaus, whereas the third group are **granite inliers**.

Anticlinal/Synclinal Hills: This group of hills has developed only in Kharga. They are small anticlines and synclines plunging in one or two directions with axes extending in a NNE-SSW direction in the northern part and in an ENE-SSE direction in the southern part of the Depression. These hills rise between 45 and 100 m above the surrounding surface of the ground. Their length varies between 0.5 and 5.0 km and their breadth between 0.25 and 2.0 km. Except for **Gabal Tarawan** which is a synclinal hill, all other hills are conformable with anticlinal structure. Representing a reversed topography, Gabal Tarawan is characterized by its Palaeocene white chalk core which is surrounded by circular ridges composed of the Exogyra beds of the Maastrichtian period (Embabi 1967). Due to quarrying, the chalk core of **Gabal Tarawan** is exposed to gradual reduction. Lying just to the south of **Gabal El-Ter**, this synclinal hill is the first one in a series of anticlinal hills along the central structural line in the northern part of the Depression (Fig. 11.1). The largest hill along this line is **Gabal El-Qarn** (Horn) whose highest point is about 90 m above the surrounding surface of the ground and extends 5.0 km from north to south and 2.0 km from east to west. Other hills of the same origin occur at the northeast corner of the eastern scarp (**Gabal El-Yabisa** and **Gabal El-Aguze**), **Gabal El-Nadura**, and **Gabal El-Qalaa** along the line of Tarawan El-Qarn, and at the southern part (**Gabal Dush**, **Gabal El-Dibabah** and **Gabal Qarn "Paris"**). Due to the presence of small bounding faults, most of these folded hills exhibit asymmetrical sides.

The Outlier Hills and Mesas: This group is concentrated in Kharga, since there is only one outlier in Dakhla, which is a small mesa called **Gabal Edminston**, 472 m a.s.l (Fig. 11.2). Most of these hills and mesas are located in the northern part of Kharga. Two of these hills (**Gabal Ghanima** and **Gabal Umm El-Ghanaiem**) lie just to the west of the eastern escarpment, whereas **Gabal Taref**, **Gabal El-Sheikh**, and **Gabal El-Ter** lie in the central part of the Depression, about 30 km to the south of the northern escarpment (Fig. 11.1). The size and form of these hills are controlled by local structures. In the case of Gabal Taref, which is the largest outlier (10 × 3 km), due to its location in the path of sand flow of the main dune belt in Kharga, dune sand climbs up its northern slope to its chalk top, then descends on its southern slopes. This process has resulted in the development of two features. The first is that climbing and descending dunes cover the northern and southern slopes of this mesa. The second feature is that moving sand has dissected the top surface of Gabal Taref into yardangs similar to those developed on the plateau surface lying to the north of the Depression.

The Granite Inliers: The third group of hills are the **granite inliers**, which project from the Nubian formations, and are concentrated in the extreme south of the Kharga Lowland. The most famous and prominent ones are the three hills that are named **Gabal Abu Bayan El-Bahari** (164 m), **Gabal Abu Bayan El-Wastani** (250 m), and **Gabal Abu Bayan El-Qibli** (270 m). Their relative height varies between 60 and 75 m above the surrounding ground. Their width varies between 1.0 and 3.5 km, whereas their length ranges between 1.0 and 5.0 km. Their slopes are characterized by gradual convex rather than angular change in contrast to the tilted sandstone overlying beds (Fig. 11.11).

Fig. 11.11 Granite hill overlain by dipping Nubian sandstone in south Kharga

11.6 Conclusion

Taking into consideration its geological aspects and location, the Kharga-Dakhla Depression is the oldest and most southern Depression in the Western Desert. It started to develop after the retreat of the Lower Eocene Sea northward from approximately 50 Ma. It is expected that the development was initiated by erosion of the southern margins of geologic formations along the contact of beds. Under the attacks of erosion and the initiation of cuesta escarpments, retreat of cuesta fronts continued in different directions according to local geologic structures, leaving behind several blocks as outlier hills and mesas. The farthest outlier detached from the Eocene Plateau is the hill called Barqat El-Shab (370 m), lying at about 60 km south of the edge of the Sinn El-Kaddab Plateau. This indicates that the escarpments of the plateau retreated northward approximately 60 km in 50 Ma, i.e., 1.2 km each million years. It also indicates that the present outliers lying at the floor of the Kharga-Dakhla Depression near the edges of the plateaus (Edminston in Dakhla, Taref, El-Ter, Ghanima, and Um El-Ghaniem in Kharga) few were only detached in the last million years. The final stage in the development of this Depression can be determined by the age of the oldest superficial sediments laid down on the escarpment slopes. The oldest tufa deposits found on Kharga escarpments were formed >450 ka ago (Sultan et al. 1997; Hamdan et al. 1999), indicating that the escarpment alignment and present-day position were acquired from then. Although no data is available for Dakhla and Abu Tartur, the age of Kharga can be applied based on the two sub-regions of the Kharga-Dakhla Depression, since they were developing under the same climatic conditions.

As the escarpments were retreating, the floor of the Kharga-Dakhla Depression was developing as an erosional undulating peneplain, with local variations due to differences in geologic characteristics and some local climatic aspects. More resistant areas to erosion will continue to rise at a higher level than the surrounding plain as hills or mesas of different sizes and origins. Also, as escarpments are retreating and the Depression Floor is expanding, local erosional and depositional forms continue to develop on the surfaces of the Depression under changing climatic conditions. During relatively wet phases, wadis developed on the steep escarpments, and when runoff reached the foot slopes, it changed to a braided flow depositing its coarse load at the piedmont surfaces as fans and gravel sheets. When runoff continued for a further distance, it ended in low-lying basins, forming temporary lakes (playas) depositing the rest of its load, which was mostly composed of fine sediments. During these relatively wet phases, the water impregnated with calcium carbonate along the escarpments and characterized by slow flow tufa, were deposited as sheets or terraces. During arid climatic phases, wind action dominated the scene, deflated fine material from fluvial sediments, and eroded the dried up sediments of the playas, creating erosion surfaces which were buried by newer sediments during other wet phases. Yardangs seem to develop during the present-day arid phase, since their orientation is conformable with the direction of the prevailing winds. Sands of dune belts of southern section of Ghard Abu Moharik in Kharga and those of Farafra Sand Sea in Dakhla were brought over by wind during arid phases.

With the exposure of Nubian sandstone at the floor of the Kharga-Dakhla Depression during the retreat of the escarpments, underground water was discharged naturally from the near-surface Nubian aquifer as springs. With the onset of arid conditions, people roaming in the Western Desert found in this Depression and other mega-depressions a permanent water source. Gradually, the number of settlers increased. The oldest artifacts found in Kharga and Dakhla can be traced back to the Acheulian period of Upper Palaeolithic (Caton-Thompson 1952; Kleindienst et al. 1999).

References

Ashour M (1974) Abu Tartur Plateau, the Western Desert of Egypt. Dissertation, Ain Shams University

Ashour MM, Embabi NS, Donner J, AbuZeid KA (2005) Geomorphology and Quaternary geology of Abu El Egl Playa, Western Desert of Egypt. Bul Egypt Geogr Soc 78:1–26

Bagnold RA (1939) An expedition to the Gilf Kebir and Uwienat. Geogr J 93:281–313

Ball J (1900) Kharga Oasis: its topography and geology. Geological Survey Department, Cairo, Egypt

Beadnell H (1909) An Egyptian Oasis (Kharga). John Murray, London

Brook G, Embabi N, Ashour M, Edwards R, Cheng H, Cowart J, Dabous A (2003) Quaternary environmental change in the Western Desert of Egypt: evidence from cave speleothems, spring tufa and playa sediments. Z Geomorph NF, Supl. 131:59–87

Brookes IA (1989) Early Holocene basinal sediments, Dakhleh Oasis region, south central Egypt. Quat Res 12:139–152

Brookes IA (1993) Geomorphology and Quaternary geology of the Dakhla Oasis Region, Egypt. Quat Sci Rev 12:529–552

Busche D (2001) Early Quaternary landslides of the Sahara and their significance for geomorphic and climatic history. J Arid Environ 49:429–448

Caton-Thompson G (1952) Kharga Oasis in prehistory. Athelon Press, London

Caton-Thompson G, Gardner E (1932) The Prehistoric geography of Kharga Oasis. Geogr J LXXX: 369–409

Churcher C, Kleindienst M, Schwarcz H (1999) Fuanal remains from a Middle Pleistocene lacustrine marl in Dakhleh Oasis, Egypt: Palaeoenvironmnetal reconstructions. Palaeogeogr Palaeoclimatol Palaeoecol 154:301–312

Cooke R, Warren A, Goudie A (1993) Wind erosion. In: Cooke R, Warren A, Goudie A (eds) Desert geomorphology. UCL Press, London, pp 290–306

Donner J, Ashour MM, Brook GA, Embabi NS (2015) The Quaternary history of the Western Desert of Egypt as recorded in the Abu El-Egl playa. Bull Soc Geog d'Egypte 88:1–18

Embabi NS (1967) A geomorphologic study of Kharga Oases Depression, the Western Desert, Egypt. Dissertation, Bristol University

Embabi NS (1968–1969) The semi-playa deposits in the Kharga Oases Depression. Bull Soc Géog d'Égypte 41–42:73–87

Embabi NS (1999) Playas of the Western Desert, Egypt. Ann Acad Sci Fennicae, Geologica-Geographica 160:5–47

Embabi NS (2004) Geomorphology of Egypt: landforms and evolution, vol I. The Egyptian Geographical Society, Cairo, The Nile Valley and the Western Desert

Goudie A, Stockes S, Cook J, Samieh S, El-Rashidi O (1999) Yardang landforms from Kharga Oasis, south-western Egypt. Z Geomorph N F Suppl-Bd 116: 97–112, Aeolian Geomorphology: Papers from 14th international Conference on Aeolian Research, Oxford, 1998

Hamdan G (1960) The personality of Egypt, part I: The physical personality, (in Arabic) Alam El-Kotob, Cairo

Hamdan M (1987) Geomorphology and Quaternary geology of Umm El-Dabadib Area, Kharga Oasis. Dissertation, Cairo University

Hamdan M, Mahmoud, A, Hassan F (1999) Geology of the tufa deposits in the north-eastern corner of Kharga Oasis, Western Desert, Egypt. In: Hafez, A (ed) in Proceedings of the 4th International Conference on the Geology of the Arab World (GAW4), Cairo University, Cairo, 1998

Haynes C (1980) Geological evidence of pluvial climates in the Nabta area of the Western Desert, Egypt. In: Wendorf F, Schild R (eds) Prehistory of the Eastern Sahara. Academic Press, New York, pp 353–371

Hermina M (1990) The surroundings of Kharga, Dakhla and Farafra Oases. In: Said R (ed) The geology of Egypt. Balkema, Rotterdam, pp 259–292

Issawi B (1971) Geology of Darb El-Arbain, Western Desert. Ann Geol Surv Egypt 1:53–92

Issawi B, El-Hinnawi M (1984) Contributions to the study of the playas in Southeastern part of the Western Desert of Egypt. In: Angela E Close (ed) Cattle Keepers in the Eastern Sahara: the Neolithic of Bir Kiseiba. Southern Methodist University Dallas, pp 41–48

Kleindienst M, Schwarcz H, Nicoll K, Churcher C, Frizano J, Giegengack R, Wiseman M (1996) Pleistocene geochronology and paleoclimates at Dakhla Oasis and Kharga Oasis, Western Desert, Egypt based upon uranium-thorium determinations from water-laid tufa. Nyame Akuma, Bul. Soc. Africanist Archaeologists 46: 96

Kleindienst M, Churcher C, McDonald M, Schwarcz H (1999) Geography, geology, geochronology and geoarchaeology of the Dakhleh Oasis region: An interim report 1. In: Churcher C, Mills A (eds) Dakhlah Oasis Project: Monograph 2, Reports from the Survey of the Dakhlah Oasis, Western Desert of Egypt. 1977–1987, Oxbow

Mahmoud AA, El-Gohary AM, Abou El Anwar EA (2002) Geology of Quatewrnary Tufa deposits in North-East of the Dakhla Oasis, Wester Desert, Egypt. Sedimentol Egypt 10:89–101

McCauley JF, Grolier MJ, Breed CS (1977) Yardangs of Peru and other desert regions. US Geol Survey, Interagency Report: Astrogeology 81

McCauley JF, Breed C, Schaber CV, Haynes CV, Issawi B, Grolier MJ, El-Kilani A (1986) Palaeodrainages of the Eastern Sahara. The radar rivers revisited (SIR-A/B Implications for a Mid-Tertiary Trans-African Drainage System) IEEE Trans. Geosci Remote Sens GE 24:624–648

McKenzie J (1993) Chemical stratigraphy of lacustrine sequences deposited during the latest Pleistocene Periods in the Eastern Sahara. In: Wendorf F, Schild R, Close A, Associates (eds) Egypt during the Last Interglacial: The Middle Paleolithic of BirTarfawi and Bir Sahara East. Plenium Press, New York, pp 106–114

Said R (1980) The Quaternary sediments of the south Western Desert of Egypt: an overview. appendix I. In: Wendorf F, Schild R (eds) Prehistory of the Eastern Sahara. Academic Press, New York, pp 281–289

Said R (1990) Quaternary. In: Said R (ed) The geology of Egypt. Balkema, Rotterdam, pp 487–507

Shata A (1953) New light on the structural development of the Western Desert of Egypt. Bull Institut du Desert d'Égypte 3:101–106

Shata A (1959) Geological problems related to the ground water supply of some desert areas of Egypt. Bull Soc Géogr d'Égypte 32:248–262

Schild R, Wendorf F (1984) Lithostratigraphy of Holocene Lakes along the Kiseiba Scarp. In Angela E Close (ed) Cattle Keepers of the Eastern Sahara: The Neolithic of Bir Kiseiba. Southern Methodist University, Dallas, pp 9–40

Sultan M, Struchio N, Hassan F, Hamdan M, Mahmoud A, El-Alfy Z, Stien T (1997) Precipitation source inferred from stable isotopic composition of Pleistocene groundwater & carbonate deposits in the Western Desert of Egypt. Quatern Res 48:29–37

Ward AW, Greeley R (1984) Evolution of the yardangs at Rogers Lake, California. Bull Geol Soc Am 95:829–837

Wendorf F, Schild R, Close A, Associates (1993) Egypt during the last interglacial: the middle Palaeolithic of Bir Tarfawi and Bir Sahara East. Plenium Press, New York

12 The Fayum Depression

Abstract

Among other mega-depressions of the Western Desert of Egypt, the Fayum Depression is the nearest to the Nile Valley. Only a 1.5–10 km wide divide separates them and facilitates the entrance of the Nile water into the Depression. This makes Fayum unique among other mega-depressions in the Western Desert. The Depression is excavated in a wide dome where it is composed of Eocene and Oligocene strata that are capped by volcanic materials. Miocene sediments are absent, since the Fayum dome was higher than the level of the Miocene Sea. Pliocene deposits are all continental and fluvial in origin. The Pleistocene deposits are all of Nile origin as the Nile waters entered into the Depression via the Hawara Channel in Early Palaeolithic, which resulted in the development of several lakes. With the intervention of man during the Ptolemaic Period, water entering the Depression was regulated according to the needs of agriculture, resulting in the shrinkage of the lake until it stood at its present-day level (45 m b.s.l). Since then, new forms have appeared at the Depression floor. Beside the lake beaches, the inverted wadis on the Nile-Fayum divide, and the Hawara Channel, which came into being during pre-Ptolemaic times, new forms were created; such are Qarun Lake, the Fayum Delta built of the Nile Silt, and the canyon-like wadis.

Keywords

Fayum Depression • Hawara channel • Bahr Youssef • Qarun Lake • Gabal Qatrani • Pliocene inverted wadis

12.1 Introduction

This Depression is unique among other Depressions of the Western Desert in that the Nile waters obtained an access to the Depression through the Hawara channel, which breaches the Nile-Fayum divide, converting it into a lake of considerable dimensions in certain intervals of the Quaternary. Thus, it has affinities with the closed mega-depressions on the one hand, and with provinces of the Nile Valley and Delta, which are fed with Nile waters, on the other hand. The Depression is bound from the north by successive cuesta escarpments (Fig. 12.1), which rise up to about 350 m a.s.l and run in an east-west direction, but they change their direction to the southwest and to the south, delimiting Wadi El-Rayan from these two directions. The lowest point at the Depression floor is 45 m below sea level and lies at Qarun Lake. It is separated from the Nile Valley in the east by a narrow divide (1.5–10 km wide) that extends from NNE to SSW, rising up

N.S. Embabi, *Landscapes and Landforms of Egypt*, World Geomorphological Landscapes,
https://doi.org/10.1007/978-3-319-65661-8_12

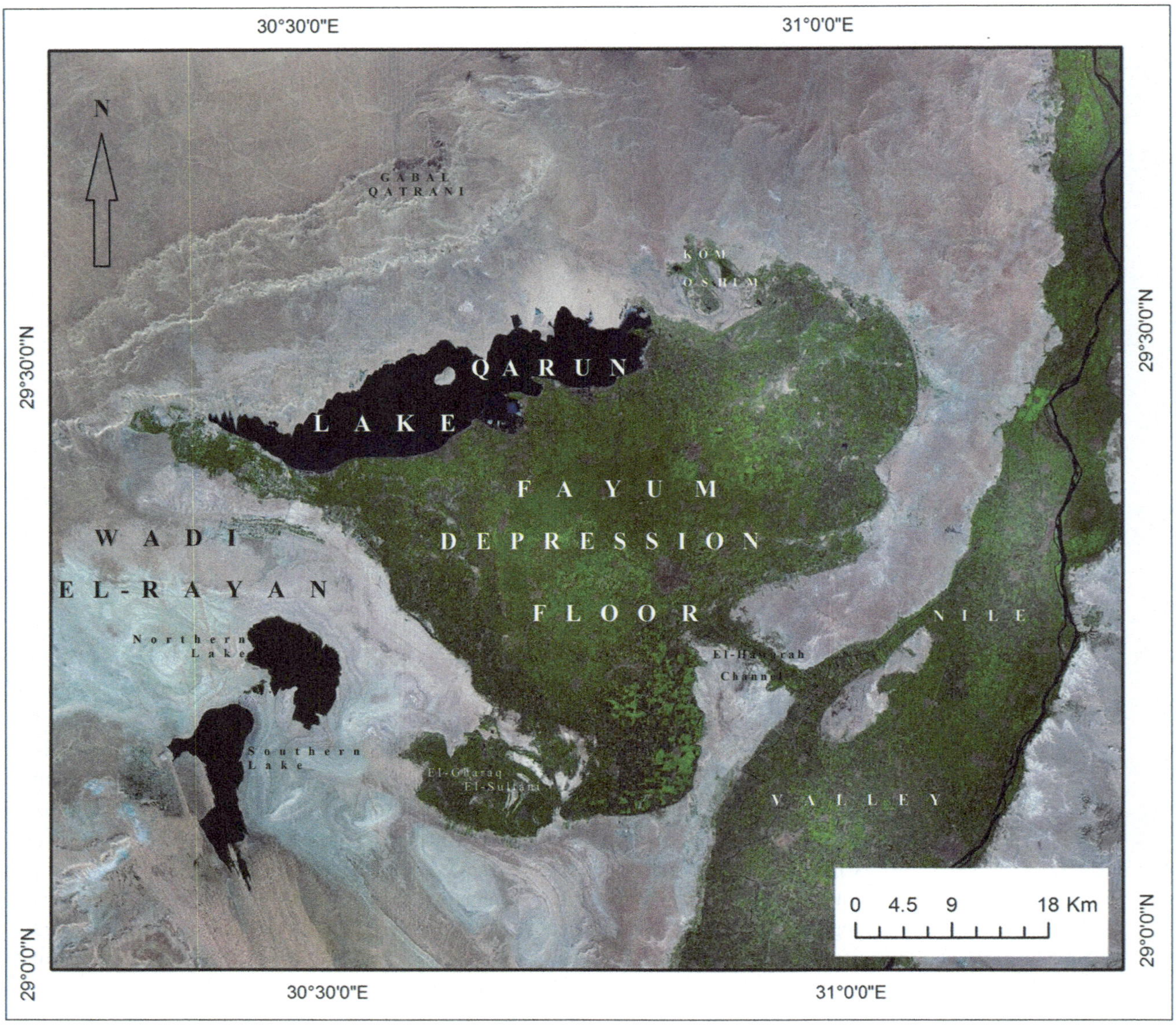

Fig. 12.1 The Fayum Depression and its environs (*Source* Landsat-TM Image)

to about 160 m asl. In the west and southwest, the Depression is separated from Wadi El-Rayan by faint relief that rises about 30–45 m asl. Characteristic landforms of the **Fayum Depression** are the bounding cuesta escarpments, the Hawara Channel, the inverted wadis along the Nile-Fayum divide, the lake beaches, the Fayum Delta, Qarun Lake, and the canyons. Before dealing with these forms, a brief summary on the geology of the Fayum Depression is presented.

12.2 Geology

This Depression is composed of a succession of sedimentary rocks belonging mainly to the Eocene and Oligocene. These rocks are overlain by some younger deposits of Pliocene, Pleistocene, and Holocene age. In fact, the Depression is excavated in the Middle Eocene formation known as "Mokattam Group," composed mainly of limestone, some shale, and sand intercalations. They are exposed at margins of the Depression and the lower sections of the sides of the deep wadis (canyons/ravines) at the Depression floor to the south of **Qarun Lake**. To the north of this lake, the upper part of the Eocene known locally as "Qasr El-Sagha Formation" forms the lowest escarpment followed by a series of scarps that are composed of "the Oligocene Gabal Qatrani Formation," separated by relatively flat land. This formation is composed of unconsolidated gravels and sands, variegated green and red clay, and limestone at frequent intervals, with fresh water fossils. Many of these sediments are point bar deposits of one or several major rivers, which flowed in Early Oligocene times from the highlands in the east toward a large southerly embayment of Tethys Sea (Bown and

Kraus 1988). The highest escarpment is called "**Gabal Qatrani,**" which is capped by a series of four basalt flows, the oldest of which is dated at 31 ± 1.0 million years (Simons and Rasmussen 1990).

The Miocene sediments are absent in Fayum. Most probably, the area of the Depression was structurally high as a wide anticline of Middle Eocene carbonate sediments dipping outward with low angles (Abul-Haggag 1967).These conditions facilitated the work of chemical processes by the rain water during the Miocene, converting the high anticline to a topographic low. This explains the presence of Pliocene sediments along the northern and eastern margins of the Fayum Depression. Pliocene deposits are of fluvial continental origin, especially those on the Nile-Fayum divide representing the relics of wadi beds of the drainage system cutting across this divide (Sandford and Arkel 1929).

Most of the Pleistocene deposits resulted from the entrance of the Nile waters into the Fayum Depression since the Early Palaeolithic. They occupy the Depression floor to the south of **Qarun Lake** and occur in several localities along its margins. In fact, this advent made the Fayum different from other mega-depressions in the Western Desert. The entrance of the Nile water not only caused the development of several lakes, but also contributed to the sedimentary infill of different origin, especially the Nile silt. This made it possible to cultivate the floor of Fayum Depression, converting it from a desert Depression to an integral part of the Nile system. Forms resulting from all these events will be discussed in the following section.

Tectonically, the Depression occupies the area of a major wide anticline, whose beds dip outward in all directions. This high anticline was acting as an obstacle that prevented the Miocene and Oligocene sediments to be laid down in Fayum. Oligocene sediments appear only in the extreme northern peripheries of the Depression, i.e., the northern margins of the anticline. It seems that this structure covered the adjacent **Wadi El-Rayan**, since the northern escarpment of Fayum extends westward acting also as northern boundary for Wadi El-Rayan. Small faults represent the second tectonic feature in the Depression. The Geologic Map shows that they spread all over the area of the Depression. Such faults were one of the significant factors that facilitated the work of chemical processes that acted on the anticline during the Oligocene and Miocene.

12.3 The Hawara Channel

The **Hawara Channel** is the water gap through which the Nile water—via Bahr Youssef—is transferred to the Fayum Depression. It cuts across the Eocene Nile-Fayum divide and extends for about 10 km from NW to SE, with a varying width of 1.0–2.0 km. This channel is filled with ancient clastics of marl, gravel, and sand which are topped by recent Nile mud. Since the surrounding surface is 23 m above sea level, and the original depth of the channel was 40 m below the surrounding surface of the ridge, the level of the original depth of the channel was −17 m. The geological cross section of the channel shows that the lower 25 m are much narrower (only 100 m wide) compared to the upper part (1100 m), with two terraces (20 and 10 m) separating the lower and upper parts (Fig. 12.2). It was thought first that

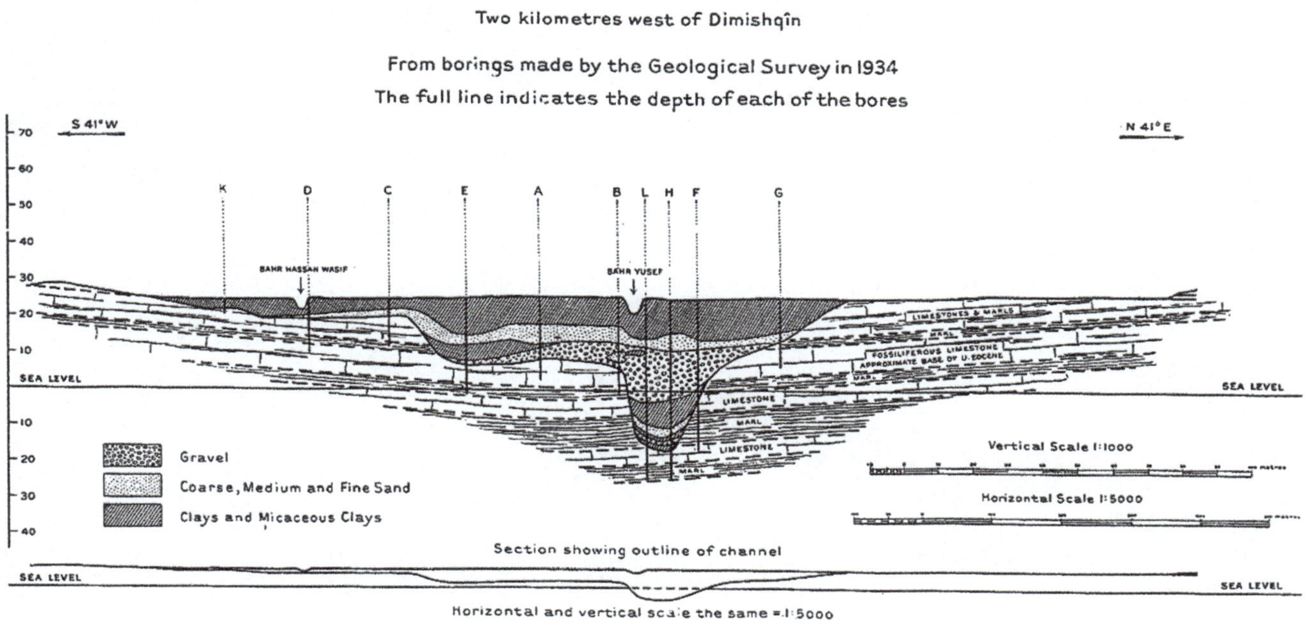

Fig. 12.2 Geological cross section across the Hawara Channel (*Source* Little 1935–1936)

this channel was developed as a gully which—under the erosive power of water flow—retreated backward until it breached the Nile-Fayum divide, allowing flood water, when reaching the height of the channel, to enter the Depression (Ball 1939). Abul-Haggag (1967) postulated that the Hawara channel developed by two rivers flowing into two opposite directions along a straight line, one toward Fayum and the other toward the Nile Valley, starting from a point approximately in the middle part of the divide. Due to the big difference in the base level of these two streams, the one that flowed toward Fayum (steeper and more powerful) was able to capture the other stream, by the action of headwater erosion, diverting its waters to Fayum and allowing the Nile waters to enter the Depression.

The work of Said (1981, 1993) confirmed the conclusions of Little (1935–1936) that the earliest connection of the Fayum Depression with the Nile seems to be that of the **Prenile** whose high water filled the Depression to an elevation of 44–46 m above sea level. According to the Geological Map of Egypt (scale 1: 500,000), relics of the Prenile deposits currently encircle the Depression at a constant elevation on all sides, especially along its NW rim. Archaeological and sedimentological evidence made it possible to conclude that the Prenile ceased to flow some 400,000–450,000 years ago, i.e., in the Middle Pleistocene, that is Early Paleolithic (Said 1993). Since the sand and gravel of the 43 m terrace of the Prenile are devoid of any archaeological material, it can be said that the connection of Fayum with the Prenile occurred before the appearance of the Paleolithic man in the Fayum or at the advent of the Paleolithic Period.

12.4 Bounding Escarpments

All maps, aerial photographs, and space images show that the Fayum Depression is bound from the north by several cuesta escarpments. Each one is separated from the upper one by a nearly flat surface, a few kilometers wide. The highest one rises to 300–350 m a.s.l. It is characterized by a steep face looking southwards toward the Fayum Depression, and a nearly flat back slope (Fig. 12.3). The escarpment is compound, since it is composed of several smaller ones. This escarpment is covered by the Gabal Qatrani basalt. Figure (12.3) also shows that the upper section of the escarpment is a vertical cliff, which is undermined. Several rock blocks were detached from the cliff lying with other smaller fallen rock fragments on the surface of the lower section of the slope of the escarpment. This case is similar to that one seen on the escarpments of Kharga-Dakhla

Fig. 12.3 The Northern Escarpment (1st from the north), showing the *top vertical*, undermined cliffs, fallen blocks from the undermined cliffs, and the nearly flat plateau

Depression (Chap. 11). Below this main escarpment, there are several others built of Oligocene and Eocene formations, of which the Eocene one called Qasr El-Sagha scarp is the most prominent one, rising to ~100 m a.s.l. The exposure of the Lower Eocene Formation of Qasr El-Sagha contains some large (1–2 m in diameter) rounded siliceous concretions lying to the north of Qarun Lake, representing a characteristic micro-feature when approaching the cultivated land along Cairo-Fayum Highway. This feature was first reported by Beadnell (1905). Two other forms characterize this region of escarpments. The first is the shallow wadis of the drainage nets cutting the escarpments from north to south. None of the wadis cuts all the escarpments from N to S, as most of them are restricted to the step where they started. The second feature is the aeolian sand accumulated as a narrow strip along the northern shores of Lake Qarun. Some of the sand is blown into the Lake forming short and small tongues into the Lake (Fig. 12.1).

12.5 Inverted Wadis

Along the Nile-Fayum divide, separate small exposures of Pliocene fluviatile gravelly deposits are capping several hills such as Gabal El-Naaloon and Gabal El-Rus. Sandford and Arkel (1929) interpreted these rounded fluviatile deposits as representing the remnants of previous beds of wadis, flowing eastward toward the Pliocene Nile Gulf. Abul-Haggag (1967) observed that the lower part of these deposits is derived from the Eocene limestone, while the upper section of the deposits is derived from the Oligocene clastics and silicified woods. In the meantime, slipped Eocene limestone blocks were found at the base of the lower part of these fluviatile sediments similar to those which were found in the Nile Valley, indicating that they slipped down from the sides of the Pliocene wadis. These characteristics indicate that the source area of these deposits is to the west of the present exposures, that is the Fayum area, and that the wadis flowed eastward to the Nile Gulf (Abul-Haggag 1967). This conclusion confirms that reached by Sandford and Arkel (1929).

Since these wadis were developed during the Early Pliocene, it can be concluded that they were exposed to modifications by erosion during Early Pleistocene. Dissection and modifications in slope direction of the Nile-Fayum divide made it possible to develop two drainage systems, one toward the Fayum and the second toward the Nile Valley. Most probably, the two streams which formed the Hawara channel were part of these drainage systems during Early Pleistocene. This means that the Hawara channel is younger than the Pliocene wadis of El-Naaloon and El-Rus.

12.6 Lake Beaches and Ancient Lakes

The presence of **lake-beaches** and lacustrine sediments at different levels in the Fayum Depression indicates that the connection with the Nile was interrupted and resumed several times since the first time in the Middle Pleistocene. As a characteristic form and as favoured sites for settlement by early man, several researchers (Sandford and Arkel 1929; Caton-Thompson and Gardner 1934; Ball 1939; Said et al. 1972; Wendorf and Schild 1976; Hassan 1986) studied these beaches. The most remarkable beach is the one locally called the **"Gisr El-Hadid"** (the railway embankment) which extends—uninterrupted—for a distance of about 50 km along the western side of the Fayum Depression from Madinat Madi in the south to the northwest tip of Lake Qarun. In fact, this beach is a bank with multiple parallel ridges (Little 1935–1936). The highest, in general, is 23 m, but has a range of 22–25 m a.s.l. The Geological Map of Egypt (Beni Suef Sheet, No NH 36 SW, scale 1: 500,000) shows that Gisr El-Hadid is composed of two separate embankments: The western embankment is composed of Prenile deposits, whereas the eastern embankment is composed of the Neonile sediments. Said (1981) postulates that the first break through of the Nile water into the Depression occurred most probably in the later part of the Prenile times. According to Ball (1939), the western bank belongs to Late Paleolithic Period, whereas the eastern is much younger, although both beaches have about the same height (23 m a.s.l). In a recent study (Donner 2011), two samples (a sand sample 110 cm below the surface and a sample of *Lymnaealagostis* shells) were collected from the southern part of Gisr El-Hadid in 2006. The dating of the samples—sand by OSL and shells by radiocarbon—gave the same age which is 40,000 yr bp. The Prenile (Paleolithic) beach has suffered more from erosion and denudation, and is more dissected by drainage lines than the Neonile (Holocene) bank. Regarding composition, Ball (1939) stated that the more recent beach is much more sandy, less consolidated, and its gravelly covering consists of flat worn fragments of fissile limestone with a much smaller proportion of dark flint pebbles than occurs in the gravel of the older beach. Although no implements have yet been discovered, fragments of pottery have been found embedded in the materials of the **Neonile beach**, indicating its relatively recent age. According to Ball (1939), this is the beach of **Lake Moeris** when it had reached its average maximum level of about 20 m a.s.l when Herodotus visited Egypt 450 BC.

One of the other prominent beaches, which appears on the topographic maps (scale 1: 10,000) is that one locally known as **Shatt (beach) El-Edwah** after the village of the same name in the area to the northeast of Fayum city. This beach extends in an E-W direction for about 10 km and rises up to 22–24 m a.s.l, with a varying width of 100–250 m. According to the Geological Map (scale 1: 500,000) (Beni Suef Sheet), its material, which is of loose gravel and sand, belongs to the Neonile deposits. Remnants of other lake beaches are also found in three localities. The first lies along the eastern margins of the Depression at a level of 42–44 m (Little 1935–1936; Ball 1939), which is probably the oldest (Prenile) one. Ball (1939) postulated that this beach appears to have been piled up by the action of storms along the shores of a lake, which once filled the whole Depression to a height of about 40 m a.s.l. Little (1935–1936) and Ball (1939) also found along the flanks of the **Hawara Channel**, remains of terraces composed of similar deposits exist up to 42 m a.s.l, from which they concluded that the formation of the 42–44 m lake resulted from the Nile water flowed through the Hawara Channel to the Fayum Depression. The second locality lies to the north of **Lake Qarun** at levels 22, 18 and −2 m. as small hills. In this latter locality, the beaches are mostly composed of the Neonile sediments. Along the eastern margins of the Depression, beaches with levels varying between 30 and 38 m a.s.l have been reported by Sandford and Arkel (1929) and belong to Paleolithic and Neolithic Periods.

Ancient lake beaches in the Fayum indicate that several lakes were developed in the Depression since the entrance of the Prenile water. This was due to the connection and disconnection between the Nile and Fayum, which occurred several times. The largest and oldest lake reached a level of about 40 a.s.l in Acheulean times, covering an area of at least 2800 km^2, and containing more than 100 km^3 of water (Ball 1939). On the basis of sedimentological and archaeological data, Ball (1939) and Caton-Thompson and Gardner (1934) inferred that from the Middle Palaeolithic (Mousterian), the level of the lake began to fall down in a descending series to 34, 28 m, and 26–22 m a.s.l. Following an interval in which the lake fell to a level of 5 m below sea level and the Nile connection became practically cut-off by a bank and shoals at the Hawara entrance, the lake rose to 18 m a.s.l at the beginning of the Neolithic period (about 8000 BC). Later work (Said et al. 1972; Wendorf and Schild 1976; Hassan 1986) recognized four stages in lake development during the Holocene, since the Neonile water entered the Depression. These stages are called Paleo-Moeris, Pre-Moeris, Proto-Moeris, and Moeris (Fig. 12.4). This figure summarizes age and level of these former lakes (Hassan 1986). The first **Palaeo-Moeris Lake** developed at about 7715 ± 45 BP, which is almost 1300 year younger compared to that of Said et al. (1972). Its elevation cannot be ascertained due to the removal of an unknown thickness of its material, but it certainly reached a level that was higher than 11–12 m asl (Said et al. 1972; Wendorf and Schild 1976; Hassan 1986). The **Pre-Moeris Lake** reached an elevation of 15–19 m asl at about 6100 BC, whereas the third **Proto-Moeris Lake** reached a level of at least 19–24 m a.s.l at about 5200 BC. The level of this third lake fell to below 13 m a.s.l about 4000 BC and then constantly aggraded until the fourth **Moeris Lake** attained its maximum level (23 m) during the Old Kingdom times. Later, artificial control of water entering Fayum resulted in gradual shrinkage of the lake area until it reached its present-day level, area, and shape.

12.7 Lake (Birket) Qarun

The **Lake Qarun** is the largest permanent one in the Western Desert, with a maximum length of 45 km from east to west and a maximum breadth of 10 km from north to south, covering an area of ~200 km^2. It is considered as the shrunken remnant of **Lake Moeris** of antiquity (Shafei 1960; Said 1993). It occupies the lowest part of the Fayum Depression where the level of the Lake lies approximately 45 m below sea level, whereas the lowest point in its floor lies at 52.9 m b.s.l. This lake is artificial since the excess drainage water feeds it after irrigating the cultivated land, and where water is disposed by evaporation from the lake

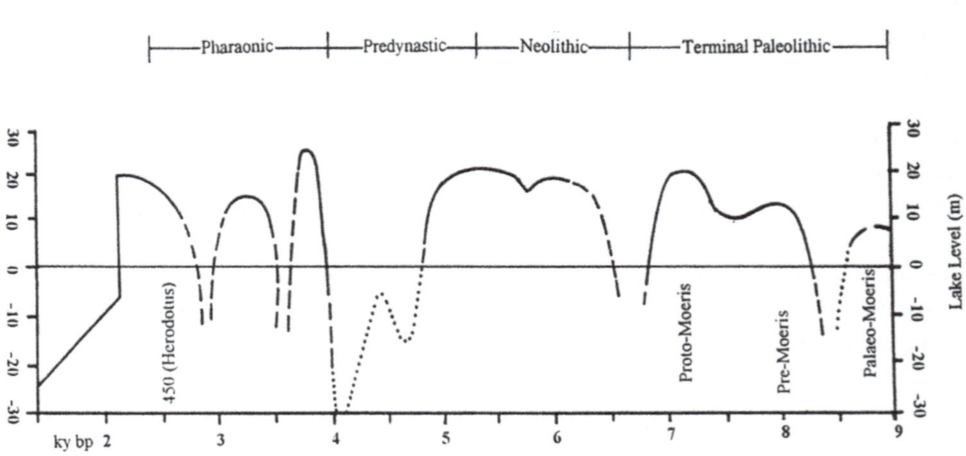

Fig. 12.4 Variations in the level of historical Moreis Lake and its prehistoric ancestors in Fayum (*Source* Modified and redrawn from Hassan 1986)

surface. The average drainage water entering the lake is about 350 million m^3 year, carrying some 60,000 tons of salts. This quantity of water is about 20% of the annual total irrigation water entering the Depression during the period 1928–1935 (Ball 1939). Due to solar evaporation, the lake salinity is increasing by time. Average proportion of total dissolved solids contained in the lake water was 11.2/litre in 1906 and became 25.4/litre between 1928 and 1930 (Ball 1939). More recently, total dissolved salts in the Lake are estimated to be 2.8/litre (Gupta and Abd El-Hamid 2003). Most probably, this decrease in salt content is due to artificial salt extraction

Due to the narrowing of its breadth, in its central section, the Qarun Lake is divided into two basins: a small, shallow eastern one and a larger, deeper western basin. Within its basin, several islands developed, of which the Qarn island is the largest (about 2.0 km^2). According to the Geological Map of Egypt (scale 1: 500,000), this island is formed of marine limestone which belongs to Wadi El-Rayan Formation of Mokattam Group (Middle Eocene), covered in places with Neonile sediments. The lake shores are irregular and are characterized by several small lagoons and headlands. In general, the northern shores are more irregular and steeper than the southern ones. One of the main causes of high irregularity of the north-western coastline is the deposition of sand by northern wind as tongues into this shallow part of the lake(Azab 2001). Locally, the headlands are called "*khashm,*" representing rocky topographic highs especially along the northern shores, or depositional promontories at the mouths of the main drains, which drain excess irrigation water into the lake.

12.8 The Fayum Delta

In the previous lakes, which covered the floor of the Fayum Depression, Nile sediments were deposited since the entrance of the Prenile water. In fact, these deposits which are topped by recent Nile silt successively built up a small delta as Bahr Youssef entered previous lakes. This **Fayum delta** takes a fan-like shape, which coincides with the pattern of branching of Bahr Youssef after entering the Depression. The surface of the delta decreases gradually from 18 m a.s.l near Madinat El-Fayum to about sea level at the northern margins of the delta, terminating with a relatively steep front. The relative height of this front varies between 2.0 and 5.0 m. Most probably, the front of the delta represents the face of the fore-sets of the Delta beds. In spite of the general slope toward the north and northwest, the delta surface is spotted with small closed or semi-closed basins 2–3 m deep, reflecting the original (subsurface) relief of the Depression floor before the deposition of the Delta sediments.

12.9 Canyon-like Wadis

The most characteristic feature of the Fayum Delta is the **canyon-like wadis**, which are incised in the delta sediments downward to the Eocene limestone (Ravine beds) in some localities (Fig. 12.5). Due to the fact that these wadis are relatively deep compared to their width, they are locally called "***khors***", i.e., canyons. They are—in fact—part of the irrigation and drainage system of the Fayum, which originated on the surface of the Fayum Delta after the recession of the "**Moeris**" **Lake**. As the topographic maps show, the Bahr

Fig. 12.5 A field photograph showing the features of a drain canyon at the floor of Fayum Depression

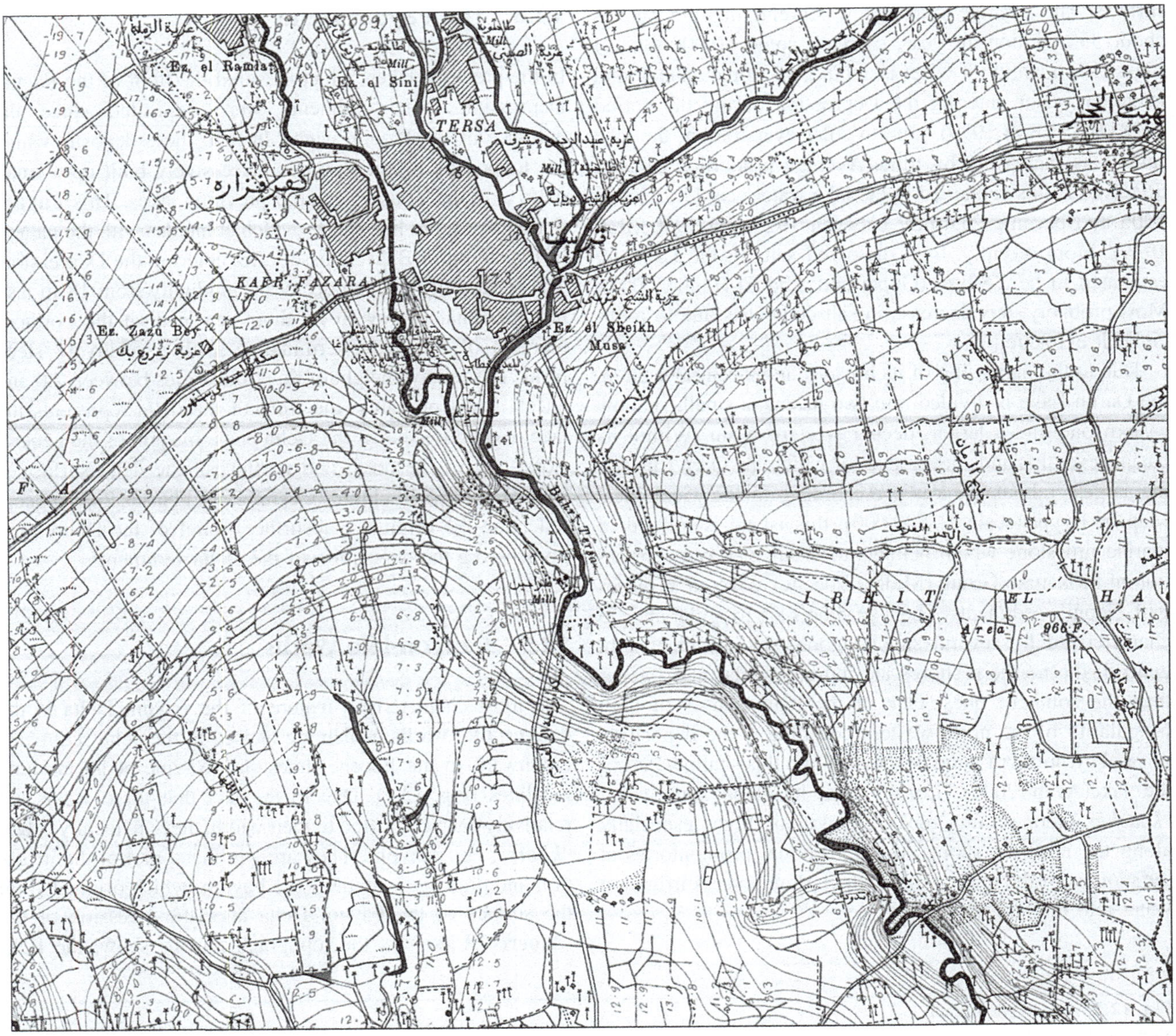

Fig. 12.6 Extract of topographic map, scale 1: 10,000 showing the finger-like canyons

Youssef divides when reaching Madinat El-Fayum into several branches in a fan-like manner, which ends in Qarun Lake. In the meantime, some canyons dissect the Delta front, which is not connected with the branching system of Bahr Youssef.

The canyons can be divided into three groups: **the Bats**, **El-Wadi**, and the **Delta wadis**. The Bats and El-Wadi are two main, relatively long (30 and 40 km), canyons, used as drains for excess irrigation water, encircling the **Fayum Delta** from the east and west, respectively. Along the sides of these two main wadis, bird's foot, small finger-like canyons join them as tributaries (Fig. 12.6). The two wadis cross some of the small basins, which characterize the surface of the Delta. In these sections as in their lower parts near **Lake Qarun**, the wadis become shallow with low gradients. Therefore, depth and width vary considerably along the two wadis. In some sections, depth is 28.6 m, and width is 150 m. In some other localities, especially in their lower sections near the lake, the wadi sides disappear, and only the channels of the two drains can be seen. Other prominent canyons (the third group) dissect the Delta front; some of them are as deep and narrow as the two previous large ones such as Khor Bahr Tersa whose depth reaches 20 m in some sections.

All these geological and geomorphologic characteristics indicate that the canyons of the Fayum Delta were the last features that developed in Fayum. This development started after the last recession of **Lake "Moeris"** by the process of headward erosion from the steep edge of the delta front. At the same time, the draining of the surface of the Delta into the main wadis permitted the development of the finger-like tributaries.

12.10 Development of the Depression

The development of Fayum can be divided into two stages. The first is the one during which the Depression came into being. The second stage started with the entrance of Nile water into the Depression.

12.10.1 The First Stage

As the Eocene Sea retreated northward, the Fayum anticline was exposed to subaerial erosion, mainly water action, since all evidence indicates that Egypt enjoyed wet climate at least from the Late Eocene. This wet climate continued during the Oligocene, Miocene, and Pliocene, and since the anticline is composed of calcareous sediments, chemical processes (karstification) played a major role in converting the Fayum anticline to lowland by dissolving the carbonate beds. In the meantime, surface runoff flowed in a radial pattern in all direction, forming consequent drainage lines on the back-slopes of the dipping beds of the anticline. These small drainage lines acted as tributaries for the adjoining large drainage systems as the Oligocene one, and as the Nile Valley in Late Miocene and Pliocene. The presence of Pliocene fluvial sediments along the eastern margins indicates that fluvial action shaped the Depression area. As the Depression deepened, new slopes developed all-around its margins, and new drainage lines developed toward the center of the Depression as can be seen today. These conditions continued during the Lower Pleistocene, but changed periodically according to climatic changes from wet to dry conditions. By the Middle Pleistocene, the Hawara channel started breaching the Nile-Fayum divide by river capture, but without the entrance of the Nile water into Fayum. Therefore, by the Middle Pleistocene, the Fayum Depression was more or less complete as any other desert one in the Western Desert but as an inverted topography.

12.10.2 The Second Stage

This stage started with the entrance of Nile water into the Fayum Depression during Early Palaeolithic through the Hawara Channel. This represents a characteristic feature in the history of development of Fayum, since all forms developed at the Depression floor were a result of this advent. The first is the formation of a lake covering the whole floor of the Depression to a level of 40–43 m above sea level (Little 1935–1936; Abul-Haggag 1967; Said 1993). This level of the Palaeolithic Lake—as called by Abul-Haggag (1967)—is based upon the presence of a 43 m sand and gravel terrace on the NE side of the Depression found and described by Little (1935–1936). This sand and gravel were also found in the Hawara Channel at depths 17–10 m a.s.l and was taken as evidence for the retreat of the Prenile (Said 1993).

The connection with the Nile was then severed for a long period, since the Niles that followed the Prenile during Middle and Late Palaeolithic were low and erratic; therefore, they were not able to overflow the Nile-Fayum divide through the Hawara Channel to the Fayum Depression (Said 1993). The connection was possible again by the "Neolithic Lake" period as named by Abul-Haggag (1967). Since then, the connection has been severed and re-established several times, until it became regulated by man in Ptolemaic times (Said 1993).

The intervention of man in regulating the water that could enter Fayum represents the onset of a third stage in the development of the Depression, associated with the development of new forms. Firstly, no more lakes were developed covering the whole floor of the Depression. The present **Qarun lake** is kept alive all-around the year to accommodate the water drained from the cultivated lands lying to the south of the lake. The level of this solar lake is kept constant at 45 m below sea level by solar evaporation. Recently, under the pressure of population increase and the need for more land to cultivate, new drainage lakes were established in Wadi El-Rayan by an artificial canal and tunnel extending from El-Gharaq El-Sultani in the south of Fayum to **Wadi El-Rayan** (Fig. 12.1). This is because this excess water would inundate the cultivated land surrounding Lake Qarun if not drained outside Fayum Depression.

12.11 Conclusion

The Fayum Depression is a unique Depression among the mega-depressions of the Western Desert, because it reflects certain features, not repeated in any other Depression. Due to its relatively close location to the Nile Valley, the Nile waters entered the Depression. Since the advent of access by Nile water to the Depression, it was transformed from a desert Depression to one with Nile affinities. The entrance of the Nile water is the crucial factor that created other unique features of Fayum.

The Nile waters formed the largest lake developed in the Western Desert—at least since the Pleistocene. Due to changes in the Nile water discharge, the level of the lake changed accordingly. During some periods, the Nile water stopped reaching the Depression, and the lake dried up. Therefore, it can be said that several lakes developed and not only one. Around the shores of these lakes, beaches were formed by the action of water movements. The Nile water carried silt to the Depression. The Nile silt was deposited in the lake as a local delta. On the delta front, deep canyon-like wadis developed in the final stage of formation.

The presence of these lakes in the Depression made life possible for prehistoric man since the Palaeolithic. With the cultivation of the fertile soils developed from the Nile silt during periods of shrinkage of lakes, it was found necessary to control and regulate the Nile water discharge into the Depression. Since then, no natural lakes have been formed, and only one artificial lake (Lake Qarun), occupying the lowest part, is kept alive through draining the water from cultivated fields into it. With the expansion in reclaimed cultivated land, more Nile water is allowed to enter Fayum, but this resulted in an increase in the drained water from cultivated fields. If this water is to be drained in Lake Qarun, its level will rise, and water will inundate the fields around the lake. This is why excess water has been drained since the early 1970s into the nearby Wadi El-Rayan, creating two large lakes occupying two relatively large basins. This emphasizes the impact of the Nile on development of new forms in the Fayum region.

References

Abul-Haggag Y (1967) The geomorphologic aspects of Fayum Depression (in Arabic). Ann Fac Arts, Ain Shams Univ 10:69–116

Azab M (2001) Geological and geomorphologic studies on the area around Lake Qarun, Fayum, Egypt, using remote sensing data and geographic information systems. Dissertation, Al-Azhar University

Ball J (1939) Contribution to the geography of Egypt. Survey and Mines Department, Ministry of Finance Egypt, Government Press, Cairo

Beadnell H (1905) Topography and geology of Fayum Province. Survey of Egypt, Cairo

Bown T, Kraus M (1988) Geology and palaeoenvironment of the Oligocene Jebel Qatrani Formation and adjacent rocks, Fayum Province, Egypt. US Geol Survey Prof Paper No. 1452

Caton-Thompson G, Gardner E (1934) The desert Fayum. Royal Anthropological Institute. London

Donner J (2011) Studies in the Quaternary of the Western Desert of Egypt. Annals Academiae Scientriarum Fennicae, Geologica-Geographica, 170: 42p

Gupta G, Abd El-Hamid Z (2003) Water quality of Lake Qarun, Egypt. Int J Environ Stud 60:651–657

Hassan F (1986) Holocene lakes and prehistoric settlements of the Western Fayum, Egypt. J Archaeol Sci 13:483–501

Little O (1935–1936) Recent geological work in the Fayum and in the adjoining portion of the Nile Valley. Bull Institut d'Égypte 18: 201–240

Said R (1981) The geological evolution of the River Nile. Springer, New York

Said R (1993) The River Nile: geology, hydrology and utilization. Pergamon Press, Oxford

Said R, Albritton C, Wendorf F, Schild R, Kobusiewicz M (1972) A preliminary report on the Holocene geology and archaeology of the Northern Fayum Desert. In: Reeves C (ed) Playa Lake Symposium, vol. 4. Lubbock, Icassalsm, pp 41–61

Sandford K, Arkel W (1929) Paleolithic man and the Nile-Faiyum divide: a study of the region during Pliocene and Pleistocene times. Orient Institute, Chicago University, vol. X

Shafei A (1960) Lake Moeris and Lahun (Mi-wer and Ro-ûn), the great Nile control project executed by the ancient Egyptians. Bull Soc Géogr d'Égypte 33: 187–217

Simons E, Rasmussen D (1990) Vertebrate paleontology of Fayum: history of research, faunal review, and future prospects. In: Said R (ed) Geology of Egypt, The Geology of Egypt, Amsterdam, pp 627–638

Wendorf F, Schild R (1976) Archaeology and Pleistocene stratigraphy of the Northern Fayum Depression. In: Wendorf F, Schild R. Prehistory of the Nile Valley, Part 12, Academic Press, New York, pp 155–228

Part V
Sand Seas and Dune Fields

13 The Great Sand Sea

Abstract

By an area of more than 100,000 km^2, the Great Sand Sea is the largest one in Egypt. Lying in the western part of the Western Desert, it occupies a huge basin that extends from the Gilf Kebir Plateau in the south, to Siwa Depression in the north. Linear dunes dominate all-over this sand sea. Characteristics of this sand sea are analyzed such as dune types and patterns, morphometric aspects, in addition to sand properties: grain size, minerals, micro-features, and source(s) of sand. Age and mode of development of the Great Sand Sea are also discussed within this chapter.

Keywords

The Great Sand Sea • Linear dunes • Barchans • Barchanoids • Whalebacks • Sief

13.1 Introduction

The term "sand sea" is often used because the mega-dunes resemble the waves of the ocean (Besler 2008). However, the term is also used for large plains without distinct dunes, e.g., the Selima Sand Sea. There are 6 sand seas and 10 dune fields in Egypt (Fig. 13.1) according to the definition of Embabi (1998, 2000). Here, two criteria are taken into consideration: the total area and the dune coverage. The minimum area of a sand sea is 5000 km^2 with dune coverage of at least 50%, whereas dune fields cover smaller areas with dune coverage less than 50%. A dune field, nevertheless, may cover an area exceeding the minimum area of a sand sea, but with less than 50% dune coverage.

Except for the North Sinai Sand, all other sand seas lie in the Western Desert (Fig. 13.1). Also, four out of the ten dune fields are developed in the Western Desert. In the Eastern Desert, dune fields are found in the NE (El-Khanka Dune Field) and in the extreme SE (the four fields of El-Hebal Dunes, West El-Hebal Field, Delta Wadi Kraf Field, and El- A'llaqi Field). El-A'llaqi Dune Field was discovered by the author to the south of **Wadi El-Allaqi**, by using Google images as means of investigation. This is the first time to record El-A'llaqi Dune Field. Thus, most of the sand accumulations in Egypt are located in the Western Desert. This is explained by the fact that among other factors, huge quantities of fluvial sediments were laid down in various parts in the Western Desert (Embabi 2000; Bubenzer and Bolten 2008; Besler 2008). In sum, sand seas (with about 24%) and dune fields (with about 4%) cover about 28% of the total area of Egypt (Table 13.1).

N.S. Embabi, *Landscapes and Landforms of Egypt*, World Geomorphological Landscapes,
https://doi.org/10.1007/978-3-319-65661-8_13

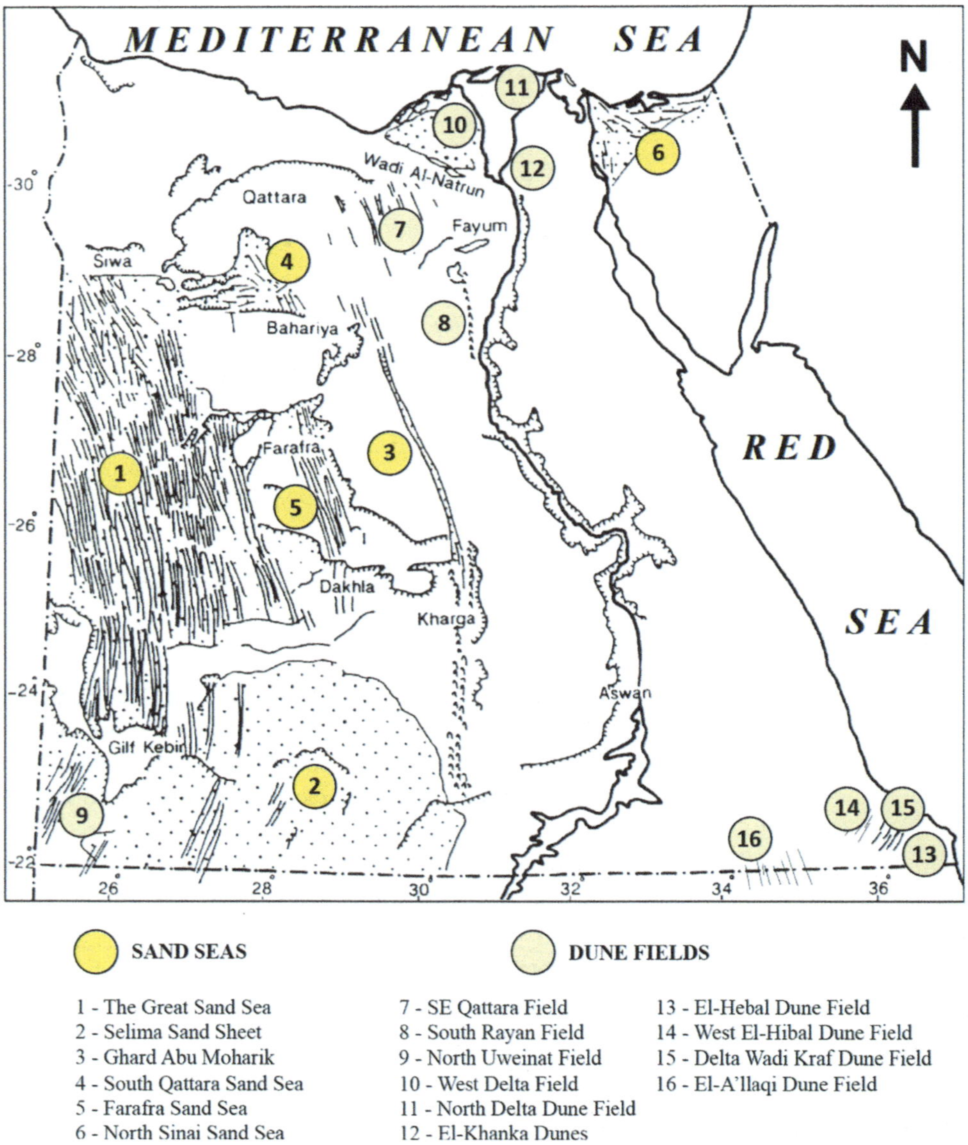

Fig. 13.1 Distribution of sand seas and dune fields in Egypt

13.2 General Characteristics of the Great Sand Sea

This sand sea acquired its name after Rohlfs who mentioned that there is a Great Sand Sea to the west of Dakhla and Farafra (Ball 1927). Beadnell (1910) also asserted that the entire country to the west of the 27th meridian E. and lying to the south of the Miocene escarpment is buried under a "Great Sand Sea"; Bagnold (1931) called it "The Libyan Sand Sea." In fact, the **Great Sand Sea** is not confined to Egypt alone, but it extends into Libya in the west. In Egypt, it extends for about 600 km from the southern margins of Siwa Depression in the north to the northern scarps of the Gilf Kebir and Abu Ras plateaus in the south, with varying width between 60 and 200 km from east to west (Fig. 13.1). Along its western margins between latitudes 26° and 25° N, dunes are straightened, trimmed, and truncated by fluvial channels heading from the south. Along these margins, alluvial deposits appear as if they are pushing the sand dunes and preventing them from advancing southward. Hence, it can be inferred that fluvial deposits are inter-fingering with sand layers at dune plinths. With an area of more than 100,000 km^2 (Table 13.1), the Great Sand Sea is the seventh largest Sand Sea in Africa and the 15th largest Sand Sea in the world (Wilson 1973).

Spot heights and ancient drainage lines indicate that the Great Sand Sea occupies a huge basin sloping in general toward the north. Topographic maps (scale 1:250,000) show that the height of inter-dune areas diminishes from 750 m a.s.l just to the north of the Gilf Kebir and Abu Ras plateaus to 50–25 m a.s. l just to the south of Siwa Depression in the north. Drainage lines dissecting, the northern parts of the Gilf Kebir and Abu Ras plateaus such as Wadi Abdel-Melik, slope toward the basin of the Great Sand Sea. Fluvial gravel is exposed in inter-dune

Table 13.1 Area of sand seas and dune fields in Egypt

Name	a Total Area (Km^2)	b Dune Area (Km^2)	c 2/1 (%)
SAND SEAS			
1-Great Sand Sea[a]	114,400	85,600	74.6
2-Selima Sand Sheet[a]	63,200	56,000	88.6
3-South Qattara Sand Sea[a]	10,400	6800	65.4
4-Farafra Sand Sea[a]	10,300	7000	68.0
5-Ghard Abu Moharik[a]	19,545	10,830	55.4
6-North Sinai Sand Sea	13,600	12,600	72.6
Sum 1–6	**238,145**	**180,610**	**75.8**
DUNE FIELDS			
7-East and Southeast Qattara	28,754	6127	21.3
8-North Uweinat[a]	8000	2800	35.0
9-West Delta[a]	4400	3200	72.7
10-South Rayan[b]	1200	375	31.1
11-El-Khanka Dunes	857	622	72.6
12-N Delta Coastal Field	80	70	87.5
13-El-Hebal Dunes	400	320	80.0
14-West El-Hebal Dunes	350	275	78.5
15-Delta Wadi Kraf Dunes	150	100	66.7
16-El-A'llaqi Dune Field	500	300	60.0
Sum 7–16	**44 691**	**14 189**	**31.4**
Total	**282 836**	**194 799**	**68.9**

[a]The Geological Map of Egypt 1986–1987
[b]Topographic maps (scale 1: 250,000, 1: 50,000) 1987
[c]TM Photomaps, (scale 1:250,000) and Google Earth 2014)

areas, especially in the southern parts of the Great Sand Sea where dune density is relatively low. This is confirmed by field studies carried out in the southern parts of the Great Sand Sea (Besler 2008; Pachur and Röper 1984; Haynes 1982). The river that drained into this huge basin is called the Gilf River and was developed during the Oligo-Miocene period after the rifting of the Red Sea (Haynes 1982; Albriton et al., 1990; Issawi and McCauley 1993) as a part of a regional drainage system called the Ancient River Nile system. There, fluvial deposits in the form of alluvial fans and playas were overridden by dunes. Worthy to be noted is that, if the sand filling of the huge basin below the draa linear dunes can be calculated, the sand budget will be immense to the extent of making the Great Sand Sea the largest one in area and volume of sand in Egypt.

13.3 Dune Forms and Patterns

13.3.1 Dune Forms

Three dune forms are recognized in the Great Sand Sea. The first is the linear/longitudinal forms, the second comprises the transverse forms, and the third is the star dunes. The first has been known since the early studies on dunes in the Egyptian deserts in the beginning of the twentieth century (Beadnell 1910; Ball 1927; Bagnold 1931; Kadar, 1934). The second comprises barchans, barchanoids, and transverse dunes. Of this second group, barcans and barchanoids and the star dunes are recorded in the present study for the first time due to the availability of high-resolution images using Google Earth. Local variations are summarized in (Table 13.2) and will be analyzed in the following section.

13.3.1.1 Linear Form

All previous studies revealed that the linear form is the dominant shape in all parts of the Great Sand Sea (Bagnold 1931, 1933, 1941; Embabi 1998; Besler 2008). Nevertheless, local variations in the linear shape were recorded in previous studies as well. The compound form which is composed of several straight ridges or with broad rippled surface (Figs. 13.2, **13.3**) is predominant throughout the sand sea. Recent active **arcuate dunes** are detected in the northern part overriding an ancient fixed transverse dune system (Fig. **13.4**). This transverse system was not mentioned in any previous study on the Great Sand Sea. To note, system changes to barchans–barchanoids in several tracts in

Table 13.2 Morphometric characteristics of dunes in the Great Sand Sea *Source* Space Atlas of Egypt, (scale 1: 250,000) National Authority of Remote Sensing and Space Sciences, Academy of Scientific research, Cairo, 1990; Topographic maps scale (1: 250,000)

Latitude (degree)	Pattern/Form	Orientation (degree)	Wavelength (km)	Length (km)	Width (km)
29° 15′–28° 50′	Network and Branching			1-5	
	1-Meso-dunes	1–360°	–	2-21–5	1.0–0.1
	2-Secondary Dune	2–10°		3- 2–2.5	
	3-Barchans	3–170°	–	3–0.35	
	4- Star dunes	–	–		–
28° 50′–28° 00′	Parallel				
	1-Arcuate				
	2-Hooked	245°–350°	1–5	5–12	1.0–2.0
	3-Linear (sief)				
	4-Whalebacks				
28° 00′–27° 00′	Parallel				
	1-Linear (sief)	1–330°			
	2-Whalebacks with ripples	2–345 –350°	2–5	10–25	1.0
27° 00′–26° 00′	Parallel				
	1-Linear Compound	1–360°	1–5	15–60	1.0–2.0
	2-Whalebacks	2–345°–350°			
26° 00′–25° 00′	1-Parallel				
	2-Branching	350° –360°			
	Linear (sief) Whalebacks		1–10	15–120	1.0–2.0
25° 00′–24° 00′	1-Parallel				
	2-Branching	1–350°–360°			
	3-Y-junction Linear (seif) Whalebacks	2–35°–40°	–	45–120	1.0–3.0

the same region (Fig. 13.5), which indicates a change in wind environment during periods of dune activity.

Distinction is made between the sharp-crested recent linear **"seif" ridges** and the broad-crested dunes. Due to low convexity of their cross section and their linear extension, they were first called **"whalebacks"** by Bagnold (1931, 1933). They are characterized by the absence of sharp crests and collapsing fronts (Fig. 13.5a). This type of linear dunes is recorded along the eastern margins of the Great Sand Sea in the vicinity of Abu Minqar Depression (Embabi 1998). The recent sharp-crested active top dunes are called **"Silk"** by Besler (2008) and are developed on the backs of the whalebacks (called '**Draa**' by Besler 2008) (Fig. 13.5b, c), or on an ancient fixed dune system as mentioned earlier in this section. The whalebacks are rippled and/or undulatory. These ripples or undulations are conjoined on a recent surface layer that has been deposited recently on the surfaces of the whalebacks. The mode of development of the **whalebacks** is still poorly known. Bagnold (1941) suggested that these non-mobile linear ridges grew so huge to the extent that they no longer move. Trenches made by El-Baz (1992) in some of them revealed that the internal structure is horizontal rather than dipping, a characteristic, suggesting that they represent remnant sand ridges left from erosion of inter-dune sand layers. However, this postulation needs further investigation and verification since orientation of the whalebacks is the same as that of active ridges and conformable with the prevailing N-NW winds.

According to Haynes (1982), the whalebacks had been active dunes from the end of the Middle Paleolithic pluvial to the beginning of the Neolithic pluvial period 10 ka BP. These whale backs stabilized by vegetation under a semiarid climate from 10 ka to 5 ka BP. Haynes (1982) concluded that the low, rounded to flattened transverse profile of a typical whaleback is probably the result of the combined action of bioturbation, human occupation, pedogenesis, and slope wash. Another characteristic observed in the field and reported in some previous studies (Haynes 1982; Kröplin1993) is that the whalebacks are covered by a surface layer of fine pebbles and coarse sand as a lag shield that leads to the stability of the ridges. In the meantime, the plinths of the linear dunes are known to be composed of low dipping sand laminae of 8° or less toward the

Fig. 13.2 N–S parallel straight, compound linear dunes in the western part of the Great Sand Sea. Note the western side of the sea is checked by the ground of the plateaus of this region (*Source* Google Earth 2012)

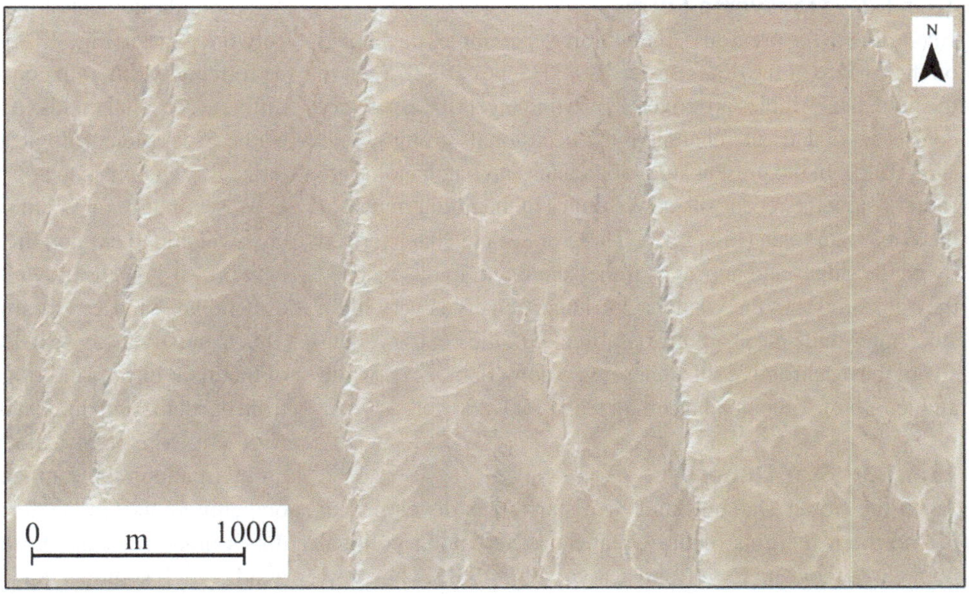

Fig. 13.3 Holocene silk linear dunes developed atop Pleistocene draa characterized by transverse very large ripples (*Source* Google Earth 2012)

inter-dune areas (Lancaster 1995). Therefore, these whale-backs are pre-Holocene linear dunes subjected to some consolidation during wet phases of the Holocene, then modified by wind deflation during the arid phases of the Pleistocene and Holocene. During these arid phases, the crests of the dunes were removed probably by the force and high velocity of the wind and only the plinths were left, where fine grains were deflated from the surface layer, leaving the coarse grains as a lag shield.

Length of the main dune ridges varies between 50 km and 120 km. This length increases southward, where the longest dunes, whose length exceeds 100 km, are present

Fig. 13.4 Curved recent silk dunes developed atop rippled ancient system in the northern part of the Great Sand Sea (*Source* Google Earth 2012)

between latitudes 26° and 24° N. Investigations made by Kadar (1934) northward the Gilf Kebir Plateau revealed that the longest linear dune is 140 km and the highest is 100 m. Dune width varies between 100 m and 3.0 km (see Table 13.2). Height varies between 20 m and 70 m.

13.3.1.2 Transverse Forms

According to the previous studies, transverse ripples developed on the surface of the draa (Besler 2008; Haynes 1982). Seen in the field, these ripples are covered by a thin layer of coarse/very coarse sand and are changing by occasional strong winds (Fig. 13.6). In the present study, barchans and barchanoids are found below the recent silk active dunes in the northern part of the Great Sand Sea (Fig. 13.7). These dunes are characterized by gentle slopes and appear as if they are fixed. In this respect, they are similar to the whalebacks. Dune density is very high, since inter-dune areas are nearly absent. On the contrary, density of the recent active silk dunes is very low (Fig. 13.7), width and length of some barchans may reach 300–350 m.

13.3.1.3 Star Dunes

As in the South Qattara Sand Sea, star dunes developed in the northern margins of the Great Sand Sea along political boundary between Egypt and Libya (Fig. 13.8). The presence of star dunes as mentioned in Chap. 10 (Sect. 10.4.2.1) indicates that there is a local multi-directional wind regime. Therefore, it can be inferred that the region which extends eastward from the northern peripheries of the Great Sand Sea to the South Qattara Sand Sea is characterized by a multi-directional wind regime.

13.3.2 Dune Pattern

The pattern of the dune changes from a network and a **branching pattern** in the northern part just south of Siwa Depression (Fig. 13.9), to a branching pattern in the southern part to the north of the Gilf Kebir (Fig. 13.10), and a parallel one throughout most of the Great Sand Sea (Table 13.3). The network pattern (Fig. 13.11) was interpreted in other deserts to be formed by different coexisting winds (Warren 1988), or by different seasonal winds (Aufrere 1935). Analysis of wind data of the Siwa Meteorological Station showed that it is characterized by the presence of N and NE principal directions and E–W secondary one (Embabi 1995). Although no wind data are available to explain the parallel pattern to the south of the network, it is well known that such pattern develops in unidirectional wind environment, with occasional crosswinds. This is mostly the case in the Great Sand Sea. The orientation of the main linear sand ridges varying between 330° and 360° from north to south indicates that the prevailing winds are from N-NW to N. Notably, the branching pattern is present in the northern and in the southern parts of this sand sea. However, branching dunes in the north open upwind, which means that dunes unite, whereas those in the southern part open downwind, which signifies that dunes branch in this direction.

Dunes in both types of branching are known in the geomorphologic literature as **"Y-Junction" pattern** (Pye and Tsoar 1990; Thomas 1997). Branching in the northern part of the Great Sand Sea was interpreted to be the result of dune convergence and unification under the effect of local ground slope in the downwind direction (Embabi 1995). Branching in the southern part is mostly due to the

Fig. 13.5 Field photographs of linear dunes at the eastern margins (Abu Minqar area) of the Great Sand Sea: **a** Whaleback ridges, **b** side view of a sharp-crested dune with peaks and saddles, overriding a whaleback ridge, **c** a sinuous recent sharp-crested dune, showing the convexity of one side and concavity of the other side, where the upper segment is occupied by a slip face

modifications in wind direction under the effect of the high grounds of the northern flanks of the Gilf Kebir. These branching dunes override the linear ridges that lie to the west of the branches. This could also imply that these secondary branches are more recent than the main larger ridges, which look as if they are fixed.

13.4 Sand Properties

Sand properties were treated in the previous studies with the same degree of morphology importance and other aspects of sand dunes due to their considerable impact on the overall aspects. Sand properties cover grain size, surface texture, and minerals.

Fig. 13.6 Very large ripples developed by strong winds on the eastern side of a linear dune in Abu Minqar area

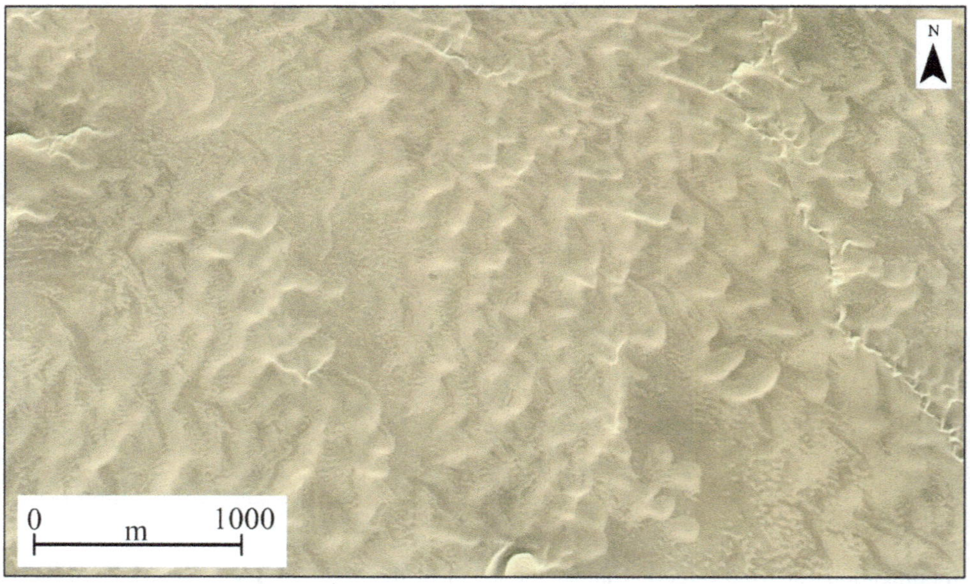

Fig. 13.7 Ancient barchans and barchanoinds in the NW part of the Great Sand Sea (*Source* Google Earth 2012)

13.4.1 Grain Size

Previous studies on grain size of sands of the Great Sand Sea were carried by El-Baz et al. (1979), = Embabi (2000) and Besler (2008). Of these studies, Besler's study (2008) is the most comprehensive one, since it covered nearly the whole sand sea, from latitude 24° 16′ N to latitude 29° 00′ N, and from longitude 25° 34′ E to longitude 27° 20′ E, and she analyzed 105 samples as well. Findings of previous studies showed that nearly all samples are composed of sands in the medium to fine fraction. In Besler's study (2008), the size ranges between 0.47 mm (medium) at latitude 26° 43′ N and longitude 26° 11′ E and 0.17 mm (fine) at latitude 24°16′ N and longitude 25° 39′ E. This characteristic is in accordance with the results of previous studies either on the Egyptian sand seas (Embabi 2000) or on other sand seas in other deserts (Besler 2002). A few samples are composed of coarse sands with mean grain size in the range of −1.25 ø (2.1 mm) to −0.74 ø (0.6 mm). The coarse sand represents the lag particles left behind as winds remove finer grains

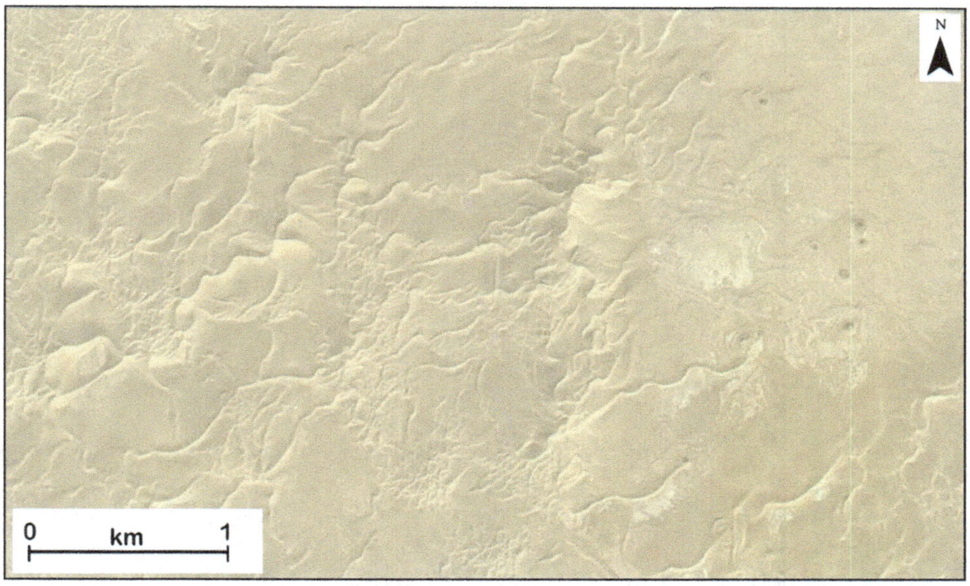

Fig. 13.8 Star dunes at the northern margins of the Great Sand Sea along both sides of the Egyptian—Libyan Borders (*Source* Google Earth 2012)

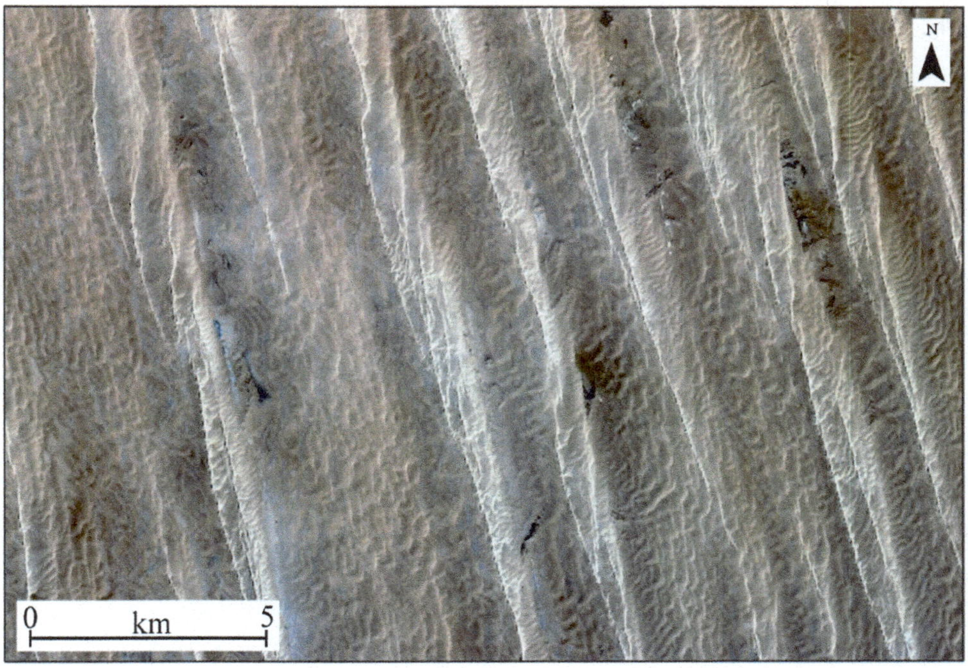

Fig. 13.9 Branching dunes in the central of the Great Sand Sea (*Source* Google Earth 2012)

from the plinths of draa ridges. All samples are unimodally distributed, except those of the plinths that are bimodally distributed due to the presence of coarse sand.

13.4.2 Surface Texture

Scanning electron microscopy (SEM) of the surface textures features of quartz sand grains provides valuable information about the processes that acted on sand grains during transportation and deposition. This is why all studies on dune sands have given this aspect special attention—since the early seventies of the twentieth century. However, previous studies on the sands of the Great Sand Sea are very few (El-Baz and Prestel 1982; Besler 2008;).

Features recognized on sand grains reflect mechanical and chemical actions on their surfaces. Mechanical features include Eolian crescents, holes, and cracks (Besler 2008). El-Baz and Prestel (1982) recorded pits on the coating surfaces, with grooves and cracks that are probably caused by Eolian abrasion. As for chemical features, these developed either through solution or the precipitation of dissolved material. Although there are other coatings (iron oxides, for example), kaolinite composes much of the coatings on

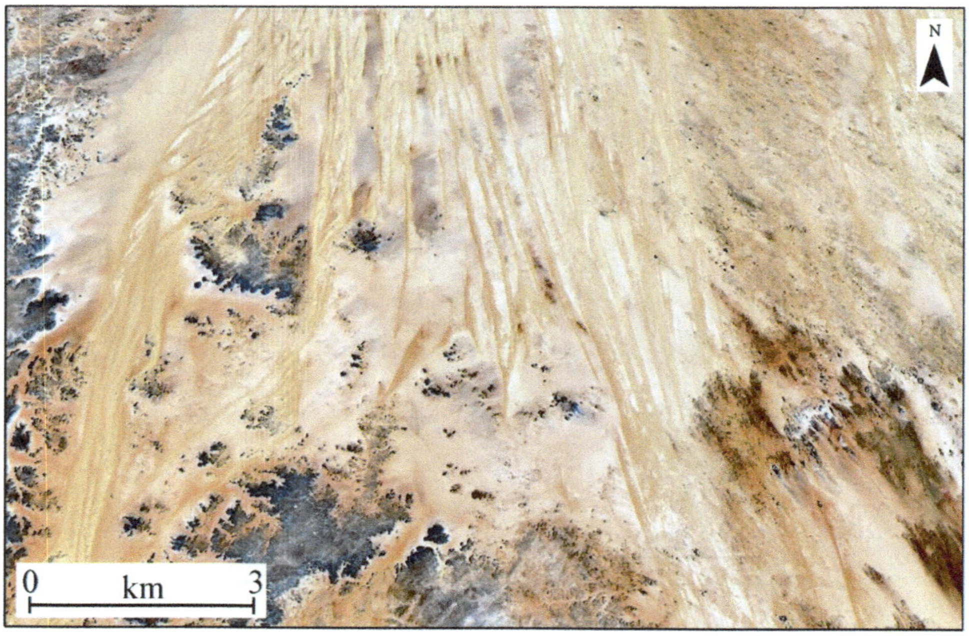

Fig. 13.10 Branching dunes in the north of the Gilf Kebir Plateau (*Source* Google Earth 2012)

Table 13.3 OSL data of the Great Sand Sea (Besler 2008)

	West Transection			Middle Transection			East Transection		
	Position	Depth (m)	Age (Ka)	Position	Depth (m)	Age (ka)	Position	Depth (m)	Age (ka)
I	N 26° 51'E 25° 36'	2	17.1 ± 2.7	N 27° 12'E 26° 02'	2	11.1 ± 1.8	N 27°19'E 26° 32'		
		3	18.1 ± 1.6		3	12.7 ± 1.8			
		4	20.4 ± 3.1		4	15.9 ± 2.1		4	12.5 ± 3.0
II	N 25° 42'E 25° 40'	2	20.4 ± 3.8	N 25° 24'E 26° 36'	2	21.3 ± 3.0			
		3	21.3 ± 2.9		2	21.3 ± 3.0			
		4	21.9 ± 4.5						
		5	22.8 ± 3,8						
III	N 25° 07'E 25° 37'	2	11.8 ± 3.0	N 25° 25'E26° 36'			N 25°25'E 27° 05'	1.0	0.24 ± 0.05
		3	11.5 ± 2.1		3.5	9.8 ± 1.6		1.4	0.28 ± 0.06
								1.5	0.25 ± 0.07
								1.8	0.24 ± 0.06
								2.0	5.00 ± 1.05

I Transverse draa. *II* Longitudinal draa, Western Flank. *III* Longitudinal draa, Eastern Flank

quartz grains collected from the floor of Wadi Bakht, which dissects the eastern side of the Gilf Kebir Plateau (El-Baz and Brestel 1982). Solution Depression and silica precipitation are recorded on several sand grains of draa in the eastern side of the Great Sand Sea. Some of the micro-features found on the sand grains of the Great Sand Sea were also found on the sand grains of the sandstone of the Gilf Kebir (Besler 2008). Due to their high significance, the percentage of the ultra-stable minerals of zircon, tourmaline, and rutile (ZTR) were extracted from the non-opaque samples (29 samples) by Besler (2008) and were found to be 54.1%, which is very near to the percentage (58.2%) of a previous sample (Embabi 2000). This relatively high percentage is taken to indicate the maturity of sediments (Embabi 2000). The occurrence of ZTR in high percentage is linked with the recycling of detritus grains and the abundance of zircon (average 22.3%). Such findings endorse the assumption that the minerals are reworked from older sediments (Sadiek et al. 1995). From this relationship, it can be inferred that sands of the Great Sand Sea have been

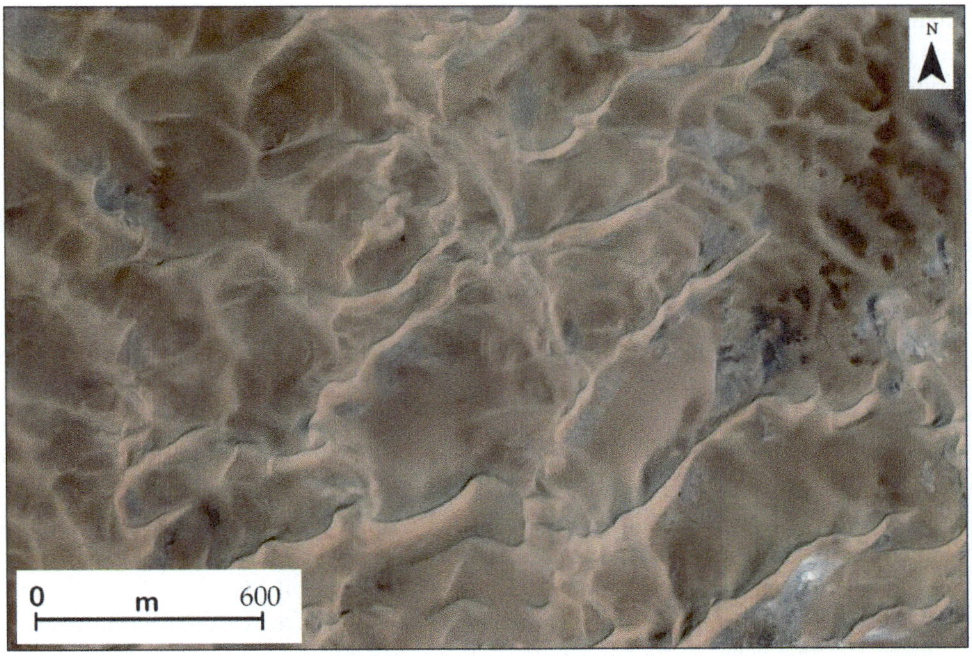

Fig. 13.11 Network pattern composed of recent silk dunes in the northern margins of the Great Sand Sea (*Source* Google Earth 2012)

recycled/or reworked from older sediments, most likely from the sandstone of the southern plateaus of the Gilf Kebir and Abu Ras.

13.4.3 Reddened Sand

Reddened sand has been observed in various deserts in other continents. Although the origin of red color is controversial, the fact that the red color increases with distance from the sand source is well established in many deserts such as western Libya (Walker 1979) and Namib (Breed et al. 1979). On Landsat-TM images, the dunes of the Great Sand Sea manifest increased reddening in the southern direction. Since the dunes in the southern margins of this sand sea fall near the end of sand transport cycle in Egypt, which is generally from north to south, the sand grains are presumed to show thicker reddened coatings. As for the reddening in Wadi Bakht, it is due to kaolinite and hematite coating (El-Baz and Brestel 1982). In the western transection of the Great Sand Sea, the reddish color of the surface sand, most probably, has been inherited from the red Gilf Kebir sandstone clay coatings observed under SEM (Besler 2008).

13.4.4 Minerals

Mineral composition with emphasis on heavy minerals of Eolian sediments represents a significant aspect in the study of dune sand, since it sheds light on the source of sand. All previous studies on Egyptian sand seas and dune fields informed that their sands are composed mainly of quartz with small amounts of feldspar and calcareous materials. The total heavy mineral content in the Great Sand Sea comes close to 0.5%, varying between 0.27% and 1.0%. These local variations were attributed to differences in sand sources or to different wind velocities during draa formation (Besler 2008). The heavy mineral fraction is divided into opaque and non-opaque minerals. The most frequent non-opaque minerals are zircon, tourmaline, rutile, epidotes, pyroxenes, staurolite, and garnet. Besler (2008) highlighted that the heavy mineral spectra as well as the individual frequencies are rather similar in all samples; therefore, nothing can be deduced with regard to different sand sources. Only the samples of Siwa region in the north diverge from the general pattern. Extremely high zircon contents and the presence of pyroxene and spinel (the latter does not occur anywhere else) indicate that the Moghra Sandstone was a sand source for the younger northern dune field (Besler 2008).

13.5 Source of Sand

Until recently, most authors adopted the views of Beadnell (1910) that the sands composing the dunes of the Western Desert of Egypt must have been derived from the clastic sediments in the north. Ball (1927) agreed with the views of Beadnell (1910) that sands of dunes originate from the great Depressions that stretch from Siwa to Wadi El-Natrun, where clastic sediments from the Miocene and Oligocene spread out and are inexhaustible. Recently, however, El-Baz and Prestel (1982) postulated from their observations of fine drainage lines leading northwards toward the Mediterranean and eastward to the Nile that the sands of the Western Desert originate from the Nubian Sandstone and other sediments in the southern part of this desert. According to them, the sand was most likely transported to the northern part of Egypt by the Nile and subsidiary drainage lines during past wetter periods. When the climate became drier, an Eolian regime sat up and sands were transported southward by northern winds.

This postulation of El-Baz and Prestel (1982) fits at least partly with that of Beadnell (1910) and Ball (1927), since it is now accepted and believed that the Oligocene and Miocene sediments were transported and deposited by drainage systems originating from the south and east (Issawi and McCauley 1993). It is also well established that the huge drainage system predating the Nile (called Ur-Nil by Blankenhorn) brought huge quantities of fluvial materials (mainly quartz) and formed an enormous delta to the west of the Nile Delta and north of the Fayum Depression. The sediments of this delta are known geologically as "Moghra Formation." This delta was composed of Late Eocene, Oligocene, and Early-to-Middle Miocene sediments. The source of drainage lines of this system was in the east (Red Sea Mountains) and south on the plateaus of the Western Desert (Salem 1976; Abdel-Rahman and El-Baz 1979; Issawi and McCauley 1993)). In this context, Issawi and McCauley (1993) suggested that a drainage system developed during the Oligocene was directed westward and was integrated with those of the Gilf Kebir, terminating into the basin of the Great Sand Sea or probably flowed further north and ended into the Mediterranean. This system is called "the Gilf Drainage System." Topographic maps show all wadis dissecting the northern parts of the sandstone plateaus of the Gilf Kebir and Abu Ras and terminated into the huge basin of the Great Sand Sea. It is assumed that the sediments laid down in the basin of the Great Sand Sea contained a high percentage of quartz grains of sand size, since these sediments were originated from the igneous and metamorphic rocks from the Red Sea Mountains and from the sandstone plateaus of the Gilf Kebir and Abu Ras. These sediments were reworked during the arid phases of the Pleistocene and Holocene and contributed to the sand body of the Great Sand Sea.

Sources of sand seas and dune fields are classified as local/internal and external (Pye and Tsoar 1990). In the case of the Great Sand Sea, local sources were derived from the ancient fluvial sediments which were laid down in its basin during the pluvial periods of the Tertiary and Quaternary. Due to the location of the Great Sand Sea to the south of Siwa Depression, it was deduced that the eroded material from it has contributed to the huge quantities of sand of this sand sea (Embabi 2000). This inference is confirmed by the finding of some minerals as mentioned in Sect. 13.4.4, indicating an input of Moghra sandstone in the northern part of the Great Sand Sea, coming with northerly winds (Besler 2008). Therefore, Moghra sediments represent an external source for the sands of the Great Sand Sea.

13.6 Age and Development

13.6.1 Age

Age determination of dune sands is done by luminescence, OSL (Optically Stimulated Luminescence) and ^{14}C techniques. Before using these methods, age used to be estimated according to available geological, geomorphologic, and archaeological evidences. In some instances, no age determination/estimation is/was made due to the inaccessibility of the sand sea/dune field, or due to the high costs of the laboratory analysis of luminescence and OSL. This was the case for most of sand seas and dune fields in Egypt.

13.6.1.1 Age Estimation

According to the available geologic, geomorphologic, and archaeological evidences, three **cycles of dune development** are envisaged (Embabi 2000). The first cycle is the oldest one, and it occurred during the arid phase of the Middle Pleistocene, which started 700,000 years ago and ended 200,000–300,000 BP. Such estimation is based on archaeological evidence found in inter-draa corridors of the Great Sand Sea. Acheulean-type artifacts were also spotted strewn on the surfaces of inter-dune areas (Haynes 1982; Pachur and Röper 1984). This means that the Eolian forms between the dune ridges were deposited during the pre-Acheulean times, i.e., before 200,000–300,000 BP. Due to the thickness (>200 m) of the sub-dune layers in the Great Sand Sea, there is a possibility of an earlier cycle than this "assumed" first one. This earlier cycle might have occurred during the hyper-arid period which converted Egypt into a veritable desert and covered the first million years of the Early Pleistocene (Said 1981).

The second cycle of dune development extended from 35,000 or earlier to 10,000 years ago. According to Haynes (1982), there is no evidence of human occupation, but there is abundant evidence of Eolian activity during that period. This arid period extended backward to 60,000 years ago, according to new evidences collected from lake deposits and human occupation in the Western Desert of Egypt (Wendorf et al. 1993). During this arid period from the end of the Middle Paleolithic Pluvial Period to the beginning of the Neolithic period, the whaleback dunes were active dune ridges (Haynes 1982; Kröpelin 1993). A similar interpretation is voiced by Besler (1998) based on the study of the archaeological and sedimentological evidence. It was concluded that the draa bodies are older than the artifacts lying upon their flanks (7000–9000 BP) and are older than 10,000 bp because only during the Pleistocene glacial periods the winds were strong enough for draa formation (Besler 1998).

The third and last cycle of dune development of the Great Sand Sea extended from the Middle Holocene to the present time. According to field observations and the dating of a number of samples, it was inferred that draa reactivation occurred during the Middle to Late Holocene (Beslar 1998). Active silks (sharp-crested "seif" ridges) on the draa (whalebacks) are the product of modern activity of wind on dunes in the Great Sand Sea (Besler 1998), since the cover of shifting sands was found to be 200–300 years only, according to OSL dating as will be seen in the following Sect. (13.6.1.2).

13.6.1.2 Age Determination

The results of OSL and luminescence dating carried out by Besler (2008) are recorded in (Table 13.3).

This table shows that age increases, in general, with depth, where the oldest sample is 22.8 ± 3.8 ka taken at a maximum depth of 5 m, and the youngest sample is 240 ± 50 years, taken at a depth of 1.0 m. This gives a clear indication that depth is an important factor in determining age and that older ages are expected to be at depths larger than 5.0 m. Consequently, the recorded age at a depth of 5.0 m does not represent the actual age at which the Great Sand Sea started to develop. Therefore, samples from the depth of more than 5 m are required for collection and analysis to draw a real picture for the age Great Sand Sea.

13.6.2 Development of the Great Sand Sea

According to the results of four interdisciplinary expeditions to the Great Sand Sea (Besler 2008; Bubenzer and Besler 2010), luminescence dating, and sediment analyses as granulometry, scanning electron, microscopy, X-ray diffractometry, and salinity, Besler (2008) suggested the following landscape development. Before and during the last glacial maximum LGM (>20 ka cal BP), the longitudinal draa were modeled by strong trade winds. Because of the geomorphologic conditions, in particular the flat underlying terrain, the determining morphometric and sedimentological factors, and the enhanced pressure gradient that give rise to higher wind velocities, it is likely that the draa formation took place due to helical roll-vortices. This inference is supported by the relatively coarse mean grain size of the draa sands that indicates higher wind speeds. In the northern part of the Great Sand Sea, strong westerlies lead, most likely, to the formation of transverse draa in the time span between 20 and 11 ka cal BP. At the end of the Pleistocene (12–10 ka cal BP), the strong and dry westerlies pushed sand further to the south and deposited it on the eastern slopes of the draa. Therefore, the western flanks of the draa are older in age of sand deposition near the surface in comparison with the eastern flanks accumulation. A possible interpretation for that is the erosion during that time span. During the Holocene humid phase or Holocene humid optimum ($\sim$ 10.6–7.3 ka cal. BP), runoff after annual precipitation amounts of 50–100 mm, and intermediate short deflation phases (Bubenzer and Hilgers 2003; Bubenzer et al. 2007; Besler 2008) shaped the draa into the present whalebacks (after Bagnold 1941). Runoff (from the slightly consolidated draa surfaces and seepage at their bases) supplied water for shallow episodic or seasonal ponds and lakes in shallow Depressions of the inter-draa corridors. These conditions allowed humans to live there as hunters and gatherers. Due to weathering during the Holocene humid phase, these so-called playa sediments show red colors and higher silt contents in the southern Great Sand Sea in comparison with the draa and dune sediments.

Sand-size fraction seems to have been incorporated into the modern dunes, because patches of red sands are noticed in many places of the investigated areas of the southern Great Sand Sea. After approximately 7 ka cal BP, the modern hyper-aridity started and since about 5 ka cal BP more and more sand was reactivated. As a result of the weaker winds, the draa bodies remained stable, and only 1–4 silks are found on top, built by a bimodal wind regime. Silk heights vary between 5 and 30 m, whereas the underlying draa have average heights of 50 m. The eastern part of the southern Great Sand Sea shows considerable progression of the Holocene modern dunes south of the Pleistocene draa for about 60 km. Bubenzer et al. (2007) showed that, although the resolution of the OSL data is up to ten times lower than that of the radiocarbon data, the former can still provide alternative and more compatible ages than ^{14}C data and can expand chronologies.

Such depiction of the development of the Great Sand Sea, although is acceptable scientifically, does not draw the complete picture. One reason behind such a case is that such depiction does not give any information on the period predating the last glacial period. In fact, this period is the longest in the history of the Great Sand Sea and therefore should not be ignored. Thus, the development presented by Besler (2008) covers- the last 20,000–25,000 years only.

13.7 Conclusion

The Great Sand Sea is not only the largest one in Egypt because it occupies an area that equals more than one-tenth of the total area of the country, but also because it accommodates a huge quantity of sand, either exposed as dunes or as a sand wedge below the dunes laid down in a mega-depression. To emphasize, the Great Sand Sea extends from the Gilf Kebir Plateau in the south to Siwa Depression in the north. Linear dunes of varying shapes and patterns dominate the landscape all-over the sand sea. In this respect, the Great Sand Sea is unique among other Egyptian sand seas. A possible reason for this is the presence of huge quantities of sand that inhibits the development of barchans. In other sand seas, barchans are formed in the distal sections and marginal areas where sand supply diminishes, (e.g., Ghard Abu Moharik Sand Sea), see Chap. 14. The presence of a faint barchanoid form in a few localities, where some of it is covered by the adjacent linear dunes, indicates that it is older than the adjacent sharp-crested linear dunes. It also suggests that such faint barchanoid has been covered by linear dunes in other localities. Due to the long history and variations in wind regime and pre-topography along the 600-km N–S axis, compound linear dunes and the presence of all types of dune patterns characterize the Great Sand Sea. These two characteristics highlight the difference and the individuality of the Great Sand Sea from other sand seas in Egypt.

Sands of the Great Sand Sea are originated from two sources. The first is local from the fluvial materials which were deposited in the previous mega-depression now occupied by this sand sea. The second source is external, from Siwa Depression. To add, mineral composition is the most significant aspect in sand properties, since it sheds light on origin and age. It was found that the percentage of the ultra-stable minerals of zircon, tourmaline, and rutile reaches about 60% of the total heavy minerals. This relatively high percentage implies the maturity of sediments and that the sands passed several cycles. It was also found that the percentage of zircon is relatively high, indicating that the minerals are reworked from older sediments. It can be inferred from these relationships that the sands of Great Sand Sea were exposed to several cycles and were reworked from older sediments/rocks, such as the sandstones of the Gilf Kebir and Abu Ras Plateaus and igneous rocks of the Red Sea Mountains.

Available archaeological and geomorphological evidence indicates that the dunes of the Great Sand Sea started to develop in Acheulian times, 200,000–300,000 years ago. Sand below the dunes was laid down most probably before that time during the arid phase of Middle Pleistocene, from about 700,000 to the beginning of dune development. Age determination by OSL was limited to the upper 5 meters, and this is why the determined age of sand at this depth (about 23,000 years, which is during the last glacial maximum) does not represent the actual age of the Great Sand Sea. This lack of information makes it necessary to collect samples from deeper horizons to accurately determine the age of the Great Sand Sea. In the meantime, the young ages (240–280 years) of the surface layers of sand at depths 1–2 m denote that only this layer is modeled nowadays.

References

Abdel-Rahman M, El-Baz F (1979) Deposition of a probable ansectral delta of the River Nile. In: El-Baz F, Warner D M (ed) Appllo-SoyuzTest Project II. Earth observation and photography, NASA, Washington

Albritton CC, Brookes JE, Issawi B, Swedan A (1990) Origin of the Qattara Depression. Egypt. Geol. Soc. Amer. Bull. 102:952–960

Aufrere L (1935) Essai sur les dunes du Sahara Algerien. Geogr Ann 18:481–550

Bagnold R (1931) Journeys in the Libyan Desert 1929 and 1930. The Geographical Journal 78:13–39

Bagnold R (1933) A further journey through the Libyan Desert. The Geographical Journal 82:103–129

Bagnold R (1941) The physics of blown sand and desert dunes. Methuen, London

Ball J (1927) Problems of the Libyan Desert.Geog Jour LXX: 21–38, 105–128 and 209–224

Beadnell H (1910) The sand dunes of the Libyan Desert. The Geographical Journal 35:379–395

Besler H (1998) Periods of aeolian activity and stability in the Great Sand Sea in Egypt. ZentralblattfürGeologie und Paläontologie, Tile 1(1997):23–39

Besler H (2002) The Great Sand Sea (Egypt) during the late Pleistocene and the Holocene. Zeit fur Geomorphologie N F. 127:1–19

Besler H (2008) The great sand sea in egypt: formation, dynamics and environmental change—A sediment analytical approach. Elsevier, Amsterdam

Breed CS, Fryberger SC, Andrews S, McCauley C, Lennartz F, Gebel D, and Horstman K (1979) Regional studies of sand seas using Landsat(ERTS) imagery. In: A study of Global Sand Seas (Editor: E. D. McKee). US Geological Survey Professional Paper 1952:305–397

Bubenzer O, Besler H, Hilgers A (2007) Filling the gap: OSL data expanding ^{14}C chronology of Late Quaternary Environmental change in the Libyan Desert. Quatern Int 175:41–52

Bubenzer O, Hilgers A (2003) Luminescence dating of Holocene playa sediments of the Egyptian southern Plateau Western Desert, Egypt. Quaternary Science Reviews 22(10-13):1077–1084

Bubenzer O, Bolten A (2008) Reconstructing palaeodrainage-systems by digital elevation data. In: Besler H (ed) The Great Sand Sea in Egypt—formation, dynamics and environmental change-A sediment analytical approach, Dev Sedimentol 59:39–46

Bubenzer O, Besler H (2010) Sands as archives of environmental change: examples from Egypt, Sudan and Namibia. In: Möhlig W, Bubenzer O, Menz G (eds) Towards Interdisciplinarity. Experiences of the Longterm ACACIA Project. Topics in Interdisciplinary African Studies, Cologne 15:23–48

El-Baz F (1992) Origin and evolution of sand seas in the Great Sahara and implications to petroleum and ground water exploration. In: Sadek A (ed) Proceedings of the 1st. Conference on the Geology of the Arab World, Cairo University, Cairo, 1992

El-Baz F, Breed C, Grolier M, McCauley J (1979) Eolian features in the Western Desert and some implications to Mars. J Geophys Research 84:8205–8221

El-Baz F, Prestel D (1982) Coatings on sand grains from Southwestern Egypt. In: el Baz F, Maxwell T (eds) Desert Landforms of Southwest Egypt: A Basis for Comparison with Mars. NASA, Washington DC p, pp 175–188

Embabi NS (1995) Types and patterns of sand dunes in Egypt. Bull Soc Geogr d'Egypte 68:57–90

Embabi NS (1998) Sand seas of the Western Desert of Egypt. In: Quaternary Deserts and climatic changes. Al-Sharhan A, Glennie K, Whittle G & Kendall C (eds) Proceedings of the International Conference on Quaternary Deserts and Climatic Change, Al-Ain, UAE 9-11 Dec 1995. A A Balkema, Rotterdam: 405–509

Embabi N (2000) Sand dunes in Egypt. In: Soliman S (ed) Sedimentary Geology of Egypt: Applications and Economics, part 1. Special Publication, TheSedimentological Society of Egypt, Cairo, pp 179–183

Google Earth (2014) Egypt Images Latitudes 22-32 N and Longitudes 22-32 E, http://www.google.com

Haynes C (1982) Great Sand Sea and Selima Sand Sheet. Geomorphology and Desertification, Science 217:629–632

Issawi B, McCauley JF (1993) The Cenozoic landscape of Egypt and its river systems. Ann Geol Survey of Egypt 19:357–384

Kadar L (1934) A study of the sand sea in the Lybian Desert. Geogr J 83:470–478

Kröplin S (1993) Geomorpholgy, landscape evolution and paleo-climates of Southwest Egypt. Catena Supplement 26:31–65

Lancaster N (1995) Geomorphology of desert dunes. Routledge, New York

O Bubenzer, A Hilgers, (2003) Luminescence dating of Holocene playa sediments of the Egyptian southern Plateau Western Desert, Egypt. Quaternary Science Reviews 22 (10-13):1077-1084

Pachur H, Röper H (1984) The Libyan "Western Desert" and Northern Sudan during the Late Pleistocene and Holocene. In: Klitzsch E, Said R, Schrank E (eds) Research in Egypt and Sudan, Band 50, Results of the Special Research Project in Arid Areas, Period 1981–1984, Reihe A, Berlin 50:249–284

Pye K, Tsoar T (1990) Aeolian sand and sand dunes. Unwin Hyman Ltd, London

Sadiek K, Rashed M, Mansour M (1995) Textural characteristics and genesis of barchan dune deposits, Kharga Oasis. Egypt. Egypt Journal of Geology 39:599–629

Said R (1981) The Geological Evolution of the River Nile. Springer-Verlag, New York

Salem R (1976) Evolution of Eocene-Miocene sedimentation patterns in parts of northern Egypt. Amer. Assoc. Petrol. Geol. Bull. 60: 34–64

Thomas D (ed) (1997) Arid zone Geomorphology: Process, Forms and Change in Drylands. 2nd Edition. John Wiley & Sons Ltd Chichester

Walker T (1979) Red colour in dune sands. In McKee E (ed) A Study of Global Sand Seas. US Geological Survey Professional Paper 1052:61–82

Warren A (1988) The dunes of Wahiba Sands. in the Scientific Report of the Royal Geographical Society's Oman Wahiba Sands Project, 1985, Journal of Oman Studies, Diwan of Royal Court, Muscat, Oman, Special Report No 3:131–161

Wendorf F, Schild R, Close A, Associates (1993) Egypt during the last interglacial: The Middle Palaeolithic of BirTarfawi and Bir Sahara East. Plenium Press, New York

Wilson I (1973) Ergs. Sedimentary Geology 10:77–1

Ghard Abu Moharik Sand Sea

14

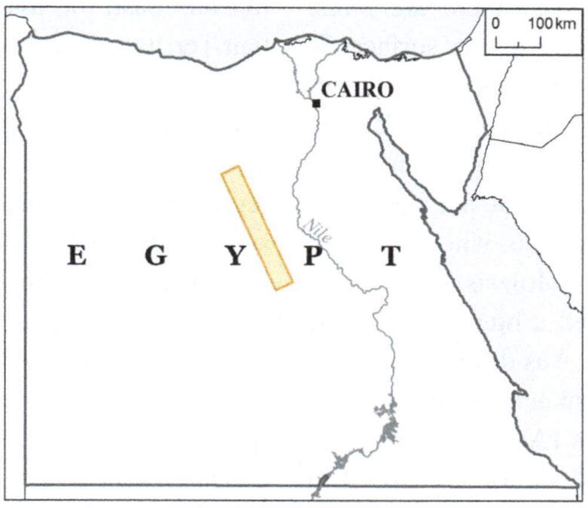

Abstract

Ghard Abu Moharik (GAM) is an unusual linear sand sea in the Western Desert of Egypt; it extends about 800 km from its proximal head northeast of Bahariya Oasis to the Egypt–Sudan border. Its width varies from 50 km at latitude 22° 15′ to 12.5 km at latitude 29°. GAM comprises three geomorphic divisions: the northern section, the central section, and the southern section. Despite the variance in dune density, the northern and middle sections comprise mainly linear dunes, whereas the southern section in Kharga Depression and the southern Plain contain mostly barchans. Mega-ripples and barchans and barchanoids with north to northwest facing slip faces characterize the central section. GAM is composed mainly of medium-to-fine quartz sands (2–3 Ø). Analysis of mineral composition indicated that the sands of the central section passed fewer cycles than those of the southern section. The texture of the grain surface exhibit effects of mechanical and chemical processes in recent and in ancient environments.

Keywords

Sand seas • Abu Moharik • Egypt • Dune forms • Barchan movement

14.1 Location and Dimensions

Ghard is an Egyptian term for any sand dune irrespective its form and/or size. It is applied to the Ghard Abu Moharik in a similar way to the term "Erg" which indicates a linear dune and is applied to some sand seas in North Africa. GAM lies in the center of the Western Desert, extending for 800 km from NE of Bahariya Oasis at 29° N latitude, across the Libyan Plateau for ~350 km, and another 450 km to the Egyptian-Sudanese border at latitude 22° N. The width of the Ghard varies greatly along its axis; 12.5–15 km along its northern section, 40–43 km in the 'central section,' and 20–50 km at the southern section, with an average width of 29 km. The linear shape distinguishes GAM from other sand seas in Egypt (Embabi 1998). Landsat-TM images show that the GAM comprises three geomorphic/sedimentary zones: the N section (upwind), the central section, and the S section (downwind). Since dune forms and thickness of sand change in any sand sea, Porter (1986) calls these three sections: Fore-Erg, Erg Body, and Back-Erg, respectively. The area of GAM is ~19,545 km^2, with a sand coverage of 55.4%, making it the third largest sand sea in Egypt, after the Great Sand Sea and Selima Sand Sheet.

14.2 Geology and Topography

GAM developed in the Quaternary period on the Eocene limestone of the Libyan Plateau (Western Desert) and across the Cretaceous sandstones of the Kharga Depression and the

N.S. Embabi, *Landscapes and Landforms of Egypt*, World Geomorphological Landscapes,
https://doi.org/10.1007/978-3-319-65661-8_14

southern Plain. Local relief has had a noticeable effect on its development and character. This 'basal' topography encompasses two major divisions: (i) the Eocene plateau and (ii) the Kharga Depression and the southern Plain. Although the plateau rises up to 200–300 m a.s.l, there are small depressions and residual hills of a former higher surface. In the central section of GAM, where some wadis are developed, the surrounding plateau surface slopes eastward. Such a case has led to the assumption that the main chain of the dunes partially occupies the trunk valley of a Paleogene–Neogene drainage system across the plateau, which predates the development of the Nile Valley (Mojzsis and El-Baz 1992). This drainage system developed a huge delta to the north of Fayum Depression. This delta was discovered at the end of the nineteenth century by Blankenhorn, but its size was only known from Landsat images (Abdel-Rahman and El-Baz 1979). The sediments of this delta are composed of sand and gravel that were deposited during the Late Eocene–Middle Miocene period (Salem 1976). The sources of these sediments are the Red Sea Mountains, the plateaus of the Eastern Desert, and the central and southern parts of the Western Desert (Said 1990; Salem 1976). The sand and gravel of this delta were a probable source for the sand of GAM (Ball 1927; Ali 1999).

14.3 Climatic Conditions

The analysis of present-day climatic condition (Chap. 3) showed that except for a narrow strip along the Mediterranean coast, the hyper-arid climate overwhelms the Egyptian Territory with cold/warm winters and hot summers, very low rainfall, and very high evapotranspiration. In a previous study on the geomorphology of GAM (Embabi et al. 2012), it was found from the analysis of wind conditions in the region (Taking Bahariya Meteorological Station as representing wind environment on the plateau and Kharga Station as representing wind conditions in the southern section of GAM), that the percentage of sand-moving winds is much lower in Bahariya (7.68%) than in Kharga (30.67%). Such findings indicate a low-energy wind environment in the region of the northern and central sections of GAM (the plateau surface), and an intermediate-wind energy environment in the region of the southern section (Kharga Depression and southern Plain), according to the classification of wind environments by Fryberger (1979). Another finding was that winds are characterized by a "wide unimodal" distribution in Kharga, where approximately 77.5% of all sand-moving winds ($\geq$ 11–16 knots) are from 315°–014°, and about 56% are from the same directions in Bahariya. The effect of this difference in wind environment on the rate

Fig. 14.1 Branching dunes of the northern section of GAM (*Source* Google Earth 2012)

of dune movement on both the Plateau and Kharga Depression is to be discussed.

14.4 Previous Investigations

It is pivotal to highlight the fact that geomorphologic studies of the GAM began about a century ago. Beadnell (1910) mapped the distribution of sand dunes in the Western Desert. In one of his descriptions, he mapped the GAM as a wide, unbroken, linear dune, trending NNW-SSE across the Libyan Plateau, with the linear dunes of Kharga Depression as a N-S continuation. In fact, Beadnell was the first to measure dune movement in Egypt. Harding King (1918) studied the slope form of six barchans in Kharga, which was similar to a study of a larger sample by Embabi (1976–1977). Ball (1927) estimated a mean southward advance of 10 m / yr for the southern section of GAM, which led to an estimate of 35,000 years for its development. Hume (1934) stated that sand forms in the Libyan Desert are composed of a series of dune ranges, and that the most notable of these dune lines is the Abu Moharik series to the east of oases. Gifford et al. (1979) and El-Baz (1992) viewed GAM as two independent divisions: Bahariya dunes and Kharga dunes. The first comprised three dune belts: Ghard Abu Moharik, Ghard Ghorabi, and Ghard El-Qazun, while the Kharga barchans were topographically separated. Sadiek et al. (1995) studied the textural properties and genesis of barchan sand in Kharga Depression. Ali (1999) carried out a comprehensive geomorphologic study on the dunes of the extreme north of the

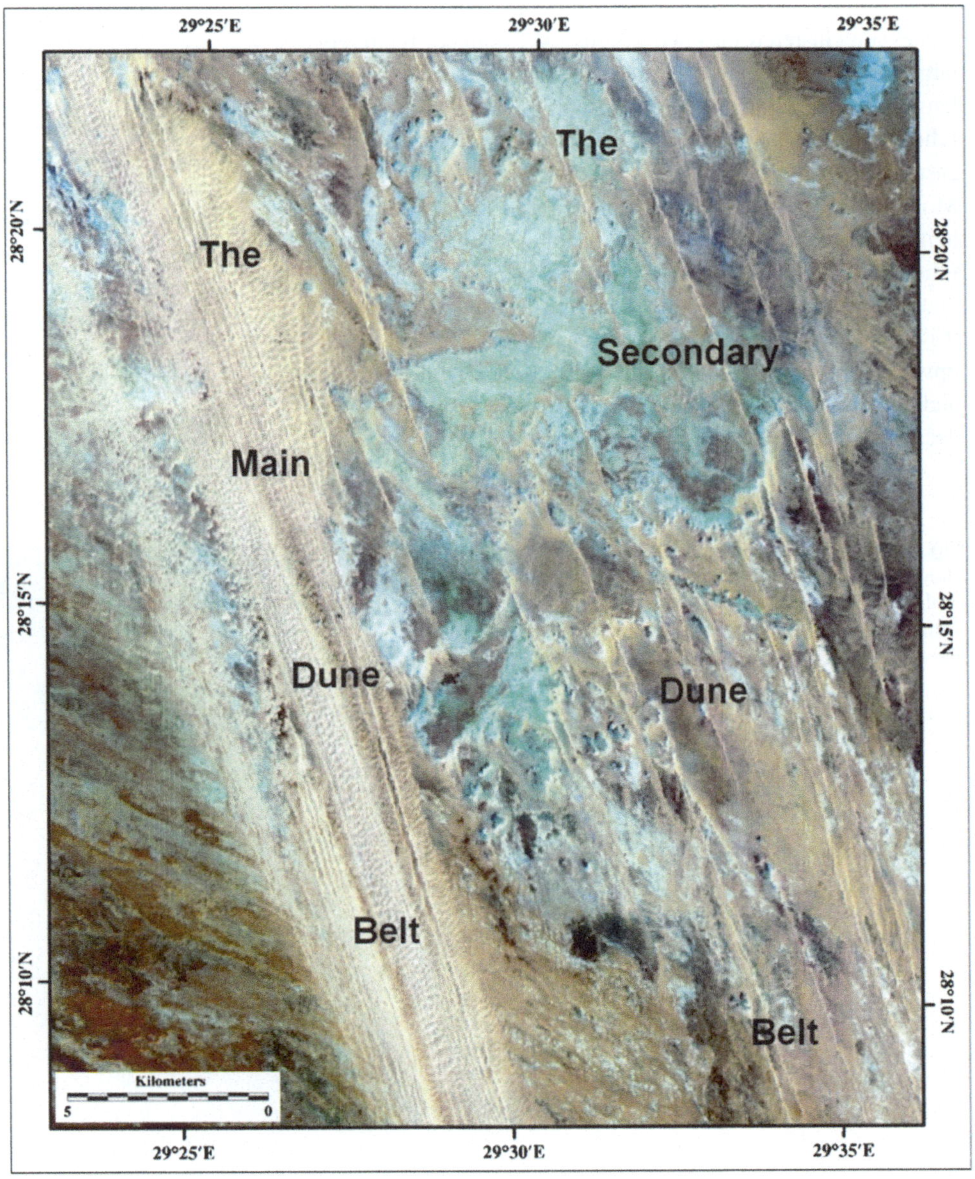

Fig. 14.2 A Landsat-TM image showing the two dune chains of the central Section of GAM

northern section of GAM. He called them "The linear dunes to NE of Bahariya Depression." The studies of Embabi (1982, 1986–1987, 1991, 1995, 1998, 2000, 2004) covered several geomorphologic aspects of GAM, topically under barchans, dune movement in Kharga, or widely through studies of the Egyptian sand dunes and sand seas. The author carried out the first thorough study of GAM and has published it recently (Embabi et al. 2012).

14.5 Dune Forms of GAM

14.5.1 The Northern Section (Fore-Erg)

The 'northern Section' of the GAM is composed of isolated small-to-large linear dunes developed at the leeward side of some outliers and escarpments trending nearly NW-SE. Dunes are clustered into two groups: (i) Chains of relatively large dunes to the west called **"Bahariya Dunes"**, with lengths from 32 km (**Ghard Williams**), to 71 km (**Ghard Ghorabi**), and heights 30–45 m (Ali 1999), and (ii) An eastern group of relatively small dunes, called "Abu Moharik Dunes" (Ali 1999), with lengths 1.5–8.5 km, and heights 5–16 m. Some of the large dunes such as **Ghard Ghorabi** and **Ghard East Ghorabi** contain several ridges that converge and unite downwind in a branching pattern (Fig. 14.1), or with "symmetrical Y-junctions" which open upwind (Thomas 1997; Pye and Tsoar 1990). Dune number and junction angle decrease downwind, resulting in a decreased width of inter-dune corridors.

14.5.2 The Central Section 'Erg Body'

Regarding the central section, it extends approximately from 28° 30′ N, SSE to 26° N, for a distance of ~280 km. The Erg body comprises high-density dunes, with predominant linear dunes, arranged in a primary chain, and a secondary discontinuous belt to the east (Fig. 14.2). The primary chain also contains barchans and barchanoid belts, as well as mega-ripples. The linear dunes lie parallel to each other with/without inter-dune corridors (Fig. 14.2). These linear dunes have sinuous, sharp crests, with slip faces looking east and overriding wide plinths (Figs. 14.3). The length of the Erg body varies from a few to tens of kilometers. In the center, between linear ridges, there is a wide belt (2–3 km) composed of coalesced transverse dunes with slip faces that take a network pattern. It is thought that this belt occupies the lowest part of a widely large wadi bed of a former Paleogene–Neogene drainage system developed on the plateau surface as suggested in a previous study (Mojzsis and El-Baz 1992). Inter-dune areas are occupied by transverse dunes, sometimes barchanoid, appearing as sand giant waves.

The barchans and barchanoids appear along the western side of the Erg body at 27° 40′ N (Fig. 14.4). Their number and density increase downwind gradually to the southern limit of the Erg body, where linear dunes disappear at 26° 30′ N. Barchans are characterized by (i) organization into several parallel chains and (ii) organization of barchan chains into two size groups: small and large barchans. Morphologic features suggest that these two groups represent different generations of barchans. To begin with, chains of large barchans (where horn-to-horn = 0.5–1.0 km, with

Fig. 14.3 A field photograph showing two parallel linear dunes in the central Section of GAM, with sharp crests and slip faces on the eastern side of the ridges

Fig. 14.4 Barchans of the western margins of the central Section of Ghard Abu Moharik (note the growth of the eastern horn more than the western one) *Source* Google Earth 2014

smooth slopes, short horns, and mostly without a slip face) are older than the belts of small dunes (width = 100–250 m). (iii) The eastern horn is longer in many small, younger barchans. Some of the long horns are connected to form small linear dunes, as noticed by Bagnold (1941) and verified by Embabi (1991, 2004) in both UAE and Egypt.

Mega-ripples (wave length ~0.25–0.5 km) are recognized on aerial photographs and space images along the eastern side of the primary dune chain and in inter-dune corridors. These mega-ripples are organized as rows parallel to the linear dunes. In some localities, these giant ripples are overridden by some linear dunes, indicating an older age for the ripples than the top dunes. The ripples have slip faces on the north, contrary to barchans and barchanoids. Space images (Google Earth) and aerial photographs show that the windward slopes of mega-ripples are darker, suggestive of a surface layer of coarse sand. Field observation proved the validity of this inference (Fig. 14.5). Such mega-ripples

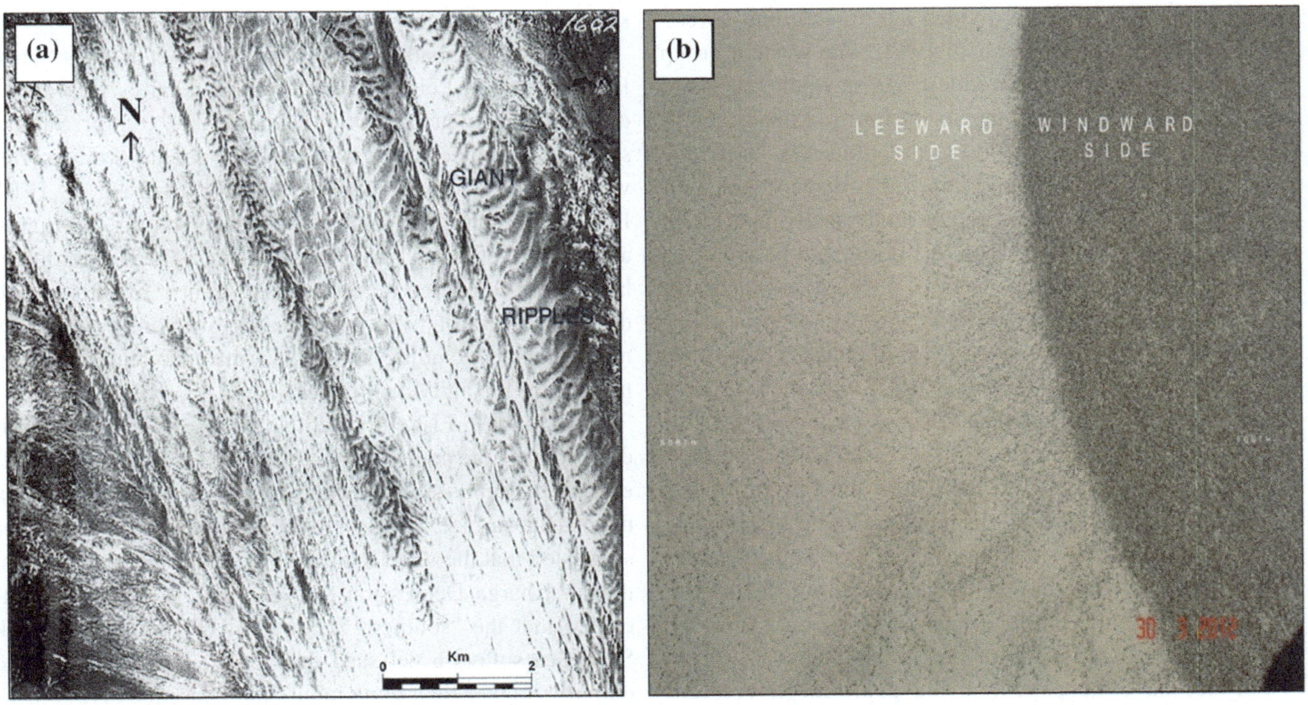

Fig. 14.5 Giant ripples of the central Section of GAM. *a* An aerial photograph, where the Ripples appear as rows at the eastern region. *b* Field photograph, where the winward side appears darker then the leeward, due to the concentration of very coarse sand grains

could have been developed in an earlier period under the influence of strong, southerly wind, with heavy sand flux.

The eastern dune belt of the GAM comprises isolated linear dunes in the lee of small outliers along the sand stream. Dune length varies from less than a km to 10's km, depending on the size of outlier obstacles and the sand supply from the north (Fig. 14.2). The dunes have sinuous sharp crests that override wide plinths. This eastern belt also contains sand-filled depressions (Fig. 14.2) which act as sand traps. Barchans are absent along this eastern belt, probably due to the coarse texture of local relief which prevents their development.

14.5.3 The Southern Section 'Back-Erg'

The 'Back-Erg' starts several kilometers on the plateau to the north of Kharga Depression (latitude 26° N), and extends southward into Kharga and the southern Plain to the Egypt–Sudan border at latitude 22° N. This southern extension, ~450 km long, is the longest and widest section in GAM Sand Sea. It is stamped by low dune density and the predominance of barchan and barchanoid chains as well as sand sheets. Dunes in this section are organized into chains oriented in an N-S direction, except for this part on the plateau, where dunes trend toward SSE. Due to local differences in landforms, relief, and wind regime, barchans are organized into two different patterns: the linear pattern and the proto-mega barchans pattern.

Linear barchan belts are the dominant dune pattern in Kharga Depression and are divided into three groups: western, central, and eastern (Embabi 1982). The western group of belts is the longest and widest. As a whole, the belt, composed of barchan and barchanoid forms, extends southward, uninterrupted for ~200 km to latitude 24° N. To the south of Kharga and southern Plain, dunes disperse and their density decreases, but dunes become larger and complex with several slip faces, with width up to 1.0 km, and height up to 30 m. By time, embryonic slip faces develop and their size and number rise. This characteristic was explained in previous studies by the amalgamation of small dunes, which move faster, with larger ones (Embabi 1982; Besler 2002a). In some cases, these large barchans become a source of sand for more younger small barchans. The extreme eastern belt, which runs at and parallel to the feet of the eastern escarpment of Kharga Depression, dismantles (due to topographic obstacles) into several small barchan chains. Some of these chains penetrate some closed basins, which are filled with excess water coming from the High Dam Lake through Tushka Spillway, and were trapped in these lakes.

The second dune pattern is the **proto-mega crescent belts**, composed of numerous, small, closely spaced barchans (Fig. 14.5), arranged in a mega-barchanoid shape and has formed leeward of low escarpments in the extreme south of GAM. These crescent belts are reported to have developed similarly in deserts of UAE (Embabi 1991; Cooke et al. 1993) and the Indian Thar Desert (Kar 1990). The width of belts is 2–6 km. As a result of the decrease in their width downwind, these belts become more elongated. Their development is attributed to a wide unimodal wind regime (Karr 1990), which prevails mostly in the south of the Western Desert.

In the south of Kharga and southern plain, a sand sheet covers the surface beneath the dunes as a result of coarse sand from the exposed sandstone bedrock.

14.6 Barchan Dune Displacement

Barchan movement is fundamental to the development, age, and prediction of GAM's future position and form(s). Previous studies have emphasized dune movement, comparing dune displacement in the downwind direction (Beadnell 1910, Embabi 1979, 1986–1987; Stockes et al. 1999; Hassan and El-Leithy 2004; Embabi et al. 2012). One study only (Embabi et al. 2012) added another aspect, which is the side movement. Here, both the downwind and lateral movements are discussed. Almost all previous studies focused on barchans of the Kharga Depression, whereas only two studies concerned barchans on the plateau (Hassan and El-Leithy 2004; Embabi et al. 2012). This is because dunes of the plateau are not accessible like those of Kharga. Sequential satellite images have permitted attempts to measure dune movement in both regions.

Studies by Embabi (1986–1987), Stokes et al. (1999), and Hassan and El-Leithy (2004) in the Kharga Depression determined similar rates of movement 9.0–9.9 m/yr. Sample size was adequate, size varied, and the annual average rates were calculated over long periods. In contrast, in 1910, Beadnell's pioneering study sample covered only five dunes, and only one year's movement was measured, recording a higher rate compared to more recent studies. Embabi (1979) drew a sample of twenty-five dunes over only one year's movement, concluding similar anomalously high rates of movement for individual dunes.

Hassan and El-Leithy (2004) worked on twenty-four dunes during the period 1984–2000, and Embabi et al. (2012) examined twenty-six dunes during the period 1987–2001 on the plateau surface. This low movement rate of 4.0–5.8 m / yr suggested that the wind regime of the plateau is weaker than that in Kharga Depression (see Sect. 14.3). It can be concluded that the "southern Section" of Ghard Abu Moharik Sand Sea suffers a net sand deficiency. The fact that dune output is higher than dune input and, consequently, is justified by the fact that dune density will decrease and GAM length will be shorter by time.

Fig. 14.6 Barchan belts in the extreme south of GAM, where dunes are organized in a Barchanoid pattern (*Source* Landsat-TM image)

The analysis of **side displacement** of the same sample of barchans on the plateau surface during the same period (1987–2001) revealed that displacement is mainly eastward. This eastward displacement is due to the effect of westerly winds (Wind categories 285°–314° and 315°–344° in Bahariya). This inference is reinforced by the fact that some barchans developed longer eastern horns than the western ones under the effect of westerly winds. A question to be posed is whether this eastward displacement is a trend in dune dynamics on the plateau surface or not? This is still unidentified, due to insufficient investigations on this point. However, if this is a trend in dune dynamics on the plateau surface, it can be concluded that there will be a shift in the 'Erg Body' of GAM eastward in the future, and consequently an eastward shift in the 'southern Section'.

14.7 Sand Properties

Previous studies on sand properties of GAM were conducted by Embabi (1982); Sadiek et al. (1995); Ali (1999); Besler (2002c); Embabi et al. (2012). These studies examine grain size, surface texture, and mineralogical composition. Samples were collected from linear dunes, barchans, and sand sheets. The samples covered the three sections of GAM. These previous studies collectively give an accurate picture for sand properties of GAM, representing a good basis for the following analysis of the sand characteristics of GAM.

14.7.1 Grain Size

Dunes of the three sections of GAM comprise mainly medium and fine sands (2–3 Ø), skewed toward FS. Mean size (Mz) lies between 2 and 2.15 Ø, in agreement with the global main dune sand fraction, which is 250–125 µm / 2–3 Ø (Besler 2002b).

Grain size in GAM dunes is unimodal, with Sorting (SD) varying between well-to-moderately sorted (0.48–0.62). Furthermore, the size of the GAM dune sand particle is almost symmetrically skewed (Sk—0.001 and 0.1), and mesokurtic to leptokurtic (K—1.01–1.17). Samples with coarse sand (1–2 mm / 0 Ø–1 Ø) were found in "southern section" and northern section. Such samples represent the sand sheets, which predominate and cover wide areas in the southern Plain. Therefore, their grain size distribution is bimodal, with a low peak for the coarse fraction. Sand in these sheets is poorly sorted (1.46), very coarsely skewed (0.36), and mesokurtic. In the northern Section, coarse sands were found at the plinths of linear dunes, where they are modeled into large ripples with different patterns. The mega-ripple rows, along the eastern side of the main dune belt of GAM body, are composed of coarse sands, with bimodal distribution, albeit with a high peak for the coarse fraction. With its very small percentage, coarse sand of the barchan dunes at the southern Plain was found along the base of the windward slope, a characteristic recorded in the Kharga barchans in previous studies (Embabi 1982; Sadiek et al. 1995). This may be due to the abundance of coarse sands produced by the disintegration of coarse-grained sandstone beds in the southern Plain.

14.7.2 Surface Texture

Scanning electron microscopy (SEM) of the surface textures' features of quartz sand grains provides valuable information about the process that acted on sand grains during transportation and deposition. **Surface textures of dune sands** of the "Erg Body" and the southern Plain were studied by Embabi et al. (2012), while Ali (1999) studied the "northern Section," and the "southern Section" by Sadiek et al. (1995). Features found on the surface of grains were photographed, with enlargements varying between 75 and 2000 times (Fig. 14.7). These features reflect the effect of mechanical and chemical processes of recent and paleo-environments. Recent mechanical features, mainly saltation impact fractures and striations and breakage due to collision, are usually superimposed on surface features formed earlier. Paleo-mechanical features include linear ridges, small and large depressions, and upturned plates. Chemical features comprise triangular solution pits, large deep depressions, and linear depressions. Silica is the most common chemical precipitate on the grains examined. Most grains, therefore, illustrate multiple surface textures, some reflecting a history of environmental change.

14.7.3 Mineralogical Composition

Dune sands of GAM are composed mainly of quartz. This confirms the results of previous studies on sand seas and dune fields in Egypt (Embabi 2000). Other light minerals are minor constituents, mostly feldspars. In the southern Section, quartz is commonly monocrystalline with minor polycrystalline grains (Sadiek et al. 1995). The predominance of monocrystalline quartz indicates multi-cycling, whereby most polycrystalline quartz and rock fragments were eliminated during transportation, due to their lower stability. Heavy minerals of dune sands comprise opaque and non-opaque minerals. Opaques include ilmenite, magnetite, and hematite. The percentage of opaques in the 'Erg Body' is lower (48.7%) than that in the 'southern Section' (76%), mainly due to local additions from surrounding bedrocks.

Non-opaque minerals are varied, ultra-stable zircon, tourmaline, and rutile (ZTR) being significant, indicating the

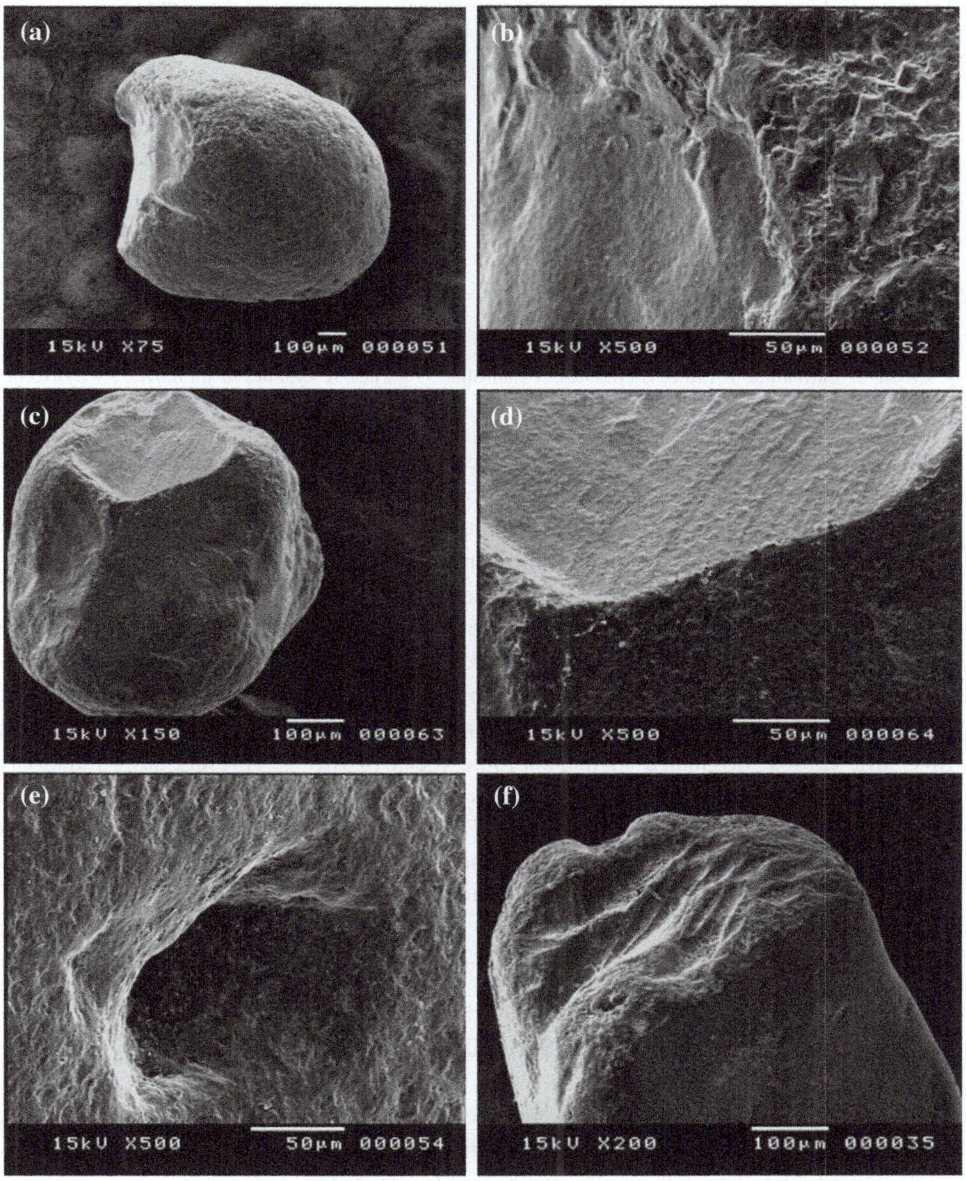

Fig. 14.7 Microphotographs of some quartz sand grains of the "Abu Moharik Body". (a) surface of a rounded quartz grain of multi-cycle erosion; (b) smoothly broken surface, and a rough surface of the silica precipitate; (c) surface of quartz grain showing two cycles of wind action represented by smooth surface and a depression on the left side; (d) an enlarged photo showing the etched and abraded surface of the top of the sand grain; (e) a deep depression where silica covers its sides and the surrounding surface; (f) different types of pits and depressions, of which the linear ones are the most conspicuous

Table 14.1 Average heavy mineral distribution of very fine sand fraction of dunes of GAM

Location	Opaque	Non-opaque	Non-opaque minerals (%)								
			Z	T	R	E	S	G	K	M	Py
Central section[(a)]:	48.71	51.29									
Dune sands			15.8	5.3	15.8	36.8	10.5	2.6	2.6	2.6	7.9
Giant ripples:			15.8	25	-		-			-	-
Windward side			15.0			60.0	-				
Leeward side			15.0			70.0		10.0	5.0		
Southern section[(b)]	76	24	37	4	22	10	7	4	2	2.2	2

(a) Embabi et al. (2012) (b) Sadiek et al. (1995)
Z zircon, *T* tourmaline, *R* rutile, *E* epidotes, *S* staurolite, *G* garnet, *K* kyanite, *M* monazite, *Py* pyroxene

maturity of recycled sediments. The total percentage of ZTR minerals of the 'Erg Body' is lower than that of the 'southern Section' (Table 14.1), suggesting that sands of the 'Erg Body' are less mature. Another significant non-opaque mineral is epidote. The percentage for the epidote mineral in the 'Erg Body' is three times more than that of the 'southern Section' (36.8–10%, respectively). The percentage of the same mineral is also very high in the giant ripples of the Erg Body (60–70%). Epidote originates in metamorphic rocks. In GAM, epidote originated in metamorphic rocks of the Eastern Desert, recycled in fluvial sediments of the Oligo–Miocene, prenile drainage system on the Eocene Plateau (Said 1990). Preponderance of epidote in the Erg Body is due to proximity of the Neogene source sediments.

14.8 Conclusion

14.8.1 Forms and Pattern of Sand Dunes

Analysis of dune forms and types along the three sections of GAM revealed three unique characteristics. The first is its linear shape, where its length ($\sim$800 km) is almost 16 times its widest area (50 km). This shape developed under the effect of pre-topography, which will be discussed later (see Sect. 14.8.3). The second characteristic is the predominance of the crescent form, a characteristic that is not reported from any other sand sea in Egypt or in other deserts of the world. It is likely that the Qatar Dune Field, which developed in the southern half of Qatar Peninsula, in the Arabian Gulf is the only field composed of barchans (Embabi and Ashour 1993). The 'southern Section', which extends for $\sim$450 km (more than half the length of the Ghard) and covers 11,000 Km^2 (also more than half the total area of the Ghard), can be considered a barchan section, as other dune forms formed under the effect of local relief occupy relatively small areas. In the 'central Section', the crescent shape developed side by side the linear form, but at their western margins. This unique characteristic could be explained by a decrease in sand supply in the 'southern Section' and along the western margins of the 'central Section'.

The second attribute is the presence of **mega-ripples** in the "central Section" with NW-NNW orientation, which is opposite to the main trend (SSE-S) of GAM. This opposite orientation cannot be explained by present-day wind environment, where expected sand drift from the southern directions is very small according to climatic conditions analysis (Embabi et al. 2012). Most probably, these forms were developed in an earlier period with wind conditions characterized by strong southerly winds blowing with higher frequencies than those of the present-day. Though the characteristics of this form can be easily recognized from high-resolution space images on Google Earth, other aspects need field investigation.

14.8.2 Source of Sand

From its location, orientation, dune movement, sand mineralogy, and wind data, it can be inferred that GAM is built of sands mainly from source(s), where fluvio-marine sediments are exposed in northern Egypt (Salem 1976). These sediments are rich in quartz sands, between 2 and 3 Ø. However, small additions cannot be ignored, especially coarse quartz grains and carbonates from local plateau sources and from Kharga Depression. Quartz grain surface textures are both relict and recent. Relicts, predominantly deep, triangular etch pits, are inherited from pre-Erg environments, such as marine inshore or sabkhas. These environments would have prevailed in the delta(s) of the Oligo–Miocene drainage system across the Western Desert, composed of cross-bedded sandstone, sand, gravel, and pebbles of quartz, chert, flint, and quartzite, derived from the Red Sea Mountains, and partly from the Nubian Sandstone (Upper Cretaceous) and the Eocene rocks to the south (Said 1990). These outcrops exist today on the northern fringes of the Fayum Depression and the Qattara escarpment (Salem 1976; Abdel-Rahman and El-Baz 1979; Said 1990).

Heavy minerals of dune sands in Kharga were more likely contributed from Upper Cretaceous sandstone bedrock flooring the Depression (Sadiek et al. 1995). Although Ali (1999) reached a similar conclusion, he suggested that the sand of the 'northern Section' of GAM is not directly derived from the Oligo–Miocene sediments, but from the linear dunes of the E and SE Qattara Dune Field that lies to the north of GAM and to the south of the Qattara northern escarpment. The sands, therefore, had been twice recycled, as confirmed by the predominance of high-stability monocrystalline quartz in the 'southern Section', and of features such as upturned plates, silica precipitates, deep depressions, wind striations, and impact fractures, developed in several environmental cycle(s).

Local bedrock and superficial deposits provide minor sources for the sands of GAM. The Nubian Group sandstones—forming the bedrock floor of the Kharga Depression and the southern Plain contain coarse sands are a possible source of the coarse fraction of the sand sheets of the "southern Section." These beds might have also contributed to the higher percentage of opaque minerals in the "southern Section" than in the "Erg Body." Satellite images and the Geological Map of Egypt also reveal Oligocene fluvial sediments, rich in quartz sands, preserved as hillocks and inverted wadis on the Western Plateau. These were brought by streams from the Eastern Desert before Late Miocene development of the Nile Valley (Said 1990). These Paleogene–Neogene fluvial sediments were the source of fine-to-coarse sands in the mega-ripples and dunes of GAM. Exposed limestone beds on the plateau, where the 'northern Section' and the 'Erg Body' extend, formed sources of carbonates in GAM sands (14.5–22.3% HCl-soluble). Precipitates on quartz grains also originated from limestone (Ali 1995).

In conclusion, the source of the quartz sands of Ghard Abu Moharik Sand Sea lies in both the primary crystalline rocks east of the Nile valley and secondary Nubian Group sandstones south of the Eocene carbonate Plateau.

14.8.3 Mode of Development of GAM

The linear form of GAM began as separate, small, linear dunes in the lee of outliers and escarpments over the plateau. These linear dunes grew and coalesced, forming larger dunes and then the sand sea. An early hypothesis suggested that the 'southern Section' of this Sand Sea advanced southward (Ball 1927), but it is more probable that the three sections of GAM developed separately and contemporaneously. Since the main sources of sand are the Oligo–Miocene sediments in northern Egypt, and the Resultant Drift Direction (RDD) is toward the SSE; linear dunes and barchans grew and coalesced also in that direction. On the other hand, the presence of some sand forms pointing NW, probably developed in an earlier period indicate that wind regime changed with time.

Local relief influenced dune form and density along the three sections of the sand sea. Mesas and north-facing escarpments, which obstruct sand flow, led to the accumulation of sand as climbing and falling dunes. Small depressions everywhere acted as sand traps. A larger scale case may have been the 'Erg Body' where linear dunes developed in a wide, Paleogene–Neogene wadi (Mojzsis and El-Baz 1992). This might explain the longitudinal form of the Erg section of GAM. Likely similar cases are the small basins lying in the sand stream of the secondary dune belt at the eastern margin of the 'Erg Body'.

14.8.4 Age of GAM

Absolute dating of the sands was not possible, for more than one reason; (i) barchan migration ensures frequent exposure of sands to optical bleaching and (ii) the difficulty of sampling linear dunes, due to the limited accessibility to the required heavy machinery. To illustrate, sampling of dunes in the Great Sand Sea reached depths of 5 meters only (Besler 2002b), which is a depth where continuous exposure and re-burial of sand is normal. Results of analysis showed merely the age at that depth (22.4 ka BP), not the age of the bulk of the dune forms, which requires sampling depths of ~ 100 m.

Dating of GAM in this study is estimated from rates of barchan movement. Embabi (1986)–1987, Stokes et al. (1999), and Hassan and El-Leithy (2004), by different, though reliable means, calculated similar rates of dune movement in Kharga, 9–9.9 m/yr. It has already been concluded—according to evidences cited in Sect. 14.8.3—that GAM developed from coeval separate sections. With length of the 'southern Section' at 450 km, and annual movement 9–9.9 m/yr, the southernmost barchan dune in this Sand Sea took about 45,000–50,000 years to reach the Egyptian-Sudanese border from the northern escarpment of Kharga Depression. Since the three sections of GAM originated contemporaneously, it can be said that the last phase of the development of GAM began 45–50 Ka during an arid period within Marine Isotope Stages 2, 3, and 4. However, there could be an earlier phase of dune development during which those forms trending NW-NNW were formed. This earlier arid phase most likely corresponds to the Marine Isotope Stage 6, from ~120 to ~200 ka. Hence, the critical need for field investigations is to verify all these inferences.

This aforementioned estimate is similar to earlier ones, which have found evidence of a hyper-arid period from 65–60 ka to 12–11 ka, (Wendorf et al. 1993; Szabo et al. 1995). In the Kharga Depression, a minimum age of 45 ± 2 ka bp for a speleothem indicates a major dry interval since that time in the Western Desert (Sultan et al. 1997). Moreover, sand samples from depths 2–5 m along the western flanks of a linear dune (draa) in the Great Sand Sea yielded OSL ages 20.8–22.4 ka BP (Besler 2002b). The interior of the 'draa' should be considerably older. The Late Pleistocene hyper-arid period is recognized in the Abu Ballas Depression where the Holocene lake deposits overlie aerodynamically shaped relief in the sandstone bedrock (Pachur and Röper 1984).

Although the results of previous studies support the age estimate reached on the basis of rate of dune movement as mentioned above, this age does not heed any change in wind direction. The same applies to the Holocene wet period. The study of Donner and Embabi (2000) showed that in the Western Desert, the orientation of elongated yardangs in playas and the direction of wind grooves in rock outcrops indicate that the prevailing winds in northern Egypt were westerly during the pre-Holocene phase of cold, arid climate. Only winter winds are westerlies at present in that area. To the south, in Kharga and Dakhla, no similar change in wind regime has been detected, where the prevailing winds were northerly even in pre-Holocene times, as they are at present (Brookes 2003). However, to what extent the westerly winds of the end of the Pleistocene in northern Egypt affected the Ghard Abu Moharik is unknown. As stated by Besler (2002b), although in the southern part of the Great Sand Sea, the paleo-westerlies beveled the tops of the draa, depositing sands on the eastern slopes, the shapes of the 'draa' were not altered, because this activity was only brief. The same may hold for GAM.

As for the effect of the wet period of the Holocene on the rate of dune movement, it can be deduced from the results of previous studies. Evidence from Djara Cave, on the Libyan Plateau at 27° 24′ N, 29° 38′ E, in the northern part of the 'Erg Body', shows that rainfall during the mid-Holocene was never more than about 100 mm/yr (Gehlen et al. 2002). Mud curls and coarse sands in playa sediments of Early Holocene age in various parts of the Western Desert signify that the climate was characterized by: short rainy and long dry seasons, and the rainfall was quite low (Embabi 1999; Swezey 2001). Therefore, it can be concluded that the relatively wet periods of the Holocene appear not to have had a significant effect on eolian activity and dune movement of GAM.

References

Abdel-Rahman M, El-Baz F (1979) Detection of a probable ancestral delta of the Nile River. In: El-Baz, Warner D (eds) Appolo-Soyuz Test Project. Vol II, Earth Observations and Photography, NASA, Washington DC, p 511–520

Ali A (1999) Geomorphology of the linear sand dunes NE of bahariya depression, the Western Desert, Egypt (in Arabic). The Arab Geographical Journal 34:323–365

Bagnold R (1941) The physics of blown sand and desert dunes. Methuen, London

Ball J (1927) Problems of the Libyan Desert. Geog Jour LXX: 21–38, 105–128 and 209–224

Ball J (1939) Contribution to the Geography of Egypt. Survy and Mines Department, Ministry of Finance Egypt, Government Press, Cairo

Beadnell H (1910) The sand dunes of the Libyan Desert. The Geographical Journal 35:379–395

Besler H (2002a) Complex barchans in the Libyan Desert: dune traps or overtaking solitons? Z. Geomorph. N. F. Suppl. 126:59–74

Besler H (2002b) The Great Sand Sea (Egypt) during the late Pleistocene and the Holocene. Zeit fur Geomorphologie N. E. Suppl. Bd 126:59–74

Besler H (2002c) The Great Sand Sea (Egypt) during the Pleistocene and the Holocene. Z. Geomorph N. F. Suppl. Bd 127:1–19

Brookes I (2003) Geomorphic indicators of Holocene winds in Egypt's Western Desert. Geomorphology 56:155–166

Cooke R, Warren A, Goudie A (1993) Desert Geomorphology. UCL Press, London

Donner J, Embabi N (2000) The significance of yardangs and ventifacted rock outcrops in the reconstruction of changes in the Late Quaternary wind regime in the Western Desert of Egypt. Quaternaire 11:179–185

El-Baz F (1992) Origin and evolution of sand seas in the Great Sahara and implications to petroleum and ground water exploration. In: Sadek A (ed) Proceedings of the 1st. Conference on the Geology of the Arab World, Cairo University, Cairo, 1992

El-Baz F, Breed C, Grolier M, McCauley J (1979) Eolian features in the Western Desert and some implications to Mars J Geophys Research 84:8205–8221

El-Gammal E, Cherif O, El-Aleem E (2005) Genetic development of Ghard Abu Moharik dune field, Western Desert, Egypt. In: Ali A (ed) Proceedings of a Symposium on the Role of Scientific Research in Combating Desertification and Fixation of Mobile Sand Dunes, Sultan Qabus University, Muscat, 2005

Embabi N (1976–1977) Slope form of barchan dunes in Kharga and Dakhla Depressions. Bulletin de la Societé de Géographie d'Égypte 49–50:13–28

Embabi N (1979) Barchan dune movement & its effect on economic development in the Kharga Oases Depression, the Western Desert, Egypt (in Arabic). J. Middle East Research Center 6:51–84

Embabi N (1986–1987) Dune movement in the Kharga and Dakhla Oases Depressions, the Western Desert, Egypt. Bulletin de la Société de Géographie d'Égypte, 59-60:35–70

Embabi N (1982) Barchans of The Kharga Depression. In: El-Baz F, Maxwell T (eds) Desert landforms of Southwest Egypt: A basis for comparison with Mars. NASA, Washington DC, pp 141–155

Embabi N (1991) Dune types and patterns in the United Arab Emirates using Landsat-TM data. Proceedings of the 24th International Symposium on Remote Sensing of Environment, 27–31 May 1991, Rio de Janeiro 2:695–909

Embabi N (1995) Types and patterns of sand dunes in Egypt. Bulletin de la Societé de Géographie d'Égypte 68:57–89

Embabi N (1998) Sand seas of the Western Desert of Egypt. In: Al-Sharhan A, Glennie K, Whittle G Kendall C (eds) Proceedings of the International Conference on Quaternary Deserts and Climatic Change, Al-Ain, Rotterdam, Balkema, 1995

Embabi N (1999) Playas of the Western Desert. Egypt. Ann Acad Sci Fennicae, Geologica-Geographica 160:5–47

Embabi N (2000) Sand dunes in Egypt. In: Soliman S (ed) Sedimentary Geology of Egypt: Applications and Economics, part 1. Special Publication, The Sedimentological Society of Egypt, Cairo, pp 179–183

Embabi N (2004) Geomorpholgy of Egypt. Vol. I, the Nile Valley and the Western Desert. Special Publication, The Egyptian Geographical Society, Cairo

Embabi N, Ashour M (1993) Barchan dunes in Qatar. Journal of Arid Environment 25:49–69

Embabi N, Mostafa A, Mahmoud A, Azab M (2012) Geomorphology of Ghard Abu Moharik, Western Desert. Egypt. Bull Soc d'Géographie d'Égypte 85:1–28

Fitzpatrick K, Summerson C (1971) Some observations on electron micrographs of quartz sand grains. The Ohio Journal of Science 71:106–119

Folk R, Ward W (1957) Brazos River bar: A study on the significance of grain size parameters. J Sediment Petrol 27:3–26

Fryberger S (1979) Dune forms and wind regime. In: McKee E (ed) A Study of Global Sand Seas. US Geol Survey Prof, Washington DC 1052:137–170

Gehlen B, Kinderman K, Linstädter J, Riemer H (2002) The Holocene occupation of Eastern Sahara: Regional chronologies and supra-regional developments in four areas of the absolute desert. Africa Prehistorica, Heinrich-Barth-Institute, Köln 14:85–116

Gifford A, Warner D, El-Baz F (1979) Orbital Observations of sand distribution in the Western Desert of Egypt, Apollo-Soyuz Test Project, Summary Science Report. In: El-Baz F, Warner D (eds) Vol. II: Earth Observations and Photography. NASA, Washington, DC, p 219–236

Google Earth (2014) Egypt Images Latitudes 22-32 N and Longitudes 22-32 E, http://www.google.com

Hamdan G (1980) The personality of Egypt, part I: The Physical personality, (in Arabic) A'alm El-Kotob, Cairo

Harding K (1918) Study of a dune belt. The Geographical Journal 51:16–33

Hassan O, El-Leithy B (2004) Monitoring of sand dunes migration for developing mitigative measures in El-Kharga Depression, Western Desert. Egypt. Egypt, Journal of Remote Sensing and Space Sciences 7:71–88

Hume, W. F., 1934, Distribution of sand deposits in Egypt, Chapter III in Geology of Egypt, Vol. I: The surface features of Egypt, their determining causes and relation to geological structure, Survey of Egypt, Government Press, Ministry of Finance, Cairo, pp. 29–81

Kar A (1990) The megabarchanoids of the Thar: their environment, morphology, and relationship with longitudinal dunes. The Geographical Journal 156:51–61

Krinsley D, Donahue J (1968) Environmental interpretation of sand grain surface textures by electron microscopy. Bulletin of Geological Society of America 79:743–748

Krinsley D, Doornkamp J (1973) Atlas of quartz sand surface features. Cambridge

Maxwell T (1982) Sand sheet and lag deposits in south Western Desert. In: El-Baz F, Maxwell T (eds) Desert Landforms of Southwest Egypt: A Basis for Comparison with Mars. NASA, Washington DC, pp p157–173

Meteorological Authority of Egypt (1975) Climatological Normals of Egypt. Cairo

Mojzsis S, El-Baz F (1992) The Bahariya River: A palaeodrainage system in the Western Desert of Egypt. In: Sadek A (ed) Proceedings of the First Conference on the Geology of The Arab World, Cairo University, Cairo

Pye K, Tsoar H (1990) Aeolian sand and sand dunes. Uniwin Hyman Ltd, London

Pachur H, Röper H (1984) The Libyan "Western Desert" and Northern Sudan during the Late Pleistocene and Holocene. In: Klitzsch E, Said R, Schrank E (eds) Research in Egypt and Sudan, Band 50, Results of the Special Research Project in Arid Areas, Period 1981–1984, Reihe A, Berlin, 50:249–284

Porter ML (1986) Sedimentary record of erg migration. Geology 14 (6):497

Sadiek K, Rashed M, Mansour M (1995) Textural characteristics and genesis of barchan dune deposits, Kharga Oasis. Egypt. Egypt Journal of Geology 39:599–629

Said R (1990) Cenozoic. In: Said R (ed) The Geology of Egypt. Balkema, Rotterdam, pp 451–486

Salem R (1976) Evolution of Eocene—Miocene sedimentation patterns in parts of northern Egypt. Amer Assoc Petrol Geol Bull 60:34–64

Stokes S, Goudie A, Ballard J, Gifford C, Sameih S, Embabi N, El-Rashidi O (1999) Accurate dune displacement and morphometric data using kinematic GPS. Zeitschrift für Geomorphologie N E, Suppl.-Bd. 116, Aeolian Geomorphology, Papers from the 4th International Conference on Aeolian Research, Oxford, 1998

Sultan M, Struchio N, Hassan F, Hamdan M, Mahmoud A, El-Alfy Z, Stien T (1997) Precipitation source inferred from stable isotopic composition of Pleistocene groundwater & carbonate deposits in the Western Desert of Egypt. Quatern Res 48:29–37

Sweezy C (2001) Eolian sediment responses to late Quaternary climate changes: temporal and spatial patterns in the Sahara. Palaeogeogr Palaeoclimatol Palaeoecol 167:119–155

Szabo B, Haynes C, Maxwell T (1995) Age of Quaternary pluvial episodes determined by uranium-series and radiocarbon dating of lacustrine deposits of Eastern Sahara. Palaeogeography, Palaeoclimatology and Palaeoecology 113:227–242

Thomas D (1997) Sand seas and aeolian bedforms. In: Thomas D (ed) Arid Zone Geomorphology: Processes, Form and Change in Drylands, 2nd edn. John Wiley, Chichester, pp 413–412

UNESCO (1977) World Distribution of Arid Regions. Paris

Wendorf F, Schild R, Close A, Associates (1993) Egypt during the last interglacial: The Middle Palaeolithic of Bir Tarfawi and Bir Sahara East. Plenium Press, New York

Wilson I (1973) Ergs. Sedimentary Geology. 10:77–71

North Sinai Sand Sea

15

Abstract

The North Sinai Sand Sea dominates the landscape of the northern part of Sinai; occupying an area of ~13,600 km. This sand sea has very limited extensions in both the Negev Desert in the east and in the plains to the west of Suez Canal. The sand sea has the most complex dune system among Egyptian sand seas. It is composed of two superimposed generations of dunes; one of them is composed of ancient fixed dunes, overtopped by a recent active dunes. Nearly all known dune forms and patterns are represented in this sand sea. Sand and dune movement represents a major hazard to development projects in the region of Sinai. Dune orientation varies greatly from one locality to another due to variations in wind regime locally as well as seasonal ones.

Keywords

North Sinai Sand Sea • Active dunes • Fixed dunes • Drowned dunes • Linear dunes • Crescent dunes

15.1 Location and Dimensions

The North Sinai Sand Sea extends from the Mediterranean coast in the north to the southern slopes of the dome-like hills and mountains of Gabal Maghara, Gabal Yeleg, Gabal Um Khisheib, and Gabal Halal in the south, and from the eastern borders of Egypt to the Suez Canal in the west. In fact, the Canal separates the western extension lying at present in the eastern margins of the Nile Delta Plain, from the main body of the sand sea in Sinai. Also, this sand sea extends across the eastern borders to the Negev Desert, and into the Bardawil Lagoon, where some drowned dunes rise above water level. Directly, in the east of the Canal, the dunes extend southward to the latitude of the Suez City (Fig. 15.1). The North Sinai Sand Sea is about 120 km from east to west, and between 30 and 120 km from north to south, with a total area of ~13,600 km^2 (Embabi 2000).

According to the type, form, pattern, and orientation of dunes, this sand sea can be divided into three main sections: The first section covers the northern part where linear dunes are the dominant form and are oriented in a nearly east–west direction. Coastal dunes along the coastal strip from the Bardawil Lagoon to Rafah at the northeastern borders represent a distinguished type in this division. The second section stretches across to the area to the east of the Bitter Lakes; this is where parallel N–S compound linear dunes and barchans of various sizes and orientations are developed. This field extends southward down to the latitude of Suez City (Fig. 15.2). Boundaries between these two divisions are gradual and not sharp. However, variations in type, form,

N.S. Embabi, *Landscapes and Landforms of Egypt*, World Geomorphological Landscapes,
https://doi.org/10.1007/978-3-319-65661-8_15

Fig. 15.1 Landsat-TM image showing the North Sinai Sand Sea

and pattern of dunes can be recognized, as will be obvious in the following analysis. The third section lies to the south of Gabal El Magharah, where barchans are forming a small dune field trending toward east-northeast (Fig. 15.3).

15.2 Previous Studies

This sand sea was extensively studied, and no other sand sea in Egypt received such attention. Nevertheless,query this sand sea has not been covered by a single comprehensive study. Previous work covered dune morphology and formation, dune movement and dynamics, sand properties, and controlling factors (wind regime, local relief, etc.) in various areas in the sand sea. Examples include the area to the west of Wadi El Arish (Tsoar 1974; Khidr 2006), the north western part of the sand sea (El-Banna 2004), sediment characteristics at the southern fringes of Bardawil Lagoon (Abdalla 2005), the northern part from the Suez Canal in the west to the international borders in the east (Hereher 2000), linear dunes west of Suez Canal (El-Disouki 2000), surface features of sand grains (Ali and Ashour 2000), sand dunes around Bardawil Lagoon (Assal 1999), sand forms to the east of Bardawil Lagoon (Abdalla 2002), eolian sand forms in North Sinai (Attia 1999), barchans in the lower section of Wadi El-Masagid (El-Disouki 1988), dune forms and sediment characteristics in various areas in the sand sea (Misak and Attia 1983), and development of dunes and morphology in El Arish region and the southeastern Mediterranean coast (Tsoar 1974, 1983, 1990). In addition to these studies, there are several others, and reference will be made to them where relevant.

Fig. 15.2 Landsat-TM image showing the various trends of linear dune patterns in the Dune Field of the eastern part of North Sinai Sand Sea

15.3 Dune Types, Forms, and Patterns

15.3.1 Types of Dunes

Field investigations and visual interpretation of Landsat-TM space images, Google Earth images, aerial photographs, and topographic maps, and previous studies (Tsoar 1974; Embabi 2000; Khidr 2006), revealed the presence of two main types of dunes in the three sections of North Sinai Sand Sea: (1) Fixed dunes, and (2) Active dunes.

Fixed Dunes: This type does not move because its sand is consolidated. Some forms of this type can be recognized on large-scale maps (scale 1:5000) and aerial photographs (scale 1:10,000). Such recognition is made feasible as most of them are modified and dissected due to exposure to erosional processes (Fig. 15.4), of which wind and water action are the most effective. **Fixed dunes** are vegetated and appear on aerial photographs as faint forms with flattened crests, whereas they are represented by contour lines on detailed topographic maps. Forms recognized are linear and barchans. This type of dunes can be classified into two groups: (a) **Continental fixed dunes**, and (b) **Drowned dunes**.

Continental fixed dunes spread all over the North Sinai Sand Sea. Most of these dunes are exposed on the surface of ground and are partially vegetated. Active dunes hide a relatively small percentage of the surfaces of the fixed dunes

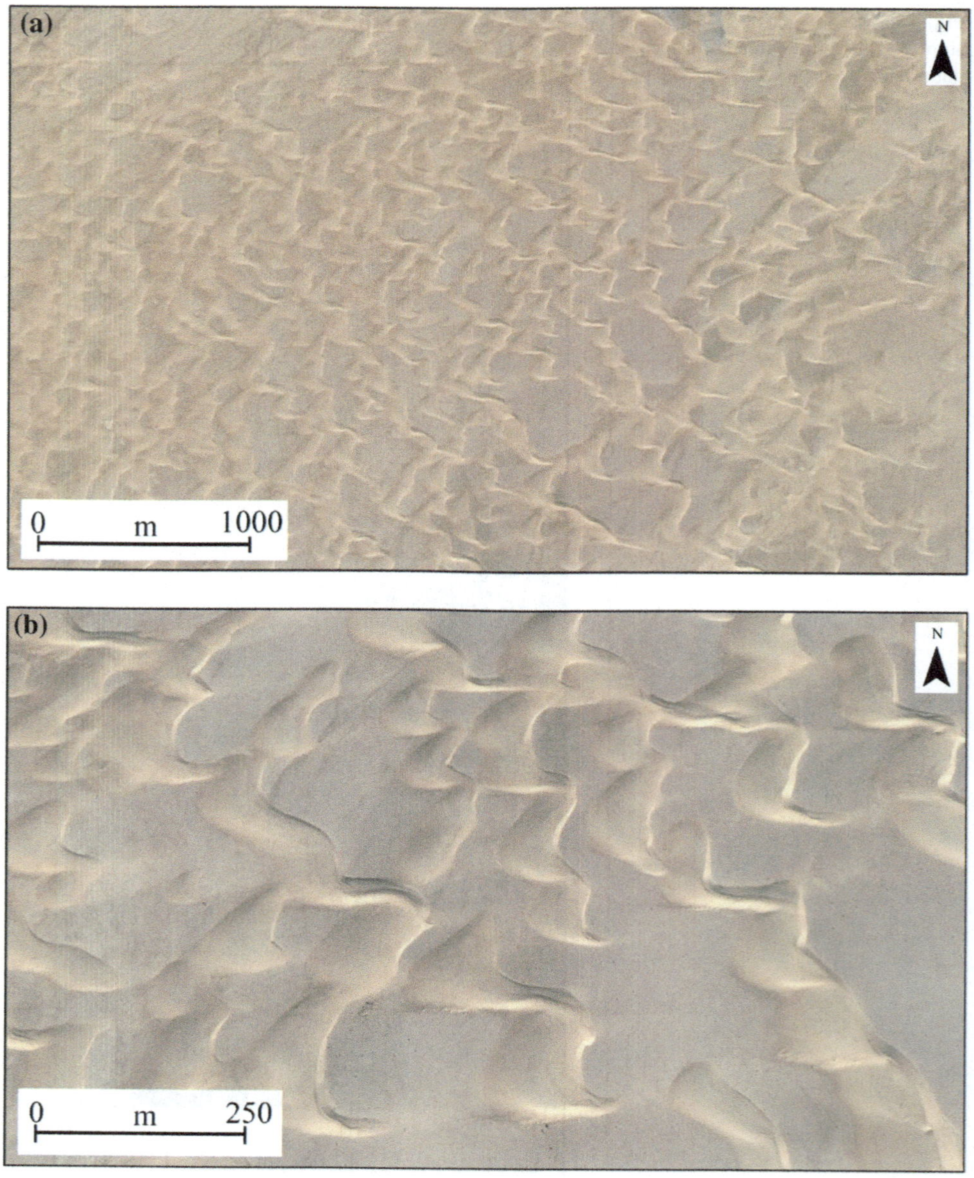

Fig. 15.3 Barchan Field between G. Maghara and G. Yeleq (*Source* Google Earth, 2014)

(Fig. 15.4). According to previous studies, these fixed dunes developed during the Late Pleistocene (Tsoar 1974; Pye and Tsoar 1990; El-Disouki 2000). Their age will be reviewed in Sect. 15.7.2.

Drowned fixed dunes: Space images (Google Earth 2014) and topographic maps (scale 1:5000) show the presence of some small islands in Bardawil Lagoon taking the form of dunes (Fig. 15.5). In addition, there are other fixed dunes on the surfaces of the coastal sabkhas. Such dune islands represent the most elevated sections of some dunes in a wider area drowned completely by seawater. To explain, the Mediterranean Sea level rose after this generation of dunes developed which caused its submergence. Available evidence shows that this rise in sea level occurred about 5000 years BP (Kidson 1982). It is likely that this submergence happened after the stabilization of dunes by cementation during the last pluvial period of the Holocene that occurred between 9500 and 5500 bp (Embabi 2004; Kuper 2006). Therefore, it can be inferred that there is an extension of the North Sinai Sand Sea into Bardawil Lagoon, and most probably into the Mediterranean Sea itself.

Active Dunes is a type of dunes composed of loose sand and exposed to wind action(s). Due to removal and deposition of sand on both sides of dunes, changes occur; dunes grow and in some instances move in the downwind direction. They developed mainly atop fixed dunes (Fig. 15.4), but they do not cover the whole surface of fixed dunes. As a result, it is inferred that the sand of active dunes is mostly originated from fixed dunes. This type of dunes can be

Fig. 15.4 Field photograph showing fixed dunes overtopped by active dunes

Fig. 15.5 Drowned dunes in the NE corner of Bardawil Lagoon (*Source* Google Earth, 2014)

divided into two categories: (a) **continental dunes** and (b) **coastal dunes**.

Continental active dunes spread all over the North Sinai Sand Sea, except for the coastal dunes strip. They take several forms of which the linear form is the most common. Other forms are barchans, transverse, and lee dunes.

Coastal active dunes exist in two areas. The first is a zone of active coastal dunes parallel to the shoreline in a narrow strip, up to 6 km wide, which extends from the eastern margin of Bardawil Lagoon to Tel Aviv in Israel. The second area is the sand barrier which separates Bardawil Lagoon from the Mediterranean, especially in the locality

Fig. 15.6 Field photograph for the highest coastal dune El-Qals Dune

known as El-Qals, where one of the dune rises up to 57 m asl (Fig. 15.6). Coastal dunes in Sinai do not develop specific form(s). Since active coastal dunes throughout the world are known to be recent, it was only after 6500–3000 year BP that the sea reached its present level after the last rapid post-glacial period. Through the analysis of human remains, it was concluded that eolian sand started to encroach on the coastal plain of Sinai between seventh and ninth centuries AD. Therefore, the maximum age of the southeastern Mediterranean coastal dunes is about 1000 years (Tsoar 1990).

15.3.2 Dune Forms and Morphometric Characteristics

The dominant dune forms in the North Sinai Sand Sea are the linear, crescent, and transverse dunes, of which the linear form is the most widespread as can be seen on TM-Landsat, Google Earth images, and aerial photographs, and as indicated by previous studies (Khidr 2006; Hereher 2000; Assal 1999; Tsoar 1974). Lee dunes developed on slopes of the structural ridges of the dome-like hills and mountains, which border this sand sea from the south. Dune orientation differs greatly from one locality to another. However, four main trends can be recognized on aerial photographs/TM-Landsat images: N-S, E-W, NE-SW, and NW-SE. These variations in dune orientation are explained by wind regime system, which changes seasonally and locally. Morphometric characteristics of these dune forms will be discussed in the following section, covering dimensions and cross section of linear dunes and barchans only, since other forms (lee and transverse dunes) have limited distribution.

Linear Dunes: There are several measurements on the dimensions of **active linear dunes** (Embabi 1995; Attia 1999; El-Disouki 2000; El-Banna 2004; Khidr 2006). Table (15.1) summarizes the results of these studies. Dune length varies between 45 m and ~25 km, the dune width ranges from 8 to 800 m, and its height from 2.4 to 64.8 m. As shown by previous studies, most of the linear dunes are relatively small ones (except for the relatively high heights and lengths of some dunes). These dimensions show that the linear dunes of North Sinai Sand Sea are smaller than those of other sand seas in Egypt (Embabi 2000), where some dunes extend for several tens of kilometers up to 100 km in the Great Sand Sea and Ghard Abu Moharik Sand Sea (see Chaps. 13 and 14). This is probably due to the fact that the active dunes are the most recent dune system in North Sinai Sand Sea that developed with the onset of the present arid climate (about 6 ka), like the active tops (silk) of the draas in the Great Sand Sea (Besler 2002, 2008). This relatively recent age did not allow linear dunes to grow and/or coalesce into longer ones. Some previous studies showed that there is a weak positive relationship between the dimensions of linear dunes (El-Disouki 2000). This indicates that the parameters of linear dunes do not grow at a similar rate like those of barchans (Embabi 1978, 2000), due to the effect bidirectional wind system, which allows a higher elongation rate than that of other parameters.

Crescent Dunes: Dimensions of **crescent dunes** were studied in several localities in the North Sinai Sand Sea (El-Kayali 1984; El-Disouki 1988; Attia 1999; Akl 2002;

Table 15.1 Statistical parameters of linear dunes in different localities in North Sinai Sand Sea (**a**) El-Disouki 2000; (**b**) Topographic Maps, scale 25,000; (**c**) Khidr 2006; Attia 1999; (**d**) Embabi 1995)

Locality	Sample size	Length (km)	Width (m)	Height (m)
East Suez Canal[a]	50	1.800–15.00	90–500	3.0–45.0
East Suez Canal[b]	20	0.500–07.00	20–050	5.0–26.0
West Wadi El Arish[c]	90	0.068–24.75	08–813	3.0–58.0
Northern Strip West Wadi El Arish (33°46′–32° 52′ E, 31° 07′–31° 01′N)[d]	425	0.045–02.92	10–555	2.4–64.8
Several areas[e]	30	1.00–20.00	100–200	3.0–55.0

Measurements based on topographic maps

Table 15.2 Crescent dunes dimensions (meters) in the North Sinai Sand Sea

Locality	Length	Width*	Height	Eastern Horn	Western Horn	Orientation
West Wadi El Arish[a]	142. 6	256	16.3	119	234	305°
Various Areas[b]	137.4	133	12.4	Average = 105.8		–
Wadi El-Masagid[c]	36.7	49	4.2	–	–	–
The Mediterranean Coastal Plain[d]	–	–	2.5–25.5	–	–	–
South G. el Magharah[e]	43.2	40.8	2.5–7.5	11.2	9.5	74.5°

[a]Khidr 2006; [b]Attia 1999; [c]El-Disouki 1988; [d]El-Kayali 1984; [e]Present study
*Distance from horn to horn. Measurements based on topographic maps

Khidr 2006). Table (15.2) summarizes the dimensions of crescent dunes according to some of these studies as well as the present one. The table reveals that the dimensions of the barchans of Sinai do not differ from other normal barchans, not only in Egypt, but also in other deserts of the world (Embabi and Ashour 1993). However, barchans with widths (distance between horns) and lengths reaching 2–4 km (mega-barchans) were recorded along the southern margins of the Sand Sea to the north of Gabal El Magharah and Gabal El-Risha, around Lat. 30° 30′ and Long. 33° (El-Kayali 1984).

Cross sections of linear and crescent dunes in North Sinai Sand Sea show that the main features of these forms are similar to those in other sand seas and fields, not only in Egypt (Embabi 1976–1977; Ali 1993) but also in other deserts (El-Disouki 1988, 2000; Lancaster 1995; Attia 1999; El-Banna 2004; Khidr 2006). Both forms of dunes have two opposing, asymmetrical sides of different form. In crescent dunes, the windward side takes the form of concave-convex, while the leeward side is straight. Linear dunes have a straight-concave windward side, while the other side is convex. On crescent dunes, however, the cross section remains unchangeable as the dune reached the mature stage, but some seasonal minor modifications occur on the crest due to the effect of opposing wind. On the contrary, the form of both sides on linear dunes is changeable due to seasonal changes in wind direction, as will be seen in the section of dune dynamics.

15.3.3 Dune Patterns

Any given dune does not only change from one form to another, but also the same form might take different patterns. This is due to many factors; beside wind regime, other controlling factors of **dune pattern** are local relief and slope direction. Due to the predominance of linear dunes and barchans, patterns of these two forms only will be discussed in the following subsection.

Linear dunes are organized in different ways/patterns. In each pattern, spacing and angles of dune lines are the critical aspects that govern the organization of dunes in each pattern. These patterns are categorized as follows: parallel, curved parallel, braided flat topped, branching (Y-junction), and network patterns.

Straight/Sinuous parallel pattern with sharp crests belongs to the active dune-type modeled under the present arid conditions. They also develop when vegetation is removed due to overgrazing from the surface of vegetated dunes along the Sinai/Negev border (Pye and Tsoar 1990). Sinuous crest lines, along which peaks and saddles are formed intermittently, characterize dunes (Embabi 1995; Pye and Tsoar 1990). **Straight parallel pattern** exists in the east of Bitter Lakes Field (Fig. 15.2), but dunes are compound since each one is composed of several ridges forming bundles of straight linear dunes. Under bidirectional wind regimes beating the dune obliquely, this pattern of dunes is shaped (Bagnold 1941; Tsoar 1974, 1983). This was verified by studies conducted on the dynamics

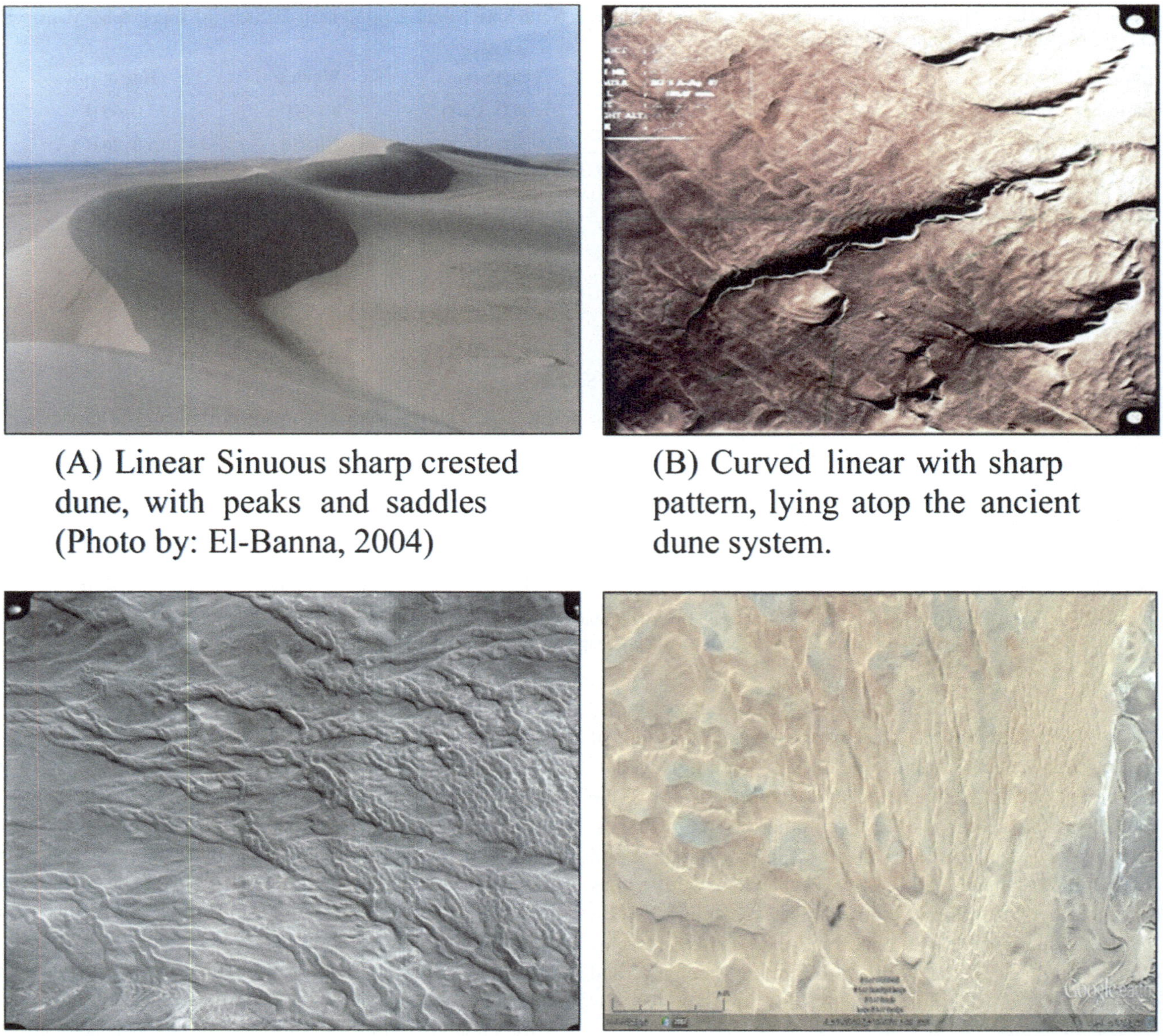

(A) Linear Sinuous sharp crested dune, with peaks and saddles (Photo by: El-Banna, 2004)

(B) Curved linear with sharp pattern, lying atop the ancient dune system.

(C) Braided flat-topped Pattern.

(D) Branching Pattern.

Fig. 15.7 Patterns of linear dunes in North Sinai Sand Sea

of sand removal and deposition on both sides of linear dunes along their longitudinal axis.

Curved parallel linear dunes developed in several areas in the North Sinai Sand Sea. Each dune is composed of several coalesced crescentic units that take curved shape. In cross section, all units developed two opposing sides, of which the lee side takes the form of a slip face (Fig. 15.7b).

Braided flat-topped pattern exists in the southern part of the sand sea near the western foot slopes of Gabal Um Khesheib. Each dune in this pattern is composed of several secondary ones that branch and unite (Fig. 15.7c). The whole system might take a network or branching pattern. All dunes are characterized by flat surfaces, with variable width between 100 and 200 m. Notably, this pattern was not reported in any previous study on sand dunes of northern Sinai; therefore, no interpretation for its development is available. However, it can be deduced that these dunes belong either to an ancient generation and have been flattened by erosional processes after being consolidated, or they belong to the present-day active generation and have been flattened by water action after periodic torrential rain.

Branching (Y-Junction) Pattern has developed in several areas in this sand sea, such as the area west of Wadi El Arish (Khidr 2006). Dunes in this pattern may branch or unite. They form what has been previously called Y-junctions before continuing as a single ridge (Greeley and Iversen 1985; Pye and Tsaor 1990). According to the orientation of branching, two types of branching dunes were recognized (Topographic Maps scale 1:5000; Khidr 2006). The first type is oriented toward the upwind direction, whereas the branching of the second type is oriented downwind (Fig. 15.7). Coalescence of linear dunes has been attributed to a deflection of the extremity of the ridge by crosswinds during the elongation process (Tsoar 1989). In another previous study (Embabi 1995), it was suggested that local relief plays a major role in merging or branching linear dunes. Dunes merge (upwind junction) when the surface of the ground slopes downwind, initiating convergence of the streamlines in the same direction. On the contrary, dunes branch in the downwind direction, where the surface of the ground rises up leading to wind divergence in the downwind direction.

Network Pattern exists in a few localities, and on the ancient fixed dune generation. It is composed primarily of linear dunes linked to secondary smaller ones. This pattern is recorded in the dune field to the east of Suez Canal, and in the southern margins of the sand sea.

Patterns of Crescent Dunes: This form of dunes takes two patterns. In the first pattern, barchans are organized into longitudinal belts that extend in the downwind direction. This occurs in the downwind section of lee dunes, and linear dunes. In this pattern, dunes grow larger as they move in the downwind direction. Also, inter-dune areas become wider. These two characteristics are the result of the coalescence of small dunes that move faster than the larger ones (Finkel 1959; Embabi 1967). In the second pattern, barchans are clustered without any clear system (Fig. 15.3). Mostly, this is due to the effect of topography, where development of barchans took place in a low-lying /basin-like area. This is the case in Sinai, where dunes were formed in the low-lying area between Gabal Maghara and Gabal Yeleq. A similar case was recorded in Kharga Depression, where barchans are trapped in a basin-like area to the east of Gabal El-Qarn in the central part of the Depression (Embabi 1967, 1982).

15.4 Dune Dynamics

The interaction between effective winds and dune sands leads to changes in the characteristics of dune form. As for all dunes in world deserts, the size of dune sands in Sinai varies between medium and fine. The wind system (direction and velocity) represents the most effective factor in dune dynamics. Active (dynamic) dunes in Sinai are exposed to winds that seasonally change direction and velocity, resulting in changes in the rate of both sand removals and deposition from one side to the other side. **Dune dynamics** result in the growth of dunes, elongation of linear dunes and movement of barchans.

15.4.1 Sand Removal and Deposition on Linear Dunes

Recent studies (Tsoar 1983; Tsoar and Yaalon 1983) confirmed that the development of linear dunes occurs under bidirectional wind regimes as deduced by several workers from the internal structure of linear dunes. This internal structure is characterized by two opposing dipping sets of sand layers (Bagnold 1941). In this regime, the dune is subjected to daily/seasonal bidirectional winds, with the wind hitting the dune sideways alike. Therefore, sand transported across the crest does not leave the dune, but rather moves down its length. On the windward flank, the sand is transported parallel to wind direction. Since linear dunes crests are meandering, the angle of wind approach fluctuates along the dune. As the wind system that forms linear dunes is bidirectional, zones of net erosion and net deposition develop on both sides of the dune. These dynamic actions have been observed first by Tsoar (1978) and were confirmed by three field measurements. The first one was carried out by Tsoar in 1983; the second one was in 2001/2002 in the vicinity of Bir El-Abd and Wadi el Gady

Table 15.3 Annual elongation rate (m) of some linear dunes in North Sinai Sand Sea

Sample size	Khidr (2006)[(a)]	El-Banna (2004)[(b)]	Akl (2002)[(c)]	El-Disouki (2000)[(d)]
1	27	12.3	3.9	Sample Size = 50 dunes
3		6.7	3.6	Range of Elongation = 5–15 m
4		5.1	5.5	
5		9.6	3.2	
6		9.6		

[a] West Wadi El Arish, during the period March 1, 2003–February 16, 2004; [b] East Suez Canal, during the period 1958–1998; [c] East Suez Canal, during the period 1984–2001; [d] East Suez Canal, during the period January–December 1998

(El-Banna 2004), whereas the third one was in the western section to the west of Wadi El Arish during the period March 1, 2003–February 16, 2004 (Khidr 2006). Due to sand removal and deposition, the linear dunes elongate in the downwind direction and the cross section changes and shifts laterally.

Elongation: Winds passing obliquely across linear dune crests are deflected and blow parallel to the dune axis on lee side, as observed in Sinai (Tsoar 1983) and in the field in Farafra Depression in the Western Desert by the author (13 April 2002). The deflected movement of wind along the lee side carries sand downwind to the far end of the dune resulting in the elongation of the linear dune.

Table (15.3) summarizes the results of four studies on the annual rate of elongation of some linear dunes in various localities in the North Sinai Sand Sea. The table shows that rate of elongation of linear dunes differs not only from one area to another, but also from one dune to another in the same area. These variations in the rate of elongation were attributed partially to the rate of sand movement along the lee flanks. This movement depends on the angle of incidence between the wind direction and the crest line (Tsoar 1983).

Previous studies (Bagnold 1941; El-Baz et al. 1980; Haynes 1980; Ali 1993) conducted in the Western Desert of Egypt revealed similar findings. These studies concluded that the annual elongation rates of linear dunes are discordant from one chain to another. Therefore, it can be deduced that linear dunes elongate at various rates according to local conditions of wind system and sand supply.

Cross Section Changes and Lateral Shift in a linear dune profile at a fixed point arise from additions of sand (deposition) resulting in an increase in volume. These cross section changes and lateral shifts could also be the result of a reduction in volume (erosion) as a result of the longitudinal movement. Recent measurements have revealed several aspects that characterize the changes in the cross section. First, removal and deposition occur on both sides of dune as recorded by Tsoar (1983), and El-Banna (2004) during the period May 2001–May 2002, and by Khidr (2006) during the period March 2003–February 2004. This was on account of daily and seasonal changes in wind direction and velocity. Khidr (2006) found that removal and deposition occur at the same point according to changes in wind characteristics. Second, due to this exchange in wind activity on both sides of the dune, lateral shift of the dune body is nearly negligible (El-Banna 2004; Khidr 2006). Additions of sand change periodically according to wind direction and velocity when occurring, especially at the top section of the leeward side. Slip faces develop due to sand deposition by wind blowing from the reverse direction in this top section of the leeward side. Moreover, changes in slope form can be clearly noticed during spring and summer months under the effect of relatively strong winds from southwest and northeast (Khidr 2006).

15.4.2 Dynamics of Crescent Dunes

Dynamics of Crescent dunes are divided into two aspects: dune advancement in the downwind direction, and changes of dune morphology.

Movement of Crescent dunes was measured in the field on three small samples in region of El Arish City (Tsoar 1974). The second measurement involved five dunes (June 2001–June 2002) in the region to the east of Suez Canal (Akl 2002). The third measurement covered seven dunes along the coastal plain during the period November 12, 1981–August 15, 1982 (El-Kayali 1984). A fourth measurement is made on two series of space images (1984–2001) in the region to the east of Suez Canal (Akl 2002). Table (15.4) summarizes the results of the field measurements studies.

In spite of the small size of the four samples and the short period of field measurements (one year), the table shows

Table 15.4 Results of measurements on annual movement of barchan dunes, the North Sinai Sand Sea

Sample size	Dune height (m)				Total advance (m)				Summer advance (%)		Winter advance (%)	
	Tsoar (1974)	El-Kayali (1984)	Akl (2002)	El-Banna (2004)	Tsoar (1974)	El-Kayali (1984)	Akl (2002)	El-Banna (2004)	Tsoar (1974)	Akl (2002)	Tsoar (1974)	Akl (2002)
1	1.9	4.5	–	–	13.1	16.0	9.6	6.8	0.9	41.7	99.1	58.3
2	1.0	5.2	–	–	6.4	15.5	11.9	2.7	6.9	36.1	93.1	63.9
3	4.0	6.3	–	–	6.2	14.5	15.3	3.0	10.7	32.7	89.3	67.3
4	–	10.0	–	–	–	14.0	12.9	4.5	–	36.4	–	63.6
5	–	14.9	–	–	–	13.2	17.1	5.0	–	39.8	–	60.2
6	–	15.2	–	–	–	13.1		4.2	–	–	–	–
7	–	16.5	–	–	–	10.7		3.3	–	–	–	–
8	–	–	–	–	–	–		1.4	–	–	–	–
9	–	–	–	–	–	–		1.1	–	–	–	–

significant variations in dune advance. Tsoar (1974) as well as authors of other previous studies attributed these variations to differences in dune size, both in Egypt and in other deserts of the world. Therefore, it can be inferred that the same rule holds for the results of Akl (2002) in the region to the east of the Suez Canal. The same table also reveals that in both studies by Tsoar (1974) and Akl (2002) most of the movement occurs in winter, especially in the coastal area of El Arish City. In winter, differences between both areas dune advances could be due to incongruity in wind energy. Tsoar (1974) found that most barchans advance in El Arish region results from winter storm winds. However, regional variations in wind energy will be discussed in the section of factors affecting the development of the North Sinai Sand Sea.

It was found from the two previous studies of Tsoar (1974) and of Akl (2002) that there are some differences in the direction of dune advance. In El Arish region, dunes move eastward (Tsoar 1974), whereas in the region to the east of Suez Canal, barchans advance in the southeast direction (Akl 2002). This dissimilarity can be interpreted by regional variations in wind direction. Also, seasonal variations in wind direction in the region of El Arish have a significant effect on dune movement. In the summer, north wind increases and a southward movement of sand is noticeable (Tsoar 1974). This causes every dune to advance southward. Measurements made on a dune of this type over a year showed that the southward summer movement is one-third of the principle advance eastward in winter (Tsoar 1974). This seasonal change represents a characteristic unique to the direction of dune movement in the region of El Arish. No similar change was recorded in other areas, according to previous studies

In addition to these two field studies of barchans advance, Akl (2002) made another study on a larger sample of dunes (32 dunes). The study involved measuring the displacement of barchans on two sets of Landsat-TM images taken in 1984 and 2001. The study revealed that barchans move southward and eastward, indicating that the dunes are exposed to two effective wind directions: the northwestern and northern ones. The annual average dune shift is higher toward the east (4.63 m) compared to that toward the south (3.95 m). This is because the NW winds are stronger and more frequent than the northern winds. It was also noticed that small dunes became larger during their movement. This means that sand deposition overweighed sand removal on these dunes and that there is a continuous sand supply from the northern upwind region.

In the study of El-Kayali (1984), dune height was measured twice, once at the beginning of the survey on November 12, 1981 and the second at the end of the survey in August 15, 1982. From these measurements, it was found that all dunes (seven dunes) increased in height by 0.8–1.4 m during this period. This result complies with the results of Akl (2002) that dunes grow during their movement.

Other attempts to investigate barchans advance in North Sinai were carried out on some dunes close to El-Sadat Village (31° 01′ N, 33° 06′ E) during the period (May 1997–May 1998). In another study (Assal 1999), field measurements made every 3 months during that year revealed an annual rate of 6–7 m advance for dune up to 4 m high. The number of dunes studied was not mentioned.

Changes in dune morphology occur on the surface of dune sides by removal and deposition of sand due to wind action, thus barchans develop under unidirectional wind regime. Another aspect of barchans morphology is that their components (horns, width, length, etc.) are symmetrical. Slope form is concave-convex on the windward side and is straight on the leeward side. Since the dune is composed of loose sand, its morphological aspects are susceptible to any change in wind direction and speed, as well as sand supply.

In an attempt to investigate the changes that occur in the morphology of barchans dunes, a field survey was carried out on five dunes lying to the east of Suez Canal (Akl 2002). Using a total station, a survey was conducted in June 2001 and another one in June 2002.

These surveys concluded that all dunes grew in size. Yet, the eastern horn grew at a greater rate than the western one under the effect of the strong NW wind. There was also seasonal change in the position of dune crest, especially in winter, due to the effect of a reverse wind that comes from the south. Dune growth is recorded also on the seven dunes of the southern part of the sand sea, where their movement was measured during the November 12, 1982–August 15, 1982 (El-Kayali 1984).

In addition, two attempts were made to investigate changes that occur on the morphology of linear dunes. The first attempt was in the area to the west of Wadi El Arish (Khidr 2006); there, a field survey was conducted on a single dune during the period March 1, 2003–February 16, 2004. The second attempt examined six dunes in the dune field lying to the east of the Suez Canal; it involved a field survey (May 6, 2001–May 6, 2002), and interpretation of aerial photographs. The first survey made by Khidr (2006) revealed significant findings on linear dunes dynamics in the North Sinai Sand Sea. The first finding is that only minor local changes occurred regarding the orientation of the dune axis. This corresponds to the stability of wind regime, which is characterized only by minor seasonal changes. Another finding is that the slip face might shift from one side to the other one and re-shift again to its original place. This occurs under the effect of temporary local reverse strong wind, which removes sand from one side to the other. Also, it was found that this dune grew in length and width. Its length

Table 15.5 Mean grain size (M_z) and sorting coefficient (σ_1) in some areas of Sinai Sand Sea (Ø)

Area/Author	West El Arish (Tsoar 1974)	Various areas in the sand sea (Misak and Attia 1983)	The coastal plain (El-Kayali 1984)	Around Bardawil Lake (Assal 1999)	North Sinai Sand Sea (Attia 1999)	Northern part of the sand sea (Hereher 2000)	NW corner of the sand sea (El-Banna 2004)	Southern margins of Bardawil Lake (Abdalla 2005)	West Wadi El Arish (Khidr 2006)
M_z	(1) 2.27 (2) 2.21	(3) 2.05 (4) 1.84 (5a) 1.16 (5b) 2.02–1.78 (6a) 2.12 (6b) 1.59	2.2	(7) 2.19 (8) 2.09	1.81	(7) 0.44* (8) 0.47*	(9) 2.08 (10) 1.95 (11) 2.19	1.51	1.65
σ_1	(1) 0.44 (2) 0.61	(3) 0.56 (4) 0.57 (5a) 1.11 (5b) 0.73–0.6 (6a) 0.57 (6b) 0.64	0.34–0.88	(7) 0.53 (8) 0.60	0.57	(7) 2.01* (8)1.95 *	(9) 0.47 (10) 0.53 (11) 0.49	0.19	0.72

(1) Coastal samples, (2) Continental samples, (3) Risan Aneiza–El Arish area, (4) El Arish—El-Tina Bay area, (5a) Western Fringes of east El-Qantara—Ismailia area, (5b) Eastern Sand Dunes of El-Qantara—Ismailia area, (6a) Gifgafa-El-Hamma Depression, (6b) South El-Tih Depression, (7) Linear dunes, (8) Crescentic dunes, (9) Windward side, (10) Leeward side, (11) Crest
* Calculated by the author

increased from 1312 to 1339 m, which means that it grew by about 2%, whereas width increased from 36 to 41 m, which shows that it widened by (about 13.9%).

Measuring the thickness of sand removal and deposition on both sides of the dune along its axis, Khidr (2006) found that the dune can be divided into two sections: the upwind section where sand removal exceeds deposition, and the downwind section where deposition exceeds removal. This process might explain the growth of dune in the downwind direction.

15.5 Physical and Mineralogical Aspects of Dune Sands

15.5.1 Physical Aspects

Grain Size was studied by several researchers/scientists. Each study covers a certain locality/various localities according to the aim and purposes of the study. Table (15.5) summarizes the results of these studies. It is clear that Sinai sands vary in size between the medium and fine fractions. Local differences in sand size can be explained by variations in the size of grains in the source region. Standard deviation indicates that sands in most areas are moderately sorted (0. 50–1.0). Well/very well sorted sands (<0.50) are present in areas where samples are composed of fine sand, whereas poorly sorted samples (>1.0) characterize coarse sand. These characteristics are similar to those found not only in other sand seas in Egypt (Embabi 2000), but also in other sand seas/dune fields in other deserts (Embabi and Ashour 1993). Such findings suggest that sands of the North Sinai Sand Sea reached the mature stage.

Previous studies (Misak and Attia 1983; Attia 1999; Khidr 2006) also revealed that there are some differences in grain size composition between different forms and between the various parts of dunes. The percentage of coarse sands was higher in sand sheets (18.1–35.5%) and sand ripples (33.87–46.5%) than in sand dunes (1.61–7.9%). Coarse sand grains are necessary for the development of sand sheets and ripples. The coarse sand constitutes the lag particles left behind as winds remove finer grains from the surface layer of the sand sheets, and these protect the underneath sand layers from being eroded. In the meantime, windward sides of dune are coarser than dune crests and leeward sides. This is due to wind selection of medium-fine grains, which leaves coarse sand as lag grains. In other words, when wind transports sand grains, there is a tendency for carrying finer grains than coarse grains. The result is the removal of finer particles leaving the coarse grains behind. These differences were also found in other deserts (Embabi 1982; Embabi and Ashour 1990).

Surface texture of sand grains shows the ancient environments in which dune sand is originated. Hence, many studies showed great interest in examining surface textures (Attia 1999; Ali and Ashour 2000; El-Banna 2004; Khidr

2006). These studies found that quartz sand grains were well rounded but with low sphericity, a characteristic that highlights the effect of wind abrasion. Nevertheless, the most significant mechanical/chemical surface features recognized by these previous studies are the following: the upturned plates, Depressions of various forms, V-shaped pits, and silica precipitates filling some Depressions in some samples or covering most of the surface of grains in some other samples. These features signify desert environmental conditions and a sabkha/coastal environment.

15.5.2 Mineralogical Composition

Although most (+90%) dune sands are composed of quartz, some grains have different composition. Analysis of mineral composition was carried on sands from various localities (Tsoar 1974; Misak and Attia 1983; Assal 1999; Attia 1999; Hereher 2000; El-Banna 2004; Khidr 2006). The most significant characteristics of mineral composition are that most of sand grains are composed of light minerals, of which quartz constitutes 90–97%. Heavy minerals are higher in the sands of coastal dunes than in those of continental dunes (7.2 and 2.4%, respectively) (Tsoar 1974). The difference in percentages owes to the fact that the bulk specific gravity of the heavy mineral is higher than that of the light mineral. According to previous studies, nearly all heavy minerals such as epidote, rutile, tourmaline, kyanite, hornblendes, garnet, olivine, andalusite, hypersthenes, augite, amphiboles are present in all localities, but with diverse percentages due to changes in the source of sand supply. The study of Khidr (2006) revealed one significant finding, namely the abundance of epidote in nearly all analyzed samples collected from the area to the west of Wadi El Arish (53.39% out of the total non-opaque minerals). In nearly half of the samples (28 in total), the percentage varied between 63 and 75%. In addition, the study of El-Banna (2004) revealed a high percentage for epidote, but is not as high (Average: 30.86%, Range: 20.4–43.7%) when compared to the samples of Khidr (2006). In the coastal strip around Bardawil Lagoon, Assal (1999) found that epidote is present in all samples, but in relatively smaller percentage (Average: 15.69%, Range: 7.95–17.02%) than reported by Khidr (2006) and El-Banna (2004). Hereher (2000) discovered similar low percentages of epidote (Average: 16.5%, Range: 10.3–25.6%). Epidote is derived from metamorphic rocks, or from sedimentary rocks, in which it constitutes one of the mineral elements. Such rocks do not exist in the nearby areas, but in the far south of Sinai—in NE Africa in the Red Sea Mountains, and in the Nubian Sandstone in Southern Egypt. This means that the source of epidote is external. The study of the minerals of the dunes of northern Sinai (Misak and Attia 1983) depended on studying the distribution of heavy material patterns. It was concluded that the possible sources of the sands of the area between Risan Aneiza and El Arish, and the area east of El-Qantara-El-Shatt are igneous and metamorphic rocks. As for the area between El Arish and El-Tina Bay, sands are derived from several sources including Nubian sandstone, metamorphic rocks, and preexisting sediments. According to Khidr (2006), the mineralogical analysis of sand samples (28 sample) revealed the presence of one mineral (Hypersthenes) in most samples, but in very low percentages, which is one of the heavy minerals in the Nile sediments in Egypt according to the study of Shukri (1950). Also, augite was found in most samples in very low percentages (El-Banna 2004; Khidr 2006) and that its grains have a shape and color similar to those described by Shukri (1950) from the recent Nile sediments.

15.6 Factors Affecting the Development of the North Sinai Sand Sea

The most essential factor affecting the development of sand dunes is the wind system prevailing in any region. Other factors contributing to the development of dunes in North Sinai are sand supply, climatic changes, sea level oscillations, and the offshore marine current.

15.6.1 Wind System

Previous works studied the effect of wind system on sand dunes in North Sinai (Hereher 2000; El-Banna 2004; Khidr 2006). Also, Tsoar (1974) pointed out to some characteristics of the wind system in Sinai. Wind data from seven meteorological stations was analyzed. Though meteorological stations differ in duration of wind data registration (1–27 years), the picture drawn from the analysis of wind data is satisfactory for the present study.

The wind system of Sinai is characterized by several aspects:

1. Complex wind regime: as winds blow from all directions with various durations and strength. Wind direction and speed change not only seasonally, but also from one locality to another (Fig. 15.8).
2. The western region of North Sinai is characterized by higher percentages of effective sand-moving winds (>11 knots) than the middle and eastern regions (Table 15.6). This suggests that wind energy decreases from west to east.
3. Percentages of effective sand-moving winds are higher during winter and spring than during summer and autumn.
4. Two major directions of effective winds predominate in North Sinai: northwest and southwest. The first direction

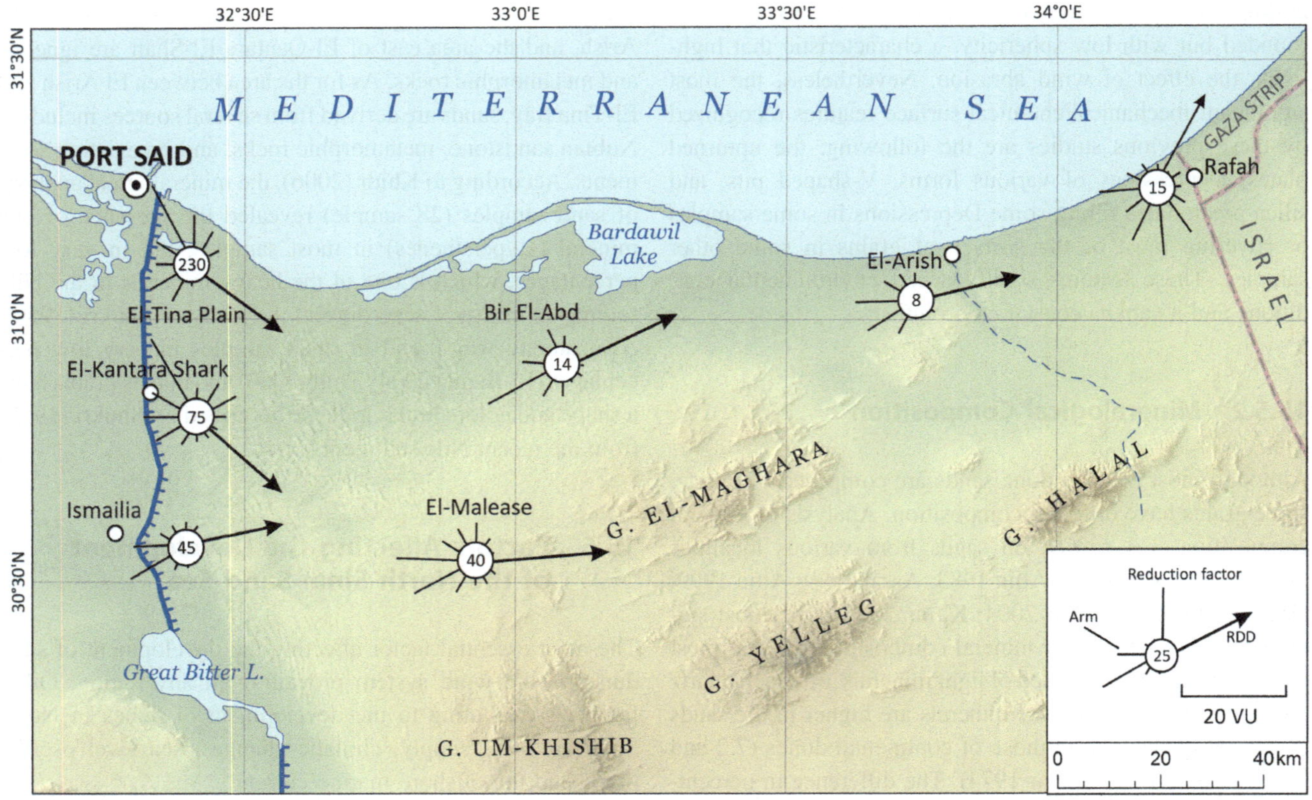

Fig. 15.8 Annual sand roses in northern Sinai (*Source* Hereher 2000)

Table 15.6 Percentage of wind occurrences according to speed in North Sinai

	Port Said[b]	El-Kantara[a]	Ismailia[a]	El-Malease[a]	Bir El-Abd[a]	El Arish[b]	Rafah[a]
Effective winds (>11 knots)	35.7	35	13	20	6	7.1	6
Non-effective winds (<11 knots)	63.1	64	70	74	60	84.2	86
Calm & variable	1.2	1	17	6	34	8.7	8
Total	100	100	100	100	100	100	100

[a]Hereher (2000); and [b]Climatological normals of ARE, up to 1975 (1980)

occurs during summer and spring, while the second direction prevails during autumn and winter.

5. The frequency of the NW effective winds decreases eastward, where the southerly and southwesterly effective winds replace them.

15.6.2 Coastal Offshore Current

There is an offshore current moving eastward along the Egyptian coasts of the Mediterranean Sea (Tsoar 1990), and it is called the East Mediterranean Current. This current transports Nile sediments eastward, from the mouths of the present-day (Rosseta and Damiatta) and ancient Nile Branches (such as the Pelusaic), to the coasts of Sinai. Northern winds carry sand deposited along the Sinai beaches to the inland. Needless to say, the bulk of sand reaching Sinai beaches decreased drastically after the construction of the High Dam in Aswan, far in the south, resulting in the deposition of Nile sediments in the basin of Lake Nasser.

15.6.3 Climatic Changes

There are active dunes and fixed dunes, and the active ones override the fixed dunes in the North Sinai Sand Sea. These fixed dunes had developed during Late Pleistocene, about 18,000 years BP according to previous studies (Sarnthein 1978; Pye and Tsoar 1990). Stabilization of the dunes, by cementation, occurred during the pluvial period of Early–Mid Holocene that occurred between 9500 and 5500 BP

(Embabi 2004; Kuper 2006). Climatic changes were responsible for the availability of fluvial sediments in the surroundings of Sinai Sand Sea, where winds deflated sand fraction necessary for the development of dunes exist. During the pluvials of the Pleistocene, huge quantities of fluvial materials rested in internal basins, flats, wadi beds, and fans Sinai (EGSMA, NARSS, UNESCO, UNDP 2004).

15.6.4 Sea Level Changes

Since the Mediterranean Sea borders Sinai from the north, it is expected that any changes in sea level will affect one aspect or another of the North Sinai Sand Sea. Available evidence confirms that the level of the Mediterranean Sea changed several times during the Pleistocene. Surface area of this sand sea increased/decreased according to sea level changes. During the last glacial maximum, sea level dropped by $\sim$100–130 m and the Sinai coastline advanced northward by $\sim$50 km. Increased aridity and desertification in the northern Negev are possible results (Roskin et al. 2011). For the same reason, this could be applied to the case in northern Sinai. Since this sand sea developed during the Late Pleistocene at the last glacial maximum (18,000 BP), it is inferred that this sand sea used to extend northward to cover a larger area than its present one. As the sea level rose, the area decreased until it approached its present-day status during the last 3–4 thousand years. The presence of drowned dunes, at least in Bardawil Lagoon, confirms such inference.

15.6.5 Sand Availability

Eolian sand is available in nature; it could be found mixed with other sediments deposited in various environments. Availability of sand to wind action depends on the location of sediment basins and localities relative to wind direction. In North Sinai Sand Sea, eolian sand is available along the beaches of the northern coasts of Sinai, and in fluvial, playa, and sabkha deposits on the mainland of Sinai. From these environments, winds from different directions with various speeds deflate sand size sediments, recycle, and model them into eolian sand forms.

15.7 Source of Sand and the Geologic Age of Dunes

15.7.1 Source of Sand

Analysis of all available data reveals that sands of the North Sinai Sand Sea have several sources. This conclusion emerges from nearly all previous studies that tackled North Sinai Sand Dunes (Shukri 1950; Shukri and Philip 1959; Misak and Attia 1983; El-Kayali 1984; Dabbour et al. 1989; Shata 1995; Attia 1999; Hereher 2000; El-Banna 2004; Khidr 2006). Available evidence suggests that sand of North Sinai originated from two main environments: (1) subaqueous (marine, coastal, beach, sabkhas); and (2) continental (fluvial, playa, weathering products) among which the Nile sediments are the most significant. Other sources are local materials, crystalline basement rocks, and Nubian Sandstone.

15.7.2 Age of Dunes

Several past studies tackled the age of dunes. It was concluded by Neev et al. (1987) that the small erg of sand dunes south the coastline between El Arish and the Suez Canal is younger than Late Pleistocene, since the age of the underlying chalky freshwater limestone is approximately 35,000 year bp. According to previous studies (Sarnthein 1978; Pye and Tsoar 1990), the fixed dunes have developed since 18,000 years bp. The stabilization of dunes by cementation occurred during the pluvial period of Early–Mid Holocene between 9500 and 5500 BP (Embabi 2004; Kuper 2006).

15.8 Conclusion

The North Sinai Sand Sea is not the only sand sea outside the Western Desert, but also the most complex sand sea in Egypt. This is due to three factors. The first factor is the landscape of north Sinai, which is characterized by the presence of a vast plain bound from the north by the Mediterranean Sea and by the anticlinal hills and mountains of the Syrian Arcs from the south. The second is sand availability from both local and outside sources. Sand of local provenance is derived from the sediments eroded from the surrounding hills and mountains and laid down in bahadas and playas. The outside sources come from the Nile sediments which were carried out by the coastal current and laid down on the Mediterranean Sea shores of Sinai. The third factor is the complex wind regime in northern Sinai. Wind blows from various directions, with varying velocities. Wind direction and speed change not only seasonally, but also from one locality to another. This complex wind regime resulted in a complex dune system, with nearly all types, forms, and patterns of dunes.

The presence of two generations of dunes suggests the development of two separate sand seas. The first is the one which is characterized by fixed dunes. The second sand sea is composed of active dunes and is developed atop the fixed dunes of the older sand sea. This assumption is based on a fact which is the presence of a surface of unconformity separating the two sand seas. This surface indicates that there was a break in dune development, and the dunes of the older sand sea were exposed to erosion during an unknown time span before the development of the more recent sand sea. This aspect needs further field investigation and age determination for the sands of both sand seas.

References

Abdalla A (2002) Sand forms in the east of Bardawil Lake. Bull Fac Arts, University of Zagazig, (In Arabic), Special Issue

Abdalla A (2005) Textural properties of dune sands, south Lake Bardawil. In: Sinai N, Proceedings of the Workshop on Environment and Development in Egyptian Deserts (in Arabic), Cairo University, Cairo

Akl MT (2002) Dune movement in the area to the east of the Suez Canal, and its effect on human activities (in Arabic). A Special Issue of the Journal (The Humanities) of the Faculty of Arts (Damanhur City), Alexandria University Branch in Damanhur City

Ali A (1993) The geomorphology of sand dunes of south and southeast Qattara depression. Dissertation, Ain Shams University, Cairo

Ali A, Ashour M (2000) Analysis of surface features of dune sands in North Sinai Sand Sea (in Arabic). Arab Geogr J 36:33–58

Assal E (1999) Sedimentological studies on the Quaternary sand dunes and sabkhas, Northern Sinai, Egypt. Dissertation, University of Mansoura

Attia N (1999) Aeolian sand forms in North Sinai: A geomorphologic study. Dissertation, Ain Shams University

Bagnold R (1941) The physics of blown sand and desert dunes. Methuen, London

Besler H (2002) The Great Sand Sea (Egypt) during the late Pleistocene and the Holocene. Zeit fur Geomorphol N E Suppl-Bd 126:59–74

Besler H (2008) The Great Sand Sea in Egypt: formation, dynamics and environmental change—A sediment analytical approach. Elsevier, Amsterdam

Dabbour G, Morsy M, Kamel A (1989) Radioactivity and heavy economic minerals of some Quaternary sediments at El Arish beach, North Sinai. Egypt. Ann Geol Surv of Egypt 16:51–56

EGSMA, NARSS, UNESCO, UNDP (2004) Geological maps of Sinai, scale 1: 250,000, Cairo

El-Banna M (2004) Geological studies emphasizing the morphology and dynamics of sand dunes and their environmental impacts on the reclamation and development areas in northwest Sinai, Egypt. Dissertation, Cairo University

El-Baz F, Boulos L, Breed C, Dardir A, Dowidar H, El-Etr H, Embabi N, Grolier M, Haynes V, Ibrahim M, Issawi B, Maxwell T, McCauley J, Moustafa M, Yousif M (1980) A journey to Gilf Kebir and Uweinat. Geogr J 146:51–93

El-Disouki S (1988) Morphometric analysis of Barchan dunes in the lower section of Wadi el-El-Masagid, northern Sinai (in Arabic). Arab Geogr J 20:137–159

El-Disouki S (2000) Linear Dunes east of Suez Canal: A geomorphologic analysis (in Arabic). Arab Geogr J 35:231–280

El-Kayali M (1984) The northern coastal plain of Sinai Peninsula, A geomorphologic Study (in Arabic). Dissertation, Ain Shams University

Embabi N (1967) A geomorphologic study of Kharga Oases Depression, the Western Desert, Egypt. Dissertation, Bristol University

Embabi N (1976–77) Slope form of barchan dunes in Kharga and Dakhla depressions. Bulletin de la Societé de Géographie d'Égypte 49–50:13–28

Embabi N (1978) Statistical relationship between barchans shape dimensions. Egypt Comput Sci J 1:1–15

Embabi N (1982) Barchans of The Kharga depression. In: El-Baz F, Maxwell T (eds) Desert landforms of Southwest Egypt: A basis for comparison with Mars. NASA, Washington, DC, pp 141–155

Embabi N (1990) Dune movement in the Kharga and Dakhla Oases Depression, the Western Desert, Egypt. In: el-El-Baz F, El-Tayeb I, Hassan M (eds) Proceedings of the International Workshop on Sand Transport and Desertification in Arid Lands, Khartoum, Singapore, 1985

Embabi N (1995) Types and patterns of sand dunes in Egypt. Bulletin de la Societé de Géographie d'Égypte 68:57–89

Embabi N (2000) Sand dunes in Egypt. In: Soliman S (ed) Sedimentary Geology of Egypt: Applications and Economics-1. Special Publication, The Sedimentological Society of Egypt, Cairo, pp 179–183

Embabi N (2004) Geomorphology of Egypt. The Nile Valley and the Western Desert. Special Publication, The Egyptian Geographical Society, Cairo, Vol. I

Embabi N, Ashour M (1993) Barchan dunes in Qatar. J Arid Env 25:49–69

Finkel H (1959) The Barchans of Southern Peru. J Geol 67:614–647

Google Earth (2014) Egypt Images Latitudes 22-32 N and Longitudes 22-32 E, http://www.google.com

Greeley R, Iversen J (1985) Wind as a geological process. Cambridge University Press, Cambridge

Haynes V (1980) Quaternary geology and archaeological investigations. In El-Baz et al (ed) A journey to Gilf Kebir and Uweinat. Part II, Geogr J 146:59–63

Hereher M (2000) A study of sand dune fields in north Sinai, and the environmental risk assessment of Aeolian processes. Dissertation, Mansura University

Khidr M (2006) Geomorphology of sand forms to the west of Wadi El-Arish, and their hazards (in Arabic). Dissertation, Ain Shams University

Kidson C (1982) Sea level changes in the Holocene. Quat Sci Rev 1:121–151

Kuper R (2006) After 5000 BC: The Libyan desert in transition. CR Pale 5:409–419

Lancaster N (1995) Geomorphology of desert dunes. Routledge, London

Misak R, Attia S (1983) On the sand dunes of in Sinai Peninsula. Egypt J Geol 27:115–131

Neev D, Baker N, Emery K (1987) Mediterranean coasts of Israel and Sinai. Taylor & Francis, New York

Pye K, Tsoar T (1990) Aeolian sand and sand dunes. Unwin Hyman Ltd, London

Roskin J, Tsoar H, Porat N, Blumberg D (2011) Palaeoclimate interpretations of Late Pleistocene vegetated linear dune mobilization episodes: evidence from the northwestern Negev dunefield, Israel. Quatern Sci Rev 30:3364–3380

Sarnthein M (1978) Sand deserts during glacial maximum and climatic optimum. Nature 272:143–146

Shata A (1995) Comparative sedimentological and mineralogical studies on aeolian and alluvial soils in North Sinai. Egypt. Ann Agric Sci, Ain Shams Univ., Cairo 40:475–482

Shukri N (1950) The mineralogy of some Nile sediments. Q J Geol Soc, London 105:5–11

Shukri N, Philip G (1959) The mineralogy of some recent deposits in el-El-Arish-Ghaza area. Bull Fac Sci, Cairo Univ 35:1–17

Tsoar H (1974) Desert dunes morphology and dynamics, El Arish, Northern Sinai. Zeit Geomorphologie, Suppl Bd 20:41–61

Tsoar H (1983) Dynamic processes acting on longitudinal (seif) dune. Sedimentol 30:567–578

Tsoar H (1989) Linear dunes—forms and formation. Prog Phys Geogr 13:507–528

Tsoar H (1990) Trends in the development of sand dunes along the South-eastern Mediterranean coast. Catena Suppl 18:51–60

Tsoar H, Yaalon D (1983) Deflection of sand movement on a sinuous longitudinal (seif) dune: use of fluorescent dye as tracer. Sed Geol 36:25–39

16 Dune Fields of SE Eastern Desert of Egypt

Abstract

There are four small dune fields (150–500 km^2) in SE Eastern Desert of Egypt, of which only El-Hebal Dune Field is known since the early years of the twentieth century. This field is composed of 14 longitudinal compound units, with an average length of 10 km and average width of 0.50 km. Each of these units is composed of sinuous, parallel linear dunes. The second dune field of delta Wadi Kraf is composed of vegetated barchans and barchanoids, and nebkhas. There are also some active lee dunes, which developed in the lee side of fixed dunes. Due to the spread of nebkhas all over the area of this field, this is a field of nebkhas and dunes. The third field lies just to the west of El-Hebal Field. In this field, sand intrudes the foothills of the Red Sea Mountains and is deposited as lee dunes in the lee side of numerous small hills which characterize the foothills. The fourth dune field lies to the south of Wadi El-Allaqi, and crosses the Egyptian-Sudanese borders. It is discovered from Google Earth. It is a field of small linear and crescentic dunes.

Keywords

El-Hebal dunes • Linear dunes • Crescent dunes • Lee dunes • Vegetated dunes • Wadi kraf • The Eastern Desert

16.1 Introduction

Four dune fields are developed in the SE corner of the Eastern Desert. Till recently, only two fields were known in this remote region (Embabi 2000), but by using the latest high-resolution images provided by Google Earth (2014), another two small fields were discovered in this region. The first is composed of fixed barchans, barchanoids, and nebkhas on the northern part of the delta of Wadi Kraf (Fig. 16.1). The other field lies to the south of the central part of Wadi El-Allaqi and is composed of belts of small linear and crescentic dunes which extend into the Sudan. Due to its location, it will be called in the present study **"El-Allaqi Dune Field"** (Fig. 16.2). The **dune field of El-Hebal** (I-Hubal as written on topographic maps) was first mentioned by Ball (1912) in treating blown sand in South-Eastern Desert, but greatly missed from Ball's study was the information on dune characteristics. In fact, this dune field was discovered by Ball while conducting the topographic survey of this corner of the Eastern Desert in the early years of the twentieth century. As a result, dunes of this field appear on topographic maps (scale 1: 500,000). El-Hebal (The Ropes) is a local name for the linear dunes of this field that lies at a few kilometers to the SE of the **delta of Wadi Kraf**, while the second field lies to the west of Wadi Kraf, where its dunes intrude the foot slopes of the

N.S. Embabi, *Landscapes and Landforms of Egypt*, World Geomorphological Landscapes,
https://doi.org/10.1007/978-3-319-65661-8_16

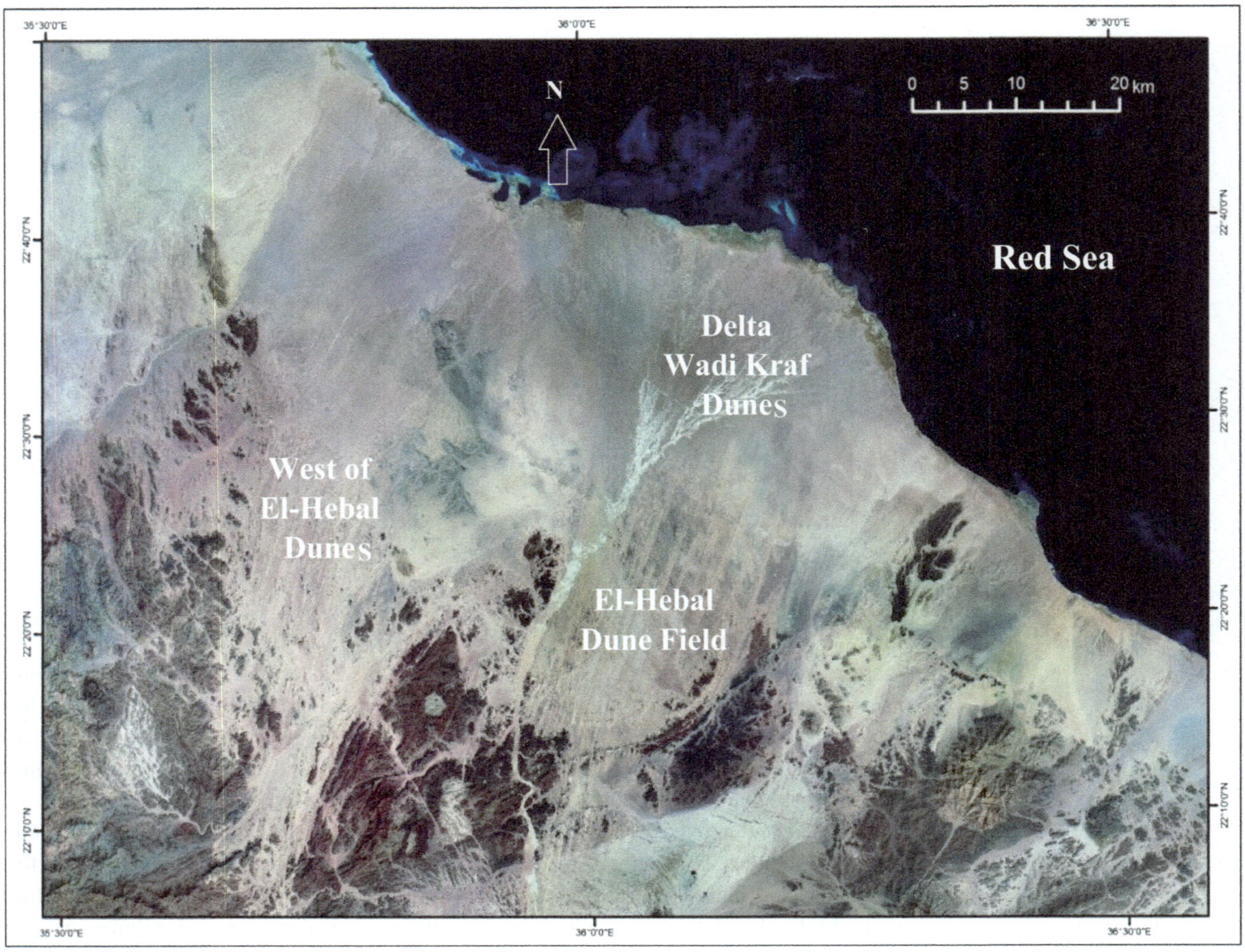

Fig. 16.1 Landsat-TM image showing the location of the three dune fields of El-Hebal, West El-Hebal, and Delta Wadi Kraf Dune Fields

Fig. 16.2 The location of El-A'llaqi Dune Field (*Source* Google Earth, 2015)

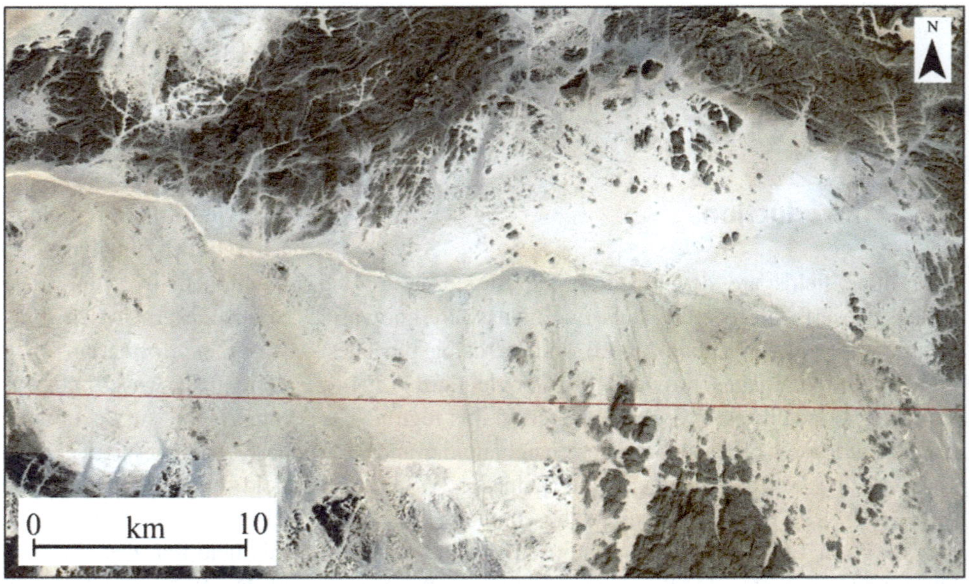

Red Sea Mountains from the NE. Since it lies to the west of El-Hebal Field, it is called here **"West El-Hebal Dune Field"** (Fig. 16.1). The third field overrides the northern part of the delta of Wadi Kraf; therefore, it will be called in this study **"Delta Wadi Kraf Dune Field"** (Fig. 16.1). In the following sections, the geomorphologic aspect of each field will be presented depending on data collected from space images of Landsat-TM and Google Earth, 2014, and topographic maps(scale 1: 50,000) as well as some field investigations conducted during a scientific trip in 1997.

16.2 El-Hebal Dune Field

16.2.1 General Characteristics

Covering an area of ~400 km^2, the dune lines of this field developed on the bahada plain and the foothills of the Red Sea Mountains. This plain is covered by a sand sheet, hence the name "Sandy plain." It rises gradually from about 50 m a.s.l at the upwind section of dunes to about 250–350 m a.s.l at the downwind section of the dune field. The dune field starts at ~20 km from the coastline of the Red Sea and extends inland for ~25 km. It starts at latitudes 22° 28′/22° 26′ N, and reaches latitude 22° 15′ N. Another characteristic is that drainage lines start braiding at the foothills of the Red Sea Mountains, where they are cut by dune ridges. Nevertheless, some recent drainages are checked by dune lines. Only a few of these drainage lines cross the Bahada plain and reach the coastline, forming deltas such as that of **Wadi Kraf**.

Regarding the orientation of dune lines, it ranges between 20° and 28°, with a mean of 24° from the north, (Table 16.1). This orientation is also recorded in the Uweinat Dune Field and the dune belts of the Southern Plain in the Western Desert, which lies to the south of Latitude 23° N (Embabi 2000). Analysis of wind data of East Uweinat Meteorological station, (which lies in the southern part of the Western Desert). Showed that about 40% of the total wind frequency comes from 014°–44° direction (Embabi 2004). Therefore, the orientation of the dune lines in El-Hebal Field can be explained by referring to a similar wind regime, which prevails in the southern part of the Western Desert. Thus, it is inferred that this regime dominates the southern part of Egypt to the south of latitude 23° N.

16.2.2 Dune Forms

El-Hebal Dune Field is composed mainly of 14 longitudinal parallel units as shown on Landsat-TM and Google Earth images, and topographic maps of (scale 1: 50,000) (Fig. 16.3). Table 16.1 shows that the length of these units is relatively short, varying between 3.2 and 18.7 km, with an average of 10 km. They have relatively wide breadth, ranging between 0.3 and 0.8 km, with an average of 0.49 km. In detail, this dune field is composed of four dune forms: linear, barchans and barchanoids, transverse and lee dunes.

16.2.2.1 Linear Dunes

Each of the 14 longitudinal units is composed of several sinuous linear ridges, running parallel to each other (Fig. 16.4). Generally, each unit is composed of 5–8 sharp-crested, sinuous linear ridges. Although some of them

Table 16.1 Morphological characteristics of linear dune units of El-Hebal Dune Field (Measurements made on Landsat-TM images)

Serial	Length (km)	Width (km)	Orientation (Degrees)
1	4.0	0.50	28
2	14.0	0.40	22
3	4.9	0.30	28
4	11.8	0.60	27
5	4.2	0.45	28
6	10.2	0.55	28
7	11.0	0.45	27
8	17.3	0.65	25
9	14.2	0.50	25
10	18.7	0.60	24
11	11.0	0.40	24
12	13.0	0.35	20
13	3.2	0.30	20
14	5.8	0.80	20
Average	10.0	0.49	24

Fig. 16.3 Landsat-TM image showing El-Hebal Dune Field to the east of Wadi Kraf

are short and unite with others, the ridges are organized in a parallel pattern. Meanwhile, spacing of these sinuous linear dunes (average 100 m) does not change along any of the main compound dunes in El-Hebal Field. Despite the existence of other compound longitudinal dunes that are composed of several linear ridges in other sand seas and dune fields in Egypt, the occurrence of the parallel pattern of the secondary ridges in all main compound units is a unique characteristic to **El-Hebal Dune Field**, which urges for more comprehensive interpretation. The most probable

Fig. 16.4 Space image showing the sinuous linear dunes and the lee dunes in El-Hebal Dune Field (*Source* Google Earth, 2015)

interpretation for such exceptional characteristic is derived from visual interpretation of the high-resolution images of Google Earth. These images illustrate that the upwind sections of the main compound units are composed of several parallel lee dunes, which developed in the lee side of the bedrock ridges. These rock ridges, which run obliquely to the main trend of the dunes, are a typical feature in this area. The lee dunes grow by time and extend in downwind direction which is SSW. The combination of both the bedrock ridges and the development of parallel lee dunes beside each other, and the availability of sand supply explain why all compound units in El-Hebal dunes are composed of sinuous linear dunes.

Sinuosity of the secondary linear dunes is another aspect which characterizes El-Hebal Dune Field. These dunes developed sharp crests, and a straight slip face on one side and a convex slope on the other side (Fig. 16.5). This cross section may be reversed not only from one locality to another along the dune, but also at the same locality. Linear dunes develop in bidirectional wind regime and/or wide uni-modal regimes (Thomas 1997). This wind regime normally creates zones of net erosion and net deposition on both sides of dunes. Therefore, peaks and saddles develop along the dune, which advance downwind in the resultant drift direction (Tsoar 1978). The outcome of this process is the sinuosity of dunes.

Fig. 16.5 Field photograph of the sharp crests of El-Hebal Dunes

16.2.2.2 Lee Dunes

In inter-dune areas between the main units, short **lee dunes** (0.2–1.0 km) developed at the leeward side of low-lying ridges (mostly dykes) that extend at right angles to the dominant wind direction which is from NNE as indicated from the orientation of the longitudinal units (Fig. 16.4). The main bundles of El-Hebal Dune Fields were originally lee dunes that grew by time and extended southward, and each group of the neighboring lee dunes formed the longitudinal bodies of the field. This mode of development is a unique characteristic of El-Hebal Field (see Sect. 16.2.2.1).

16.2.2.3 Barchans and Barchanoids

Barchans and barchanoids are formed along the downwind section of the linear dunes (Figs. 16.6 and 16.7). As indicated by previous studies (Embabi 2012, 2000; Lancaster 1995), the availability of sand in these downwind sections is relatively limited. Due to this shrinkage in sand supply, linear dunes dismantle into crescentic ridges and small sand hillocks, which develop by time into barchans and barchanoids. Despite having a slip face and windward slope, barchans of El-Hebal are characterized by some aspects which make them different from other barchans in other dune fields/sand seas in Egypt.

Barchans of El-Hebal Field have three characteristics. The first is that most of them are semi-circular/circular in form, and the second is the very short horns of barchans or without horns. The third characteristic is that they are smaller comparatively to barchans developed in other Egyptian sand seas. It is likely that as the linear dunes extend southward, they move on surfaces that are getting higher, which means sand flow is dramatically decelerated. Consequently, short horns and circular-like dunes are developed (Fig. 16.6). Measurements made on a sample of barchans (100 dunes) showed that the average length and width is 50 m. In Kharga and Dakhla, average width for a sample of 50 barchans in each of them is 173–228 m (Embabi 1995).

16.3 West El-Hebal Dune Field

The field intrudes into the foothills of the **Red Sea Mountains** more than El-Hebal Field does. The surface of the ground rises up from 140–150 m a.s.l on the bahada plain to 350–400 m a.s.l. Therefore, this field is developed on a rough surface where numerous small hills occur. This is why almost all dunes in this field are **lee dunes**, which developed in the leeward side of these hills. The field covers an area irregular in shape of ~350 km^2, with a maximum length of ~26 km from north to south, and a decreasing E-W breadth, from ~22 km in the upwind section to 4–5 km in the downwind section.

Lee dunes are commonly short, as they do not exceed several hundreds of meters in most cases. Nonetheless, some of them may reach 2–10 km, taking the form of linear dunes. The orientation of this dune field is NNE (20°–25°) and is similar to that of the El-Hebal Dunes. Other dune forms are transverse dunes which are common in the northern part of the field where the ground surface is generally smooth. Climbing dunes on wadi sides and mountain slopes occur wherever orientation of wadis and slopes is favorable for their development.

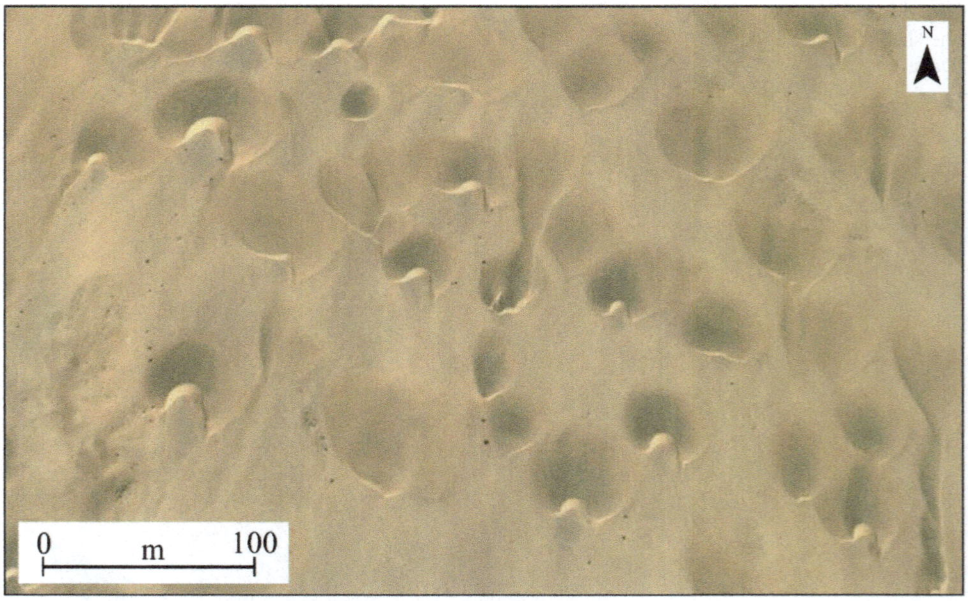

Fig. 16.6 Small domal-shaped crescent dunes at the downwind section of two linear dunes of El-Hebal Field. Note the short horns, the small lee dunes and the small slip face of the dunes (*Source* Google Earth, 2015)

Fig. 16.7 Field photograph for the barchanoids of El-Hebal Dunes

16.4 Delta Wadi Kraf Dune Field

This field differs from the previous ones in three main facets. Firstly, it is composed of **barchans** and barchanoids, second its dunes are vegetated and fixed, and the third facet is that most of its surface is covered by nebkhas (Figs. 16.8 and 16.9). The field covers most of the northern part of the delta of Wadi Kraf, with an area of ~400 km^2, with a maximum E-W length of ~20 km, and a maximum N-S breadth of ~21 km. The deltaic plain rises gradually from 5–10 m a.s.l near the Red Sea coast to 40–45 m a.s.l at the inland extension of the dune field. Apart from the dunes, this plain is covered by a sand sheet.

There are five dune forms in this field. The first is the crescentic while the second is the barchanoid, and the third is the linear. As for the fourth, it is lee dunes, and the fifth is nebkhas. Notably, lee dunes are the only active forms. They grew on the lee sides of other dunes in the downwind section of the field. A clear reason for this is the fact that sand of these lee dunes is the product of wind erosion of other fixed dunes lying northward. These lee dunes are generally short (100–200 m in length), but the length reaches 300–400 m in rare instances (Fig. 16.8). As for barchanoids, they are more prevalent than barchans in this dune field. Most barchanoids are coalesced barchans, due to the increase in sand supply, which forms curved rows parallel to each other with a

Fig. 16.8 Vegetated and lee dunes of the Delta of Wadi Kraf Dune Field (*Source* Google Earth, 2015)

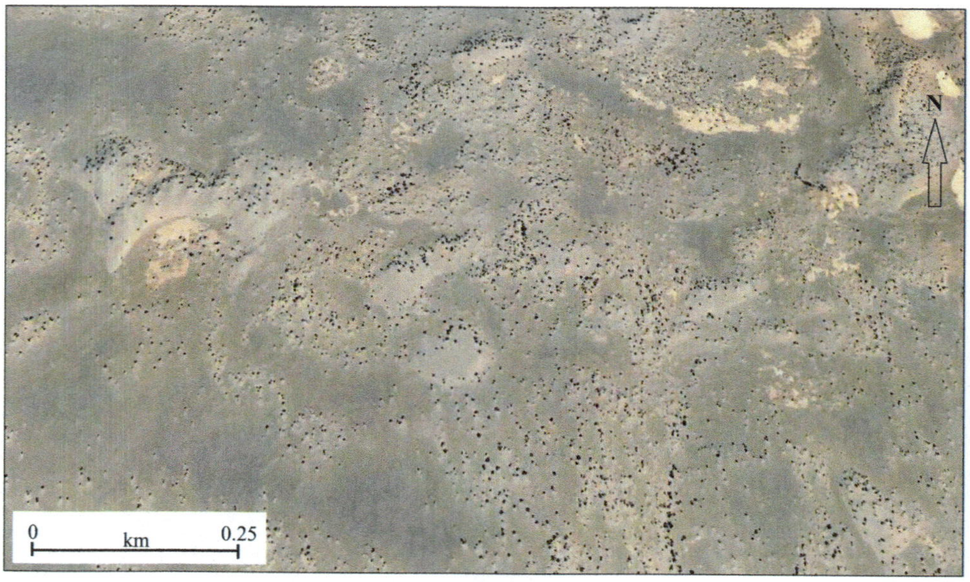

Fig. 16.9 Fixed dunes and nebkhas of the Delta Wadi Kraf Dune Field (*Source* Google Earth, 2015)

general orientation NNE-SSW. In some cases, barchanoid ridges resemble linear dunes, especially those with sharp crests. Heights of barchans, barchanoids, and linear dunes seem to be low not exceeding a few meters as faces of some forms are short. Fixed dunes are exposed to sand removal by wind (resulting in the reduction of their heights).

In spite of their tiny forms, **nebkhas** are the most widely spread sand forms in this dune field (Fig. 16.9). This is a result of the prevalence of small shrubs on the floor of this part of the deltaic plain of Wadi Kraf. A small dune developed around each shrub, forming a short tail at most of them. The presence of nebkhas distinguishes this field from other fields in this group of Dune Fields of SE Eastern Desert and from other dune fields and sand seas in Egypt.

16.5 El-Allaqi Dune Field

This field is not reported in any previous study and is discovered during visual examination of Google Earth images of the southeastern region of the Eastern Desert. It lies to the south of the central section of Wadi El-Allaqi where it takes an E-W orientation and extends into the Sudan (Fig. 16.2). It has an irregular shape, with a maximum E-W length of ~26 km, a N-S maximum extension of ~21 km, and an area of ~450–500 km^2.

Dune density decreases southward, since sand deposition and dunes start to develop just to the south of Wadi El-Allaqi (Fig. 16.10). This is due to two factors. First, dunes do not grow by the same rate, so that spacing increases in the downwind

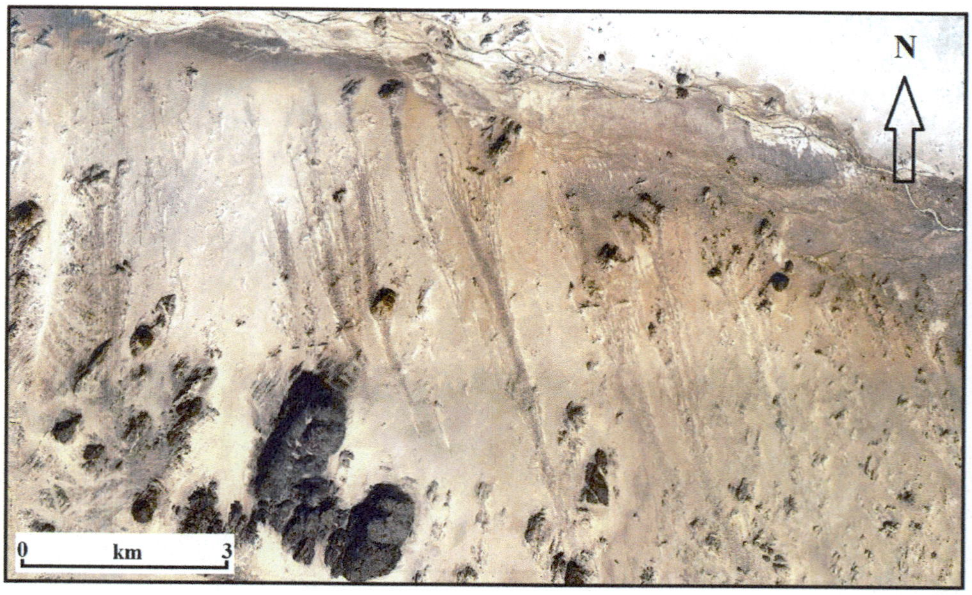

Fig. 16.10 Compound, branching linear dunes, with low density in El-A'llaqi Dune Field (*Source* Google Earth, 2015)

Fig. 16.11 Small linear and domal dunes with a very high density in the eastern part of El-A'llaqi Dune Field (*Source* Google Earth, 2015)

direction and density decreases. Second, decrease in sand supply as distance increase from the source of sand which is Wadi El-Allaqi. However, in some localities, density either remains high or increases (Figs. 16.11). In these two cases, dunes are either fed by additional local sand supplies from wadis crossing this locality (Fig. 16.11), or trapped in a basin-like area.

Nearly all forms and patterns of dunes are present in El-Allaqi Dune Field. Simple/compound linear developed in all parts of this field. The largest linear dune extends for ~9.0 km. All large linear dunes are sinuous and compound of several parallel/branching ridges (Fig. 16.10). Lee dunes also spread in this field, because of the presence of vegetation/rocky obstacles. Due to the smaller size of vegetation obstacles, lee dunes are much smaller than those of the rocky ones. Length of vegetation type dunes does not exceeds several tens of meters, whereas that of the rocky ones reaches several tens of kilometers. Barchans and domal dunes are the third characteristic form in this field. Dunes are organized in linear belts, while some others are not (Fig. 16.12). Barchans are less frequent than other dune forms due to environmental conditions which are not favorable for their development, mainly the roughness of surface and sand properties. Barchans appear in two small localities, with relatively small sizes where length/width does not exceed 40 m (Fig. 16.12). Most of barchans have no slip face, whereas the rest developed a very short slip face. With surfaces characterized by low angles, it can be inferred that these barchans are in their early stage of development, which is the embryonic stage (Coque 1962; Embabi 1967; Embabi and Ashour 1985). Google Earth images show that nearly all barchans show the development of small lee dunes at their leeward side, which indicates that

Fig. 16.12 Cluster of small barchans in El-A'llaqi Dune Field (*Source* Google Earth, 2015)

most of dunes are not moving or moving at very slow rate. These aspects of barchans are very similar to those of El-Hebal, indicating similar environmental conditions.

16.6 Sand Properties

Of the four dune fields, only four samples were collected from one of the linear dunes of El-Hebal Dune Field during a short field in 1997. Three of the samples were collected from the dune body, whereas the fourth was from the beach sand. Despite being insufficient, the analysis sheds light on the characteristics of the sands of El-Hebal dunes.

16.6.1 Grain Size

Table (16.2) summarizes the grain size characteristics of the four sand samples collected from the El-Hebal Dunes and the Wadi Kraf Beach. The grain size is fine to very fine, and the mean size is fine where it ranges between 2.4 and 2.7 ø.

This size is smaller than sand from other sand seas in Egypt, where most of them fall within the medium to fine fractions (El-Baz et al. 1979; Embabi 2000; El-Baz et al. 1979). Sorting of all samples is very good, except for the beach sample which is good. As indicated in previous studies (Lancaster 1995; Embabi 2000; Embabi et al. 2012), sorting becomes worse as the sand gets either finer or coarser. This is the case of beach sand whose sorting is worse than other samples, because it contains a much higher percentage of medium size sand fractions.

16.6.2 Mineral Composition

Table (16.3) illustrates heavy mineral composition of dune and beach sands. The majority of samples are composed of non-opaque mineral. On the contrary, opaque specimens are high in the beach sample—almost one third of it. This is because opaque minerals are heavier due to their high bulk specific gravity. The percentage of the ZTR minerals in the dune samples is small than in the beach sample. These minerals are known as the ultra-stable minerals. This characteristic indicates that the sands passed a few cycles of deposition, and consequently the dunes are relatively recent (Embabi 2000).

Taking into consideration all the previous geomorphic characteristics of dune fields and those of dune and beach sands, it can be concluded that dune sands of El-Hebal, West El-Hebal, and Delta Wadi Kraf Dune Fields were derived from beach sands. These beach sands were brought from the Red Sea Mountains and laid down as deltaic sediments by fluvial action.

Due to its location in the extreme southern part of the Eastern Desert no fieldwork was carried out on El-Allaqi

Table 16.2 Grain size of El-Hebal Dunes and beach sands in SE Eastern Desert

Sample location	Mean size (ø)	Sorting (ø)	Grain size (ø)				
			2	3	4	5	Total (%)
(1) 22°26′ 03″ N 34° 07′ 00″ E	2.7	0.32	1.20	84.40	14.3	0.10	100
(2) 22° 25′ 20″ N 36° 08′ 45″ E	2.4	0.36	1.86	82.65	17.2	0.17	100
(3) 22° 16′ 35″ N 34° 07′ 00″ E	2.7	0.33	1.40	82.51	14.7	1.39	100
(4) Beach 22° 23′ 18″ N 36° 25′ 32″ E	2.6	0.43	9.30	74.70	15.0	0.15	100

Table 16.3 Some aspects of mineral composition of El-Hebal Dunes and beach sands

Sample number	Non-opaque	Opaque	ZTR[a]
(1)	84.70	15.30	3.60
(2)	67.54	32.46	4.80
(3)	79.51	20.49	2.06
Beach sample (4)	36.71	63.29	22.98

[a]*ZTR* Zircon, Tourmaline, and Rutile as % of non-opaque totals

Dune Field no information is available about dune sands properties.

However, some information can be inferred from Google Earth images. These images show two facts. First, most of the El-A'llaqi Dune Field lies directly to the south of the southern side of Wadi El-A'llaqi. Second, only a small subfield of small lee dunes lies to the north of the Wadi El-A'llaqi on the surface of some wadi fans in the extreme eastern part of the field. From these two facts, it can be inferred that the sands of El-A'llaqi Dune Field are derived from the fluvial sediments either of Wadi El-A'llaqi or other wadis.

16.7 Conclusion

The study of these four dune fields was beyond the bounds of possibility without the space images of Landsat-TM and Google Earth. Each of the dune fields is characterized by a unique aspect. El-Hebal Dunes are composed of linear dunes that developed from lee dunes; the dunes of the field of the Delta of Wadi Kraf are vegetated and the nebkhas dominate the field. The third field which developed to the west of El-Hebal Dunes is a field of lee dunes. El-A'llaqi Dune Field is the most complicated one in this region because of the rugged topography of this locality. Wind conditions are not similar in the whole region of SE Eastern Desert. Wind regime is characterized (as inferred from dune form/pattern) by a wide uni-modal regime in the subregion of the first three contiguous fields of El-Hebal, West El-Hebal, and Delta Wadi Kraf Fields. Also, it can be inferred that winds from NNE are the most effective. In the subregion of El-A'llaqi Dune Field, it can be inferred also from dune form/pattern that wind regime is narrow uni-modal, where NNW direction is the most effective. In fact, pre-topography as well as the wind conditions and sand availability were responsible for the development and differences among the four fields. Although space images provided much information about the geomorphologic characteristics of sand forms and topography, the full picture remains incomplete and requires field investigations and further study.

References

Ball J (1912) The geography and geology of south Eastern Desert. Government Press, Cairo

Coque R (1962) La Tunisie presaharienne: etude geomorphologique. Paris

El-Baz F, Slezak MH, Maxwell TA (1979) Preliminary analysis of color variations of sand deposits in the Western Desert of Egypt, Apollo-Soyuz test project, summary science report. In: El-Baz F, Warner D (eds) Vol. II: Earth Observations and Photography. NASA, Washington, DC, pp 237–262

Embabi NS (1967) A morphological study of Kharga Oases Depression, the Western Desert, Egypt. Dissertation, Bristol University, UK

Embabi NS (1995) Dune movement in the Kharga and Dakhla Oases Depressions, the Western Desert, Egypt. In: El-Baz F, El-Tayeb IA, Hassan MHA (eds) Proceedings of the international workshop on sand transport and desertification in Arid Lands Khartoum, Sudan: 79–105, World Scientific, Singapore, 17–26 November 1985

Embabi N (2000) Sand dunes in Egypt. In: Soliman S (ed) Sedimentary geology of Egypt: applications and economics, part 1. Special Publication. The Sedimentological Society of Egypt, Cairo, pp 179–183

Embabi NS (2004) The geomorphology of Egypt: landforms and evolution, Vol. I: The Nile Valley and the Western Desert. The Egyptian Geographical Society, Cairo

Embabi NS, Ashour MM (1985) Sand dunes in Qatar Peninsula (in Arabic), vol 2. Qatar University, Doha

Embabi N, Mostafa A, Mahmoud A, Azab M (2012) Geomorphology of Ghard Abu Moharik, Western Desert. Egypt. Bull Soc d'Géographie d'Égypte 85:1–28

Google Earth (2014) Egypt Images Latitudes 22-32 N and Longitudes 22-32 E, http://www.google.com

Lancaster N (1995) Geomorphology of desert dunes. Routledge, New York

Thomas D (1997) Sand seas and aeolian bedforms. Chapter 17 In: Thomas D (ed) Arid zone geomorphology: processes, form and change in drylands, 2nd edn. Wiley, Chichester, pp 373–412

Tsoar H (1978) The dynamics of longitudinal dunes. Final Technical Report. European Research Office, US Army, London

Part VI

Mountains and Drainage Systems

17 The Red Sea and Sinai Mountains

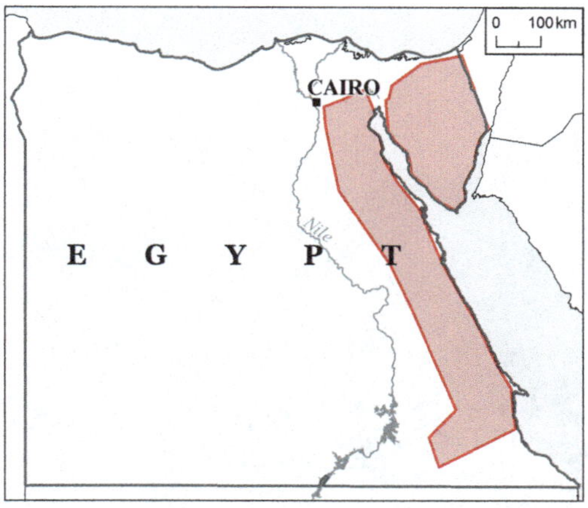

Abstract

Red Sea and Sinai Mountains are characterized by several distinctive forms which are not repeated elsewhere—except for the erosion surfaces which are found also in the Western Desert. These unique forms are as follows: dyke swarms, intermountain basins, ring complexes circular mounts, old erosion surfaces, and lacustrine terraces of Wadi Feiran in Sinai Mountains. Each of these features developed under certain structural/ environmental conditions. Their significance is derived not only from their geomorphologic characteristics, but also from their implications for past climates.

Keywords

Red Sea Mountains • Sinai Mountains • Dyke swarms • Intermountain basins • Ring complexes • Erosion surfaces • Wadi Feiran lacustrine terraces

17.1 Introduction

The Red Sea Mountains and Sinai Mountains have similar geological and geomorphologic characteristics; therefore, they will be treated together. During their long geological and geomorphic history, several landforms types developed under the control of geologic structures. The most recurrent forms are the fluvial ones as the climate was rainy tropical since the end of the Eocene (Chap. 3). The drainage networks dissect the mountains into blocks with steep slopes and V-shaped valleys (Fig. 17.1). Some of the eroded materials were laid down in intermountain basins, or as fans at the foot slopes of the mountains, or as deltas along the coasts of both Sinai and the Red Sea. By time, the sedimentary cover was removed, except for a few localities at the foot slopes of the mountains along the Red Sea coast, and in some intermountain basins. In Sinai, remnants of the sandstone cover—which were deposited atop the old erosion surface—can be clearly observed in several localities in the mountains. The dissection of the mountains by the wadis and the removal of the sedimentary cover in some places revealed the old erosion surface, which had developed a long time ago before the development of the mountains.

Time elapses and the effect of geologic structure and tectonic activity on the landscape and landforms becomes more pronounced, and new morpho-tectonic forms are created. To add, due to lithological differences, ridges of the dyke swarms appear as a characteristic form in the landscape of the Mountains. Also, ring complexes developed into their circular ridges and mounts, and became a characteristic geomorphic feature. Climatic and sea level changes were of great importance in the development of some forms. Nevertheless, it is not viable to examine all landform diversity developed in these mountains in the present chapter.

N.S. Embabi, *Landscapes and Landforms of Egypt*, World Geomorphological Landscapes,
https://doi.org/10.1007/978-3-319-65661-8_17

Fig. 17.1 Red Sea Mountains, with steep slopes, sharp peaks, and V-shaped wadis. SE part of the Eastern Desert

Therefore, characteristic forms will only be analyzed, whereas other forms will be treated in other chapters.

Forms characterizing the Red Sea and Sinai Mountains are as follows: dyke Swarms, intermountain basins, ring complexes or circular mounts, old erosion surfaces, and the lacustrine terraces of Wadi Feiran in Sinai Mountains.

17.2 Dyke Swarms

Dykes are the most widespread forms. They are narrow steeply dipping bodies, a few meters thick and several kilometers long. The basement formations in Egypt and especially in the Eastern Desert and Sinai are among the most intensely dyke-intruded regions known, and these intrusion are commonly referred to as **dyke swarms** (Meneisy 1990). In addition to their geological significance, their geomorphic importance relates to their positive and negative topographic relief. This topographic expression and their linear shape and their color make it easy to recognize the dykes and interpret from space images and aerial photographs (El-Rakaiby 1997), (Fig. 17.2). In southern Sinai, dykes vary in thickness from 0.5 to 20 m and are up to 20 km long (Iacumin et al. 1998). In the central Eastern Desert, dykes -formations either simple single ridges or parallel – 250–500 m Apart. -. As for simple dykes, they vary from a single dyke, several meters in thickness, to several thin dykes separated from one another by less than 10 m of basement rocks (El-Rakaiby 1997). Orientation of dykes varies greatly from one locality to another. In the central section of the Eastern Desert, dykes are notably oriented perpendicularly to the Red Sea NW structural trend (El-Rakaiby 1997). The same trend was found in Wadi Feiran area and Gabal Minadir area, while at Wadi El-Sheikh in southern Sinai dykes strike NE and NW (Iacumin et al. 1998).

Authors of previous studies agree that dykes have intruded the basement rocks along fractures and planes of weakness (Moawad 2008; Iacumin et al. 1998; El-Rakaiby 1997; Meneisy 1990). For example, in the area between Safaga and Quseir, dykes are considerably controlled by the geometric characteristics of fractures and/or the fault systems, and tend to be either parallel narrow straight or sinuous lines (Moawad 2008). They represent transitional stages between different types of tectonic events. These events emerged from the Late Precambrian and were rejuvenated throughout the Phanerozoic era.

The geomorphic significance of dykes is derived from their surface expressions. In general, dykes show both positive and negative topography, according to their relative resistance to the host rocks. Positive dykes take the form of sharp crests of elongated ridges due to their resistance to weathering, whereas weak dykes are eroded readily and become lower than the adjacent host rock, taking the form of an elongated trench. In fact, surface expression of both types of dykes is of low relief, but when their density increases, they develop rugged surfaces (Fig. 17.2). Despite being straight and parallel to each other, positive dykes are incised and crossed by some wadis as it can be seen in southern

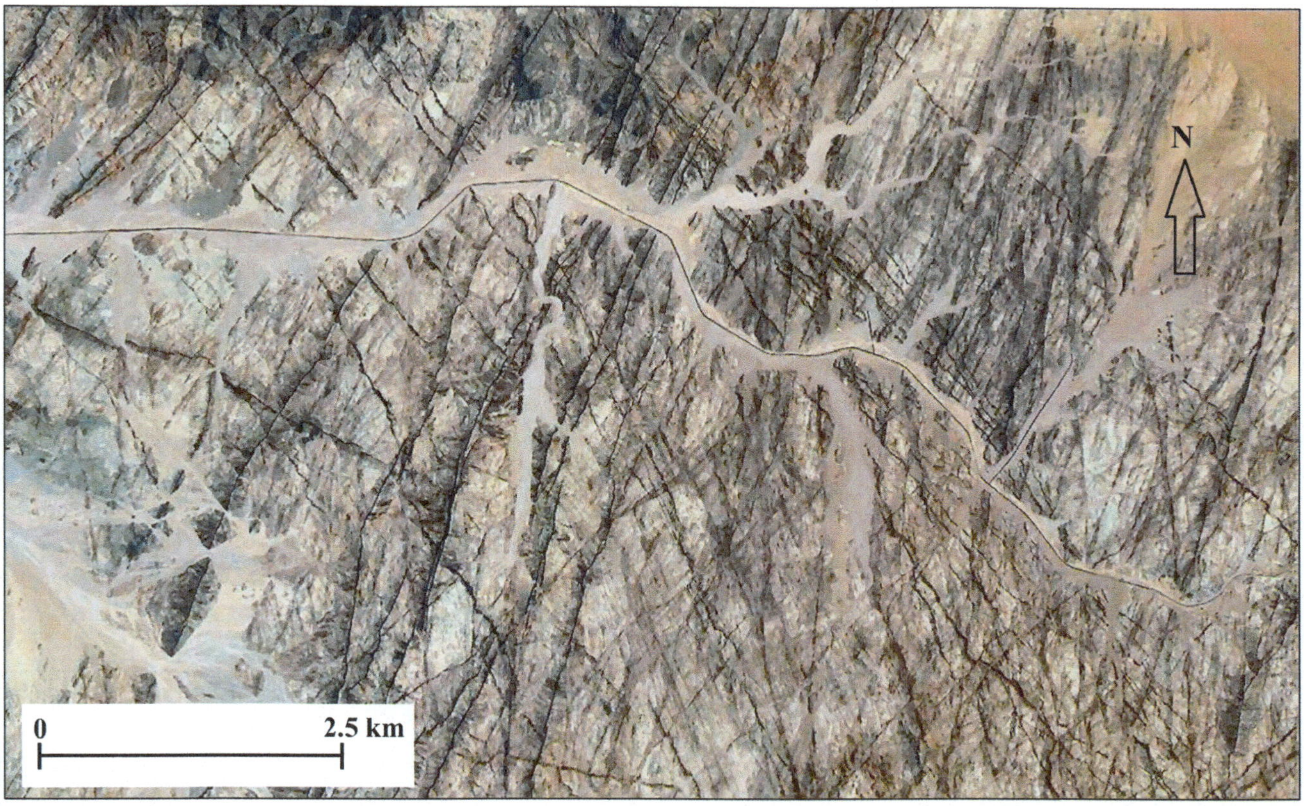

Fig. 17.2 A space image showing the landscape of dyke swarms in central Sinai Mountains at the environs of Wadi Feiran (*Source* Google Earth, 2014)

Sinai (Fig. 17.2). This indicates that these wadis are antecedent ones and have been developed on the original surface of the basement complex long before the dykes were exposed to differential erosion. Wadis, on the other hand, were able to keep their channel across the dykes due to the slow rate of dyke erosion, as well as the relatively high rate of fluvial erosion.

17.3 Intermountain Basins

Basins in the Red Sea and Sinai Mountains originated due to tectonic movements. They are delimited by faults, which is the reason why they are considered **morpho-tectonic Depressions**. In fact, topographic maps prove that these basins are semi-closed where drainage lines can easily pass in both directions i.e., in and out (Fig. 17.3a, b). The shape of these basins is the reason they are filled with fluvial sediments (Fig. 17.3a). All basins are irregular in shape, with a few semi-circular ones due to their tectonic origin, where faults bound their outer sides and/or cross them internally.

Worthy to be noted is that the exact number and size of the intermountain basins have not been specified yet, but topographic and geologic maps show that they spread all-over the Red Sea and Sinai Mountains. These maps also show that the basins vary greatly in size and density. In an attempt to estimate their number from the Geologic Map of Egypt (scale 1: 500,000) (Conoco Coral and the Egyptian General Petroleum Corporation, 1987), the number of basins (areas covered with fluvial sediments and into which drainage lines flow) is around 349 basins found in the Red Sea Mountains and 56 in the Sinai Mountains.

Table 17.1 summarizes the morphological parameters of a sample of five basins in the area lying between Safaga and Quseir as well as the divide line between the Nile Valley and Red Sea based on a study by Moawad (2008).

In addition to the above-mentioned morphological aspects of intermountain basins, space images of Sinai (Google Earth, 2014) and Moawad's (2008) study on some basins demonstrate that low-lying sedimentary hills project from the floors of nearly all large basins in spite of being rimmed by basement complex rocks (Fig. 17.3a). In most cases, these sedimentary hills comprise of remnants of the sedimentary cover. These remnants were atop the basement rocks and are kept till the present-day because they were downthrown in the tectonic basins. To consolidate, geological sections made for these hills illustrate that the whole succession is present in the basins. This indicates that basins developed in the early days of the evolution of the Red Sea

Fig. 17.3 a. A small intermountain basin in Red Sea Mountains, West Ras Banas. (Note the recent white channels which enter the basin from the west, and flow out from the east, and also the trees which grow along these channels) **b**. A field photograph showing an intermountain basin in Sinai Mountains, to the west of Nabaq Reserve on the Gulf of Aqaba. (Note the wadi cutting the rim of the basin)

Table 17.1 Morphological parameters of some inland basins in Central Eastern Desert (Moawad 2008)

Basin name	Mean length (km)	Mean width (km)	Area (km^2)	Orientation
Mohamed Rabah	11.2	3.7	27.4	NNE–SSW
Wassif	15.4	3.0	43.6	NNW–SSE
Umm al-Hawaytat	17.8	3.5	41.4	NNW–SSE
An-Ankheil	54.7	10.0	348.8	NW–SE
AbuTandab	47.3	4.0	161.8	NW–SE

Source Moawad 2008

and Sinai Mountains when the sedimentary cover was still present atop the basement complex.

In relation to the top layer of the sediment fill, it is the only one exposed. This top layer is composed of Quaternary fluvial materials, but no information is available about the underlying sediments. These underlying sediments are indicative of the geological and geomorphologic history of the Red Sea and Sinai Mountains since their development. Therefore, further investigation(s) and/or studies are recommended for possible future venues of research about underlying sediments.

17.4 Ring Complexes Circular Mounts

Examination of topographic maps (scale 1: 50,000) and Landsat-TM and Google Earth space images revealed the presence of several mounts encircled with circular ridges. Geologically, this form is known as "**Ring Complex**". Therefore, this form is conformable with its structure, and it can be called a structural landform. In Egypt, ring complexes are mostly found in the Red Sea Mountain of the Eastern Desert (Fig. 17.4), where 15 rings were recorded in this Desert (El-Ramly and Hussein 1985). Only another two ring complexes are recorded outside the Eastern Desert: the first is in Gabal Uweinat in the extreme SW the Western Desert, and the other one is in Gabal St. Katherine in Sinai.

17.4.1 Geological Aspects

Ring complexes intrude a dominantly Proterozoic basement rocks. They are circular/elliptical in plan, a few kilometers in diameter, and include a wide variety of rocks, such as granites, syenites, nepheline, gabbros, and volcanics. All ring complexes are arranged in a dominantly concentric pattern (El-Ramly and Hussein 1985); their ages range between 554 Ma for **Wadi Dib complex** and 89 Ma for **Abu Khruq complex** (ages of some rings are recorded in Fig. 17.4). Such ages indicate that their development extended from the Cambrian to the Cretaceous (Serencsits et al. 1979). During this time span, their development was episodic, where a period of 52 Ma separates each episode from the next (El-Ramly and Hussein 1985, after Lutz 1979). Analysis of the rocks of the ring complexes indicates that they developed during *several cycles* of magma activity rather than in one cycle.

17.4.2 Morphological Characteristics

Table 17.2 summarizes the parameters of some of the ring complexes of the Red Sea Mountains.

This table compares the wide variations between the parameters of the ring complexes. Mostly, these variations are structurally controlled, and role of erosion role is nearly negligible in changing their original form. One more observation is that although some of ring complexes rise more than 1000 m a.s.l, other parameters appear to be smaller than those of lower complexes (Table 17.2), since the area of the largest one is less than 50 km^2. In spite of this, they represent a unique landform type in the Red Sea Mountains, because of their circular outline and their characteristic local forms.

The most recurrent local forms are circular ridges; some of which form a complete circle all around the ring, while - others are partial ridges. Although all show erosion features, the partial ridges represent a small sector of a circle and do not exhibit signs of being exposed to removal of some parts of them. Therefore, most probably, they originated structurally in a similar manner to circular ridges. In cross section, ridges have two opposing slopes that meet at a sharp crest where the inner slope is steeper than the outer one (Fig. 17.5a). Both sides are dissected by small wadis, which feed small alluvial fans at the foot slopes. In some localities, the fans of the outer slope coalesce making depositional aprons around the ridges (Fig. 17.5b). In some ring complexes, relatively large wadis cut across the ridges making gaps in-between (Fig. 17.5a). Until the time of writing, no study has provided a clear interpretation or justification as to how these wadis cut through the ridges and crossed them? There are two possibilities: The first is that these wadis developed along fault/joint lines. The other possibility is that these wadis were anteceding the ridges, and they kept their lines by deepening while the rings were building up by the materials stemming from the molten rocks. In favor of this

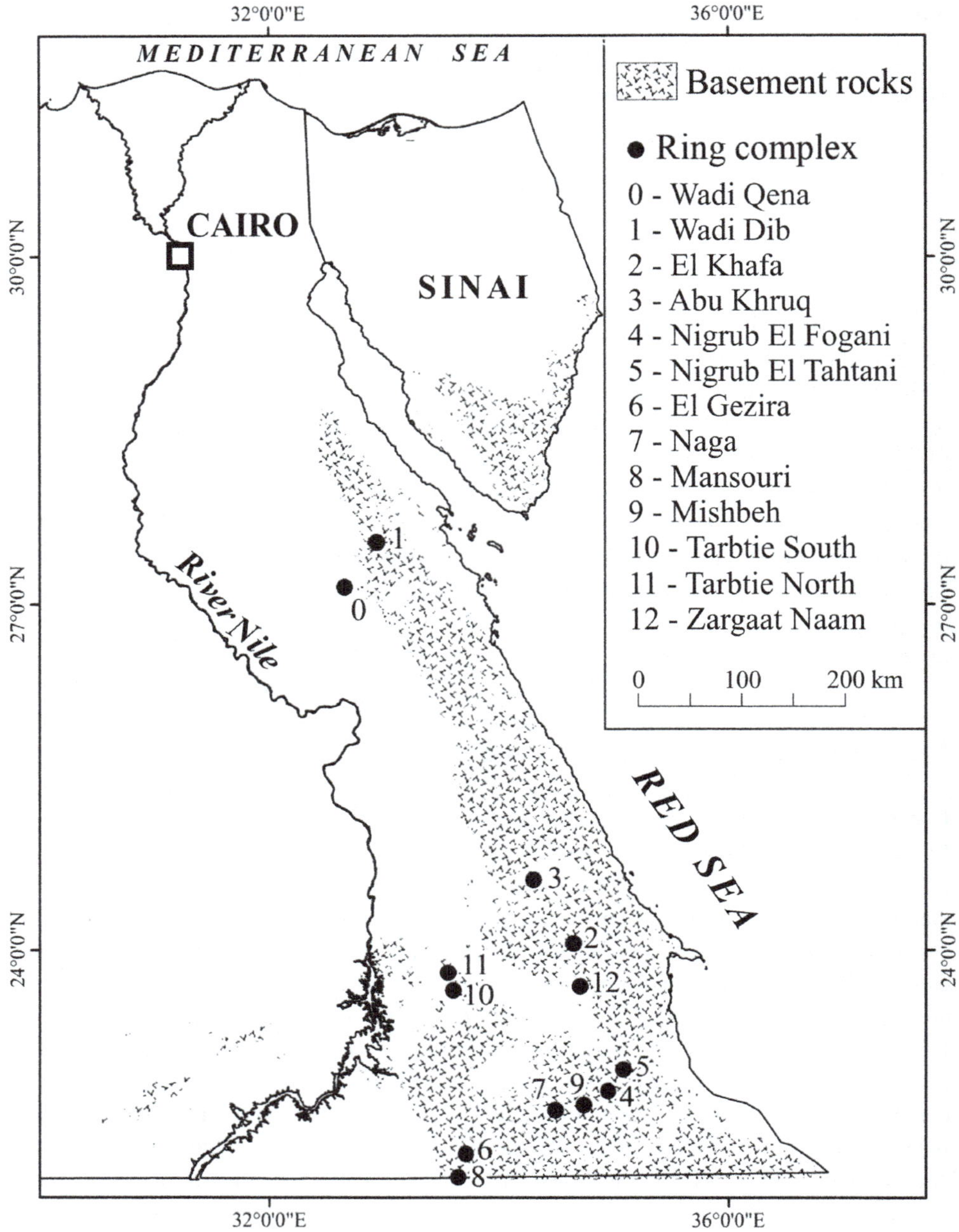

Fig. 17.4 Distribution of ring complexes in the Eastern Desert (Modified from El Ramly and Hussein 1985)

Table 17.2 Parameters of some the Ring Complexes of the Red Sea Mountains (*Source* Adam 2015, Topographic Maps, (scale 1:50,000) Google Earth, 2014, and Landsat-TM images)

Ring Complex	H	R_1	R_2	A	C	Ring Complex	H	R_1	R_2	A	C
Wadi Dib	1000	2.4	1.6	4.18	7.3	Mansuri	798	6.0	5.0	18.0	19.0
Abu Khruq	847	6.3	3.6	28.7	4.0	El-Naqah	826	6.0	3.5	23.5	18.0
El-Kahfa	1018	3.6	1.6	5.7	14.0	Mishbeh	1353	10.2	5.4	46.3	33.0
Nuqrob S	829	3.2	1.6	3.5	9.0	El-Gezeirah	483	3.5	1.2	5.9	9.6
Nuqrob N	1076	4.0	3.0	23.8	27.6	Tarabit	480	3.0	0.7	6.3	6.4

H Height (m), R_1 Radius$_1$ (km), R_2 Radius$_2$ (km), *A* Area (km^2), *C* Circumference (km)

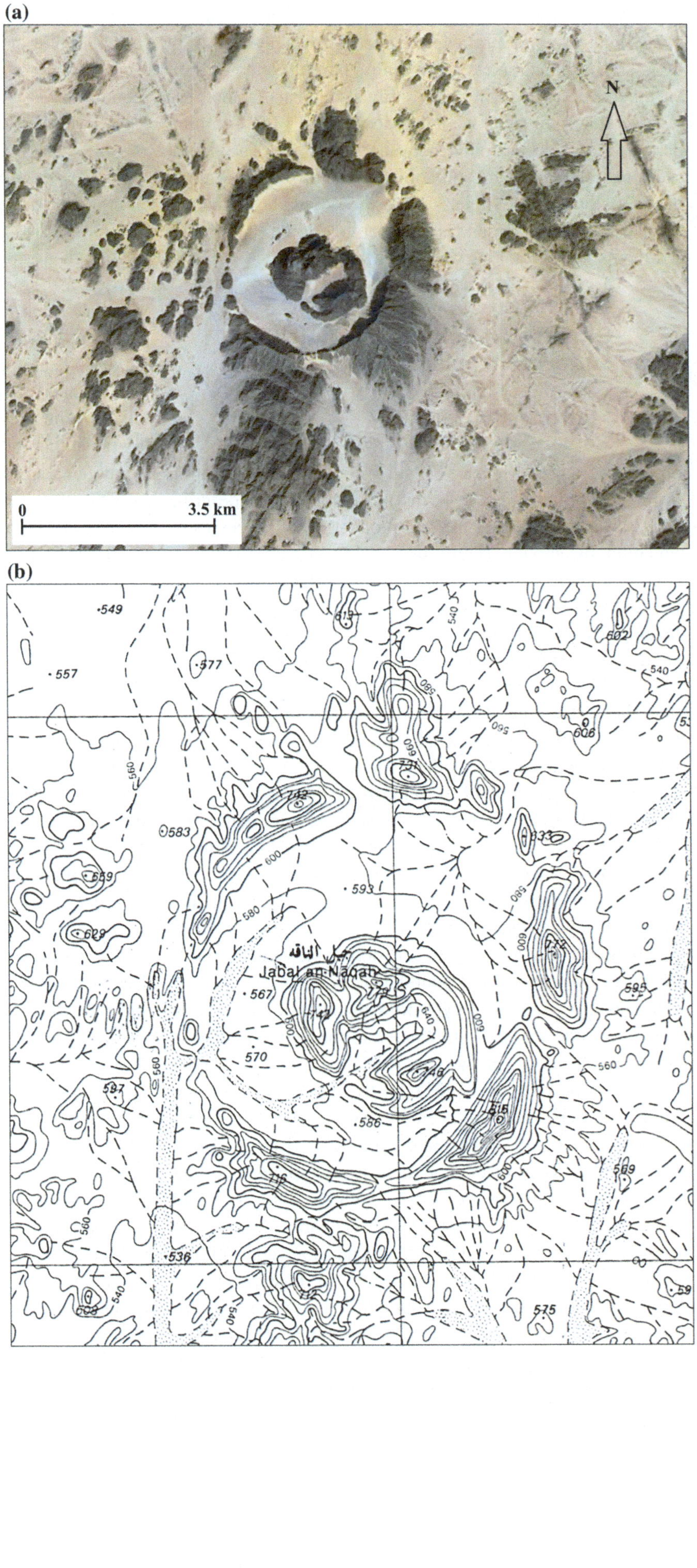

Fig. 17.5 **a**. Gabal El-Naqah nearly circular ring complex with a dissected outer ridge, with two opposing slopes (*Source* Google Earth 2014). (Note the inner slope is steeper than the outer slope, and an inner circular core, taking the form of a ridge rather than a solid block). **b**. Contour map of Gabal El-Naqah Ring Complex, Red Sea Mountains

suggestion is that each ring complex developed not in one cycle, but during several cycles (Sect. 17.4.1).

In some ring complexes, the central part takes the form of an inner Depression, in which_ fluvial deposits, sometimes with aeolian sediments, were laid down (Fig. 17.5a). Similar deposits can also be found between the ridges. The fluvial sediments develop fans and sometime small bahadas along the foot slopes of the inner side of the ridges; however, the aeolian materials spread as a sheet overlying the fluvial sediments without taking any specific form.

17.5 Old Erosion Surfaces/Planation Surfaces/Etchplains

17.5.1 Introduction

Erosion surfaces are also called planation surfaces/etchplains. These terms are employed to describe plain surfaces which are the end product of all processes of planation by erosion (Fairbridge 1968; Migoń 2004a, b). This definition denotes that erosion surfaces are plains etched by erosion processes, and passed all the stages of development (youth, maturity, and old stages) which were suggested by William Morris Davis for the cycle of erosion (Davis 1899). At the final stage, the land surface becomes a plain, but a plain that developed by erosion, but has various names: It is called by Davis "peneplain" in the normal cycle of erosion and is called by others "pediplain" in arid regions as they believe such plains developed by the coalescence of pediments (King 1953). Davis indicated one important point which is that the cycle of erosion takes a very long time to reach its final stage, probably millions/hundreds of millions of years providing that no interruption occurs during any stage of the cycle. Therefore, the Egyptian planation surfaces represent the final stage of an erosion cycle and represent previous or old peneplains/pediplains.

Planation surfaces in Egypt drew the attention of several researchers due to their significance to the geologic history and its implications on the geomorphology of Egypt. First mentioned by Clayton (1933) in the SW corner of the Western Desert, then followed by Awad (1941) in Gabal Uweinat, Awad (1951) in Sinai Mountains, Butzer (1964, 1965) on the slopes of the karstified Eocene plateau around Kurkur Oasis to the west of the Nile Valley, Philobbos and Hassan (1975) in the Eastern Desert, Embabi (2004) in the southern plain of the Western Desert, and Moawad (2008) in the area between Safaga and Quseir in the Eastern Desert. Along with these studies, the southern plain of the Western Desert is described in nearly all previous studies (Peel 1966; Said 1975) as a pediplain, since it lies in a desert with hyper-arid climate. Nonetheless, when the Space Shuttle Columbia discovered the subsurface wadis and channels in November 1981, it was realized that this southern plain went a long cycle of fluvial erosion before being buried under a surface layer of aeolian sand after the commencement of arid conditions.

17.5.2 Planation Surfaces in the Red Sea and Sinai Mountains

Although mentioned in some previous studies, no scientist carried a thorough study on planation surfaces in the Red Sea and Sinai Mountains. Works of Philobbos and Hassan (1975), and Moawad (2008) represent incomplete case studies. Therefore, it is of great necessity to conduct studies to examine this phenomenon. According to the information cited in the previous studies, and some field investigations by the author, all planation surfaces in the Red Sea and Sinai Mountains are developed on the basement rocks of varying resistance and can be classified into three groups: buried peneplains, etchplains, and exhumed surfaces.

Burried penelains—developed on the basement rocks and were covered by a sedimentary cover of Nubian Sandstone (Fig. 17.6) in Sinai Mountains (Awad 1951) and by Tertiary chalk, limestone, and carbonate sediments in Gabal Naqarah in the Red Sea Mountains (Moawad 2008). Since the sandstone succession in Sinai is of Cretaceous age, it can be assumed that the age of the peneplained surface of Sinai Mountains is Pre-Cretaceous. Despite being covered by Tertiary rocks, Moawad (2008) postulates that the buried surfaces of the Red Sea Mountains were developed during the period 680–640 Ma, since the basement rocks underwent severe peneplanation.

Etchplains: Philobbos and Hassan (1975) recorded a well-defined lateritic palaeosoil layer capping the former rocks in seven localities of this buried surface between the latitudes of Safaga and Mersa Alam on both sides of central Red Sea Mountains (Fig. 17.7a, b). This type of palaesoil indicates the surface of the basement rocks was corroded severely by deep weathering under tropical climate (Migoń 2004a, b). Accordingly, the resulting planation surface can be named "Etchplain." The peneplained surfaces were submerged by sea (where the sedimentary cover was laid down) and remained unexposed until the block was most probably uplifted during the Oligocene, when the Red Sea and the Gulf of Aqaba were developing.

The Exhumed Planation Surfaces were buried under a sedimentary sandstone cover like the other localities that remain unexposed. When the whole region was uplifted during the Oligocene as suggested previously, it was exposed to fluvial erosion which removed the sedimentary cover from several localities; thence, the old surface was exposed to erosion (Fig. 17.8). These penelaind surfaces are locally called "*Farsh*" which means a flat area, such as

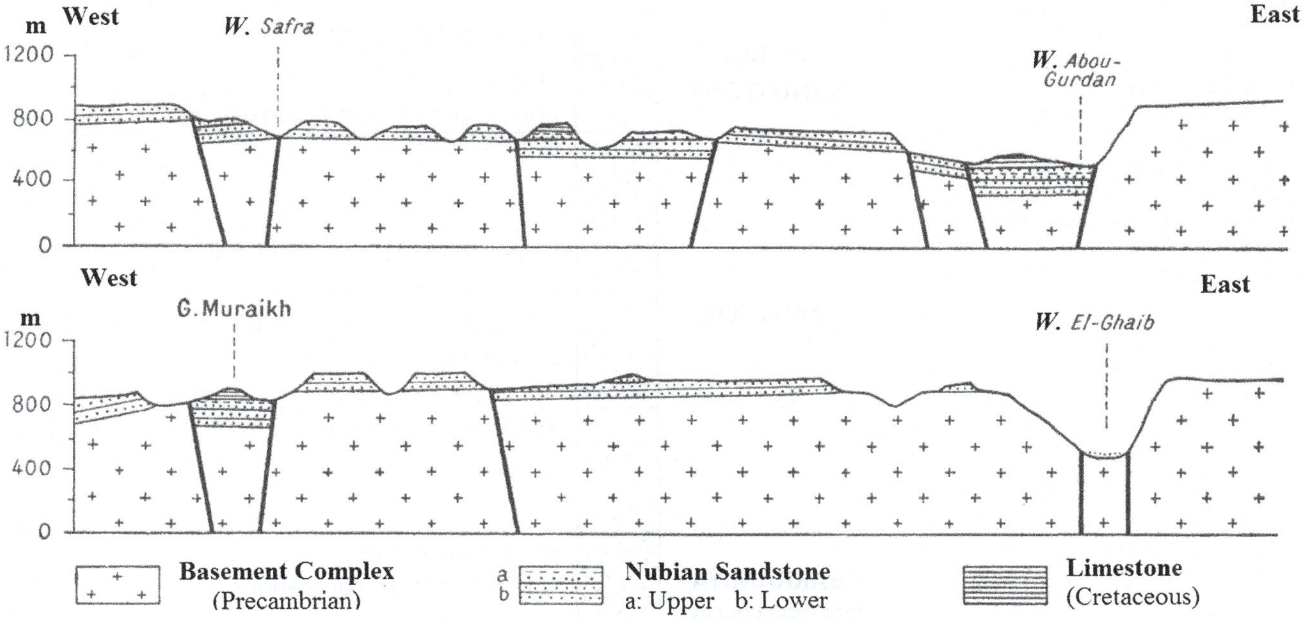

Fig. 17.6 Two geological sections across Sinai Mountains, showing the buried planation surface which developed on the Precambrian Basement Complex (*Source* Awad 1951)

Farsh Zubir, and **Farsh El-Qasab**. Since these exhumed surfaces developed on the basement rocks of the Arabo-Nubian Shield, they are the oldest erosion surfaces in Egypt. In fact, it would be no exaggeration to say they are of the oldest erosion surfaces in the world based on the givens provided.

The exhumed erosion surface was spotted in the Red Sea Mountains of Central Eastern Desert (Moawad 2008). This type of planation surfaces is exposed, attaining an elevation between 500 and 1000 m a.s.l, with an almost flat surface characterized by a low relief and occasional residual hills (Monadnocks). The surface of this peneplain truncates different rocks with various resistances to erosion and weathering (biotite, gabbro, and metavolcanics). The surface is also severely subdued, weathered, and locally affected by younger granite intrusions and morpho-tectonic Depressions. This is why Moawad (2008) considered it as an old erosion surface.

17.6 The Terraces of the Lacustrine Sediments in Wadi Feiran in Sinai Mountains

Since the beginning of the twentieth century, many geologists and geomorphologists were attracted to the unique landscape of Sinai Mountains, the terraces of the fine-grained lacustrine sediments of Wadi Feiran, and its tributaries Wadi El-Sheikh and Wadi Solaf in the heart of Sinai Mountains (Barron 1907; De Martonne 1948; Issar and Eckstein 1969; Awad 1952; Mostafa 1987; Gladfelter 1988–1989; Rögner and Smykatz-Kloss 1991; El-Sherbini 1992; Rögner and Smykatz-Kloss 1993; Rögner et al 1999; Zalamah 2003).

The terraces are distributed as patches rising 20–35 m from wadi beds along the sides of **Wadi Feiran**, **Wadi El-Sheikh**, and **Wadi Solaf**, and the lower sections of some of the tributaries of these wadis (Awad 1952; Rögner and Smykatz-Kloss 1993; Mostafa 1987). The terraces are distributed along a distance of about 40 km from Feiran Oasis (600 m asl) in the west to Watia Gorge (1100 m asl) in the east, with concentrations around the two oases of Feiran and Tarfat.

17.6.1 Sediment Properties

All previous studies described the materials of the terraces as fine-grained sediments, with yellowish color, which indicates that the **lacustrine sediments** are composed of sand, silt, and clay. This grain size composition could be seen readily at the faces of the terraces during field trips (Fig. 17.9). These sediments are made of quartz, feldspars, calcite, and clay minerals, and include zircon, tourmaline, rutile, epidote, hornblende, and iron and manganese oxides as opaque minerals (Zalamah 2003). Regarding the fine-grained sediments, aeolian and fluvial processes in an environment characterized by still-water conditions (Rögner and Smykatz-Kloss 1993) are invoked. Evidence for these still-water lacustrine conditions is that the thin layers of the fine-grained sediments were found without any disturbance, and that the coarser materials were found only at the

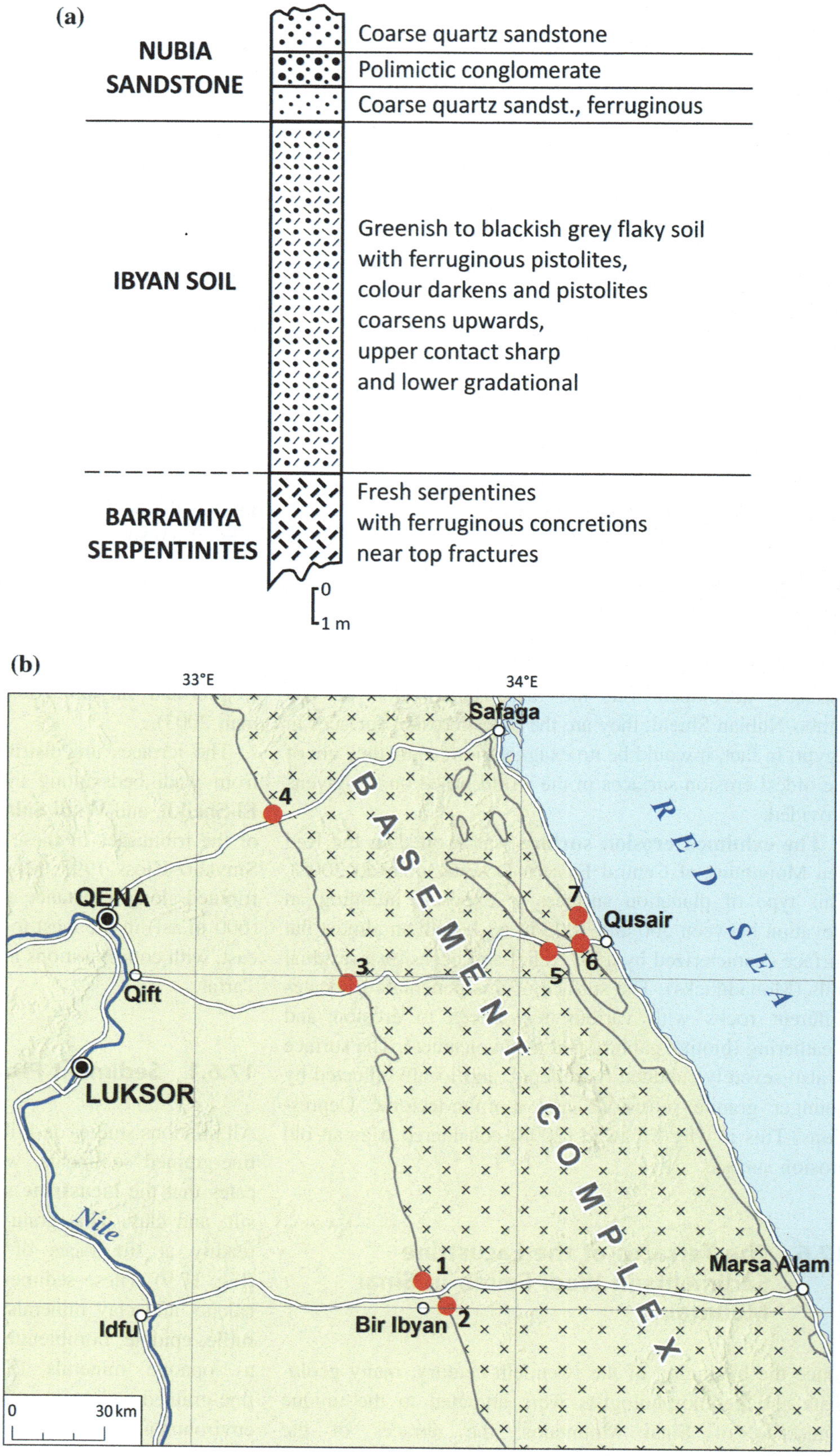

Fig. 17.7 **a** Geological Column showing the succession of formation at the type locality of the etchplain developed on the basement rocks in The Eastern Desert (see Fig. 17.7b for location). **b** Map showing the localities where lateritic palaeosoils were recorded on the buried etchplain in the Eastern Desert (*Source* Philoppos 1975)

Fig. 17.8 Exhumed planation surface in Sinai Mountains, covered in places by some Nubian sandstone remnants, NW Gulf of Aqaba

Fig. 17.9 A field photograph showing a remnant of a terrace of the Lacustrine Sediments of Wadi Feiran, central Sinai Mountains

entrance of such still-water basins (lakes). In some localities, thin layers of fine gravel were found in-between the fine sediments (Issar and Eckstein 1969): evidence of fluvial torrential processes. Because of its angularity, it could be inferred that the coarse material represents the weathering products of the basement. It is more concentrated in the lower layers of the sediments and the marginal areas, where it inter-fingers with the fine-grained sediments (Rögner and Smykatz-Kloss 1993). These sediments are generally un-fossiliferous, except for the relics of some plant roots of fresh water reeds (Issar and Eckstein 1969; Zalamah 2003). These relics of plant roots were found in some horizons, indicating a shallow periodic water body rather than a permanent deep lake. Rögner and Smykatz-Kloss (1993) suggested that the sediments accumulated in at least 11 different sedimentary basins, and the marginal parts of the basins contain material a little coarser in grain size than the central parts. The theory of the presence of several lake basins where sedimentation occurred was suggested by former researchers, of whom Awad (1952) was the first.

Some previous studies recognized decrease in grain size of sediments from gravel to sand, silt, and clay, with a top layer of plant remains, and that this change is repeated several times along the section, indicating a cyclic change in deposition and in water volume in the lake basins (Rögner and Smykatz-Kloss 1991; El-Sherbini 1992; Zalamah 2003).

17.6.2 Age

All previous studies suggested an age for the lacustrine sediments of Wadi Feiran. Suggestions varied between Late Plio/Pleistocne (Issar and Eckstein 1969), Quaternary (Awad 1952), early Quaternary for Feiran sediments, and late Quaternary for Tarfat sediments (Nir 1974). In spite of the large number of previous studies, only a few of them tried to determine the age of the sediments of Wadi Feiran by accurate methods. Using TL technique, Rogner et al (1999) found ages 11.3 ka at a depth 2.5 m, and 27 ka at a depth 25 m. Also Brook et al (in prep. 2017) using radiocarbon and OSL methods, found ages 30 ka for the lower layers and 12 ka for the upper layers. According to the citation of Mostafa (1987), personal communication with a researcher from the University of Illinois, USA, TL age dating of some samples from Wadi Feiran sediments was 35 ka and 12 ka for the lower and upper layers respectively. These three groups of ages suggest sediment deposition since about 27–35 ka and continued up 12 ka ago. Of course, deposition was not continuous all this period. Pattern of deposition will appear in the paper of Brook et al (in prep. 2017) when published.

17.6.3 Origin and Mode of Development

The question which all previous studies tried to answer is the nature of damming that occurred in Wadi Feiran bed, and its tributaries to create (a) lake(s) in these wadis. Until the time of writing, there is no final answer to such an inquiry. The difficulty in finding a conclusive answer lies in the fact that no evidence was found to indicate that there were dams in these wadis, where one or several bodies of water were created. Awad (1952) was the first to suggest the formation of several dams by the alluvial fans of the tributaries of Wadi Feiran which resulted in the formation of several lakes. This suggestion was also echoed in the study of Schick and Lekach (1988). On the other hand, Rögner and Smykatz-Kloss (1991) rejected the suggestion altogether, since the thickness of the lacustrine sediments is 30–35 m, a thickness of alluvial cones of more than 10 m would be unexpected in light of the mode of fans development. Damming by rock sliding and rock fall seems more viable to Rögner and Smykatz-Kloss (1991, 1993). For them, such possibility explains the simultaneously occurrence of lakes at the constricted sections of the wadis under the effect of a spontaneous tectonic event such as earthquake.

Dyke dams across Wadi Feiran are another suggestion for damming the Wadi and its tributaries to form (a) lakes (s), especially that this area of the Wadi is crossed by many dyke swarms (Issar and Eckstein 1969). Though both the Geological Maps of Sinai and field investigations showed the absence of any dyke of any scale projecting from the wadi beds, this suggestion was also rejected by Mostafa (1987) because he believes that the wadis cut across the dykes long before the deposition of the sediments and were eroded completely from wadi beds before the deposition of the sediments.

Although all previous studies concluded that the sedimentological environment of the fine-grained sediments is "Lacustrine," the origin of these sediments remains an unresolved problem. El-Sherbini (1992) concluded that the sediments of Wadi Feiran and its tributaries were derived from one and the same source rocks (granite and granite gneiss), and the derived sand and gravel were transported by the melt-water flowing from the mountain glaciers. This process is repeated three times during the deposition of the whole succession, at its base, middle, and upper stages due to climatic fluctuations, which were responsible for the vertical variations of the sediments from fine- to coarse-grained deposits. Two features are in favor of the glacial processes suggested by El-Sherbini (1992). The first is that the shape of the upper section of Wadi El-Sheikh is U-shaped, similar to glacial valleys of the Pleistocene Ice Age (Fig. 17.10). The second feature is that during cold

Fig. 17.10 A field photograph of the U-shaped Wadi, the upper section of Wadi El-Sheikh, Sinai Mountains

winters of today's climatic conditions, snow falls instead of rain, and in some occasions, it persists for several days. These two features constitute a high probability of the existence of an ice cap on high summits of Sinai Mountains, especially if the age dating of the layers of the sediments mentioned above, 35,000–20,000 BP is taken into consideration, which is an age that coincides with the last Ice Age. The recent study of Brook et al (in prep. 2017) found evidence to conclude that these sediments were deposited and later eroded by rivers at different times in the past. An earlier sediment sequence dates to c. 30 ka and a later one began shortly after 15 ka following partial erosion of older deposits. Dry conditions introduced loess into the valley and its tributaries, and this was later washed into the main valley where it accumulated along with coarser material. Importantly, the amount of flow in the river system was insuffient to transport the sediments out of the basin into the Gulf of Suez and so it accumulated slowly in the valley. During the Holocene wet phases, the sediments was washed out by intense rains storms brought by a more active monsoon at a time of increased northern Hemisphere solar radiation centered about 9 ka. The presence of extensive fluvial deposits in the sediments indicates that they were not deposited in a series of lake basins as suggested by previous studies. As the deposits are of variable ages, and contain sufficient amounts of fluvial material, they were not deposited in lakes produced by earthquakes—generated landslides and valley blocking as previously suggested.

Rögner and Smykatz-Kloss (1993) present an alternative speculation on the origin and development of the lacustrine sediments and terraces of Wadi Feiran and its tributaries. They believe that the fine-grained carbonaceous material is aeolian in origin, and it was transported as airborne dust from their source rocks, which lie 50 km to the west of the first occurrence of these sediments in Wadi Feiran. This airborne material is laid down on the slopes of the mountains of Sinai and partly in the wadis of southern Sinai. Then, it was transported from the slopes downward to wadi beds by fluvial action during rainy seasons. By time, the material accumulated in wadi traps (basins) under still-water conditions (lacustrine environment). Periodic drying up of the basins allowed the growth of reeds, forming the fossil roots horizons. Recently, fluvial erosion partly eroded the lacustrine sediments along the wadi bed, leaving some patches on both sides of the wadis in the form of terraces.

In this model, there are some gaps in research among which the source rocks, the process of transportation, deposition of the fine-grained sediments, and reasons for the development of the fine-grained sediments in Wadi Feiran are the most critical ones. The first issue is how aeolian processes derived the fine materials from bedrock? It is a well-established fact that prerequisite process is needed; this process is the decomposition of the carbonate rocks to derivate the fine grains, of which there was no mention. The second issue is the process of transportation and deposition of the fine material from the source rock which lies 50 km to the west of present deposits in Wadin Feiran. There is no proof for a western wind during the period of transportation; second, during transportation, there was a chance for some material to fall down in other wadis, and the question is why no development of similar lacustrine sediments occurred? The interpretation—by Rögner and Smykatz-Kloss—(1993) holds that local conditions in other wadis were not favorable for the formation of lake(s). Moreover, there is no mention to what extent the source of carbonaceous rocks provided the quartz grains, and how these grains were transported from the source rocks.

17.7 Conclusion

Although the Red Sea and Sinai Mountains differ greatly in their geomorphologic characteristics, there are some common aspects which are significant to both the geologic and geomorphologic history of Egypt and the past climates over Egypt. The first aspect is that the Old Planation Surface (OPS), which developed on the basement complex, indicates that the original Arabo-Nubian Shield was exposed to planation for a very long period. During this period, the Shield surface became a peneplain by the influence of fluvial erosion under humid climatic conditions, rather than marine planation. Two circumstances speak in favor of this conclusion. The first is the soil horizon atop the etchplain and below the sandstone. The second evidence is the highly rounded quartz gravel found in the lower sandstone layers lying atop the planation surface, or found loose on the surface of the ground after the sandstone was exposed to erosion.

During field investigations, not a single gravel grain was found flattened, which shows that marine erosion had no role in the planation of this OPS. Due to lack of evidence, it was suggested that its age is pre-deposition of Nubian Sandstone (Precambrian?), which could be more than 600 million years. It is likely that this OPS is one of the oldest erosion surfaces on earth.

Dyke Swarms (DS) and Ring Complexes (RC) are unique geological features that characterize the Red Sea and Sinai Mountains. These features are composed of intrusive rocks, which took two different forms according to the system of joints and fissures characterizing areas of development. DS spread over much wider areas than RC, because the linear structural lines which controlled their development are more frequent than the circular ones of the RC. Also, the density and orientation of structural lines controlled these aspects of DS and RC. These morpho-tectonic forms developed over a long time span, which indicates that tectonic activity is still going on as a factor in the building up of the Egyptian landscape.

Intermountain Basins (IB) are also a morpho-tectonic form. They represent depositional basins, into which fluvial sediments were carried out from the surrounding summits and hills. Their size and shape are controlled by a fault system in each locality, and this is why their morphologic shape differs from one locality to another.

The terraces of the lacustrine sediments in Wadi Feiran and some of its tributaries represent the remnants of once more widespread deposits which filled the beds of these wadis. The origin of these terraces and their development are still a matter of great controversy. However, their significance lies in their past climatic implications. The only speculation on their origin and development which can be confirmed with some evidence is the glacial origin. Firstly, the shape of upper section of Wadi El-Sheikh is a U-shaped valley, similar to those of the glaciers (Fig. 17.8). Secondly, present-day snow falls down on Sinai Mountains during cold winter. Finally, these sediments were laid down during the Last Glacial Period (20,000–35,000 yr. BP). From all these aspects, it can be hypothesized that small ice caps were present on the summits of Sinai Mountains during the LGP, and that ice as an erosion process played a local role in shaping present-day landscape of Egypt. In spite of this conclusion on the origin of sediments, the issue of the formation of (a) lake(s) where the sediments existed, seems to be unresolved. However, the study of Brook et al (in prep. 2017) concluded that the presence of extensive fluvial deposits in the sediments indicates that they were not deposited in a series of lake basins as suggested by previous studies. As the deposits are of variable age, and contain sufficient amounts of fuvial material, they were not deposited in lakes produced by earthquake—generated landslides and valley blocking as previously suggested.

References

Adam S (2015) Geomorphology of ring complexes in the Eastern Desert of Egypt (in Arabic). Dissertation. Ain Shams University, Cairo

Awad H (1941) Structure et relief de la Libye Meridionale. Annale de Geographie LIII-LIV:192–203

Awad H (1951) La Montagne du Sinai Central. Étude Morphologique. Société Royale de Géographie d'Égypte, Le Caire

Awad H (1952) Signification morphologique des depots lacustre de la Montagne de Sinai Central. Bull Soc Royale de Géographie d'Égypte XXV:23–28

Barron F (1907) The topography and geology of the Peninsula of Sinai Western Portion. Survey Department, Cairo

Brook GA, Embabi NS, Ashour MM, Yang L, Ramadan M (in prep. 2017) Age and origin of relict fine grained sediments sequences in Wadi Feiran, Sinai, Egypt.

Butzer KW (1964) Pleistocene palaeoclimates of the Kurkur Oases, Egypt. Canadian Geographer 8:125–141

Butzer KW (1965) Desert landforms at the Kurkur Oasis, Egypt. Ann. Assoc. Amer. Geographers 55:578–591

Clayton P A (1933) The western side of the Gilf Kebir, in A reconnaissance of the Gilf Kebir by the Late Sir Robert Clayton East Clayton. Geographical J LXXXI:254–259

Davis W (1899) The geographical cycle. Geographical J XIV:481–504

De Martonne E (1948) Reconnaissance Géographique au Sinai. Bull Soc Royale de Géographie d'Égypte XXII:105–136

El-Rakaiby M (1997) Remote sensing studies for dyke swarms in the Central Eastern Desert of Egypt. J African Earth Sci 24:653–656

El-Ramly M, Hussein A (1985) The ring complexes of the Eastern Desert of Egypt. J African Earth Sci 3:77–82

El-Sherbini M (1992) Composition and paleoclimate of the Pleistocene lacustrine sediments of Wadi Feiran, Sinai. Proc. 3rd Conf. Geol. Sinai Develop, Ismailia, p 153–160

Embabi NS (2004) The geomorphology of Egypt: landforms and evolution. The Nile Valley and the Western Desert, vol I. The Egyptian Geographical Society, Cairo

Fairbridge R W (ed) (1968) The encyclopedia of geomorphology. Dowden, Hutchinson & Ross. Inc

Gladfelter B (1988–1989) Late Pleistocene lakes within the Mountains of Sinai: observations at the Tarfat Oasis. Bulletin de la Société de Géographie d'Égypte 56–57:29–52

Google Earth (2014) Egypt Images Latitudes 22-32 N and Longitudes 22-32 E, http://www.google.com

Iacumin M, Marzoli A, Ellmetwally A, Piccirillo E (1998) Neoproterozoic dyke swarms from southern Sinai (Egypt): geochemistry and petrogenetic aspects. J African Earth Sciences 26:49–64

Issar A, Eckstein Y (1969) The lacustrine beds of Wadi Firan: their origin and significance. Israel J Earth Sci 18:21–27

King LC (1953) Canons of landscape evolution. Geol Soc Am Bull 64:721–752

Lutz T (1979) Strontium and oxygen isotope relations and geochemistry of Abu Khruq complex, Egypt, Implications for the petrogenesis of the alkaline rocks of the Eastern Desert, Egypt. Dissertation, University of Philadelphia

Meneisy M (1990) Volcanicity. In: Said (ed) The geology of Egypt. Balkema, Rotterdam, p 157–172

Migoń P. (2004) Etchplain. In: Andrew Goudie (ed) The Encyclopedia of Geomorphology. Routledge, London

Migoń P. (2004) Planation surfaces. In: Andrew Goudie (ed) The Encyclopedia of Geomorphology. Routledge, London

Moawad M (2008) In: VDM Verlag Dr. Mueller. K (ed) Application of remote sensing and geographic information systems in geomorphologic studies: Safaga-El-Quseir area, Red Sea Coast, Egypt as an example. Saarbrücken

Mostafa M (1987) Wadi Feiran Basin: a geomorphologic study. Dissertation (in Arabic), Ain Shams University

Nir D (1974) Lacustrine/fluviatile sediments in Feiran and Tarfat el Kudrein. Zeit fur Geomorphologie, N F SupplBd 33:32–34

Peel RF (1966) The Landscape in Aridity. Institute of British Geographers Transactions 38:1–23

Philobbos ER, Hassan KK (1975) The contribution of palaeosoil to Egyptian lithistratigraphy. Nature 253:33

Rögner K, Smykatz-Kloss K (1991) The deposition of eolian sediments in lacustrine and fluvial environments of Central Sinai (Egypt). Catena Supplement 20:75–91

Rögner k, Smykatz-Kloss W (1993) The fine-grained sediments of Wadi Firan (Sinai-Egypt): Origin and sedimentology. Z Geomorph N F Suppl Bd 88:123–139

Rögner K, Smykatz-Kloss K, and Zoller L (1999) Oberpleistozane palaoklimatische varanderungen im Zentral-Sinai (Egypten). Erdkunde Bd 33:220–330

Said R (1975) Some observations on the geomorphological evolution of the South Western Desert of Egypt and its relation to the origin of the ground water Ann. Geol. Survey of Egypt 5:67–70

Schick P, Lekach J (1988) A high magnitude flood in the Sinai desert. In: Mayer, Nash (eds) Catastrophic Flooding, pp 381–410

Serencsits C, El-Ramly M, Faul H, Foland K, Hussein A (1979) Alkaline ring complexes in Egypt: their ages, and relationship to tectonic development of the Red Sea. Ann Geol Survey of Egypt 9:102–116

Zalama A (2003) The Quaternary sediments of Wadi Feiran and Wadi Khashbi, Southern Sinai, Egypt. Sedimentol Egypt 11:291–315

The Drainage System of the Eastern Desert (with a Special Reference to Wadi Qena Basin) 18

Abstract

The drainage system of the Eastern Desert like that of Sinai is inherited from past humid climates. Three regional drainage subsystems have developed in this region under the effect of regional slopes: eastward toward the Red Sea, westward to the Nile Valley, and finally to the Mediterranean, eastern region of the Nile Delta. Only a few localities are drained internally. All types of drainage patterns are present in the Eastern Desert, of which the dendritic pattern is the most common. The size of basins is larger in the Nile Valley and the Mediterranean systems than in the Red Sea. Wadi Qena is one of the largest and unique wadis in this desert. It runs from north to south along the contact between basement rocks and the sedimentary formations. In this Wadi, four fluvial terraces, fanglomerates, have been developed along both sides of the Wadi.

Keywords

The Eastern Desert • Drainage system • Wadi qena

18.1 Introduction

As in the Sinai Peninsula, the drainage system in the Eastern Desert is inherited from past wet climates, since present-day arid and hyper-arid conditions are not favorable to generate surface water flow to develop such a drainage system. Although present-day floods lead to some changes, the drainage system remains intact and is composed of numerous basins of varying sizes. Previous studies on some selected drainage basins in the Eastern Desert (Moawad 2008) showed that the network of these basins follows the rules and laws of Horton (1945) on stream and drainage basins, indicating that there are no fundamental changes that occurred in the network characteristics under the present-day arid conditions.

18.2 Regional Drainage System

The **Eastern Desert**, as mentioned in Chap. 1, is bound from the east by the Red Sea, the Gulf of Suez, and the Isthmus of Suez, from the west by the Nile Valley and the Nile Delta, and Wadi El-Tumaylat from the north. Due to the effect of tectonics, the Red Sea and the Eastern Desert follow the same trend, which is NNW-SSE. Slopes in the Eastern Desert take three general directions: the first is toward the Red Sea and Gulf of Suez, the second is toward the Nile Valley, and the third is toward the Nile Delta Embayment and the east-delta plain. Therefore, there are three drainage systems in the Eastern Desert. The first one is the **Nile Valley-Mediterranean Drainage System**, the second one is

N.S. Embabi, *Landscapes and Landforms of Egypt*, World Geomorphological Landscapes,
https://doi.org/10.1007/978-3-319-65661-8_18

the **Red Sea Drainage System**, and the third one is the **East Delta Drainage System**. The first two systems are external, but the third is internal since most of the wadis are terminating in the gravelly fluvial plain east of the Nile Delta, and some of them in the embayment of the Delta. Internal drainage is not only restricted to the northern drainage lines, but it is also present in another two areas. The first lies to the south of Ras Gharib (see Fig. 21.2), where wadis coming from the west (W. Um Yasar, W. Ghareb, W. Garf, and W. Kharim) and east terminates in the basin of **El-Mallaha**, which lies below the sea level (2–12 m bsl). The second area with internal drainage is the Maaza Plateau that lies in the west of the Red Sea Mountains and in the north of Qena Bend of the Nile Valley. This plateau is underlain by Eocene limestone and highly karstified, where small solution Depressions receive local drainages. Therefore, this plateau is partially drained internally (see Chap. 17).

This picture of the drainage system in the Eastern Desert was attained—as late as in Late Miocene after the Nile Valley came into being. It is now well established that during the period from Late Eocene till Middle Miocene; drainage lines which were flowing westward from the Red Sea Mountains and represented the tributaries of the drainage system whose delta is lying to the north of Fayum Depression as external drainage or terminated in the Depression of the Great Sand Sea and other low lying areas in the Western Desert as internal drainage. When the Nile started to develop backward from the Hing Zone in Late Miocene, it captured all the drainage lines which flowed westward and became tributaries for the River Nile (Said 1981). The beheaded sections of the tributaries were all obliterated by erosion except for a few of them, where they are still present at the western side of the Nile Valley.

18.3 Drainage Basins

The **number of drainage basins of the Eastern Desert** has never been cited in any previous study on this region. To count the drainage basins is impossible for several reasons. The first is the great number of basins in a region with such a vast area (225,000 km^2). The second factor is that there are no detailed maps or high-resolution images/aerial photographs covering an area like that. The third factor is the human interference in modifying the earth's surface for land reclamation under the needs for new land for expansion outside the Nile Valley and the Delta due to rapid population growth. However, from the maps available constructed for various studies, some aspects of the drainage basins can be inferred.

Figure 18.1 shows the basins of the two drainage systems of the Red Sea and the Nile Valley/Mediterranean. This figure shows that the basin area of the wadis of both systems is variable, where there are small basins as well as large ones. The large basins of the Mediterranean system outnumber those of the Red Sea System. Examples of large wadis belonging to the first system are **W. Sannur**, **W. El-Asyuti**, and **W. Qena** on the Eocene Plateau and **W. Kharit**, **W. Shaait**, and **W. El-Allaqi** on **El-Ababda Plateau** in the south. Examples of wadis of the second system are **W. Gheweiba**, **Wadi Araba**, and **W. Hodein**. The figures also show that the large wadis of the Mediterranean are larger than those of the Red Sea. These two characteristics are caused by relief, where the divide lines of the Red Sea Mountains are nearer to the Red Sea than the Nile Valley, resulting in steeper slopes on the side facing the Red Sea. A second factor is that the western side of the Red Sea Mountains is bordered by low gradient surfaces of the plateaus of Maaza in the north and Ababda in the south. This was a significant factor in developing large drainage basins. A concomitant aspect is that the large basins dominate the landscape of the western side of the Red Sea Mountains. On the other hand, the small size basins represent a characteristic feature of the Red Sea System.

18.4 Drainage Patterns

In a vast region such as the Eastern Desert, it is expected to find all types of drainage patterns. The most widespread pattern is the dendritic one. Nearly all the large basins have a dendritic network, especially on the plateaus (Fig. 18.2). Some small basins also may develop a **dendritic drainage pattern** such as those in the mountainous area between Safaga and Quseir (Moawad 2008). **Parallel drainage pattern** is present in areas affected by a system of parallel faults, or steep slopes. This pattern can be seen in several localities, such as the steep western side of Wadi Qena (Fig. 18.3) and in the southern part of the Eastern Desert where some wadis are controlled by parallel faults (Fig. 18.4). Radial patterns are present on all forms which have outward slopes such as volcanic ones and ring complexes. Here, drainage lines are short and may unite with larger wadis. A centripetal pattern developed where wadis drain internally in low-lying areas, of which inter-mountain basins are the most frequent (Fig. 18.5). In some cases, it occurs in some ring complexes such as G. El-Naqah (Fig. 17.5, Chap. 17). A **braided drainage pattern** developed on very low gradient surfaces mainly those of alluvial fans and deltas. It exists also in inter-mountain basins, where the channels are transformed from single-thread form to braided features.

18.5 Some Characteristics of Geomorphic Significance

The drainage network in the Eastern Desert is characterized in certain localities by some aspects of geomorphic significance. There are some wadis which are deeply incised on the

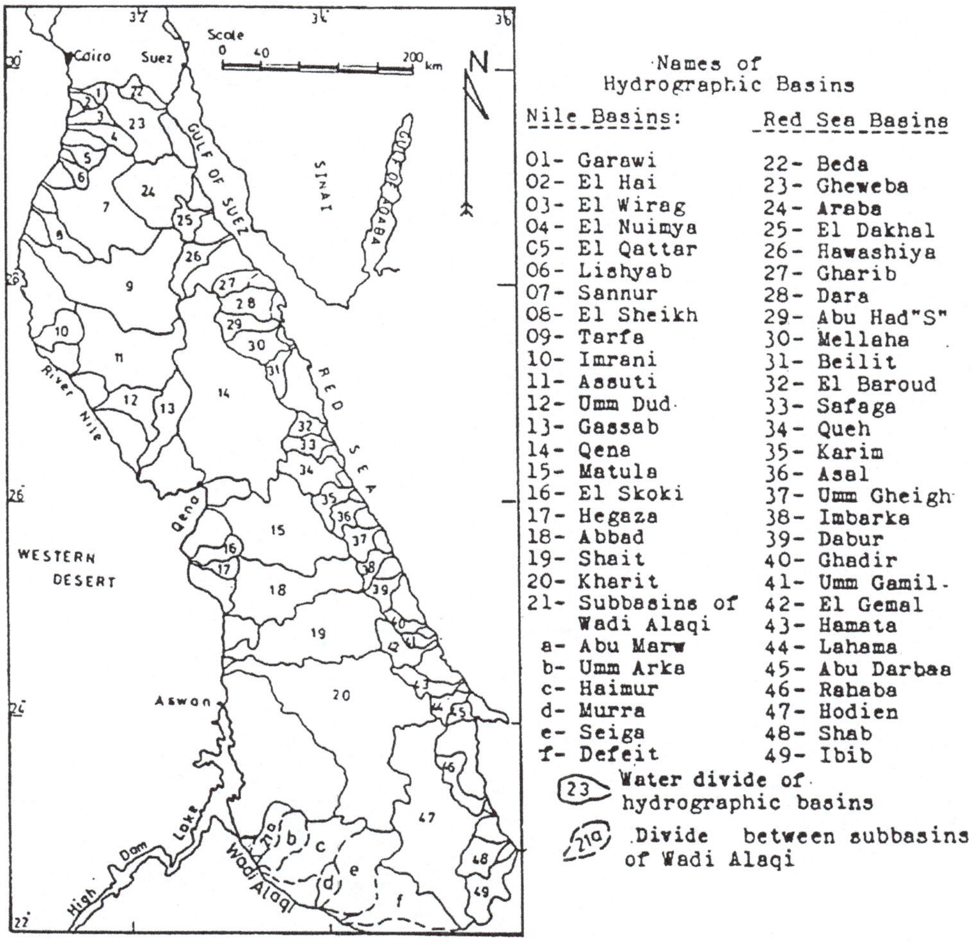

Fig. 18.1 Drainage basins in the Eastern Desert (*Source* El-Etr et al. 1997)

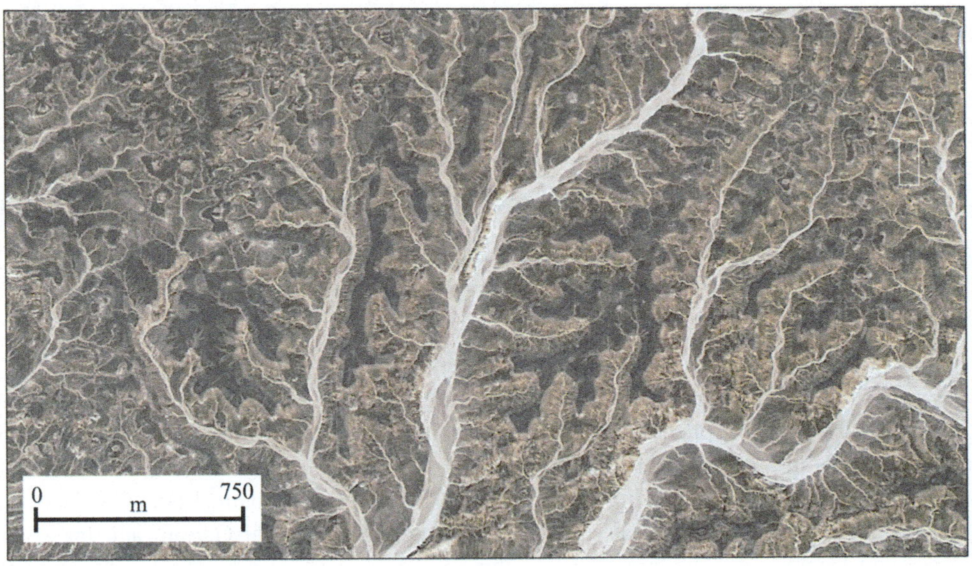

Fig. 18.2 A space image of the Eocene Plateau showing the dendritic pattern of the drainage lines dissecting the plateau (The black areas represents the remnants of the original surface) (*Source* Google Earth, 2014)

Maaza plateau to the west of Wadi Qena (Fig. 18.2). Moreover, some wadis cross the two **Ranges of Aish El-Mallaha** from west to east through narrow gorges (Fig. 18.6).

Figure 18.2 shows that all wadis in this western part of the limestone plateau of El-Maaza are incised, indicating the occurrence of tectonic uplift which resulted in an increase in the erosive power of water flow in the wadis, the deepening

Fig. 18.3 Parallel drainage pattern of the small W-E tributaries of Wadi Qena (*Source* Google Earth, 2014)

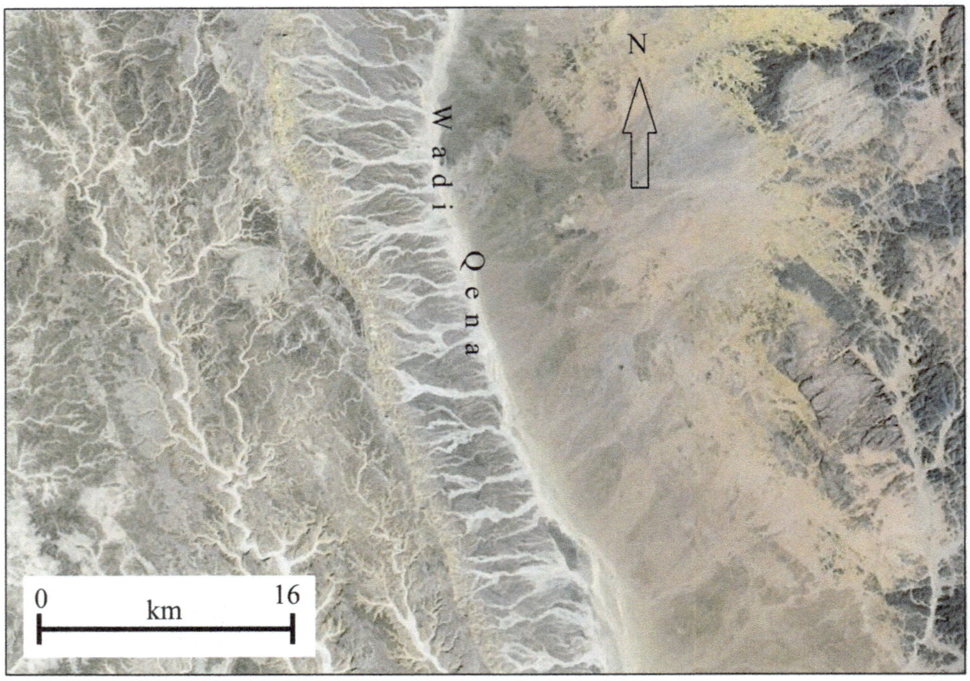

Fig. 18.4 Parallel drainage pattern developed under the effect of faults in the southern part of the Eastern Desert (*Source* Google Earth, 2014)

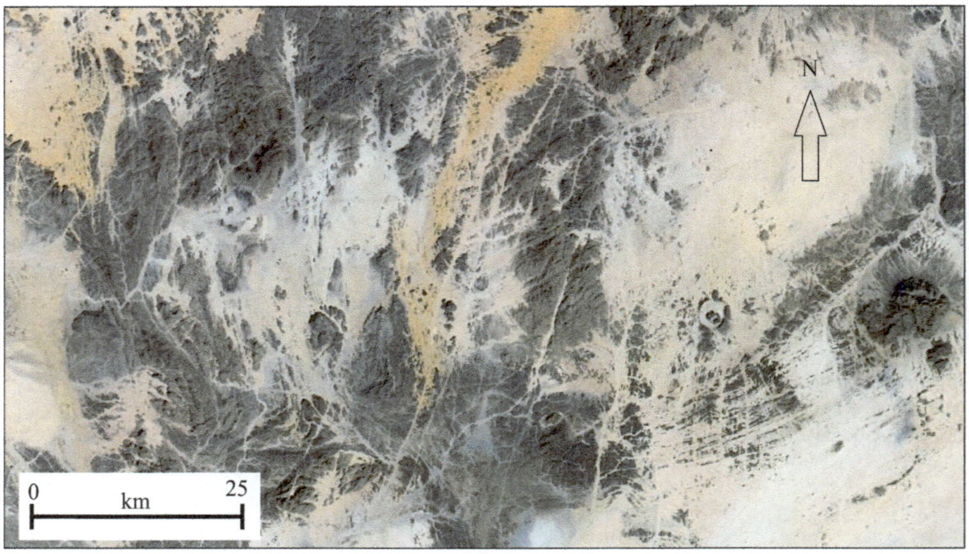

of the wadi beds, and the development of steep-sided valleys. Most probably, this uplift occurred during the Late Miocene. The second phenomenon is the crossing of four major wadis the **Aish El-Mallalaha Range** (Fig. 18.6), taking the form of a gorge. This range is composed of two sub-parallel ranges of different size and separated by a low-lying Depression. The large range is the one known as Aish El-Mallaha, whereas the small one is called Gabal Safar El-Dib in the north, Gabal Safar Abu Had and Safar El-Mallaha in the central section, and Safar El-Aish in the southern part. The main range is a faulted linear elevation of Precambrian basement complex (El-Etr and Abdel-Rahman 1975). It runs parallel to the Gulf of Suez, and extends 55 km from NNW to SSE, with a varying width of 5–10 km, and heights between 250 and 350 m with a highest point of 450 m a.s.l, whereas heights to the west and east of this range are 110–200 m a.s.l. The northern section of this range is named Homret El-Girgab, the central part is called Gabal Abu Garfan, and the southern section is called Gabal Um Dorah. The four wadis crossing Aish El-Mallaha Range are as follows from north to south: **W. Dib**, **W. Abu Had**, **W. Mallaha**, and **W. Bali**. These wadis drain mainly the northern part of the **Red Sea Mountains**, extending westward for several tens of kilometers. Their mode of development was interpreted in a previous study as antecedent wadis (El-Etr and Abdel-Rahman 1975). Most probably they

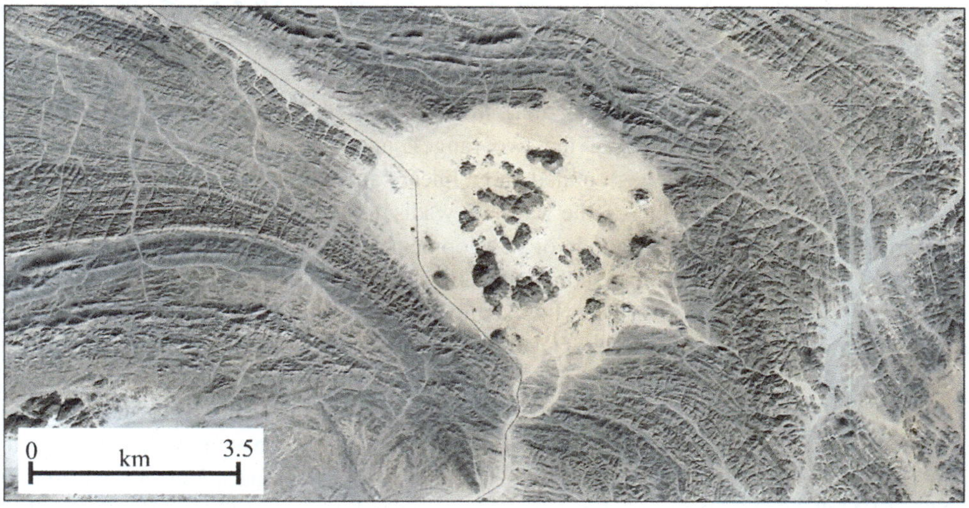

Fig. 18.5 Inter-mountain basin with centripetal drainage system in the southern part of the Eastern Desert (*Source* Google Earth, 2014)

Fig. 18.6 Landsat-TM image showing the drainage system of the two ranges of Aish-El-Mallaha

antedate the uplift of Aish El-Mallaha faulted range. They deeply incised their courses and succeeded in crossing transversely the tectonically controlled range. Another probable mode of development is that there were two opposing wadis on both sides of the two ranges, and by headward erosion the wadis were united and a linear capture occurred diverting the water of the western wadis to the eastern ones. A third probable mode of development is that these wadis were running on a sedimentary cover sloping toward the Red Sea, and over time this sedimentary cover was wiped out and the wadis were superimposed on basement igneous and metamorphic rocks of the two ranges of Aish El-Mallaha.

Topographic maps and Google space images show that these antecedent wadis are not restricted to the main range of Aish El-Mallaha, but they are developed also on the little range lying to the west of the main range (Fig. 18.6). The same wadis of Aish El-Mallaha cross this secondary range also as deep and narrow gorges in most cases. Other wadis of the same type are in the Red Sea Mountains to the north of Aish El-Mallaha Range. The most prominent wadi in this region is **Wadi Hawashiya** which drains into the Gulf of Suez 25 km to the north of Ras Ghareb Town.

18.6 Wadi Qena Basin

Wadi Qena acquired its name after the city of Qena in the Nile Valley where it debouches into the Nile River. Wadi Qena is a long, wide dry valley that begins at the southern slopes of the **South Galalah Plateau**, extending 250 km in a nearly N–S direction to the town of Qena. Its basin covers an area of ~16,000 km^2, which equals ~7.2% of the total area of the **Eastern Desert**.

It is a unique wadi among the drainage networks of the Eastern Desert, because it is characterized by certain aspects which are uncommon in other wadis. It runs southward, contrary to the direction of the wadis of this desert which are directed either W or E, and contrary to the Nile River which flows northward. It is developed along the contact line between the old basement rocks of the **Red Sea Mountains** in the east, and the more recent sedimentary cover in the west (Fig. 18.7). Therefore, it can be considered as a subsequent wadi. It accommodates features indicating that its development predated the Nile Valley, and it was captured by the Nile River and transformed to become one of its large tributaries. The Wadi is also characterized by the presence of a unique volcanic and ring structure in its central section (Bandel et al. 1987). The eastern tributaries are superimposed from a former sedimentary cover.

18.6.1 Geomorphologic Characteristics of Wadi Qena

The **Wadi Qena Basin** is fed by 43 major tributaries, of which 35 wadis issue from the Red Sea Mountains in the East and only 8 wadis from the **Maaza Plateau** in the west (Gawish 1981). The divide line between the Qena basin and other basins in the east separates the **Red Sea Drainage System** from the **Nile Drainage System**. In the west, the divide line runs along the edge of the escarpment of the Eocene Maaza Plateau. The eastern divide line along the Red Sea Mountains is higher (2187 m for **Gabal Shayeb El-Banat** in the north, and 562 m for **Gabal Nag El-Ter** in the south) than the western divide along the edge of the Maaza Plateau (580 m for **Gabal El-Sheikh Isa** in the south, and **Gabal Um Tanasib** 1110 m in the north).

Wadi Qena basin area can be divided into four morphologic units as follows:

The first unit is the eastern slopes that represent the northern part of the eastern side of Wadi Qena and the western slopes of the Red Sea Mountains, at the same time. Their height rises gradually from west to east and from north to south. These slopes are composed of igneous and metamorphic rocks, of which granite is the most widespread, where it intruded the old metamorphic rocks. One of the examples of granite forms is **Gabal Qattar** whose steep slopes appear as towers, rising abruptly ~600 m above the wadi bed. In the southern part of this unit, **Gabal Dokhan** with its steep slopes, composed of granite, extends in a north–south direction.

This unit is dissected by numerous faults that take various directions, where those taking NE–SW and N–S directions are predominant. These faults governed the development of most wadis in the eastern side of Wadi Qena, such as **W. Gisi**, **W. Um Balad**, and **W. El-Urf**.

The second unit is the eastern carbonate plateau: The southeastern section of Wadi Qena basin is composed of a small plateau of Palaeocene chalk and Eocene limestone. It extends from **Gabal Abu Had** in the north to **Gabal El-Sarai** and **Gabal El-Gir** in the south. It is dissected by a group of wadis, such as wadis Gariah, Fatirah. In fact, Gabal Abu Had represents a detached part from El-Maaza Plateau in the west.

The third unit is western escarpments of El-Maaza carbonate plateau representing the western side of Wadi Qena. They are composed of Eocene limestone and shale, and older Cretaceous rocks. They are characterized by numerous joints and faults, which run parallel along some sections of the escarpment, and are followed by courses of some wadis. The escarpments are dissected by a few large

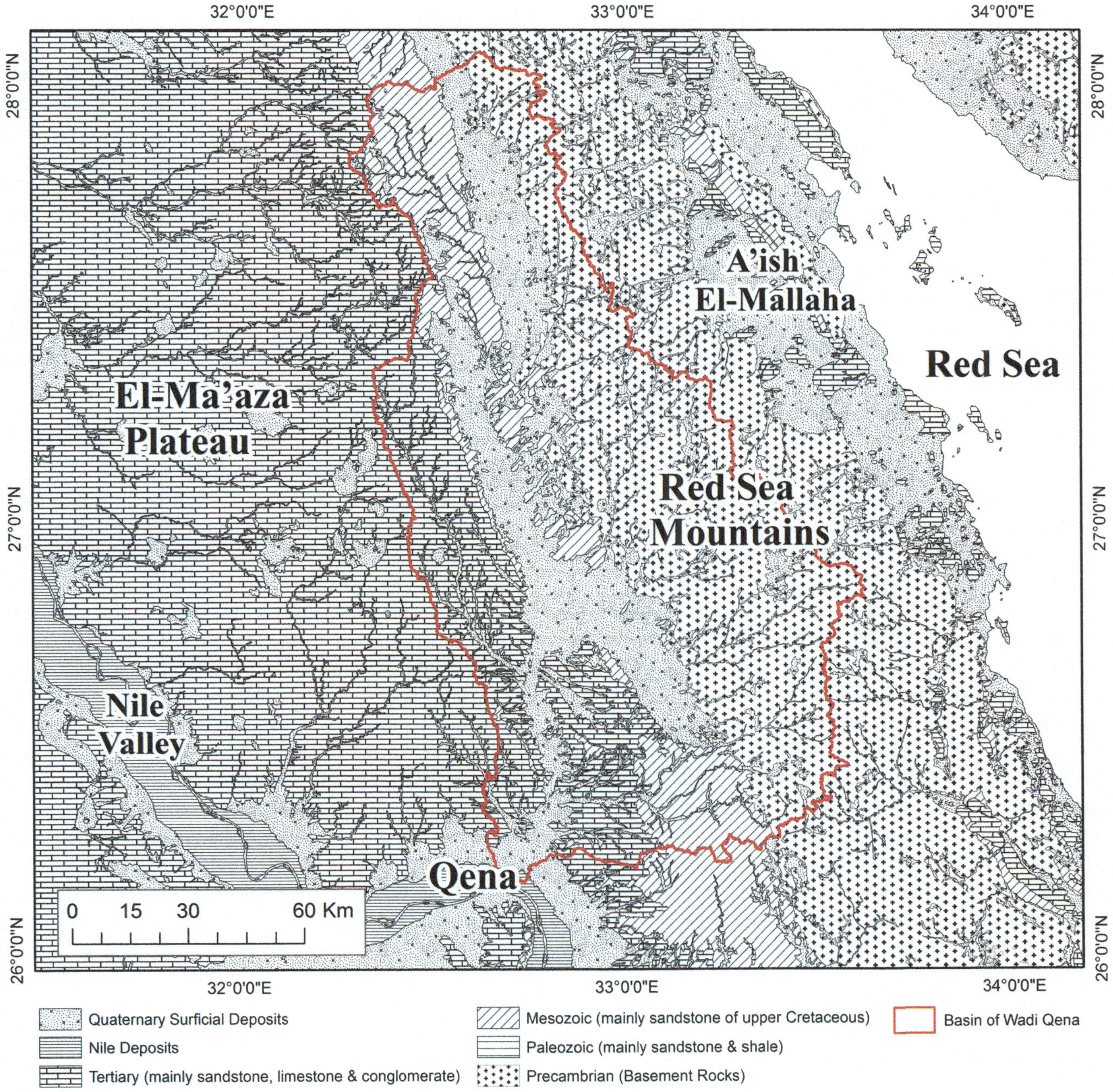

Fig. 18.7 Simplified geologic map of Wadi Qena (Simplified from the Geological Map of Egypt, scale 1: 2,000,000)

wadis from El-Maaza Plateau, which represent the main eastern tributaries of Wadi Qena, such as **W. El-Gurdi El-Soghair**, **W. Gurdi**, **W. Umm Gilu**, and **W. Shahadin**. In the meantime, they are dissected by numerous small wadis, which also join the main channel of Wadi Qena.

The fourth unit is the floor of Wadi Qena, where all tributaries meet the main channel, carrying the eroded material from east and west, and spreading them on the low lying areas of the floor. The top layer of these wadi deposits is of Holocene age. No information is available for the subsurface materials, but it is expected to be of Pleistocene. The bedrock below these surface materials varies from basement rocks in the north to sedimentary ones in the lower, southern section. Width varies, but in general, it reaches several tens of kilometers in some sections.

18.6.2 Features and Forms of Wadi Qena

18.6.2.1 Unique Features of Wadi Qena

The first feature is that Wadi Qena developed along the contact line between the old basement complex and the tilted sedimentary cover (Fig. 18.7). The sedimentary formations had covered the whole area of the Red Sea Mountains before the opening of the Red Sea and the rising of the mountains. This fact was inferred from the presence of sedimentary remnants along the eastern side of the mountains. The rising of the Red Sea Mountains resulted in the up-arching of the sedimentary cover, and the tilting of beds eastward toward the Red Sea and westward to the Nile Valley. The sedimentary cover was eroded gradually and retreated eastward and westward, exposing the basement complex rocks. Most probably, this started to occur during Late Eocene/Early Oligocene, as evidenced by the presence of Lower Eocene formation at the eastern and western sides of the Red Sea Mountains, and the presence of Early Eocene material at the base of the sediments of the Prenile huge delta lying to the north of Fayum Depression. By time, new slopes developed on the basement rocks in the east and on the retreating edge of the sedimentary cover in the west. The surfaces of the basement and necessarily water flow were sloping westward toward the Nile Valley, but the trend was interrupted by the slopes of the edge of the sedimentary cover, which was facing eastward. Consequently, water flow coming from the east and the west were forced to meet at the low-lying line that separates the eastern and western slopes. By this process, Wadi Qena developed along the contact line between the old basement rocks and the more recent sedimentary cover.

The second feature is that Wadi Qena runs from North to South. Although there are several wadis running from north to south in the Eastern Desert, they are all small tributaries of main wadis running eastward/westward. Only one main wadi in this desert has this characteristic, which is Wadi Qena with a length of ~250 km. So far, no previous study answered the question: why does this Wadi take this direction for such a long distance? From the available geologic and topographic data, it can be deduced that the tectonic rise of the region of Wadi Qena was not equal from north to south. Most probably, the northern area has been raised more than the southern part. This inference is supported by the data derived from topographic maps (scale 1: 100,000) which show that the height of the wadi floor decreases from ~600 m a.s.l at its upper reaches (latitude 28° N) to ~200 m a.s.l near its mouth in the Nile Valley. On the surface of El-Maaza Plateau, above its edge, heights also decrease from 750 to 800 m a.s.l in the north to 550–600 m a.s.l in the south. Geologic maps (scale 1: 250,000) show that basement formations are exposed on the northern half of the eastern side (28° 00′–26°45′ N), whereas the sedimentary rocks are exposed on the southern half of the same side. Here, the blocks of Gabal El-Serei, Gabal Qreiya, and Gabal Abu Had rise to 550–600 m a.s.l forming the eastern side of Wadi Qena. This indicates that the sedimentary cover was removed from the northern part of the eastern side and is still present in the southern part of this side, hiding the basement rocks because the northern area was uplifted more than the southern one.

18.6.2.2 Local Forms of Wadi Qena

Apart from the wadi tributaries, several features and forms developed along the wadi sides and its floor. They include Pliocene sediments, fanglomerates, fluvial wadi terraces, and circular structure and volcanic cone.

The Pliocene sediments are distributed along both sides of the southern section of Wadi Qena. They are composed of fluvial materials (mainly gravel and sand) that were deposited in the Pliocene Marine Gulf, which invaded the Nile Valley and the lower sections of some of its tributary wadis by the advent of the Pliocene (Said 1981, 1990). The sediments laid down became consolidated and were exposed to dissection by water action during the pluvial periods of the Pleistocene. This is why their dissected remnants are present only on the sides of Wadi Qena and some of its tributaries (e.g. W. Gilu. W. Gurdi, W. Shadan on the western side; and W. Abu Had, W. Umm Selimat on the eastern side). These Pliocene sediments have drawn the attention of several geologists since the first decades of the twentieth century (Sandford 1934; Little 1935–1936; Ball 1939), because of their significance to geologic and geomorphologic evolution not only of Wadi Qena, but also for other landforms in Egypt such as the Nile Valley. The significance of the Pliocene sediments in Wadi Qena comes from the fact that they indicate the development of the wadi in Pre-Pliocene times.

There are two possible scenarios of the development of Wadi Qena. The first is that it came into being with the Nile Valley during Late Miocene; in other words, they were contemporaneous and developed during the same period. This argument, nevertheless, lacks supporting evidence. The second probable age is Middle Miocene (~16 Ma), as was suggested by Shata (1962) and Issawi and McCauley (1993). These two studies postulate that Wadi Qena divided into two branches at the present-day site of Qena city in the Nile Valley, the first branch went west to the Western Desert, while the second continued southward to Nubia Region. During the development of the Nile Valley in Late Miocene times (~6 Ma) and its growth southward by headward erosion, the powerful Nile captured all streams coming from east/west, of which Wadi Qena was one. In this scenario Wadi Qena is older than the Nile Valley.

Terraces of Wadi Qena were studied by Sandford (1929, 1934) and since then no other study has been conducted except for the study of the sediments properties, which was part of a more comprehensive study on the geomorphology of this Wadi (Gawish 1981). Sandford (1929, 1934) recognized four terraces in Wadi Qena at levels of 100 ft. (30 m), 50 ft. (30 m), 30 ft. (9 m), and 15 m (3 m).

100 ft (30 m) terrace is present on both sides of the Wadi on a magnificent scale over a distance of ~80 km from the Nile toward the Red Sea Mountains (Sandford 1934), and on the sides of lower sections of some of the tributaries, such as wadis El-Atrash and Fatira. These terraces are dissected into small conical hills due to erosion during later periods. They are covered by gravels, some of which of originated from igneous rocks. In a few localities, boulders are a common constituent in surface material with gravels of igneous origin. The gravels with sand and clay appear as conglomerate due to consolidation. Sandford (1934) found some implements belonging to **Early Lower Paleolithic** (Chelean) Period.

50 ft. (15 m) terrace can be seen at the extreme lower section of Wadi Qena to the E/W Bir Arras, Gabal Shahadin, and along the highway of Qena-Safaga. This 50 m terrace is covered by gravels of igneous and metamorphic originated from the Red Sea Mountains, which were transported by wadis of the White Fatira and Blue Fatira. In some places, the sediments of this terrace were consolidated in a matrix of sand clay as conglomerate. Due to dissection, the conglomerate appears as gravel islands at the Wadi Qena bed. Implements found in the sediments of this terrace were recognized as Late Lower Paleolithic.

30 ft. (9 m) terrace is present on both sides along the lower and middle sections of the Wadi

Qena. But, at the western side, the terrace is dissected resulting in the removal of some parts. Sediments contain a high percentage of igneous and metamorphic gravels. Thickness of sediments reaches 2 m on average, but reaches 3.5 m at the Wadi Qena mouth. Implements in this terrace were identified as Early Middle Paleolithic.

15 ft. (3 m) terrace: extends for 2 km on both sides from mouth of the Wadi. It is covered by loose gravels, which are brought by flash floods.

Analysis of **Wadi Qena Terrace Sediments** (Gawish 1981) revealed the following characteristics:

Size distribution in all terraces is bimodal, where the high peak represents the coarse and very coarse division and is composed of large and very large gravels. The smaller peak represents the fine particles of sand (0.25–0.5 mm) and coarse sand (0.5–1.0 mm).

Analysis of average size of grains indicates a decrease from the highest terrace toward the lowest one. This characteristic is interpreted by a decrease in wadi discharge due to decrease in rainfall during the period of sediments deposition.

Sorting in the 30 and 15 m terraces is good, whereas that of terrace 9 m is medium, and that of terrace 3 m is weak. These characteristics can be interpreted by the fact that the sediments are heterogeneous, since they contain very large boulders and very large gravels (16–64 mm) and medium/coarse sands (0.25–1.0 mm). On the other hand, the weak sorting of terrace 3 m is due to the fact that its gravel was exposed to the effect of flash floods.

Gravel particles are rounded to semi-rounded. Roundness decreases from older to younger terraces. Here also, drainage discharge plays an important role in these variations in roundness. It seems that rainfall and consequently drainage discharge decreased over time during the period of deposition.

Nearly, all gravels (99%) of the highest terrace (30 m) originated from limestone, chalk, flint, and clay, whereas gravels of the other lower terraces are composed of mixed igneous and metamorphic/sedimentary materials. This characteristic indicates that during the deposition of the highest terrace in Early Lower Paleolithic, the igneous and metamorphic of the Red Sea Mountains were still covered by the sedimentary rocks, therefore, these latter rocks were the only ones available for fluvial transport. By time, the basement rocks were exposed gradually, resulting in a bed load of a mixture of igneous and metamorphic/sedimentary gravels in the lower and more recent terraces.

Fanglomerates are fans composed of consolidated gravels produced and brought from the Red Sea Mountains and the western escarpments during humid periods at the end of the Tertiary and in the Quaternary, covering much of the wadi floor and low tabular outliers (Bandel et al. 1987). The Geological Map (scale 1: 250,000) shows that **fanglomerates** cover the western side of Wadi Qena from latitude 26° 45′ N to latitude 27°30′ N without any break, except for the dissection made by small wadis issuing from the escarpments of **Maaza Plateau**. On the eastern side of the Wadi, fanglomerates cover smaller patches in the same latitudes of the western side.

A -volcanic cone and a ring structure were recorded in central wadi Qena (Bandel et al. 1987), at about latitude 27° 30′ N. The ring structure lies about 20 km to the SW of the volcanic cone. The circular dome-like structure has a diameter of about 3 km. Its north and west sides dip about 10°, south and east sides dip at 20°′ and more. The central dome has a flat top and is 150–200 m above the surrounding grounds. Some sills were recorded near its base and contain andesitic basalt, which most probably forms a solid cushion-like intrusive body within the dome and was responsible for metamorphosis above (Bandel et al. 1987). At the top of the central mount of the ring structure as well as its sides, marl and limestone have become strongly altered due to the heat induced metamorphosis. Limestone has changed into crystalline marble, and sandy layers have been molten and turned into glass-like structures. The metamorphosis has also altered the granite of the basement complex and changed it into a sandy clay-rich material (Bandel et al. 1987).

18.7 Conclusion

The drainage system in the Eastern Desert is inherited from the past humid periods, mainly during the period from Early Late Eocene to Late Pliocene. This is why some wadis developed during Prenile times and were captured by the Nile during its development by headward erosion in Late Miocene. Although Pliocene sediments cover the sides of the lower section of some wadis such as Wadi Qena, the exact time of development not yet specified. Previous studies on the Prenile drainage systems postulated that major wadis, like Qena, Assyuti, and Allaqi had extended westward during the Miocene and Oligocene before the Nile came into being.

The geologic structure and topographic expressions controlled the development of the drainage system and its characteristics in the Eastern Desert. The three major drainage systems were determined by the slope orientation of the Red Sea Mountains and the surrounding plateaus. Basin size and network pattern were also controlled by the same factors, which explains the presence of a large number of drainage basins, a wide variety in basin size and the development of all types of drainage patterns. In general, basins of the Red Sea System are smaller than those of the Mediterranean. Large basins in this latter system outnumber those in the Red Sea system.

Wadi Qena was taken as an example of the wadis of this region because it is a large unique wadi, in which one expects some features uncommon in other wadis in the Eastern Desert. Two aspects characterize Wadi Qena compared to any other wadi. It is the sole large wadi running from north to south and it is the only wadi that developed along the contact line between the basement rocks and overlying the sedimentary formation. Although the latter characteristic can be easily explained, none of the previous studies made an attempt to interpret the first feature. Erosion processes working on the surface found it easier to erode along the contact line between two different formations; therefore, sedimentary cover recessed westward at a higher rate compared to the hard crystalline rocks of the basement. Gradually, Wadi Qena appeared as a longitudinal trough. It was suggested that due to the lifting up of the northern section above the southern one, this trough acquired a southward gradient. Consequently, Wadi Qena contained flow from north to south. Several pieces of evidence are in favor of this interpretation, including the complete removal of the overlying sedimentary cover on the basement rocks in the northern upper reaches of the Wadi, whereas this sedimentary cover is still present along the eastern side in the lower section of the Wadi.

Three other features characterize Wadi Qena. They are gravel wadi terraces, fanglomerates, and a circular structure and a volcanic cone. Four terraces (30, 15, 9, and 3 m) were recognized in Wadi Qena almost 80 years ago by Sandford (1929, 1934). Through the implements, Sandford (1929) succeeded to determine the age of the higher three terraces dating them to the period Early Lower Palaeolithic–Early Middle Palaeolithic. Since then, no study has correlated these terraces with either climatic changes or sea level. This aspect remains unresolved until now and needs further investigations.

The third feature of Wadi Qena is the fanglomerates which cover the sides of Wadi Qena, especially the western side. Fanglomerates are alluvial fans composed mainly of gravels. Because the gravel is inherited from the past wet climates, they were consolidated and transformed into conglomerate. Since they have been deposited as fans, they are called fanglomerates. They have varying thickness of 1–20 m and have developed during the Pleistocene. Along the western side in the lower section just to the north of Wadi Shahadin mouth, the fanglomerates lie below the margins of the Pliocene sediments giving the impression that they are older than Pliocene material. This aspect and the consolidation of the gravel and its large thickness are all in favor of an age older than the Pleistocene.

The fourth feature is the presence of a volcanic cone and a circular structure in the floor of Wadi Qena, indicating that the Wadi has already existed before these two features came into being. The question which arises from this phenomenon is when has the volcanism/tectonics occurred? According to the analysis of volcanic phases, it was found that the phase of tertiary volcanic episode last recorded occurred during the Middle Miocene, ~15 Ma (Meneisy 1990). This means that Wadi Qena developed —at least—during Early Middle Miocene. This conclusion is in accordance with the postulation of Shata (1962) and Issawi and McCauley (1993) that Wadi Qena developed during Middle Miocene (~16 Ma).

References

Ball J (1939) Contribution to the Geography of Egypt. Survey and Mines Department, Ministry of Finance Egypt, Government Press, Cairo

Bandel K, Kuss J, Malchus N (1987) The sediments of Wadi Qena (Eastern Desert, Egypt). J Afr Earth Sci 6:427–455

El-Etr H & Associates (1997) Hazard assessment and mitigative measures of flash flooding on the Red Sea towns, Egypt, Special Publication No. 1, National Authority for Remote Sensing and Space sciences "NARSS", Cairo, Egypt

El-Etr H, Abdel Rahman M (1975) Drainage pattern and drainage lineations of Esh El-Mallaha Range, Gulf of Suez district. Proceedings 7th Arab Scientific Congress, vol. IV, Cairo

Gawish A (1981) Wadi Qena basin: a geomorphologic Study. Dissertation, Cairo University, Cairo

Google Earth (2014) Egypt Images Latitudes 22-32 N and Longitudes 22-32 E, http://www.google.com

Horton R (1945) Erosional development of streams and drainage basins, hydrophysical approach to quantitative morphology. Bull Geol Soc Amer 56:275–370

Issawi B, McCauley J (1993) The Cenozoic landscape of Egypt and its river systems. Ann Geol Survey Egypt 19:357–384

Little O (1935–1936) Recent geological work in the Fayum and in the adjoining portion of the Nile Valley. Bull Institut d'Égypte 18:201–240

Meneisy M (1990) Volcanicity. In: Said (ed) The Geology of Egypt. Balkema, Rotterdam, p 157–172

Moawad M (2008) Application of remote sensing and geographic information systems in geomorphologic studies: Safaga-El-Quseir Area, Red Sea Coast, Egypt as an example. VDM Verlag Dr Mueller. K, Saarbrücken

Said R (1981) The geological evolution of the river Nile. Springer, New York

Said R (1990) Cenozoic. In: Said R (ed) The geology of Egypt. Balkema, Rotterdam, pp 451–486

Sandford K (1929) The Pliocene deposits of Wadi Qena and the Nile valley between Luxor and Asuit. Quart J Geological Soc London 85:493–550

Sandford K (1934) Paleolithic man and the Nile valley in Upper and Middle Egypt: a study of the region during Pliocene and Pleistocene Times. Oriental Institute, Chicago, Illionis Publication No 3

Shata A (1962) The lower Nuba area, Egypt, U. A. R. Bull Soc Géogr d'Égypte 35:273–299

The Drainage System in Sinai (With a Special Reference to Wadi El-Arish)

19

Abstract

The drainage system of Sinai is inherited from past wet climates. Therefore, present-day hyper-arid climate had a negligible effect on its inherent characteristics. With the exception of the NW area that has internal drainage, Sinai drainage is external, running toward the Mediterranean Sea, Gulf of Suez, and Gulf of Aqaba. The radial pattern dominates the drainage system in Sinai. With its large area (nearly one third of Sinai), Wadi El Arish basin is the largest drainage system. Its net has a denderitic pattern, starting from the extreme southern tip of El-Egma Plateau to the Mediterranean Sea (more than 350 km). Although all its sectors are well fitted in one drainage network, they, however, differ greatly in their geomorphic characteristics.

Keywords

Sinai Peninsula • Drainage system • Wadi El Arish • Antecedent wadis • palaeo-lakes

19.1 Introduction

Sinai Peninsula as one of the geographic regions of Egypt has a unique landscape in which nearly all geologic structures and landforms are contained. During its geologic history, a drainage system developed under the past wet climates, since rainfall of the present-day arid and hyper-arid climate is not sufficient to develop a permanent and efficient drainage system. Therefore, similar to others, the drainage system of Sinai is inherited from past wet climates. The present chapter deals with a drainage system whose characteristics have not developed nowadays.

19.2 Regional Drainage Systems

Sinai Peninsula is encircled nearly from all sides by open seas, the Mediterranean Sea from the north, the Gulf of Suez from SW, and the Gulf of Aqaba from SE. This characteristic gives the impression that all of Sinai is externally drained. Furthermore, the map of Sinai also shows that this impression is almost true for all of Sinai, except for the extreme northern region of the North Sinai Sand Sea, where drainage is internal. All drainage lines dissecting the northern slopes of the folded hills and mountains disappear under the sand dunes of the North Sinai Sand Sea. This area is a

N.S. Embabi, *Landscapes and Landforms of Egypt*, World Geomorphological Landscapes,
https://doi.org/10.1007/978-3-319-65661-8_19

Depression (Kusky and El-Baz 2000), and this is why wadis were not able to reach the Mediterranean Sea. Only **Wadi El Arish** cuts through the sand dunes and succeeded to reach the sea. In the eastern part of Sinai, in the area to the north of Gulf of Aqaba, wadis drain outside this region, Therefore, it can be concluded that there are two drainage systems in Sinai: the first is external that drains most of Sinai land surface, and the second one is internal, that drains small areas in the north and in the east.

The map of Sinai also illustrates another aspect of its drainage systems, namely that the external drainage lines drain in the Mediterranean, in the Gulf of Suez, or in the Gulf of Aqaba. Wadi El Arish is the only one draining into the Mediterranean Sea, whereas the slopes of Sinai Mountains divide the drainage lines into two groups, one toward the east and another one toward the west. Although there are no clear drainage lines cutting through, the extreme NW area known as El-Tinah plain was drained externally by one of the ancient Nile branches called Pellusaic. In fact, it belongs to the Nile system, and not to Sinai systems. Therefore, the drainage lines in Sinai can be divided into three systems: the Mediterranean system, the Gulf of Suez system, and the Gulf of Aqaba system. Apart from the Mediterranean system, each of the other two systems is composed of several basins (Fig. 19.1).

19.3 Drainage Basins

Figure (19.1) shows the distribution of drainage basins in Sinai. There are 40 basins in Sinai, of which **Wadi El Arish** basin is the largest, with the total area of 21,700 km^2 (36.1% of Sinai). This area is six times as large as the area of **Wadi Watir** (3516 km^2), which is the second largest basin in Sinai. However, there are seven wadis, such as **Wadi Dahab-El-Ghaib** (2067 km^2) and **Wadi Kid** (1090 km^2) in the east, and **Wadi Gharandal** (1966 km^2) and **Wadi Feiran** (1830 km^2) in the west with catchment areas of more than 1000 km^2. In the meantime, there are several small basins, some of which are less than 50 km^2 in area. Figure (19.1) also shows that the shape of the basins tends to be elongated, due to the steepness of the mountains slopes and

Key to Hydrographic Basins

Map No.	Wadi Name	Area Km^2
A	Northern Wadi El-Arish System	
1	Upper El-Arish	16.306
1a	El-Bruk	3.345
1b	El-Ruag	6.481
1c	El-Aqabah	2.839
1d	Geraia	3.641
2	Lower Wadi El-Arish	2.749
3	Northeast coast	963
4	North coast	5.148
5	Hegayib	1.680
6	El-Hassana-El-Hema	3.549
B	Eastern Gulf of Aqaba System	
7	El-Gerafi	2.446
8	Wasit (Watir)	4.204
9	Dahab	2.684
10	Kid	1.355
11	Umm Adawi	964
C	Western Gulf of Suez System	
12	El-Qaa	3.904
13	Abu Durba	266
14	Feiran	1.717
15	Sidri	1.163
16	Baba	841
17	Tayiba	860
18	Gharandal	829
19	Wardan	1.569
20	Sudr	895
21	Lahata	603
22	El-Raha	847
23	El-Hagg	621
24	El-Giddi	703
25	Umm Khisheib	4.641
	Total Approximately	**61.000 m^3**

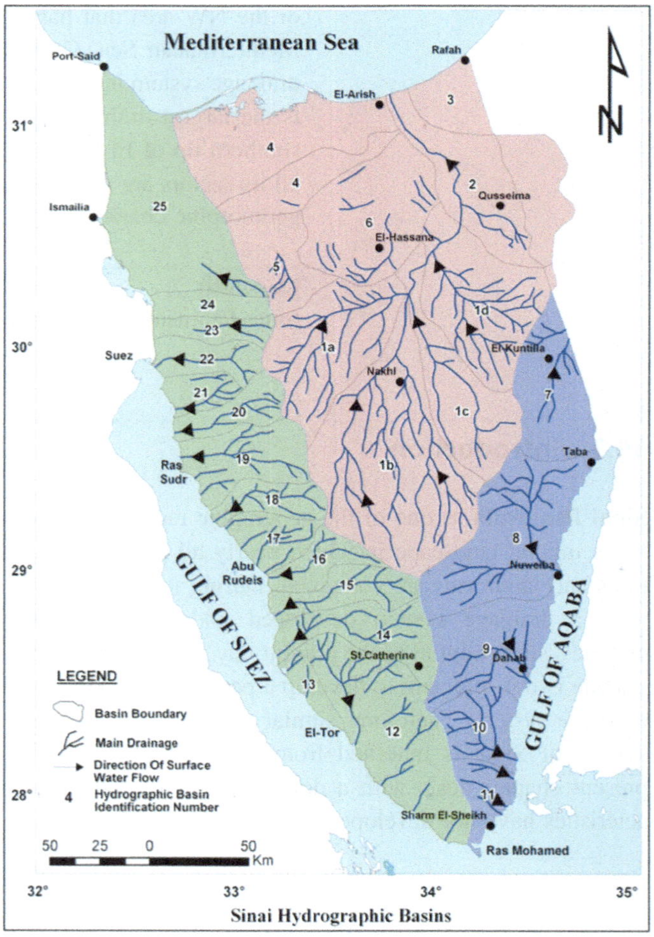

Fig. 19.1 Drainage basins and nets of Sinai Peninsula (*Source* NARSS et al. 2009)

the effect of faults which controlled the development of the wadis.

19.4 Drainage Patterns

Well-known drainage patterns are parallel, dendritic, rectangular, radial, trellis, and several others. The regional pattern is nearly radial, since—as explained in the previous subsection—wadis in Sinai drain northward, eastward, or westward, starting from the central Sinai Mountains. Therefore, it can be said that Sinai has a regional drainage pattern.

Previous studies (Mostafa 1987; Torab 1988; El-Rashidy 1994; Rashwan 1994) of some wadi basins in Sinai revealed that each basin accommodates most of the well-known drainage patterns. The common patterns are: dendritic, trellis, and radial, of which the dendritic pattern is the most common. This characteristic is in accordance with drainage patterns worldwide (Thornburry 1954). It is best exemplified by **Wadi El Arish** drainage network (Fig. 19.2). Trellis pattern is typical in the basin of **Wadi Dahab**, which drains into the Gulf of Aqaba. In this basin, there are several wadis, which developed under the control of faults. Some of these wadis developed in grabens, of which **Wadi El-Ghayeb**, a tributary of Wadi Dahab, and **Wadi Nasab** are the best representatives. Along the sides of these two "**Rift Wadis**" several parallel short wadis developed creating a trellis pattern in both. Local radial drainage pattern developed on the slopes of the dome-like hills and mountains in northern Sinai (Fig. 19.3). A parallel pattern developed on the steep slopes of landforms in various localities in Sinai, but the respective valleys are mostly short, and join larger wadis, as tributaries.

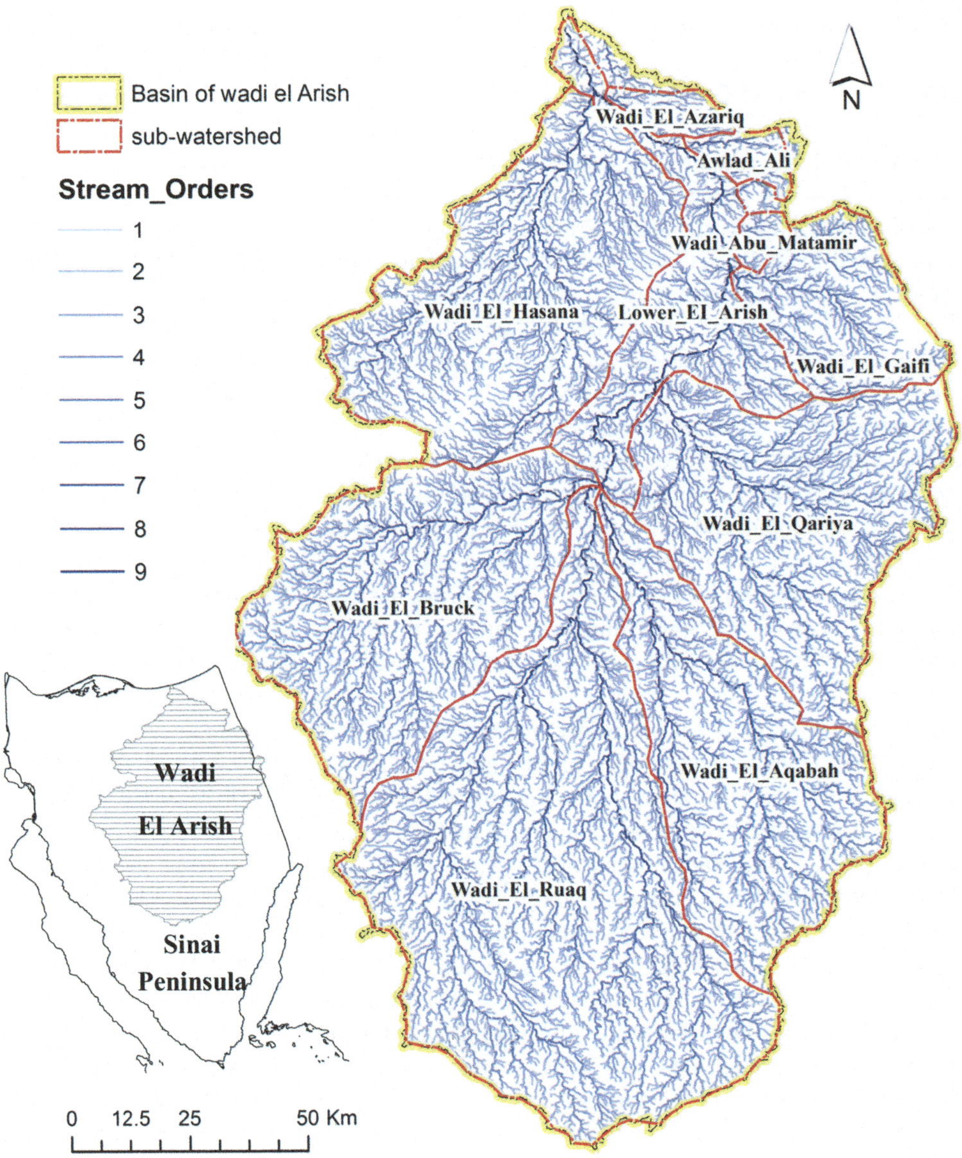

Fig. 19.2 Drainage net of Wadi El Arish (*Source* Moawad 2013)

Fig. 19.3 Space image for a small anticline hill with a radial drainage pattern (*Source* Google Earth 2014)

19.5 Wadi El Arish

19.5.1 General Characteristics

Wadi El Arish is taken to represent Sinai wadis, for several reasons. It is the largest wadi in Sinai, therefore, it is expected that all aspects of the drainage networks will be represented in its basins; and it has its own unique forms. Wadi El Arish drainage basin covers an area of about 22,000 km^2, and extends in a nearly S-N direction for about 365 km. It is the second largest wadi in the Egyptian deserts after Wadi al-Allaqi. It is composed of six major basins. Except for some small tributaries joining its lower section, which come from the **Negev desert**, its catchment area lies in Sinai. It discharges more than two thirds of Sinai waters, equivalent to about 160 million m^3. Although it is dry for most of the year, it might flow during the winter season. During flood times which follow heavy rain storms, the wadi looks like a real river, presenting a hazard to all human activities in its lower section (Hamdan 1960; Moawad 2013).

Most of Wadi El Arish headwater tributaries come from the northward gentle back-slopes of **El-Egma** and **El-Tih** Plateaus in Central Sinai at height reaching up to 1400–1600 m a.s.l. These two plateaus are composed of Cretaceous–Eocene formations: Nubian sandstones, white chalk, and limestone. The beds dip regionally northward at a very low angle. In the meantime, the beds of El-Egma Plateau are folded in a broad synclinal structure plunging northward (Awad 1982). Therefore, all drainage lines on El-Tih and El-Egma Plateaus followed the original dip slope, flowing northward in general and toward the central parts of El-Egma Plateau. Other structural aspects of El Arish basin are the folded Syrian Arcs Structures which dominate the landscape in northern Sinai, taking a general NNE-SSW orientation, and are bounded by faults taking the same trend. Several types of Quaternary sediments occur within the basin of which fluvial/lake sediments are the most widespread.

19.5.2 Geomorphologic Aspects

The geologic and topographic maps and Google Earth images show several aspects characterizing **Wadi El Arish** that are of geomorphic significance. These aspects are: (1) The fact that the wadi to the north of El-Egma Plateau narrows to form several gorges (Daika in Arabic); (2) The presence of an erosion surface of the interfluves on the surfaces El-Egma Plateau; (3) The development of two systems of entrenched meanders along the upper reaches of the Wadi on El-Egma Plateau; and (4) The presence of wide areas along the middle section—to the south of the large anticlinal mountains of Yeleq and Halal, covered with fluvial/lake deposits.

19.5.2.1 Antecedent Gorges (The Daika Form)

Wadi El Arish to the north of **El-Egma Plateau** is characterized by several gorges (Daika in Arabic), of which the one cutting the eastern part of **Gabal El-Halal** Anticline (known as **El-Daika**) is the most well known. It was concluded in a previous study (Sneh 1982) that the daikas were formed during the Late Pliocene or perhaps even in the Early Quaternary because these daikas are filled with Quaternary sediments, and no Neogene sediments were found in them. According to Sneh (1982), these gorges, which are post folding, were cut through a continuous down-cutting by a fluvial system. Some other previous studies on neo-tectonics and fluvial geomorphology of the northern Sinai Peninsula (Moustafa and Khalil 1990); Kusky and El-Baz 2000) concluded that the deformation of the Syrian Arcs in north Sinai began with minor uplifts in the Late Cretaceous, and that the major phase of uplift began in Late Oligocene to Early Miocene, concomitant with rifting of the Red Sea Basin. It is now well known that the climate of Egypt was humid since the Late Eocene, and continued to be so, at least, to the end of the Miocene (Salem 1976; Said 1981, 1990). Therefore, it is expected that El Arish drainage system came into being as early as Late Eocene, similar to other systems that developed on the Egyptian landmass, and consequently the down-cutting of daikas along the anticline structures started —during Late Eocene, and continued until Late Pliocene/Early Pleistocene. This means that the present daikas, which cross some anticlines in northern Sinai, can be classified as antecedent wadis. This conclusion appropriately agrees with that reached by—Kusky and El-Baz (2000), in which the section of Wadi El Arish that cuts through the Halal-Dalfa anticline is an antecedent stream. However, this conclusion further adds that the cutting of this gorge follows one of the faults along this section.

19.5.2.2 Entrenched Meanders in the upper reaches of Wadi El Arish

This is the first mention of such forms in the upper reaches of Wadi El Arish, because no large scale—maps, aerial photographs, or space images were available to unravel their existence. Figure (19.4) shows an example of the **entrenched meanders** in the upper sections of the tributaries of Wadi El Arish on El-Egma Plateau. It shows the presence of two systems of incised meanders: the first is a complete one, where meanders are connected together, and the second is composed of the remains of disconnected meanders. It is often inferred that incised meanders were inherited from a previous erosion cycle and are hence second-cycle features, providing evidence of rejuvenation (Thornburry 1954). Therefore, it can be assumed from the space images of Google Earth that Wadi El Arish experienced two cycles of erosion, at least. Two questions arise now: what is the cause of rejuvenation? And when did it happen?

Rejuvenation occurs due to sea level lowering, or due to tectonic land rise. Although no data is available, previous studies pointed out that tectonics started with a minor uplift in north Sinai in the Late Cretaceous and mostly ended by the Eocene, and that the major phase of uplift began in the Late Oligocene to Early Miocene. Most probably, the first rejuvenation started during the Late Cretaceous to Eocene phase while the modified meanders were entrenched and modified under the effect of erosion in a later period. The second cycle occurred during the major uplift of the Oligocene–Miocene phase during which the complete meander system developed. Most probably, minor tectonics which occurred from the Miocene to Quaternary kept this system complete until today.

19.5.2.3 Palaeo-Lakes

Geological and topographic maps of Sinai (scales 1:500,000, 1:250,000, and 100,000) show that Wadi El Arish and its tributaries on the northern part of El-Egma Plateau, and on the low-lying area to the south of G. Halal, G. Maghara, and G. Yeleq up to the northern margins of El-Egma Plateau are covered with Quaternary fluvial sediments, some of which are classified as **playa deposits**. These sediments were treated in several studies and were classified partially/wholly as **paleo-lake** beds formed during past humid climatic periods (Sneh 1982; Goldberg 1984; Kusky and El-Baz 2000; AbuBakr et al. 2013). Playa deposits including several

Fig. 19.4 Entrenched meanders at one of the upper tributaries on El-Egma Plateau (*Source* Google Earth, 2014)

tens of centimeters of thick gypsum layers from south of Gabal Yelleq have been recorded by Sneh (1982). Late Quaternary sequence from the Quseima area was studied on one of the tributaries of Wadi El Arish (Goldberg 1984). This sequence begins with early **Middle Palaeolithic** (90,000–65,000 BP) gravels, which were eroded and then buried by a new sequence of sands, clays and silts of about 33,800 BP. Some of the fine-grained sediments are interpreted as lake deposits, although many of them probably existed for short period only. Other lakes may have persisted for longer periods, as indicated by the presence of clays with ostracod, inter-bedded with horizontal silt and sand layers of Upper Palaeolithic (Kusky and El-Baz 2000).

According to—Kusky and El-Baz (2000), the palaeo lake sediments occur in wadis and low areas between anticlinal hills and mountains. This mode of development of paleo-lakes is confirmed by the study of AbuBakr et al. (2013). By using radar data, this latter study revealed the development of three paleo-lakes along Wadi El Arish within structurally controlled Depressions formed due to the effect of the anticlinal ridges of the Syrian Arc System in North Sinai.

It is possible that the wadi channels at the heads of the **palaeo-lakes** were blocked by migratory sand dunes, which are abundant in the area (Sneh 1982), on condition that eolian sand supply exceeds the transport ability of the drainage flow. This happened at the heads of some wadis in the Gilf Kebir Plateau in the Western Desert. Although some of the palae-olake beds in Wadi El Arish are associated with Depressions formed behind anticline ridges and occur up-gradient of active dune fields, there is a much a stronger correlation between palae-olakes and eroded faults than a correlation with the dunes (Kusky and El-Baz 2000). Regardless of cause of blockage of drainage flow, the huge quantities of palaeo-lake sediments may explain the reason behind of the non-existence of a delta for Wadi El Arish in the Mediterranean Sea. A question arises—what was the fate of the sediments which passed through El-Daika Gorge after its development, while wadi El Arish was transporting its load across the plains to the north of G. Halal. Most probably the channel of the Wadi became braided after being freed from **El-Daika**, depositing most of its load on the plains (at present it is covered by eolian sands and dunes). Consequently, the flow reached the Mediterranean Sea with a little quantity of sediments not sufficient for building a delta, while the offshore current removed the sediments reaching the sea. Something similar happened under present-day arid conditions, when a flood occurred in Wadi El Arish—but only for five days (February 21–26, 1975)—bringing huge quantities of sediments 500,000–600,000 m^3 to the sea, and building up a small delta (450 × 700 m). Erosion of the delta started only two weeks after its development, and only three years later it disappeared (Klein 2000).

19.5.2.4 Erosion Surface

Space images of Google Earth (2014) show the presence of wide low-gradient dark-tone tracts at the interfluves of the tributaries of Wadi El Arish on El-Egma Plateau (Fig. 19.5), suggesting that they represent an old **erosion surface**. Figure (19.5) shows that these surfaces are crossed by shallow channels; however, they are dissected by steep-sided tributaries. In some other localities, this erosion surface is highly dissected, and only a few small remnants can be spotted. Although very small/tiny, some **karst features** (cone karst hills, tiny Depressions) can be discerned on the surface of the limestone plateau. Since it is the highest surface on El-Egma Plateau, this erosion surface is projected to have developed

Fig. 19.5 Probable erosion surface (black-tone high surfaces) dissected by incised deep wadis, upper reaches of Wadi El Arish, El-Egma Plateau (*Source* Google Earth 2015)

with the beginning of El Arish drainage network, Late Eocene, and to have become a separate erosion surface during the peak of tectonic activity in the Oligocene-Miocene period. The suggested erosion surface on El-Egma Plateau needs further investigations in the field.

19.6 Conclusion

The drainage system of Sinai is simple and complex at the same time. It is almost fully external. Therefore, it drains all water that falls on the surfaces of Sinai externally into the Mediterranean Sea, the Gulf of Suez, and the Gulf of Aqaba. Due to the orientation of slopes toward the north, west, and east, all drainage lines take a regional radial pattern, a characteristic that is not existent in other regions of Egypt. The steep slopes of both eastern and western sides of Sinai Mountains resulted in the development of small drainage basins, although the western ones are a little larger than those in the east, and in any of them the basin area exceeds a few thousand square kilometers. With low-gradient surfaces toward the north, Wadi El Arish became the largest basin in all Sinai, covering more than one third of the Sinai area.

Wadi El Arish is composed of several sections that differ in their geomorphologic characteristics, due to local conditions and tectonic aspects. These characteristics are inherited from past wetter climates. The huge quantities of fluvial sediments, which cover wide areas to the south of the large anticline mountains indicates that Wadi El Arish was, most probably, draining internally in this area because this area is lower than the surroundings, and the obstruction of water flow by the steep slopes and high sides of Syrian Arcs mountains of Yeleq, Magharah, and Halal occurred. It is suggested that the gaps between these mountains allowed some drainage to get through and drain internally in the low-lying area in the NW of Sinai, but the bulk of the water and sediments was caught in this area forming lake(s) from time to time. Even, if some of the flow reached the sea, the small quantity of sediments is expected to be removed by the W-E offshore current.

References

AbuBakr M, Ghoneim E, El-Baz F, Zeneldin M, Zeid S (2013) Use of radar data to unveil the paleolakes and ancestral course of Wadi El-Arish, Sinai Peninsula, Egypt. Geomorpholgy 194:34–45

Awad H (1982) Physical Geography of Sinai Peninsula: The geomorphologic Units (in Arabic). The General Egyptian Authority for Books, Cairo, p 1–16

El-Rashidy A (1994) Wadi Gharandal Basin. A Geomorphologic Study. (in Arabic) Dissertation, Dept. of Geography, Faculty of Arts, Ain Shams University

Goldberg P (1984) Late Quaternary history of Qadesh Barnea, northeastern Sinai. Zeit Geomorph 28:193–217

Google Earth (2014) Egypt Images Latitudes 22-32 N and Longitudes 22-32 E, http://www.google.com

Hamdan G (1960) The Personality of Egypt. (In Arabic), Dar El-Hilal, Cairo

Klein M (2000) The formation and disappearance of a delta at Wadi El-Arish river mouth. Proceedings of Jerusalem Conference on the Hydrology-Geomorphology Interface: Rainfall, Floods, Sedimentation and Land Use, IAHS Publication No 261, May, 1999

Kusky T, El-Baz F (2000) Neotectonics and fluvial geomorphology of the northern Sinai Peninsula. J African Earth Sciences 31:213–235

Moawad MB (2013) Analysis of the flash flood occurred on 18 January 2010 in wadi El Arish, Egypt (a case study). Geomatics, Nat Hazards Risk. 4(3):254–274

Mostafa M (1987) Wadi Feiran Basin: A Geomorphologic Study. Dissertation, Ain Shams University

Moustafa A, Khalil M (1990) Structural characteristics and tectonic evolution of north Sinai fold belts, Chapter 20. In: Said R (ed) The Geology of Egypt. A A Balkema, Rotterdam

NARSS, UNESCO, UNDP, EGSMA (2009) Atlas of water resources in Sinai Peninsula: 36 National Authority for Remote Sensing and Space Sciences, Cairo

Rashwan Kh (1994) Wadi Dahab Basin, Sinai Peninsula: A Geomorphologic Study. (In Arabic). Dept. of Geography, Faculty of Arts, El-Minya University, El-Minya

Said R (1981) The geological evolution of the River Nile. Springer, New York

Said R (1990) Quaternary. Chapter 25 in Said R (ed) The Geology of Egypt. A A Balkema, Rotterdam

Salem R (1976) Evolution of Eocene—Miocene sedimentation patterns in parts of northern Egypt. Amer Assoc Petrol Geol Bull 60:34–64

Sneh A (1982) Drainage systems of the Quaternary in Northern Sinai with emphasis on Wadi El- Arish. Z Geomorph 26:129–185

Thornburry W (1954) Principles of geomorphology. John Wiley & Sons Inc, New York

Torab M (1988) Wadi Beda, SW Esswes City: A Geomorphologic Study. (In Arabic), Dept. of Geography, Faculty of Arts, Alexandria University, Alexandria

Part VII
Coastal Forms

Ridge-Depression Landscape Along the Mediterranean NW Coast of the Western Desert

20

Abstract

The carbonate ridges distinguish the NW coast of Egypt, since they are not repeated elsewhere in Egypt. These ridges form together with the low-lying areas, separating them from each other, a local landscape system called "The Ridge-Depression System." This system roughly extends parallel to the coastline of the Mediterranean from Alexandria in the east to Sallum city in the extreme west. It is composed of a sequence of low-relief ridges, but with local variations in their number. The height of the ridges varies between 10 and 25 m a.s.l. for the first coastal ridge and 110 m a.s.l. for the inland eighth ridge. The width of the ridges ranges between ~100 and ~10,000 m. Some ridges were found submerged in the Arab Gulf and Maryut Lake or buried by Nile sediments in NW the Nile Delta. A calcareous layer of soils covers Depressions that separate the ridges. These soils are mainly derived from the adjacent ridges. In the first and second Depressions, sabkhas and coastal lagoons were developed either due to connection with the Mediterranean, such as Matruh and Alamain lagoons, or due to seepage of seawater. The ridges are polygenetic, and their age ranges between 0.6–4.3 ka for the first coastal ridge, and 900–950 ka for ridges No. 5–8.

Keywords

Northwest coast • Ridges-Depression system • Submerged ridges • Coastal lagoons and sabkhas

20.1 Introduction

Although the surface of the coastal plain slopes gently toward the sea, the plain is characterized by a sequence of low carbonate ridges (bars), which are roughly parallel to the present coastline. These ridges are separated from each other by sabkha-lagoonal Depressions that are interrupted in parts by secondary transverse bars. This Ridge-Depression System is one of three landform systems in this coastal plain. The other two systems are the Wadi-Fan System and the Coastal-Marine System. Several researchers have given special attention to this system since the end of the nineteenth century, because it represents a unique landscape along the Mediterranean coast (Fourteau 1915). The studies of Paver and Pretorius 1954,

N.S. Embabi, *Landscapes and Landforms of Egypt*, World Geomorphological Landscapes,
https://doi.org/10.1007/978-3-319-65661-8_20

Shukri et al. (1956), Butzer (1960), Slim (1974), El-Demerdash et al. (1978), Shata (1987), Cherif et al. (1988), Hilmy et al. (1989), Lindell et al. (1991), Stanley and Hamza (1993), Wali et al. (1994), Moussa et al. (1998), El-Asmar (1998, 1994a, b), El-Asmar and Wood (2000), and Moawad (2003) are the most relevant ones to the present study. These previous studies show considerable controversy concerning, the origin, age, and the extent of the ridges. In this chapter, the geomorphologic characteristics of the ridges will be treated first, then those of the Depressions.

20.2 Distribution of the Ridges in the System

Carbonate ridges of the NW coast are present in two environments. The continental environment, where ridges are developed on the mainland of Egypt, and the second is the marine environment, where ridges are submerged by seawater. Therefore, ridges can be divided into two groups depending on their location. Most of the continental ridges are developed on land along the Mediterranean coast, whereas the second group is submerged by seawater.

20.2.1 The Continental Ridges

The geologic map of Egypt (scale 1:500,000) Sheets NH 35 NE and NH 35 NW (Conoco Coral and Egyptian General Petroleum Corporation 1987) show that the ridges extend for long distances along the coastal plain from the extreme east to the extreme west. They also show that in several localities, there are several lines of these ridges parallel to each other and to the coastline as well. Nevertheless, due to their low relief and the small scale of the map, not all the ridges, especially the small ones, appear on maps. Even the topographic sheets (scale 1:100,000, 1:50,000, and 1: 25,000) do not show the exact number of the ridges because the contour intervals do not allow the representation of the small low aridges. There is also the dissection of the ridges into several sections by wadis, which renders their recognition on maps quite difficult. Consequently, the description of local variations in the distribution and geomorphologic characteristics will mainly depend on available sources and on previous studies.

The number of bars increases—in general—as the coastal plain becomes wider and lower. The largest number of ridges that can be recognized is along the coastline of the Arab Gulf in the east and in the Sallum basin in the extreme west. Butzer (1960) and Shukri et al. (1956) mapped eight lines of these ridges with varying length and width in the Arab Gulf region (Fig. 20.1). The height of the ridges increases from about 10 m for the coastal ridge to 110 m a.s.l. for the southern one, and that the width varies from one ridge to another and even along the same ridge (Table 20.1). The widest symmetrical ridge is the fourth from the coast (5–8 km), whereas the widest ridge in general (12 km) is the eighth ridge. The longest and most continuous ridges are the first (coastal) and the second (Abu Sir) ones (Fig. 20.1), where both extending from Alexandria city and Abu Qir to the west of Alamain (about 150 km). According to Shukri et al. (1956), the coastal bar extends without nearly any break or change in height for about 500 km from Alexandria city to about 13 km east of Sallum city. The examination of the topographic sheets (scale 1:100,000 and 1:25,000) confirms this fact, except for the bulge of Sidi Barrani, where the coastal ridge disappears apart from very small occurrences. These maps (Sheets Mahatet El-Zawaida, Mersa El-Tarfaya, and Wadi El-Maktala) also indicate that the width of the first ridge changes due to the dissection by wadis, which occurred during the deposition of this ridge below present-day sea

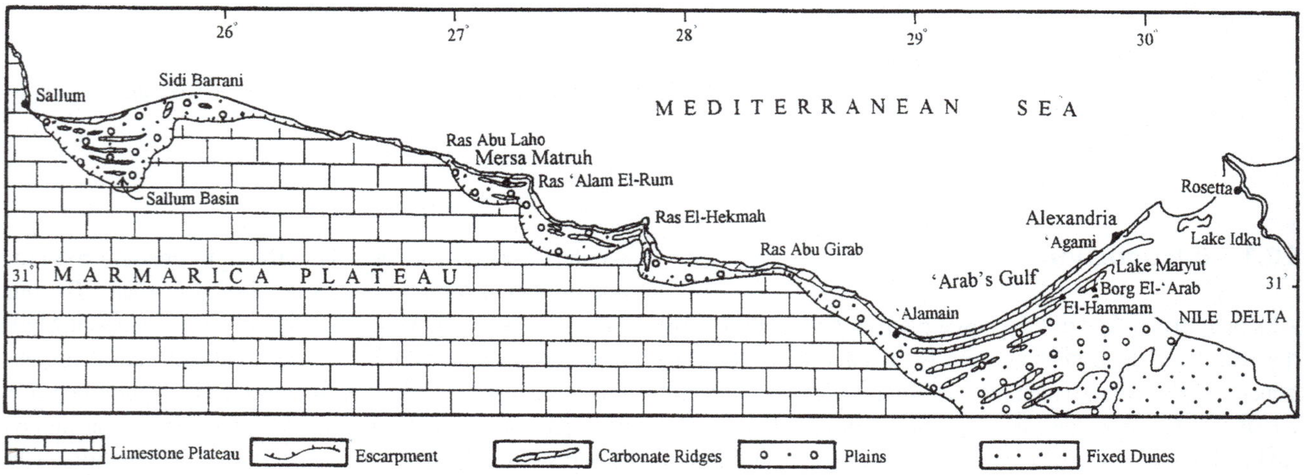

Fig. 20.1 General Features of the NW Coastal Plain of the Mediterranean Sea (*Source* Embabi 2004)

Table 20.1 Number and height (m) of the coastal ridges at several localities

The Arab Gulf[a]	El-Dabaa[b]	Fuka and Ras El-Hekma[c]	Mersa Matruh[d]	Umm El Rakham[e]	Sallum Basin[c]	Sidi Barrani[c]
1 (10)	1 (25)	1	1 (20)	1 (18)	1 (12–36)	–
2 (25)	2 (32)	2	2 (28)	2 (28)	2 (22)	–
3 (35)	3 (35)	3	3 (44)	3 (40)	3 (32)	–
4 (60)	4 (50)	4	4 (65)	4 (100)	4 (28)	–
5 (80)	5 (90)	5	5 (60–110)	–	5 (?)	–
6 (85)	–	6	–	–	6 (?)	–
7 (90)	–	7	–	–	7 (?)	–
8 (110)	–		–	–	8 (?)	–

[a]Shukri et al. (1956)
[b]El-Shami et al. (1969)
[c]Aerial photographs (scale 1: 60,000) and Geologic Maps (scale 1:500,000), Topographic Maps (scale 1:25,000)
[d]El-Shazly and Shata (1969)
[e]El-Sanussi and Shata (1969)

level. In the meantime, it seems that the third and fourth ridges are dissected into sections and are divided into several branches. All ridges in the Arab Gulf area (except for the second Abu Sir ridge) are ended abruptly in the east. Based on this characteristic, Butzer (1960) inferred that two fault lines that run at right angles to the direction of the ridges (Fig. 20.2) bound them. The orientation of these fault lines is similar to faults, which run parallel to the coastline of the headlands of the central part of the Mediterranean coast. This truncation may also be due to Nile channel incision during glacial sea level low stands that were followed by deposition of Nile silt (Sestini 1989). Another possibility is down-flexing under Delta loading. The presence of oolitic limestone and other littoral sediments in the subsurface in the NW corner of the Nile Delta in Abu Qir, Kafr El-Dawar city station, and Edfina (Attia 1954) indicates that the ridges extend in the eastward direction at least to the present Rosetta Nile Branch. It also suggests that the eastern parts of the truncated ridges were—most probably—exposed to tectonic subsidence rather than being eroded by an old Nile Branch. What consolidates this suggestion is that the sediments of the ridges extend downward more than 100 m below sea level and the surface of the Nile Delta (Attia 1954). On the other hand, one of the subsurface ridges in the NW of the Nile Delta was considered as a NE continuation of the third ridge and not for the other five ridges from 4 to 8 (Wali et al. 1994).

In the Sallum basin, there are eight ridges which run parallel to the coastline, of which the first (coastal) and the second are the most continuous and prominent ones. On the contrary, the southernmost one that lies at about 25 km from the coastline is the smallest in this basin. The coastal ridge rises in some places (Ras El-Siyadah locality) up to 36 m (Table 20.1) and runs for about 25 km without any break. As in the group of the Arab Gulf, some of the ridges have local names as in the case of the third ridge which is called Alam Barghûth.

Apart from these two localities (the Arab Gulf and the Gulf of Sallum), the number of ridges recognized by previous workers or shown on the topographic and geologic maps is less than eight. They vary between three ridges in Matruh area and seven bars in Fuka region. Even though the majority of ridges are parallel to the coastline and to each other, there are two localities where some of the branches of the ridges are perpendicular to the coastline. These two localities are in El-Dabaa area and at Ras El-Siyadah in the middle of Sallum Bay. In El-Dabaa area, geologic sheets at a (scale 1:100,000) (The Geological Survey of Egypt 1986) show that at Ras Abu Girab, a 12.0 km ridge extends in a N-S direction, whereas at Ras El-Dabaa only small ridges (2–4 km) run in a N-S direction. Moreover, in Ras El-Siyadah, several short (about two km long) ridges (including Ras El-Siyadah promontory) branch from the coastal ridges. Most of these branches are perpendicular to the main ridges. These perpendicular ridges were most likely developed by the same processes, which developed the main ridges but at favorable sites such as estuaries.

In cross section, the ridges are convex-concave in form (Fig. 20.2), with varying slope angles between 2 and 15°. The form of the ridges may be modified and changed into step-like or irregular due to the mode of development, or due to the subsequent erosion. In Matruh region, the first ridge shows a gentle slope landward and a steeper one seaward due to the effect of wave action that resulted in a vertical cliff overlooking the coastline (El-Shazly and Shata 1969). Disintegrated layer of the parent material (carbonates or aeolianites) covers most of the ridges, except for the coastal one.

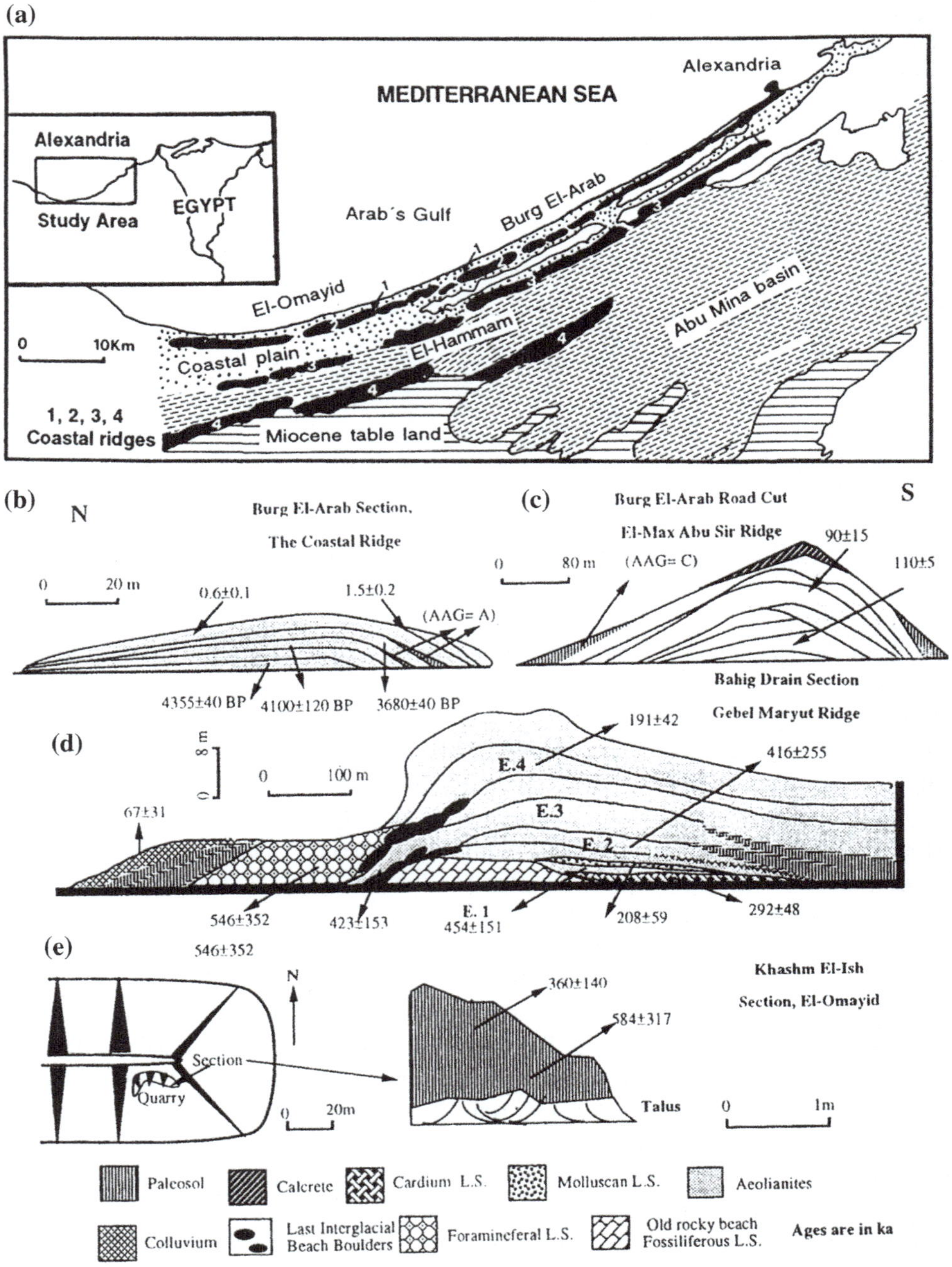

Fig. 20.2 Coastal Ridges of The Arab Gulf Region **a** Distribution Map (*Source* Butzer 1960) **b** Cross Sections and age dating results for the exposed sedimentary units constituting the first four ridges (*Source* El-Asmar and Wood 2000)

This layer becomes thicker at their foot slopes. This surface material forms a thin layer of soil, which is cultivated by locals with fig trees that produce the main cash crop in this coastal region. Older and similar layers of weathered material appearing in the stratigraphic record of the ridges are called "geosols" (El-Asmar 1994a, b), suggesting that these ridges developed during several cycles of deposition and erosion. The coastal bar is covered by loose aeolian sand, which is deposited at present by the northern winds on both sides of the bar. This aeolian sand hides the original lithified bedrock slopes of the ridge that might expose some irregularities. The coastal ridge is called in some previous works (Sundborg and Nilsson 1985) the dune ridge, since it consists of two parts: a basal, weakly cemented ridge and a relatively, thin coverage of loose windblown sand. The cemented basal part is eroded into steep cliffs along the present coastline in some localities.

20.2.2 The Submarine Ridges

This group of ridges was recognized in the offshore area in the eastern and western parts of the Arab Gulf and the Alexandria western harbor using different methods (remote sensing, echo-sounding, boring and from the British Admiralty maps). Besides, submerged ridges were traced in Lake

Maryut where small islands can be noticed projecting above water level in the western arm of the lake. Data from a set of seven borings put down offshore from Alexandria Western Harbor to El-Agami revealed the presence of a submarine (first) ridge composed of oolitic limestone, marl with shell fragments and calcareous sand. This submarine ridge lies at 0.71–15.79 m below sea level and extend parallel to the coastline (Attia 1954). Butzer (1960) recognized some submerged bars extending from Alexandria city to the longitude of El-Hammam city through a chain of islands and shoals running parallel to the coastline 2–6 m below sea level (Fig. 20.2). He also recognized other submerged bars from the Admiralty maps. One of these ridges lies 11 m below sea level at about one kilometer from Pharos Island (Alexandria city), whereas two other ridges lie 40 m below sea level, some 6 km offshore.

Recent studies which analyzed space images (Sundborg and Nilsson 1985; Lindell et al. 1991; Moussa et al. 1998; Moawad 2003) revealed the presence of submerged ridges in the western part of the Arab Gulf, and along the coasts of Ras El-Hekma and Alam El-Roum. The ridges of the Arab Gulf, consisting of recent aragonitic oolites, are the first recent finding ever to be observed in the Mediterranean (Lindell et al. 1991). These ridges are directed toward southeast, and they make an angle of 20° with the coast. Their spacing varies between 900 and 1500 m, and their width between 400 and 600 m. Seaward, they can be followed to a depth of 6–8 m and for a length of 10–15 km. Landsat images show that there is a system of narrow transverse ridges making a feather-like or branching pattern on the sides of the large ones. Three cross sections along the offshore made by echo-sounding in the intake area of the Qattara Canal Project near Alamain revealed that there are three well-developed submarine ridges with widths varying between 250 and 800 m, and heights between 0.5 and 4.0 m. Attached to the main ridges are lobe-shaped bars generated by wave induced currents (Sundborg and Nilsson 1985). Comparison of Landsat images from 1972 and 1978 showed no changes in size or position of these submarine ridges. From this fact, it was concluded that these images gave no clear indication to whether the submarine ridges are active sedimentary features, or lithified remains of terrestrial dunes across fossil coastlines (Sundborg and Nilsson 1985). On the other hand, by comparing the old Admiralty chart, MSS 1972 image and 1986 SPOT images, Lindell et al. (1991) concluded that the submarine ridges are in a hydrodynamically stable situation. This stability occurs in a location with long-shore transport and variable wind pattern during the year. Using Landsat-TM images of 1986 and 1999, Moawad (2003) recognized some submarine ridges along the coasts of Ras Alam El-Roum and Ras El-Hekma. The 1999 images show the development of two ridges parallel to present-day coastline of Ras Alam El-Roum. The first one extends for about 5 km at a distance of 500–800 m from the coastline. The second ridge-runs for a distance of 3 km at −1.4–2.0 km from the coastline. Along the coastline of Ras El-Hekma, there is a 7 km branching ridge developed at a distance of 300–900 m. Comparison of the two series of Landsat-TM images signaled the growth of the submerged ridges in the both areas, indicating that they are dynamically active sedimentary features (Moawad 2003). This conclusion is not conformable with that of Lindell et al. 1991 concerning the submerged ridges of the Arab Gulf. It is highly possible that this difference is due to the inaccuracy of the Admiralty chart.

20.3 Sedimentological Composition of the Ridges

The coastal ridges are described in previous works as calcaranite or aeolianite ridges, indicating that they are mainly composed of calcareous materials. A few detailed studies were recently carried out on sedimentological composition of these coastal ridges (Shukri et al. 1956; Selim 1974; Hassouba 1980; El-Asmar 1991; Stanley and Hamza 1992; Wali et al. 1994; El-Asmar 1994a,b; Mahmoud 1995; El-Asmar 1998; El-Asmar and Wood 2000; Shata 2000) These recent studies revealed that these ridges are composed of alterations of marine, aeolian, fluvial sediments, and palaeosoils. There are significant differences in sediment composition and age from one ridge to another, and that these differences may shed light on the origin of these ridges or can be correlated to climatic and sea level changes. Detailed studies concentrated on the first four ridges, known as the coastal, El-Max-Abu Sir, Gabal Maryut, and Khashm El-Aish ridges in Borg El-Arab and El-Hammam area along the coast of the Arab Gulf (El-Asmar 1994b, 1998; El-Asmar and Wood 2000). Some observations were made also on other ridges in the Arab Gulf region (Shukri et al. 1956). Other relevant studies were made for the first three ridges in Sallum area (Selim 1974), Matruh area (El-Shazly and Shata 1969) and Um El-Rakham area (El-Sanussi and Shata 1969). On the other hand, the deep boring put down in the northwest corner of the Nile Delta revealed the presence of some of the ridges below the Nile sediments (Attia 1954). Though no mention is made as to which ridges these subsurface bars belong; the data of the deep borings are the only available for the whole succession of sediments from top to base in any ridge (Fig. 20.3).

Most of the ridges along the northwest coast are covered with a hard brown or gray, calcareous duricrust (calcrete). This crust has been observed by Shukri et al. (1956) on top of the second (Abu Sir), third (Gabal Maryut), fourth (Khashm El-Aish), and fifth (Alam El-Khadem) bars in Borg El-Arab region. They described this crust as a solid

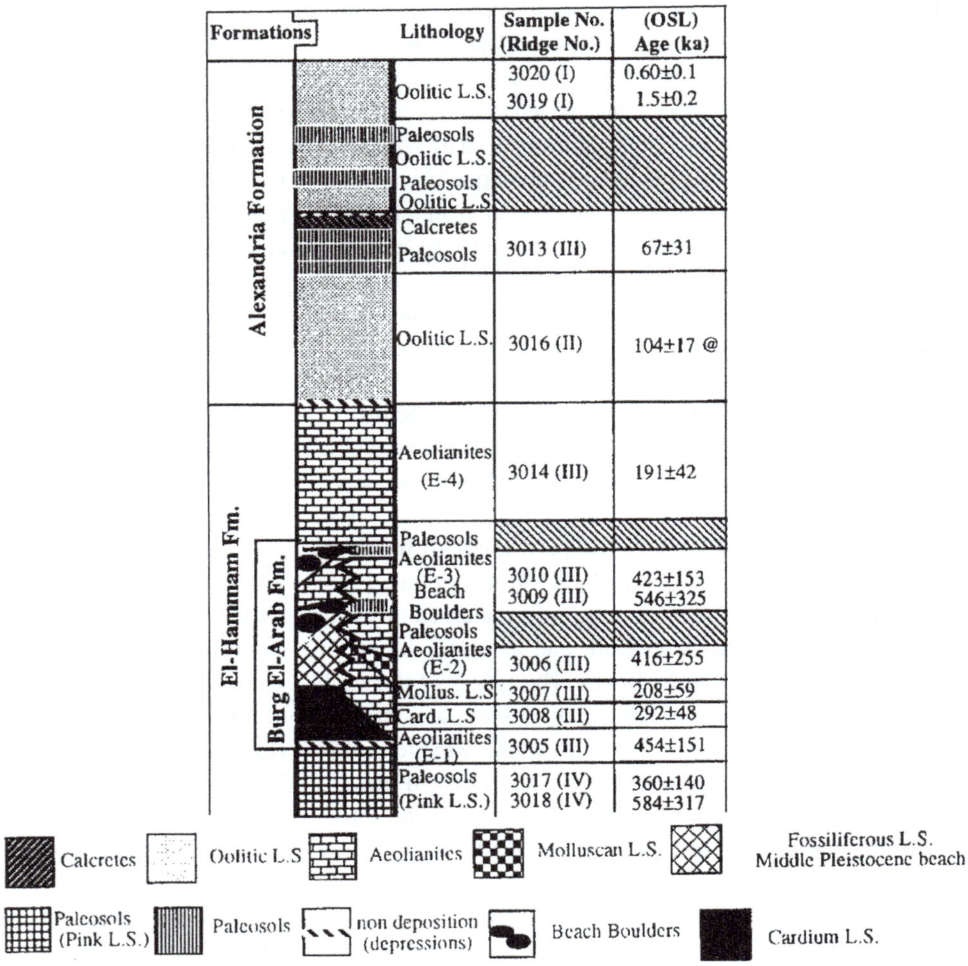

Fig. 20.3 Composite Geologic Column and OSL age results for the first four ridges in the Arab Gulf Region (*Source* El-Asmar and Wood 2000)

re-crystallized brownish top layer of a few millimeters in thickness. They also found that this crust in the older bars is not restricted to the top layer but is re-deposited in the interstitial spaces as cement or as crack filling and that it extends for a few meters below the surface. Thin calcrete crusts, that are a few to 15 cm thick, hard, light to dark brown, and thinly laminated, are repeated irregularly throughout the sequence of the second (Abu Sir) and third (Gabal Maryut) ridges (Stanley and Hamza 1992). This internal occurrence of calcrete crusts suggests intermittent exposure of ridge deposits to subaerial processes. The calcrete crust was observed also by El-Shazly and Shata (1969) in Mersa Matruh region on the second (Matruh) and third (south Matruh) ridges. These researchers described the crust, which developed on the second ridge, as hard crust with pinkish brown color and a thickness of a few millimeters. They also suggested that this crust was developed by re-crystallization caused by aridity. El-Shazly and Shata (1969) also recorded that the crust, which developed on the top of the third ridge, is thicker (few centimeters) than that of the second ridge, but has the same color. The crust was also recorded on the top of the second (El-Matareih) ridge in Umm El-Rakham area and was described as a hard crust that is composed of brown limestone (El-Sanussi and Shata 1969). A calcrete layer was found at the base of the second (Gabal Maryut) ridge in Burg El-Arab region in the artificial cut by the drainage canal near Bahig village (Hassan et al. 1986; Hegab and El-Asmar 1995). However, only a few thorough studies have been carried out on this crust (El-Shahat et al. 1987; Hegab et al. 1987; Rashed 1998) in Borg El-Arab El-Omayed region.

The main constituent mineral of the calcrete is the low Mg calcite that was derived from either the host rocks by dissolution from soil cover carbonates, or the airborne dust that may carry considerable amounts of C_aCO_3. According to Hassouba (1980), this mineral represents 87% of the total minerals. Other minerals are aragonite, quartz, feldspar, and dolomite. From these characteristics and other data mentioned earlier in this subsection, it was concluded that the calcretes were developed within the upper few meters of the Pleistocene oolitic ridges under subaerial semi-arid environment in the vadose zone (Rashed 1998).

20.3.1 The Coastal (First) Ridge

In the Arab Gulf region, the top 6.0 m of this ridge are composed of hard oolitic limestone with two horizons of palaeo-soil (Fig. 20.2). Such composition indicates that this section was deposited during a generally arid phase with two relatively short periods of wet climate (El-Asmar 1998). This upper part overlies 4 m of unconsolidated-to-weakly cemented oolitic sand in El-Hammam area (El-Asmar 1992). Along the shores of this locality, the dipping oolitic strata can be seen extending below the present sea level; there, some of the beds comprise flattened beach gravel and some small rounded gravel, an indicator of a littoral depositional environment. Ball (1939) also observed the extension of the oolitic beds below the present-day sea level without determining the locality. This type of composition was also recorded by Shukri et al. (1956) and was observed in the field in the same locality. Due to its composition of pure carbonate, Stanley and Hamza (1992) classified this coastal ridge as an ooid grain-stone ridge in which small gastropod shells were found. In addition, it is recorded on the geologic map (scale 1:500,000) (Conoco Coral 1986) as the dominant composition of all coastal ridges.

In El-Agami (Hanoville) at the extreme western margin of Alexandria region, it was found that the exposed part of this first ridge is composed of lagoonal mud of poorly sorted calcareous clayey silty oolitic sand (about 1.5 m thick). Several thin (0.1–0.7 m) units of compact to friable calcarenites (Mahmoud 1995) overlie this unit. These two layers are separated by a sharp contact that lies just below sea level and contains Greco-Roman potsherds embedded in light yellowish brown calcareous clayey sand. Based on these characteristics, it was inferred that this region was exposed to neotectonism during the period 24–20 BC (Mahmoud 1995).

In the locality of El-Ballah Touristic Village, which lies at 0.5 km to the west of El-Hammam city, a layer (0.5–1.0 m) of conglomerate of fluvial gravel was seen in the field exposed at the top of the coastal ridge. The presence of this layer confirms the conclusion derived from the presence of two palaeo-soil horizons that there were short periods of relatively wet climate. Moreover, it indicates that during the development of this ridge, some channels cut across the ridge. What consolidates this fact is the detection of the dissection of the coastal ridge by several wadis debouching into the Mediterranean Sea. In the Sallum Basin in the extreme west, the first ridge is characterized by a distinct seaward dipping cross-bedding (Selim 1974). Samples collected from this ridge are composed of 58% ooids, 9% fossils, 6% quartz grains, and 27% cement. The cement consists of low magnesium calcite mostly as grain contacts (Selim 1974). In Matruh area, the first ridge is made up of oolitic sands, which demonstrates aeolian structures such as cross-bedding, cross-lamination, and steep dipping (El-Shazly and Shata 1969). This ridge is almost covered by recent loose snow-white oolitic sand. This surface layer of aeolian sand is taken to signpost the role of wind in the partial deposition of similar material on top of ridges that demonstrates aeolian structures (El-Shazly and Shata 1969). Borings in this ridge revealed the occurrence of shelly limestone layer at the base, which contains pebbles, referring to a near-shore environment (El-Senussi and Shata 1969). In Umm El-Rakham area, the first ridge is composed of medium hard snow-white oolitic limestone, which shows conspicuous cross-bedding (El-Sanussi and Shata 1969).

The seven borings, dug offshore Alexandria coastline from El-Agami Marine Airport to the western harbor, revealed the presence of a submerged ridge that runs parallel to the coastline. This submerged ridge can be considered a submerged (eastern) extension of the first coastal ridge (Fig. 20.2). The depth of borings varied between 3.6 m (boring No. 2) and 17.3 m (El-Agami boring). They reported that the base of all borings was composed of oolitic limestone with a thickness varying between 0.7 and 4.0 m, and was overlain by several layers of shelly sand, marl, calcareous sand, and marly limestone with shell fragments. It was concluded that these deposits are of a littoral nature (Attia 1954). This submerged continuation of the first ridge was deduced from the Admiralty maps of the coasts of Alexandria (Butzer 1960). The submergence of this part of the first ridge is due to subsidence in the coastal area of Alexandria region that occurred most likely during the period 24–20 BC. To note, this period witnessed neotectonism that affected the NW region of the Nile Delta (Abdel El-Hady 1994).

20.3.2 The El-Max-Abu Sir (Second) Ridge

This ridge is composed mainly of oolitic limestone with a top layer of palaeosols, with total thickness of 4.0 m, in Borg El-Arab area (El-Asmar 1994b). This succession indicates that the climate was generally arid during the deposition with relatively short intermittent periods of moderately wet climate during which the topsoil horizons or calcrete was developed. In the Alexandria region (from Agami to Abu Qir headland), examination of several sections revealed local variations in the composition of the second ridge along short distances (Stanley and Hamza 1992). The most significant finding of this examination, is that the content of quartz that constitutes most of the insoluble residues is very high (76.6%) in the east (Abu Qir headland and Hadra). This percentage decreases markedly to the west of El-Max. Examination of thin sections revealed that quartz grains are medium to coarse grade and are angular; such findings mean that they are not likely to have been transported by wind action (Stanley and Hamza 1992). From these characteristics, it was concluded that these siliciclastic components were transported northwestwardly by River Nile distributaries, such as the Canopic and Alexandria branches (Stanley and Hamza 1992). The same conclusion was drawn from the study of the sedimentological and mineralogical characteristics of 125 m borehole sediments in Burg El-Arab Area (Shata 2000). The presence of both bands of sands and sandstone in a subsurface creamy limestone bed (Hilmy et al. 1978) as well as the presence of violet augite and brown mica(that characterizes the Nile sediments in the soils to the east of Borg El-Arab area) (El-Demerdash et al. 1978), indicate a contribution by an ancient Nile branch. It is natural to find a decrease in the percentage of terrigenous material westward due to the long distance from the source, namely the Nile water. This study also remarked that to the west of Agami, the ridge is fairly consistent in lithology where the carbonate content is almost always higher than 95%, and displays large scale low-angle cross-bedding. In the same study, abundant fish bone fragments and teeth were recorded in some horizons to the west of Agami. In Alexandria city, two deep borings were dug in this second ridge by the Municipality and Shari El-Mamun (Attia 1954). Both borings revealed alternating layers of oolitic limestone, marine sand, calcareous sand-rock, foraminiferal sand and clay of littoral origin. The bottom of these two borings was about 113 and 81 m below sea level, with their tops at 4.0 m and 8.0 a.s.l., respectively. If the succession of these two borings represents the second ridge, then it can be inferred that the area was exposed to subsidence either during or after the development of the ridge. It is most likely that this subsidence occurred during the same period (pre-Greco-Roman times) when the first ridge was submerged in Alexandria region (between El-Agmi and West Harbor). In the Sallum Basin, the second ridge is composed of oolitic limestone, with 38% ooids, 9.5% fossils, 3.5% quartz grains, and 33% calcite cement, which is impregnated with argillaceous material (Selim 1974). The Matruh second ridge comprises white medium hard oolitic limestone and calcite cement with conspicuous cross-bedding in its upper part (El-Shazly and Shata 1969). In Umm El-Rakham area, similar sedimentological characteristics to those of Matruh area were observed (El-Sanussi and Shata 1969).

20.3.3 The Gabal Maryut (Third) Ridge

A full exposure (15–36 m thick) of this ridge is located in the cut of Bahig drain, about 3.0 km to the east of Borg El-Arab city. The section in this cut is composed of alternations of marine deposits, palaeosols and aeolianites (Hegab and El-Asmar 1995; El-Asmar and Wood 2000). The marine deposits comprise several types of limestone and beach boulders (50–57 cm). These beach boulders were found at two levels, 3–4 and 6–7 m a.s.l. (Hegab and El-Asmar 1995). On the other hand, the aeolianites are composed of shell fragments, ooids and quartz grains, and other subordinated calcareous grains with the characteristic planar-tabular cross-stratification (El-Asmar and Wood 2000). According to El-Asmar (1998), the marine sediments lying between the beach boulders were deposited during the last interglacial stage when the climate was warm and the sea level was in the order of 6–8 m above the present level. These aeolianites appear to be associated with arid climatic phases alternated with the marine transgression in wet phases. The base of the ridge is composed of two beds of aeolianite and cardium limestone (Fig. 20.2). Shukri et al. (1956) stated that this ridge is rich in organic remains, which are represented by Mediterranean foraminifera with certain Atlantic and Indo-Pacific species. Calcareous algae, pelecypod and gastropod shell fragments, and enchiniod spines are also abundant. Field and petrographic study of five sections along the third ridge from Ameriya to Borg El-Arab revealed gradual changes from a terrigenous facies in the east to a carbonate-dominated facies in the west (Stanley and Hamza 1992). According to this study, ridge III has the highest terrigenous (mainly quartz) content, which extends further west than in the first and second ridges. This study also found that as in Abu Sir Ridge, ridge III is characterized by repeated occurrence of palaeosols and calcrete layers. In Matruh area, the third ridge (South Matruh ridge) is made of medium hard oolitic limestone of pinkish white color, with conspicuous cross-bedding at the upper part (El-Shazly and Shata 1969). These characteristics were also reported about Um El-Rakham area (El-Senussi and Shata 1969a). In the Sallum area, the third ridge consists mostly of interclasts of microcrystalline calcite and low percentage of quartz grains, cemented by fine to coarse dolomitized low magnesium calcite (Selim 1974).

Three deep borings were dug in NW Nile Delta near Difshu village and Edfina, with bottoms lying 168.8, 122.5, and 124.0 m below sea level. They reveal successions of alternating layers of oolitic limestone and other marine and littoral deposits with varying thickness. The bottom oolitic limestone layer in the Edfina boring reaches a thickness of 17.5 m (between 106.5 and 124.0 m below sea level).These deposits may be correlated with the third ridge. In a previous study, the thick oolitic limestone bed found at the bottom of the Edfina boring was also correlated with the third ridge (Hassouba 1980). Such thick subsurface littoral succession of the third ridge indicates that this part was exposed to slow subsidence—as that of the second ridge and the submarine part of the first ridge—due to neotectonism, which occurred 24–20 BC in NW Nile Delta (Abdel-Hady 1994).

20.3.4 The Fourth (Khashm El-Aish) Ridge

This ridge is made of oolitic hard limestone, which contains a huge number of Indo-Pacific species of foraminifera with a re-crystallized brownish top layer (Shukri et al. 1956). On top of this ridge, there is a band of limestone rich in fossils, especially in megascopic pelecypods, *Cardium edule,* and *Perinella* sp., with some broken shells. The fossils and the broken shells constitute the fauna of the bed (Shukri et al. 1956). In El-Omayid area, the top of the fourth ridge is composed of pink to pale-brownish paleosols (pink limestone), overlying the Miocene deposits with thickness up to 4.0 m (El-Asmar and Wood 2000). In the area of Matruh, the fourth ridge, which is called Alam El-Awsagi Ridge, exists at the foot of the Miocene escarpment of the plateau. It is made of oolitic limestone of pinkish white color, intermixed with shelly deposits of near-shore origin. The ridge also rests directly on the Miocene rocks (El-Shazly and Shata 1969).

20.3.5 The Fifth, Sixth, Seventh, and Eighth Ridges

The fifth ridge is called Alam El-Khadem and is made of hard limestone, which is brownish in color on top due to re-crystallization. This ridge and the older ones contain Indo-Pacific species, which refer to a connection between the Mediterranean Sea and the Red Sea in the time of formation (Shukri et al. 1956). The three ridges from six to eight are called El-Mikherta, Raqabet El-Halif, and Alam Shaltut, respectively, and are composed of oolitic limestone with re-crystallized top layer (Shukri et al. 1956).

20.4 Inter-ridges Depressions

The aeolianite ridges are separated from each other by Depressions. In some Depressions, lagoons and sabkhas are developed, whereas some others are filled with fluvial sediments or gypsum and clay. Examination of topographic maps (scale 1:100,000; 1:50,000; and 1:25,000) and geologic maps (scale 1:500,000 and 1:100,000) revealed that lagoons and sabkhas developed only in the first and second Depressions from the coastline, between the first and second ridges and the second and third ridges. In these two Depressions, the level of the surface of the ground lies at one or 2 m above/below sea level. This correlation is an indication to the relationship between these two forms and the Mediterranean Sea (Moussa et al. 1998).

20.4.1 The Geomorphologic Characteristics of the Depressions

The geomorphologic characteristics of these inter-ridge Depressions are known from a few localities only. Therefore, further studies are suggested to add knowledge about this Ridge-Depression System as it dominates the landscape of the coastal plain west of Alexandria. Nonetheless, available studies provide sufficient data on their geomorphologic aspects and mode of development. As noticed in the ridges, the Depressions were given local names either by locals or by authors who worked on them, such as El-Dekheila (first) Depression west of Bahig cut and El-Qasr (first) Depression in Matruh area.

The floor of **the Dekheila (first) Depression** is composed of 1–2 m of Holocene brownish-yellow calcareous muddy loam, gray gypsiferous sabkhas, and marsh deposits. This stratum overlies another stratum of shallow marine to lagoonal sediments composed mainly of shell debris, calcareous ooids, and pellets with few clay and quartz grains. The results of the oxygen isotope analysis of the collected fossils from this Depression signified that accumulation occurred in a shallow sea with a frequent influx of fresh water (Hegab and El-Asmar 1995). **The first Depression in Matruh region** is called the "near-shore Depression" (El-Shazly and Shata 1969). A series of eight salt lakes, of which the West Matruh Lake is the largest and deepest, occupy this Depression. The western part of this near-shore Depression is known as El-Qasr Depression and is covered with reddish brown alluvial deposits. In the extreme west, the level of the ground of this Depression becomes closer to sea level and is exposed to flooding by seawater and torrential surface flow from wadis issuing from the plateau. As a result, marshes with salt tolerant plants dominate this part of the Depression. In Umm El-Rakham region, the first Depression is called '**Umm El-Rakham Depression**' (El-Senussi and Shata 1969). Its surface rises 5 m above sea level and is covered by reddish brown alluvial calcareous deposits. Several wadis, such as Wadi Abu Mangore, Wadi Umm El-Rakham, and Wadi El-Habis, drain the southern plateau and gush in this Depression, with some fans formed at their mouths. These fans are composed of pebbles, cobbles, gravel, and boulders embedded in a matrix of calcareous loam and are considered fanglomerate by El-Senussi and Shata (1969). In the eastern part, the Depression merges gradually into Sabkhat Umm El-Rakham.

Lake Maryut arm, which is locally known as "the Marine Arm," occupies the eastern part of the second Depression in the Arab Gulf region. The width of this Depression varies between 5.0 and 2.0 km, whereas the height of its floor increases gradually westwards from a few meters below sea level in the east of Bahig to a few meters above sea level to west of El-Hammam city. To the west of El-Hammam, the floor of the Depression is covered by calcareous clay with high salt content. The study of the materials of some sections between Borg El-Arab and El-Alamain revealed the presence of gypsum occurrences of lagoonal origin at the floor of this second Depression, and that the succession of the gypsum layers is separated by marine limestone and clay (Abdallah 1967). This study also showed that the alternations of the gypsum and marine layers are separated by unconformities, signifying that marine and lagoon conditions alternated several times. Thickness of sections also varied between 3.0 m in El-Omaid and 46.5 m in Alam El-Milh (7.0 km east of El-Alamain), suggesting that the floor of this second Depression is irregular in form. Only the section of El-Hammam Gypsum Quarry reached a layer of gritty conglomerate limestone, with >1.0 cm pebbles of quartz and calcareous material. Such finding stands as a representation to the beach and foreshore deposits of the Millazian Sea (Said et al. 1956). The section of Alam El-Milh shows cavernous, karstified limestone layers, while that of El-Ghirbaniyat Gypsum Factory exposes a layer of crystalline stalactitc gypsum, which reflects the effect of karstifiction in the development of the Ridge-Depression System along the coastal plain of the Western Desert.

The second Depression in Matruh region is known as "**Matruh Depression**" (El-Shazly and Shata 1969). It is well defined to the east of Mersa Matruh city with a width of about one kilometer and is covered by a yellowish brown loamy soil and white oolitic sands, with a thickness of more than 4 m. In its western part, this Depression opens at the northern near-shore Depression and at the inland one to the south, as evident on topographic maps. In this western part, there are some deposits. These deposits are indicative of the presence of a possible ancient fresh water lake, and that the lake formed at the time when the Matruh ridge was acting as a dam, preventing water from draining northwards (El-Shazly and Shata 1969). Similar deposits were found in two small basins (Hatawa and Sidi Salim), closely behind the South Matruh ridge. The **second Depression in Umm El-Rakham** area is called El-Matareih Depression and occupies a small area in the eastern extremity of this region between El-Matareih ridge and Sidi Salim ridge and the plateau escarpment (El-Sanussi and Shata 1969). As in other Depressions, reddish brown calcareous loam, derived from the parent material of the surrounding ridges and the plateau, cover its surface.

The third Depression in El-Omaid area in the Arab Gulf region lies between the third (Maryut) and fourth (Khashm El-Aish) ridges. This Depression is covered by a succession of soil, calcareous brownish, friable soil with a thickness between a few centimeters to 6 m, sandy limestone up to 10.3 m, white-to-grayish gypsum (0.2–4.8 m). Marl also exists (0.3–2.7 m) and includes black carbonaceous matter, and fossillifereous limestone (25.6 m thick) (Abd El-Aziz et al. 1971). **The third Depression in Umm El-Rakham region** is called "Sidi Salim Plain," with an elevation 15–50 m above sea level. It lies between Sidi Salim ridge (+40 m) in the north and Ailet Housain ridge (+100 m) in the south. This plain has a rocky surface, veneered with a thin fluvial cover of gravel mixed with calcareous loam (El-Sanussi and Shata 1969). Shallow gullies of south–north orientation dissect its surface.

20.4.2 Sabkhas and Salt Marshes

These two forms cover extensive areas in both the first and second Depressions either separately or around the lagoons. Areas devoid of these two forms along the coastal plain are characterized by the absence of inter-ridges Depressions, such as the area centered by Ras Abu Girab, west of El-Dabaa city. Sabkhas are transformed into salt marshes in some localities when shallow water covers the surface of the ground permanently, which in turn allows permanent growth of some halophytes. Lagoons, sabkhas, and salt marshes witness significant seasonal changes. Due to human interference, sabkhas and marshes have disappeared along extensive tracts in the first inter-ridge Depression by the construction of tourist villages along the Mediterranean coast.

Topographic maps (scale 1:25,000 and 1:100,000) show that the largest sabkhas developed in the Sallum Basin in the first and the second inter-ridges Depressions. The first sabkha, which developed in the first inter-ridge Depression, extends for about 38 km from east to west, whereas the second sabkha extends for about 18 km. Both sabkhas have a width varying between 1.0 and 2.0 km, and are 1–2 m below/above sea level.

20.4.3 Lagoons and Salt Lakes

As called on topographic maps, lagoons or salt lakes occupy small basins in the inter-ridges Depressions. As is the case with sabkhas, only few studies were carried out on lagoons (El-Shazly and Shata 1969; Sundborg and Nilsson 1985; Moussa et al. 1998). Yet, the studied lagoons can be taken as examples of this feature. Lagoons are restricted to a few localities along the first and second inter-ridge Depression. Permanent lagoons/lakes can be spotted on topographic maps in three localities. The first locality is the Borg El-Arab area, where the Lake Maryut extends in the second Depression, taking the form of an arm. The second one is the El-Alamain area where several small lakes developed from

El-Alamain to Dabaa locality. The third locality is the Matruh area where three permanent lakes are linked to the Mediterranean Sea.

The largest permanent water body is the western arm of Lake Maryut. It occupies the extreme eastern part of the second Depression between Abu Sir Ridge and Gabal Maryut ridge. This arm extends for about 50 km, with a varying width of 0.5–5.0 km, and ends up at a point to the west of El-Hammam city, lying 0.5–2.0 m below sea level. Due to the hyper-salinity of water, this arm is called Mallahet Maryut and is used for salt extraction in several localities. Water level changes significantly from season to season, transforming the margins of the arm into sabkhas with salt crust. In this arm, 50 small (0.1–3 km long, 0.1–0.6 km wide) shallow (4–11 m) aeolianite islands project from the water surface. These islands are aligned in two lines paralleling the main ridges to the north and south of the lake. Such characteristics are (Butzer 1960) a sign of some drowned sections of the second ridge in this region (see Fig. 20.2).

In the Alamain area, small lakes developed along a distance of about 45 km in the first inter-ridge Depression from the surroundings of Alamain city in the east to Ras El-Dabaa in the west. Two of these salt lakes were studied recently (Sundborg and Nilsson 1985; Moussa et al. 1998). The first lagoon is composed of two basins with a total length of about 3.0 km, varying width of 300–700 m and a depth of 3.0 m (Moussa et al. 1998). Isolated hollows with depths of 8.0 m were also found in the lagoon bed (Sundborg and Nilsson 1985). The second lagoon is smaller and shallower, taking a circular form with a diameter of 400 m, and a depth of about 120 cm. Regarding the water level in the two lakes, it is nearly constant since evaporation from their surfaces is compensated for by the inflow of marine water (Moussa et al. 1998). The lagoons are surrounded by sabkhas from three sides as for the northern side; the lake abuts the foot slope of the coastal ridge. Despite the fact that the ionic ratios indicate that the lagoonal water is of marine origin, the salinity of water is about 2.5–3 times more than that of seawater. Analysis of sediment samples collected from lagoon bed showed that the sediments are composed of a mixture of sand (56%), silt (28%) and clay (16%). The shallowness of the small lagoon made it possible to measure sediment thickness, which is about 180 cm, whereas this was impossible in the large deeper lake (Moussa et al. 1998). In relation to sand composition, coarse sands are composed mainly of mollusk shells, ooids, aggregate grains and plant detritus, whereas medium sands are made of biogenic fragments, smaller aggregates, foraminiferal tests, ostracod shells and quartz grains. The sediments are composed of silicates (25–40%) and carbonates (60–75%). The mineral association of the non-silicate fraction comprises mainly aragonite (29–37%), calcite (32–35%) and dolomite (23–25%), (Moussa et al. 1998).

The Matruh salt lakes are formed of three interconnected basins, of which only the middle one is directly connected – through a wide opening – (1.25 km) to the Mediterranean Sea (Fig. 20.4). The three basins extend for about 7.0 km from east to west. The lakes drop in size from west to east. The western lake has a length of 4 km and a maximum width of 1.5 km. The middle one (Matruh lake) has a length of 2.0 km and a maximum width of 1.25 km. The eastern lake has a length of 1.0 km and a maximum width of 0.75 km. These three lakes are—in fact—the only ones of a series of eight lakes lying between the first and second ridges in Matruh area, which are directly connected to the Mediterranean Sea. Other lakes lying to the east of Matruh lakes are small and shallow and were transformed into

Fig. 20.4 Lakes of Marsa Matruh (*Source* Google Earth 2016)

saltpans to produce rock salt some time ago. At present, they are no longer used for salt production and are changed gradually into marshy tracts where salt tolerant plants grow (El-Shazly and Shata 1969). Depth of Matruh lakes is unknown except for the Matruh Lake, which is 5.0 m (El-Shazly and Shata 1969). The basins of the Matruh lakes may have a karstic origin since the bed of the middle lake is expected to reach the bedrock (Miocene?), which is composed mainly of limestone.

Lagoonal deposits (banded gypsum inter-bedded with dismicrites and clayey dismicrites) were found in the area from El-Gharbaniyat to El-Alamain and are indicative of the development of previous lakes in the first inter-ridge Depression (Hilmy et al. 1978). Previous studies (Abdalla 1967; Hilmy et al. 1978) concluded that these ancient "lagoonal" conditions prevailed during the mid-Pleistocene period which was characterized by climatic variations from hot/dry (gypsum development) to cold/wet (development of inter-bedded layers of dismicrites and clayey dismicrites). In the third Depression, near El-Hammam city, seven beds of gypsum were identified alternating with seven beds of clay in a quarry of gypsum, suggestive of seven periods of aridity and seven periods of wetness in between (Shukri et al. 1956).

20.5 Origin and Mode of Development of the Ridge-Depression System

The origin and mode of development of the oolitic limestone ridges of the Mediterranean Coastal Plain are a matter of a great controversy. This controversy is mostly due to lack of data or due to the misinterpretation of available data. From these conclusions, it can be argued that there are three possible modes of development: aeolian, marine and polygenetic.

Several scientists have preferred the possibility of **the aeolian origin**. Hume (1925) ascribed a dune origin to the consolidated parts of the ridges formed by the prevailing north-northwesterly winds. Ball (1939) concluded that the oolitic ridges certainly originated from the consolidation of ancient coastal sand dunes. Sandford and Arkell (1939) concluded from their discussion of the origin of the ridges that they are formed by wind action along a receding shoreline. Hilmy (1951) supported Ball's view on the non-marine origin of the ridges. Paver and Pretorius (1954), after examining the first and second ridges in Burg El-Arab area, considered these two ridges to represent old, consolidated dunes, the materials of which were derived from beach deposits.

Several other researchers concluded that these ridges developed by **marine and coastal processes**. Blanckenhorn (1901) and Butzer (1960) ascribed these ridges to a marine origin and considered them to represent bars cutting off lagoons, and that dune sand overlies the near-shore (first) ridge. Butzer (1960) considered the ridges as stretches of consolidated coastal bars separating ancient lagoon floors. Shukri et al. (1956) concluded that the ridges represent bars or spits and the Depressions separating the ridges represent lagoons in which typical lagoon deposits are present. The former conclusions depend on the following evidence:

1. The uniform extension, direction and elevation of the ridges (except in a few localities);
2. The presence of typical true oolites formed in situ;
3. The presence of Pleistocene foraminifera;
4. The presence of a conglomerate layer on top of the fourth ridge (Khashm El-Aish);
5. The presence of lagoonal deposits in the Depressions separating the ridges; and
6. The presence of several layers of gypsum in the lagoon separating the fourth and fifth ridges.

Attia (1954) concluded from the analysis of the materials of the subsurface ridges in NW Nile Delta that they are of littoral origin. Butzer (1960) suggested that the ridges and Depressions represent offshore bars and lagoons, which were formed due to sea transgressions during the Pleistocene. Based on the existence of a shelly limestone layer at the base of the first ridge in Matruh area, El-Sanussi and Shata (1969) suggested that this ridge originated as an ancient littoral dune along a receding shoreline. Ismail and Slim (1969) found that the characteristics of the microfacies associations of the three ridges in the Sallum area indicate that these ridges were deposited under agitated, shallow, littoral conditions receiving abundant terrigenous material. Selim (1974) concluded from the study of the limestone petrography of the first three ridges in the Sallum Basin that they originated as coastal barriers. Stanley and Hamza (1992) suggested that first ridge has likely been accumulated in a shallow, high-energy marine environment receiving a very low terrigenous influx. It was suggested that this first ridge developed under similar environment to that prevailing at present along the coast of west of Alexandria (Cherif et al. 1988).

Polygenetic mode of development -emerges from recent detailed studies on some ridges and in some localities. Shata (1971) suggested that these ridges have a marine and terrestrial origin, and presumably mark the position of the receding Mediterranean coast during the Pleistocene and Holocene periods. Hilmy et al. (1978) concluded from the study of the false bedding of rock exposures that the ridges developed in a littoral-continental environment of deposition. This environment was littoral with high-energy index during the first stages of deposition, and changed into continental environment during the next stages of deposition.

Sandburg and Nilsson (1985) also concluded that although a marine origin has been recognized and accepted, aeolian processes have certainly been active. Therefore, they explain that the ridges consist of at least two elements: a basal marine section formed by wave and current action as offshore bars and an upper part of aeolian origin. This conclusion was also confirmed by Moawad (2003) in some parts of the coastal ridge of Ras El-Hekma. Moawad (2003) revealed that the inter-structure of the consolidated aeolian dunes differs from that of the coastal ridge, and recognized cut-and-fill basins that indicate the action of wind erosion and deposition. Hassan et al. (1986) concluded, from the study of the sedimentological sequence exposed in the Bahig cut of Gabal Maryut (second) ridge that it represents a succession of cycles commencing with a hemicycle linked to marine transgression and followed by a terrestrial hemicycle. The formation of aeolianites and associated with sea regression characterize the latter cycle. El-Asmar (1991) observed the presence of sedimentary littoral cycles in which marine deposits alternate with aeolianites and palaeosols. Wali et al. (1994) found that the sedimentary and diagenetic features of the ridges, suggest aeolian modification of submarine bars accompanied by complex successive marine, vadose and phreatic diagenetic phases. Therefore, they concluded that the exposed ridges are composed of a lower sub-aqueous phase overlain by an upper aeolian phase. They also found that tidal channels with reverse graded bedding separate the two stages; representing a transitional period between submerged wave-accumulated sediments and sub-aerial wind-drifted movement. Stanley and Hamza (1992) found that petrographic attributes of the second ridge, (Abu Sir) indicate that most of this ridge is originally deposited under shallow marine conditions and is, as denoted by calcrete crusts, intermittently exposed. Wali et al. (1994) also suggested that the facies criteria of the third ridge (Gabal Maryut) signals its formation in shallow sub-tidal zone, with periodic emergence.

In conclusion, the vertical and horizontal facies distributions and their inferred environments suggest that the ridges represent bar and perhaps beach complexes, which were formed during sea level high-stands. Each ridge may have formed sequentially as sea level dropped or remained constant. As a result of marine regression, a phase of sub-aerial exposure followed. During this regressive period, wind transported carbonate ooids and bioclastic fragments from the shore onto the platform interior. Even though most of the material in the ridge is of marine carbonates and bioclastics, and the inland terrigenous materials also contributed to local variations in ridge composition. It can be also claimed that the terrigenous materials of the ridges are of local origin from the southern plateau. Only the extreme eastern margins of some ridges (mainly the second and third) are of outside origin from the Nile sediments. On the other side, while marine, littoral and aeolian were the main responsible processes for the development of the ridges, fluvial action, soil-forming processes and weathering were secondary factors. In the meantime, the successive marine regressions were a major element in the development of the successive series of ridges along the Mediterranean coast.

20.6 Age

Based on the available data and techniques for age dating, age of the ridges suggested by previous studies is summarized in Fig. (20.3) and Table (20.2).

Table 20.2 Age of the coastal ridges according to previous studies

Ridge Number	Age	
Submarine	500 yr. bp for the top layer of a submarine bar, 1900 yr. BP for offshore skeletal sand (Sundborg and Nilsson 1985), 800 yr. for the oolitic material of submarine ridges (Lindell et al. 1991).	
1	0.6 ± 0.1 and 1.5 ± 0.2 ka for the top layers at El-Omayid, 3680 ± 40 and 4355 ± 40 BP for lower layers at Burg El-Arab (El-Asmar and Wood 2000), 3.0 ka for coastal dune oolite (Sundborg and Nilsson 1985), post Greco-Roman (2.0 ka) for the top layers at El-Agami (Mahmoud 1995).	Ridges 1–3: Upper Pleistocene: Monasterian from 130 ka to Recent (Wali et al. 1994).
2	Paleosols at the northern flanks of the ridge: 67 ± 31 ka (El-Asmar and Wood 2000).Oolitic Limestone: 90 ± 15 ka near Burg El-Arab and 110 ± 5 ka at Abu Sir Castle (El-Asmar 1994a, b) and 104 ± 17 ka near Burg El-Arab (El-Asmar and Wood 2000).	
3	Aeolianites at Bahig drain cut: 191 ± 42 ka for top sample, 416 ± 255 and 454 ± 151 ka for two bottom samples, Marine beds at the same locality: Moluscan Limestone: 208 ± 59 ka, Foraminiferal Limestone: 546 ± 352 ka, Beach boulders: 423 ± 153 ka, Cardium Limestone: 292 ± 48 ka (El-Asmar and Wood 2000).	
4	Upper paleosols: 280 ± 75 ka (El-Asmar 1991). The whole ridge: Milazzian from 300 to 70 ka (Wali et al. 1994).	Ridges 4–8: Lower Pleistocene (Butzer 1960). Ridge 5: Millazian (Shukri et al. 1956) Ridges 6–8: Sicilian (Shukri et al. 1956)
5, 6, 7, 8	Lower Pleistocene: Sicilian from 950 to 900 ka (Wali et al. 1994).	

The petrographic characteristics and age of the submarine ridges in the Arab Gulf indicate that they consist of recent oolitic deposits and that oolitization is a still ongoing process (Sundborg and Nilsson 1985; Lindell et al. 1991). This recent contribution of oolitic deposits holds also for the coastal (first) ridge, since the age of the top layer is 0.6–0.1 ka. This finding as indicated by El-Asmar and Wood (2000) had not previously been evident, with most researchers assuming that oolitic grains had not been produced since the beginning of the Holocene. In the meantime, the fauna of the lower part of the coastal ridge suggests similar conditions to those prevailing nowadays along the coasts of Alexandria (Cherif et al. 1988). The age of the submarine ridges samples and the first ridge at Burg El-Arab was determined by radiocarbon method (Sundborg and Nilsson 1985; Lindell et al. 1991; El-Asmar and Wood 2000). As for the samples of the first ridge at El-Omayid, they were determined by the OSL method (Optically Stimulated Luminescence) (El-Asmar and Wood 2000). In addition, OSL was used in age determination of the samples from the second and third ridges, except for the oolitic limestone of the second ridge, which was analyzed using U-series method. This latter method was used to determine the age of the upper paleosols of the fourth ridge (El-Asmar and Wood 2000). Depending on their height, and supported by the fauna and foraminifera, Butzer (1960) considered that ridges four to eight lie within the range of the Lower Pleistocene eustatic level. The Millazian age was assumed for ridge five while the Sicilian for ridges six to eight, as suggested by Shukri et al. (1956). Such suggestion is based on faunal evidence. To note, Wali et al. (1994) adopted the age estimation of Shukri et al. (1956), with additions of equivalent dates for age periods according to previous studies conducted in the Mediterranean Basin. Only one study (Mahmoud 1995), used archaeological materials (Greco-Roman potsherds), in estimating the age of the top layers of the coastal ridge concluding that they developed during the last 2000 years.

20.7 Climatic Significance

Due to the rise in the thickness of aeolianites and the fall in marine deposits in the sections from Gabal Maryut ridge to the coastal ridge, it was inferred that there is a general tendency toward increasing aridity by time (El-Asmar 1994a). This conclusion is confirmed by the increase in the isotope values for both oxygen and carbon because of the decrease in the effect of fresh water diagenetic processes as well as the reduction in the intensity of pedogenesis from Gabal Maryut to the Coastal ridge (El-Asmar 1994a). Analysis of the samples, those representing the rock units constituting the first three ridges, made it possible to recognize the following six climatic phases:

a. The first youngest arid phase was about 5000–3000 years BP (Holocene), when the sea started to attain its present level and the climate was warm and dry (El-Asmar 1994a; Sundborg and Nilsson 1985).
b. A warm-wet phase characterized Early Holocene (ended 8000 years BP). During this phase, thick reddish brown geosol horizons were developed.
c. The second arid phase had an age of 90 ± 15 ka (isotope sub-stage 5c/5a) during which the aeolianites of El-Max –Abu Sir ridge and the aeolianites at the top of Gabal Maryut Ridge were deposited. The study of the foraminiferal fauna from the Gabal Marytut (El-Amriya) ridge indicated that this ridge was laid down under warmer conditions than those prevailing nowadays in the Mediterranean (Cherif et al. 1988; Shukri et al. 1956). A surface of unconformity separates the aeolianites of this arid phase from the upper geosol horizons from the following wet phase.
d. A fourth complex phase, where alternations of warm/wet to warm/dry conditions prevailed. This phase is associated with the deposition of the last interglacial marine deposits (125 ka), interglacial palaeosols and the aeolianites of Gabal Maryut ridge.
e. An arid phase followed the marine transgression, which led to the deposition of the basal *cardium* limestone with an average age of 300 ka.

When these deduced climatic phases were correlated with those deduced from the study of the playas in Egypt (Embabi 1999), it was found that there are some similarities especially in those of Late Pleistocene and Holocene. Besides the presence of two (wet and arid) major climatic phases during the Holocene, the correlation confirms the presence of a long arid phase that prevailed during Late Pleistocene in the whole Western Desert, despite the difference in the beginning of this period. This beginning is 90 ± 15 ka along the coast of the Mediterranean, whereas it is 60 ka in the southern part of the Desert (Wendorf et al. 1993).

20.8 Environmental Conditions of Ridge-Depression Development

Based on the available data and the conclusions made by the previous studies, it can be said that environmental conditions were favorable for the development of Ridge-Depression System, and these conditions are still exist. The formation of this system requires the following environmental conditions:

a. Calcium carbonate saturates the surface seawater. Marine sedimentation of calcium carbonate is also expected because of the large additions of calcium from the bedrock (Selim 1974; Sundborg and Nilsson 1985).
b. The detrital nuclei, which provides the initial surface on which aragonite precipitates, is available. In the

submarine area of the Arab Gulf, two fundamentally different types of calcareous sands prevail (Sundborg and Nilsson 1985). Skeletal sand—consisting of shells and skeletons of various calcareous organisms, such as molluscs, echinoderms, bryozoans, coralline algae and benthic foraminifera—dominates the seaward area from a depth of about 15 m and probably continues out to the shelf break. Oolitic sand, consisting of spherical grains with diameters of 0.3–1.0 mm, exists in a 5-km broad zone along the shore, to a maximum depth of about 10 m. The depth interval 10–15 m is a zone of transition. This condition is expected to dominate the whole coast of the Mediterranean of the Western Desert with local variations, since the Ridge-Depression System prevails along the whole area.

c. Agitation of the grains, which prevents the detrital grains from connecting to the sea floor and ensures that the whole surface of the grains is exposed to the saturated water (Selim 1974). The strongest winds blow from west, northwest and north mainly during winter and spring (El-Fiky 1999; Sundborg and Nilsson 1985). Wave measurements show a strong dominance from west, northwest and north with a significant peak from northwest (Sundborg and Nilsson 1985). Such winds and wave conditions are responsible for the agitation of grains.
d. Three types of microfacies were distinguished in the oolitic limestone (Pleistocene) in Umm El-Rakham area (El-Sanussi and Shata 1969). The first type is essentially composed of true oolitic grains mixed with varying quantities of superficial and pseudo-oolitic grains. The second type is composed of pellets, oolitic grains and shell fragments. The third type is essentially composed of fragments of oolitic detrital particles and shell fragments. The characteristics of these microfacies suggest shallow and warm marine conditions, with strong tidal currents, which become then affected (land phase) by extensive amounts of mud and detrital material carried out into the sea by strong flows from wadis (El-Sanussi and Shata 1969). Similar depositional conditions are inferred from the study of microfacies of the first three ridges in the Sallum area, and from the subsurface borings in NW Nile Delta. In those areas, the characteristics of the lithofacies and biofacies indicate that deposition occurred under shallow, agitated, and littoral marine environment that received abundant terrigenous material (Attia 1954; Ismail and Selim 1969).
e. Not all-marine environments were favorable for oolitic growth. This was exemplified by the formation of the third ridge in the Sallum basin. The petrographic characteristics suggested that the grains forming this ridge were derived from the nearby limestone outcrop and accumulated in shallow water environment, which was not favorable for oolitic development (Selim 1974). The same holds for the *Cardium* limestone, which forms the basal unit of Gabal Maryut (third) ridge in Borg El-Arab area.
f. In spite of the marine origin of the oolitic grains, it was found that some horizons in the ridges were deposited by wind under arid conditions. Paver and Pretorius (1954) suggested that the marine oolitic grains were washed into the beach and then transported by wind to form the present coastal dunes. Likewise, the first and second ridges formed in a similar manner, as suggested. This mode of development is proposed by El-Asmar (1998) for the aeolianite layers in the coastal Abu Sir and Gabal Maryut ridges.
g. Local disparity in micro-fauna, petrographic characteristics and depositional facies indicate some variations in local environmental conditions. The foraminiferal fauna of the Abu Sir ridge reveals that this ridge was laid down in front of a coast receiving a considerable amount of fresh water (Cherif et al. 1988). The source of such fresh water was either the old Nile Branch of the Nile Delta, which outstretched to the area of the ridge (some 50 km west of Alexandria) or rainfall, since the fauna denotes a pluvial (Abbasiya of Said 1981) phase (Cherif et al. 1988). Conglomerate intercalations in some parts of the first ridge connote intermittent fluvial deposition during the development of the ridges. Relative high percentage of quartz grains in the eastern margins of the second and third ridges indicates that the Nile contributed to the development of the eastern parts of some of the coastal ridges (Stanley and Hamza 1992).

References

Abd El-Aziz M, Gabra S, Ahmed S (1971) El-Omaid gypsum deposits. Ann Geol Survey Egypt 1:111–116

Abdallah AM (1967) Geology of some gypsum deposits in north western of Egypt, General Egyptian Organization for Geol. Research and Mining, Geological Survey Paper No. 41, Cairo

Abdel-Hady YE (1994) Role of geophysics in earth sciences with special emphasis on earthquake seismology. Bull. Institut d'Égypte 72–73:87–130

Attia MI (1954) Deposits in the Nile Valley and the Delta. Government Press, Cairo, Geol. Survey of Egypt

Ball J (1939) Contribution to the Geography of Egypt. Survey and Mines Department, Ministry of Finance Egypt, Government Press, Cairo

Blanckenhorn M (1901) Neues zur Geologie und Palaeontologie Aegyptens: III: Das Miozän. Zeitschrift der Deutschen Geologischen Gesellschaft 53:52–132

Butzer K (1960) On the Pleistocene shoelines of the Arab Gulf of Egypt. Jour Geology 69:626–637

Cherif O, El-Sheikh H, Labib F (1988) The foraminifera of the Pleistocene coastal ridges of the area to the west of Alexandria. Egypt. Revue de Paléobiologie 2:735–740

Conoco Coral and Egyptian General Petroleum Corporation (1986) The Geological Map of Egypt, 20 sheets, scale 1: 500 000, Cairo

El-Asmar H (1991) Old shorelines of the Mediterranean coastal zone of Egypt in relation with sea level changes. Dissertation. El-Mansoura University, Egypt

El-Asmar H (1992) Recognition of diagenetic environments of the Pleistocene carbonate rocks at El-Hammam, NW Mediterranean coast of Egypt using stable isotopes and electron microprobe

El-Asmar H (1994a) Aeolianite sedimentation along the northwestern coast of Egypt: evidence for Middle to Late Quaternary aridity. Quatern Sci Rev 13:699–708

El-Asmar H (1994b) Recognition of diagenetic environments of the Pleistocene carbonate rocks at El-Hammam, NW Mediterranean coast of Egypt using stable isotopes and electron microprobe. N Jb Geol Paläont Mh 1:7–22

El-Asmar H (1998) Middle and Late Quaternary palaeoclimatic evolution, northern Mediterranean coast of Egypt. In: Alsharhan S, Glennie K, Whittle G, Kendall C, Balkema A (eds) in Quaternary Deserts and Climatic. Rotterdam, Proc. Inter. Conf. Quaternary Deserts and Climatic Changes, al-Ain, UAE, 9–11 Dec. 1995, pp 261–271

El-Asmar H, Wood P (2000) Quaternary shoreline development: the northwestern coast of Egypt. Quaternary Sci Rev 19:1137–1149

El-Demerdash A, Abdelaal ShI, Fathy A, Kassim Y (1978) Mineralogy of the sand refraction criterion for profile uniformity and development in the western coastal littoral A.R.E. Bull Inst Desert 28–1:177–191

El-Fiki A (1999) Wind conditions in Egypt: a climatic study. Dissertation, Ain Shams University

El-Sanussi MY, Shata A (1969) Geomorphological and morphological aspects of Umm El Rakhem area. The Desert Inst. Bull., U.A.R.19: 31–62

El-Shahat A, Ahmed M, Hegab OA, El-Hefnawi MA (1987) Petrography and geochemistry of some calcretes from the northwestern coastal plain of Egypt. Egypt Jour Geol. 3:161–180

El-Shami M. El-Shazly A, Shata A (1969) Contribution to the geology of El-Dabaa area, Western Mediterranean littoral zone. Bulletin of the Desert Institute, U.A.R 19: 63-95

El-Shazly M, Shata A (1969) Geomorphology and pedology of Mersa Matruh area, Western Mediterranean littoral zone. Bull Desert Inst Egypt 19:1–29

Embabi N S (1999). Playas of the Western Desert, Egypt. In J. Donner (Ed.), Studies of playas in the Western Desert of Egypt. Academia Scientiarum Fennicae, 160: 5–47. Helsinki: Academia Scientiarum Fennica

Fourteau R (1915) Contribution a L'étude des depôt Nilotique, Mem. Institut d'Égypte, Le Caire 8:56–94

Hassan FA, Hegab O, El Shahat A (1986) Mediterranean littoral cycles, west Alexandria Egypt and implications for archeological exploration. Nyame Akuma 27:3–5

Hassouba H (1980) The occurrence of palygorskite in Quaternary sediments of the coastal plain of north-west Egypt. Clay Miner 77–83

Hegab OA, El-Asmar HM (1995) Last interglacial stratigraphy in the Burg El Arab region northwestern coast of Egypt. Quatern Int 29 (30):23–30

Hegab OA, El-Shahat A, Ahmad MI, El-Nagar M (1987) Clay Mineralogy of some Quaternary Calcretes in Burg El-Arab area, northwestern coastal zone. Egypt Egypt Jour Geol 31:181–196

Hilmy ME (1951) Beach sands of the Mediterranean coast in Egypt. Journal of Sedimentary Petrology 21:109–120

Hilmy ME, El-Shazly MM, Korany EA (1978) Lithostratigraphy and petrology of the Miocene and post-Miocene sediments in Borg El-Arab-El-Dabaa area. Bulletin of the Desert Institute 28:1–24

Hilmy ME, Ezz M, Ali Y, Osman A (1989) Geologic occurrence and composition of the sediments of El-Alamain lagoons along the western side of the Mediterranean coastal zone of Egypt. Bull Nat Inst Oceanography and Fisheries, Egypt 15:35–75

Hume W (1925) Geology of Egypt. Vol 1, Survey of Egypt, Government Press, Ministry of Finance, Cairo

Ismail M, Slim A (1969) Stratigraphy of the Sallum area, Western Desert. U.A.R. Bull Faculty of Science, Alexandria University 330–9:309

Lindell L, Alexanderson E, Norman J (1991) Satellite mapping of oolitic ridges in Arab Gulf. Egypt Geocarto Int 1:49–59

Mahmoud A (1995) Evidence of the Late Holocene subsidence, Northwest of Alexandria, Mediterranean Coast. Egypt Bull Soc Géogr d'Égypte 68:99–112

Moawad M (2003) The Geomorphology of the coastal Zone between Ras Abu Girab and Ras Alam El-Roum, the NW Coast of Egypt. Dissertation, Ain Shams University

Moussa A, Shata M, Draz S, El-Sayed M, El-Sabroutti M (1998) Carbonate sedimentation in coastal hypersaline lagoons of Alamain, Egypt. In: Alsharhan S, Glennie K, Whittle G, Kendall C, Balkema A (eds) in Quaternary Deserts and Climatic. Rotterdam, Proc. Inter. Conf. Quaternary Deserts and Climatic Changes, Al-Ain, UAE, 9–11 Dec. 1995, pp 43–56

Paver G, Pretorius D (1954) Report on reconnaissance hydrogeological investigations in the Western Desert coastal zone. Publication de L'Institute du Desert d'Égypte 5: 145p.

Rashed MA (1998) Calcrete on Pleistocene coastal ridges, west Alexandria. Egypt Sedimentology of Egypt 6:113–128

Said R (1981) Geological Evolution of the River Nile. Springer-Verlag, New York

Sandford K, Arkel W (1939) Paleolithic man and the Nile Valley in Lower Egypt. Chicago University Press. Oriental Institute Publication 36

Said R, Philip G, Shukri N M (1956) Post-Tyrhenian climatic fluctuations in Northern Egypt. Quaternaria 3: 167172

Selim AA (1974) Origin and lithification of the Pleistocene carbonates of the Salum area, western coastal plain of Egypt. Journ. Sedimentary Petrol 44:70–78

Sestini G (1989) Nile Delta: a review of depositional environment and geologic history, deltas: sites and traps for fossil fuel. In: Whateley MIG, Pickering KT (eds) Geological Society of America. Special Publication 41:99–127

Shata A (1971) The geomorphology, pedology and hydrology of the Mediterranean coastal desert of U.A.R. In: C. Gray (ed), Proc. Symposium on the Geology of Libya, Tripoli, April 14–19, 1969, University of Libya, Tripoli, pp 431–346

Shata M (2000) Sedimentological and mineralogical characteristics of subsurface sediments in Burg El-Arab Region, NW coast of Egypt. Bull Nat Inst Ocean Fisheries 26:1–26

Shukri NM, Philip G, Said R (1956) The geology of the Mediterranean coast between Rosetta and Bardia. II Pleistocene Sediments: Geomorphology and Microfacies 37:395–427

Stanley DJ, Hamza F (1992) Terrigenous-carbonate sedimen interface (late Quaternary) along the northwestern margin of the Nile Delta. J Coastal Res 8–1:153–171

Sundborg A, Nilsson B (eds) (1985) Qattara Hydrosolar Power Project: environmental assessment. UNGI Report No. 62, Uppsala University, Department of Physical Geography, Uppsala

The Geological Survey of Egypt (1986) Geologic Maps, scale 1:100 000 for Ras El-Dabaa, Ras Abu Girab, Ras El-Siyadah, Cairo

Wali A, Brookfield M, Schreiber B (1994) The depositional diagenetic evolution of the coastal ridges of NW Egypt. Sedimentary Geol 90:113–136

Wendorf F, Schild R, Close A, Associates (1993) Egypt during the last interglacial: the Middle Palaeolithic of Bir Tarfawi and Bir Sahara East. Plenium Press, New York

Red Sea Coastal Forms

21

Abstract

Red Sea coasts exhibit a wide variety of forms of different origins—changes in sea level have played a significant role in the development of coastal forms. Coastal forms of the Red Sea in Egypt are classified into four groups according to origin: fluvial, aeolian, marine, and sea level changes. Fluvial forms comprise micro-deltas, bahadas, and drainage channels. As for aeolian forms, they are dominated by nebkhas, while marine forms include coral reefs, sand barriers, barrier lagoons, beaches, cliffs, and platforms, as well as sabkhas. Forms resulting from sea level changes comprise estuaries, submerged deltas, submerged canyons, submarine terraces, and reef terraces. Most of these forms are small or micro-ones, since their size ranges between a few kilometers to a few hundred meters.

Keywords

Egypt • Red Sea • Micro-deltas • Coral reefs • Sabkhas • Salinas • Lagoons • Sand barriers • Reef terraces • Submerged forms

21.1 Regional Setting

Along the Red Sea and Sinai coasts, a variety of forms are developed through several processes. These forms are classified into four groups according to their origin, as follows: fluvial, aeolian, marine, and forms resulting from sea level changes.

21.2 Fluvial Forms

Deltas and bahadas are the most widespread fluvial forms along the coasts of the Red Sea as well as the Gulfs of Suez and Aqaba. This is due to the development of a big number of wadis on the slopes of the Red Sea and Sinai Mountains and plateaus which carry runoff into the sea.

21.2.1 Deltas

Examining deltas in the Red Sea, this subsection will depend on the previously available studies with the aid of space images, especially Google Earth images. Microfroms represent the most distinctive feature of deltas in this area, though their area does not exceed a few tens of square kilometers, and many of them are even less than that. In the meantime, the area of some deltas reaches several tens of square kilometers, such as **Delta Wadi Gheweiba** (74.7 km^2) at the NW coasts of the Gulf of Suez (El-Disouki 1998), and Wadi Fieran Delta (48 km^2) on the eastern coast of the Gulf of Suez (Mostafa 1987). The geomorphic study of the Red Sea coast between Safaga and Quseir (88 km) has shown that there are 12 deltas with—areas varying greatly between 2.8 km^2 for Wadi Asala and 30.25 km^2 for Wadi

N.S. Embabi, *Landscapes and Landforms of Egypt*, World Geomorphological Landscapes,
https://doi.org/10.1007/978-3-319-65661-8_21

Safaga (Moawad 2008). When the area of deltas was correlated with that of drainage basins, a positive linear relationship between these two variables was found. An impediment in studying the deltas in the Red Sea coasts is that their exact number was not reported in any previous studies; it is also difficult to count that number on maps/images. This difficulty in attaining an exact number is due to the fact that some of the deltas are coalesced/modified because of human activities which resulted in the disappearance of their deltaic characteristics. The coast of Gulf of Aqaba (length 200 km) is shown as an example of the number of deltas along the Egyptian coast. When counted on Google Earth images, the estimated number was ~ 120 deltas.

Previous studies (Mostafa 1987; El-Disouki 1998; Moawad 2008) and field investigations in Marsa Alam area found that the deltaic sediments of the Red Sea are composed of two types: the first is composed of rock blocks, boulders, gravels, while the second type comprises coarse, medium and fine sands, clay, and silt. Even though the general trend is a decrease of rock blocks, boulders, and gravel in the downstream section and an increase in the percentage of fine particle, this trend cannot be applied to all sections in the deltas. As seen in the field in the area of Marsa Alam, sediments of these deltas exhibit poor sorting since they are composed of a mixture of fine and coarse materials without stratification. This stands in contrast to other localities where sediments with variable size show some stratification. Most of the gravels are angular and sub-angular. These characteristics reflect many factors including the change in the energy of water flow over their history of development and the short distance between the source areas in the adjacent mountains and depositional area where the deltas are formed.

Local forms developed on the surfaces of the deltas of Red Sea coast can be divided into two groups: (1) drainage channels and pro-deltas and (2) sabkhas, salinas and lagoons. In the first group, the channels bifurcate at the apex of the delta, taking a braided pattern, based on the direction of slope. They also carry and spread sediments and on the surface of the delta. Some of these channels reach the sea where they deposit their load. By time, a new form develops as an extension of the original delta taking the form of a pro-delta, but with different characteristics in size, shape, and sediment composition (Fig. 21.1). The size, shape and extension of pro-deltas into the sea depend on local conditions. As an example, Delta Wadi Watir in Sinai (Fig. 21.1) exhibits three different types of pro-deltas. The first one is composed of three rather small deltas deposited by three independent wadis in SW of the main delta. Their morphologic aspects show that these small deltas developed contemporaneously with the main delta. The second type of pro-delta developed in Wadi Watir itself along the coastal front of the main delta, whereas the third type is composed of extremely small pro-deltas developed along the coastal margins of the main delta. These latter are the most recent ones and are probably developed due to powerful flash flood (s) that take place under the present climatic conditions.

The second group of forms that characterize the deltas in the Red Sea coasts comprises three forms: sabkhas, **salinas,** and lagoons. These three forms may develop side by side. Lagoons are permanent water bodies because they are connected to the open sea. Due to sea water seepage, the surrounding tracts of the lagoons are saturated with sea water and by time are transformed into sabkhas. Water movement in both direction from and to the sea and evaporation of water from the surface of the sabkhas result in an increase in water salinity to the degree that salt is precipitated on the sabkha surface, forming a temporary/permanent salt crust. In areas where water exchange between sea/lagoons with sabkha is low and evaporation rates are relatively high, thickness of the surface salt crust increases by time and becomes a characteristic feature of the sabkha surface. At this stage the sabkha is transformed into salina. This transformation between these forms may occur in a different way. Lagoons may be transformed into sabkhas if their connection with the open sea is interrupted and recharging water does not compensate the evaporated water. In this case, the lagoon becomes a sabkha, and some of its floor will be a salina. Such a case occurred in Bardawil Lake at a time when water recharge was not sufficient to compensate the evaporated water from the lake body and was transformed into sabkha. Then after making two new artificial openings, water recharge from the Mediterranean Sea was resumed and Bardawil became a lagoon again (see Chap. 22).

Of these three forms, sabkhas are the most widespread along the Red Sea coast and are the ones that occupy the widest areas when compared to the other two forms. This type of sabkha is known in the literature as coastal sabkha since it lies on the coastal fringes of the deltas and is fed by seawater. Examples of relatively large sabkhas are the one along the coastal margins of Wadi Gheweiba on the western side of the Gulf of Suez, which extends 10 km from north to south, with the varying width of 100–1000 m (El-Desouki 1998). The second example is sabkhas at the delta of Wadi Ed-dir. This sabkha measures 10 km from north to south, with a width of 4 km. The third example is the sabkha of the delta of Wadi Abu Shar, which lies 22 km to the north of Hurghada. It has N-S length of 8 km and a varying width of 1.5–5.0 km. The fourth example of coastal sabkha lies to the north of Abu Bakr headland. Its length is 10 km and its width is 1 km. This sabkha is separated from the coastline by a sand beach, which resembles a sand barrier connected to the mainland by marine deposition.

Although it is not a deltaic feature, **El-Mallaha**, which means "the salina," is worth mentioning in this section

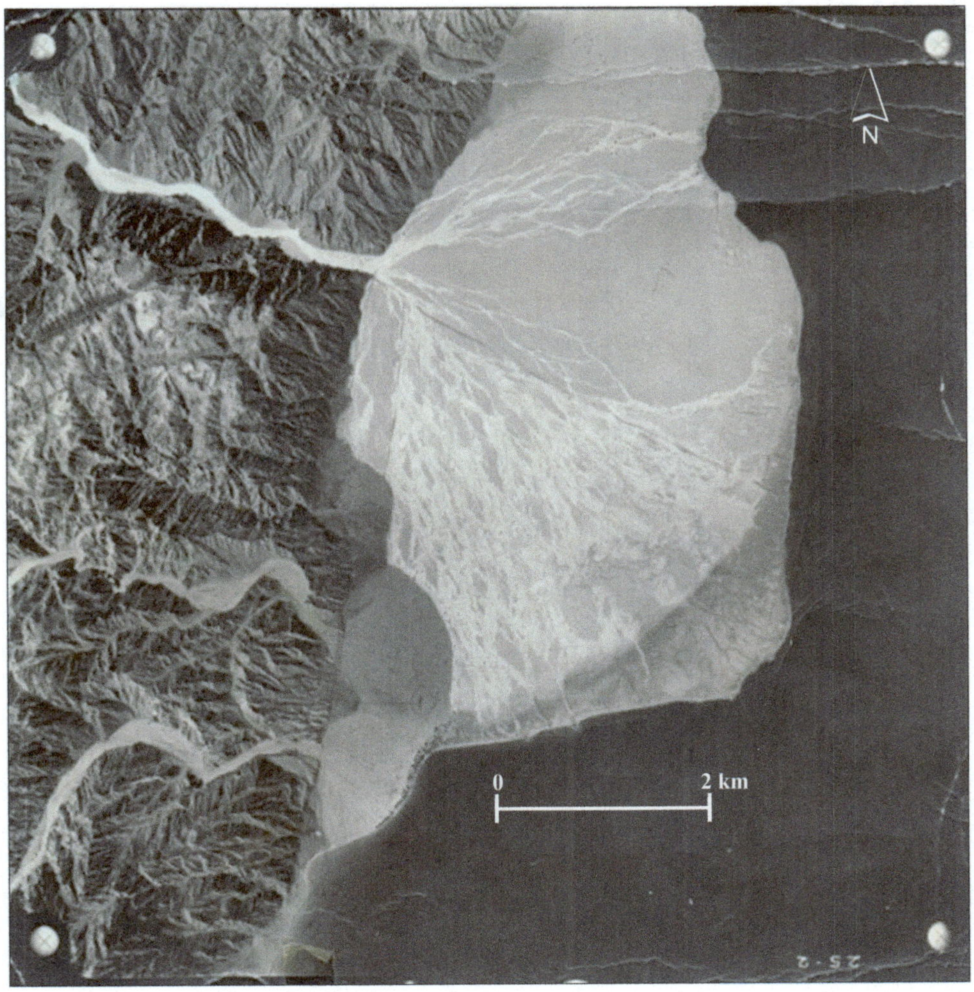

Fig. 21.1 An example of pro-deltas along the delta of Wadi Watir, Gulf of Aqaba

because it is the largest known sabkha-salina on the coasts of the Red Sea and occupies a tectonic Depression (Keheila et al. 1989). It is also an example for the transformation of previous lagoon into salina/sabkha. El-Mallaha sabkha lies between Ras Gharib and Ras Shukeir (Fig. 21.2), with a maximum length of 15 km and a maximum width of 3 km. The sabkha occupies the lowest parts of a wider Depression. This Depression is 2–4 m below sea level in the north and 11–12 m below sea level in the south; its escarpment lies 3–5 km to the west of El-Mallaha, and an eastern ridge (1–1.5 km wide) rises 11–16 m a.s.l, separating from the sea. This saline sabkha, which has a salinity of 200–280%, is fed by sea water seepage and by inland freshwater through the wadis coming from the Red Sea Mountain/Aish El-Mellaha Range. In the meantime, brine reflux moves from the sabkha to the sea, indicating water exchange between the two. Some lagoonal sediments (gypsum) and fauna (shells) were found at higher levels along the margins of the Depression. This finding suggests that this sabkha was formerly a lagoon that stood at a higher level (a few meters) than the one of the present day (Orszag-Sperber et al. 2001).

There are a few small saline lakes along the southeastern margins of El-Mallaha sabkha. The study of Orszag-Sperber et al. (2001) revealed that the surface of the lakes is 5.5 m below sea level and 1–1.5 km from the shoreline. The lakes are separated from the Gulf of Suez by a low relief ridge of Pliocene-Quaternary sediments, developed parallel to the coastline. The presence of multiple springs emerging at the floor of the lakes suggests that seawater continuously feeds the lakes, with a salinity of 42–46%.

Salinas developed in other areas along the margins of deltas at the Red Sea coasts. Salinas are reported to be present in Wadi Nakkara, Mersa Alam, Quseir El-Qadim, Wadi Sharm El-Bahari (Plaziat et al. 1995; Orszag-Sperber et al. 2001). This form changes rapidly due to local environmental conditions such as sea level changes, evaporation rates, rainfall, and seawater recharge. This is why remnants of previous Pleistocene salinas are found in the above-mentioned sites. Figure (21.2) shows that there are some salinas developed in the northern part of Sabkhat El-Mallaha, while the study of Orszag-Sperber et al. (2001) reported that the salinas are concentrated in the southern part

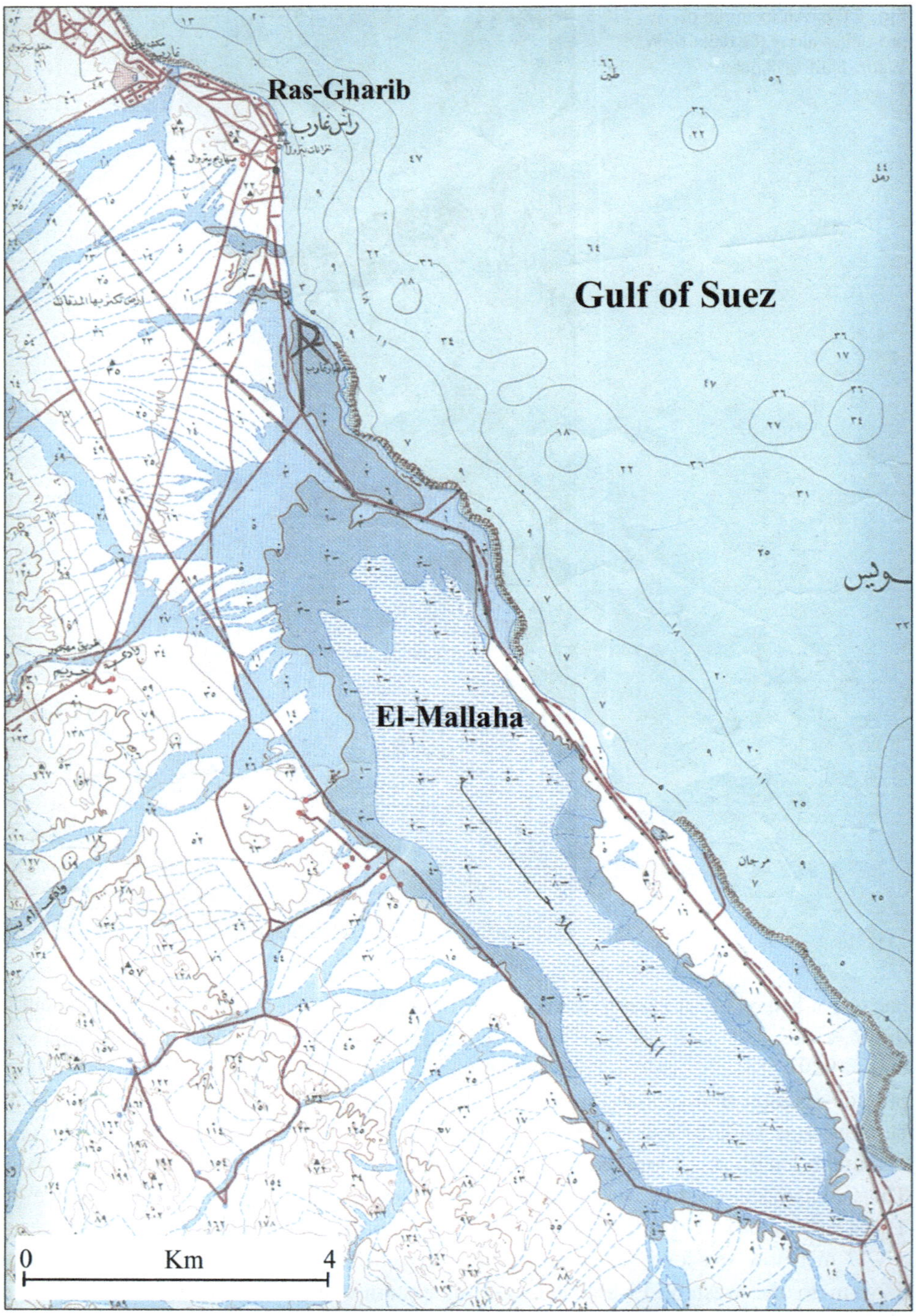

Fig. 21.2 Extract of a topographic map to show El-Mallaha Sabkha-Salina Basin at coast of the Gulf of Suez

of this sabkha, which indicates that the salinas develop and disappear in a short time.

21.2.2 Bahadas

Bahadas are developed due to the coalescence of alluvial fans and deltas. The largest bahada is in Sinai and is known as **Al Qaa plain**. It occupies the area between the Sinai Mountains and Gabal Qabaliyat in the north and the Gulf of Suez in its southern part. This plain extends for ~135 km from Wadi Feiran in the north to Ras Mohamed in the south, with a varying width between 5 and 20 km. All wadis dissecting the western slope of Sinai Mountains contributed to the development of this bahada. The wadis take an E-W orientation in the southern part of the plain, but in the northern part they shift gradually to SE; then, they debouch in a small basin with the N-S consequent Wadi El-Aawag, forming together a playa with an area of 50 km^2 (Hassan 2003). The sediments of this bahada are of Pliocene age and are composed of alternating beds of calcareous grits and sandstone. The second largest bahada in Sinai lies in the southern part of the Gulf of Aqaba. It extends from the delta of Wadi Kid to Sharm El-Sheikh, for ~50 km from N to S with a varying E-W width of 5–15 km. As indicated on the geological of Sinai, scale 1:50,000 this bahada is composed of Holocene wadi deposits.

Furthermore, there are three large bahadas along coasts of the Red Sea. The largest bahada developed just to the south of Ras Zaafarana, extending southwards to the Port of Safaga for 265 km. It is interrupted at 30 km N of Hurghada by two small linear ranges of mountains, running parallel to the coastline. The first range of mountains lies inland and is **Aish El-Mallaha** (448 m a.s.l). The second range of mountains is Gabal El-Zeit (457 m a.s.l) which directly overlooks the coastline. Although the wadis coming from the west of the Red Sea Mountains are obstructed by Aish El-Mallaha range, some of them succeeded —as antecedent wadis— to cross the range and deposit another fans to the east of this range (see Fig. 18.6). The second bahada plain is called "**El-Waddah**", which developed in the extreme 210 km of the coast, from 15 km N of Shalatin town to the Egyptian-Sudanese borders. Its width varies between 40 km at the Wadi Kraf Delta and 7.5 km at Halayib town. There are several hills of different sizes projecting from the surface of this bahada. The hills are separated from the Red Sea mountains and retreat under the effect of erosion. All wadis coming from the mountains contributed to the bahada by depositing various quantities of sediments of which some take the form of fans. The sediments are composed of fluvial wadi materials of gravel, sand, and silt, and the top layer was deposited during the Holocene. The third large bahada along the Red Sea coast developed from Mersa Himeira lying to the north of **Bernice Peninsula** for a distance of ~60 km and width of 2–10 km.

21.3 Aeolian Forms

Previous works on aeolian forms along the Red Sea coasts concede that they are depositional rather than erosional (Hume 1925; Farag 1950; Manibary 1991; Amroon 2001; Moawad 2008;). Except for **Dune Fields in SE Eastern Desert** (Chap. 16), nebkha is the dominant form Nebkhas develop when sand is trapped by vegetation clumps, as isolated mounds around individual plants. Nebkhas develop in clusters because vegetation bushes grow in clusters. Several fields of nebkhas were recorded in various localities along the Red Sea Coastal Plain, covering a total area of ~220 km^2 (Manibary 1991). The development of nebkhas in any locality depends on the growth of plant clusters, which in turn depends on the presence of water in the subsoil layer. This is why nebkhas developed at wadi beds near their mouths and on the surfaces of coastal sabkhas (see Fig. 16.6).

Depending on field measurements of the dimensions (length, width, height, slope of sides) of a sample of 50 nebkhas from various localities, Manibary (1991) found that height varies between 0.2 and 4 m, with a mean of 1.2 m. Due to local variations in sand supply, some nebkhas take the form of a mound with nearly symmetrical sides. There, sand accumulation resides on and around the plant clumps. In some other nebkhas, a lee dune developed —as a tail— in the downwind direction, with a length of 1–7 m (Manibary 1991). In a previous study on nebkhas in Kharga Depression, it was found that length of lee dune is a function of nebkha's height (Embabi 1967). Field observation on the nebkhas of the Red Sea Coastal Plain showed that the lee dunes are composed of opposing cross-bedding structure. This type of structure was found in a previous study on a nebkha at the Tehama coastal plain of the Red Sea in the Arabian Peninsula (Embabi 1976–1977). This cross-bedded structure is a result of the bifurcation of wind when it blows atop the nebkha.

Sand properties of nebkhas were found in the following manner (Manibary 1991; Amroon 2001). Medium-fine sands predominate (77%), with a mean grain size of 0.23 mm on the leeward side, while the percentage of coarse sand increases on the windward side. Sorting is medium (0.77), which suggests that the source of sand is not very far. Moreover, mineral composition varies greatly from one locality to another. In some areas (El-Aish estuary), quartz coated with red iron oxides predominates, while in another locality (Wadi El-Gimal), sands contain a high percentage of fragments, marine shells, and other marine materials which suggest a marine origin. Mineral analysis of a sand sample showed the presence of some heavy minerals such as tourmaline, pyroxene, and epidot, and some light minerals such as quartz and feldspar, which indicate a local source for sands.

21.4 Marine Forms

This group includes a wide variety of forms resulting from different marine processes acting along the coasts of the Red Sea. These processes are tide, waves, and offshore currents. Forms that characterize the coasts of the Red Sea are coral reefs, cliffs and marine platforms, sand and **gravel beaches**, and **sand barriers**.

21.4.1 Coral Reefs

Reefs occur on most of the coasts of the Red Sea and on both coasts of the Gulfs of Suez and Aqaba. They tend to be at their best in the central and northern parts of the Red Sea, on the coasts of Egypt, Sudan, and Saudi Arabia (Head 1987). All previous studies (Head 1987; Manibary 1991; Moawad 2008) divided Red Sea coral reefs into three categories: **fringing reefs**, barrier reefs, and atolls. These studies also suggest that fringing reefs are the predominant form along Red Sea coasts. Topographic maps show the presence of coral reefs along most of the Red Sea coast. This presence is expected since all the elements of the coastal environment are suitable for the growth of coral reefs. This subsection will deal with contemporary fringing reefs, which have been growing during the past 5000–7000 years (Braithwaite 1987; Moawad 2008) as a result of the rising sea level. Emerged older reefs will be discussed in Sect. 21.5.2.

Fringing reefs develop close and parallel to the shoreline. They are the most basic and most widespread type of reefs in the Red Sea (Head 1987). In the sector between Safaga and Quseir, they are usually separated from the shore by narrow, shallow lagoons owing to lack of upward and outward growth. Fringing reefs grow on platforms owing to a combination of erosion and deposition processes. The living corals form a thin veneer over platforms composed of biogenic debris, mainly of the remains of previous corals. The width of the platforms varies from tens to a hundred meters (Moawad 2008). In other sectors of the coast, fringing reefs disappear in front of wadi mouths resulting in openings in the reefs for small ships anchorage, and this is why such marine sites are called "Mersa"/"Sharm" like Mersa Alam and Sharm El-Sheikh, and several others. Hence, Mersas and Sharms make excellent natural harbors for small boats.

21.4.2 Coastal Cliffs and Marine Platforms

Coastal cliffs develop along the Red Sea coasts by the action of waves near sea level and the subsequent collapse of blocks of rocks from the hanged top. They have a vertical face or near vertical at the upper part and an undermined lower part at the base of the cliffs. The depth of the undercutting notch is 1–2 m. Some of the cliffs directly overlook the beach, while in some other localities; the cliffs are 100–200 m from the shoreline (Manibary 1991). Field observations and previous studies (Manibary 1991; Moawad 2008) showed that the cliffs are composed of reef limestone (1–4 m) and are overlain by alluvial sediments. This is because reefs emerged several times during the Quaternary as the sea level dropped during this period. This is why there are several cliffs at different heights. In the localities where ancient salinas developed, the cliffs are composed of salina sediments (mainly gypsum). Bedrock cliffs are found along the southern coastline of Ras Banas Peninsula (Fig. 21.3).

Undercutting the cliffs and the collapse of the overhang section resulted in the retreat of cliffs and the development of a **wave-cut platform** (Fig. 21.4). Width of this platform varies from a few meters to a few hundred meters in the coastal section between Safaga and El-Quseir (Moawad 2008). The wave-cut platform is usually covered with a sand layer and wadi sediments forming a sand/shingle beach. The surfaces of rocky platforms are characterized by the development of solution holes of different sizes and shapes.

Fig. 21.3 Space image showing the coastal cliffs and submerged deltas along the southern coastline of Ras Banas Peninsula. *Source* Google Earth 2014

(a)

(b)

Fig. 21.4 Two examples of Marine terraces and platforms

As wave energy is not too strong, it is suspected that rates of cliff retreat are not high (Moawad 2008).

21.4.3 Sand and Gravel Beaches

Beaches are classified into two types based on the grain size of material covering their surfaces. Sand grains cover sand beaches, whereas gravels cover the shingle beaches. Sand beaches are recorded along the Red Sea, Gulf of Suez, and Gulf of Aqaba coasts. They do not stretch for long distances along the coasts and are limited to narrow strips due to limited sand supply. Thickness of sand varies greatly according to the local environment. It attains a few centimeters at the sides of estuaries, while it is just a thin sand cover on the surfaces of some coral surfaces. According to Moawad (2008), sand size ranges from fine to medium fractions. Moawad (2008) postulates that sands of the beaches originated from shallow sea bottom and that the role of wind and supply from the wadis are very limited. On the contrary, field evidence recorded by the present author is in favor of more significant role of wadi sediments, which are laid down at their mouths along the Red Sea coasts. Sand at beaches of the delta of Wadi Kraf is derived from the sediments of the wadi and is driven inland forming the dunes of El-Hebal Dune Field (see Chap. 16). Another example is the presence of shingle beaches at some localities along the coasts of the Gulf of Aqaba, mainly at the mouths of some wadis. There, all the sediments brought by the wadis were laid down and exposed to wave action carrying the fine materials ($\geq$ sand) that were redistributed by long shore currents at protected areas. Coarse fractions ($\leq$ gravel) were left behind as lag, resulting in shingle beaches (Fig. 21.5).

Fig. 21.5 Shingle beach along the coast of the Gulf of Aqaba

21.4.4 Sand Barriers

Sand barriers include all linear sand forms that developed in parallel to the coastline (Fig. 21.6). Most of the Red Sea barriers are spits/hooks, since sand barriers without any connection with the coastline are rather rare. Most of the barriers are pointing southwards, except for a few of them which point northward, indicating that the offshore currents are moving in the same direction. Along the coasts of the Gulf of Aqaba, barriers are connected from both sides to the coastline, forming closed barrier lagoons, such as those of Abu Galoom and Nabq reserves. The only exception to this is the complex hook of Dahab, which extends southwards from the coastline for several hundred meters. Furthermore, several micro-barriers developed along the coasts of the Gulf of Suez, some of which are complex hooks. Along the Red Sea proper, the characteristics of sand barriers and linked lagoons do not differ much from those of the coasts of the Gulfs of Suez and Aqaba. The most significant difference is in the orientation of barriers, especially in the southern parts where it shifts toward the north, indicating a change in the

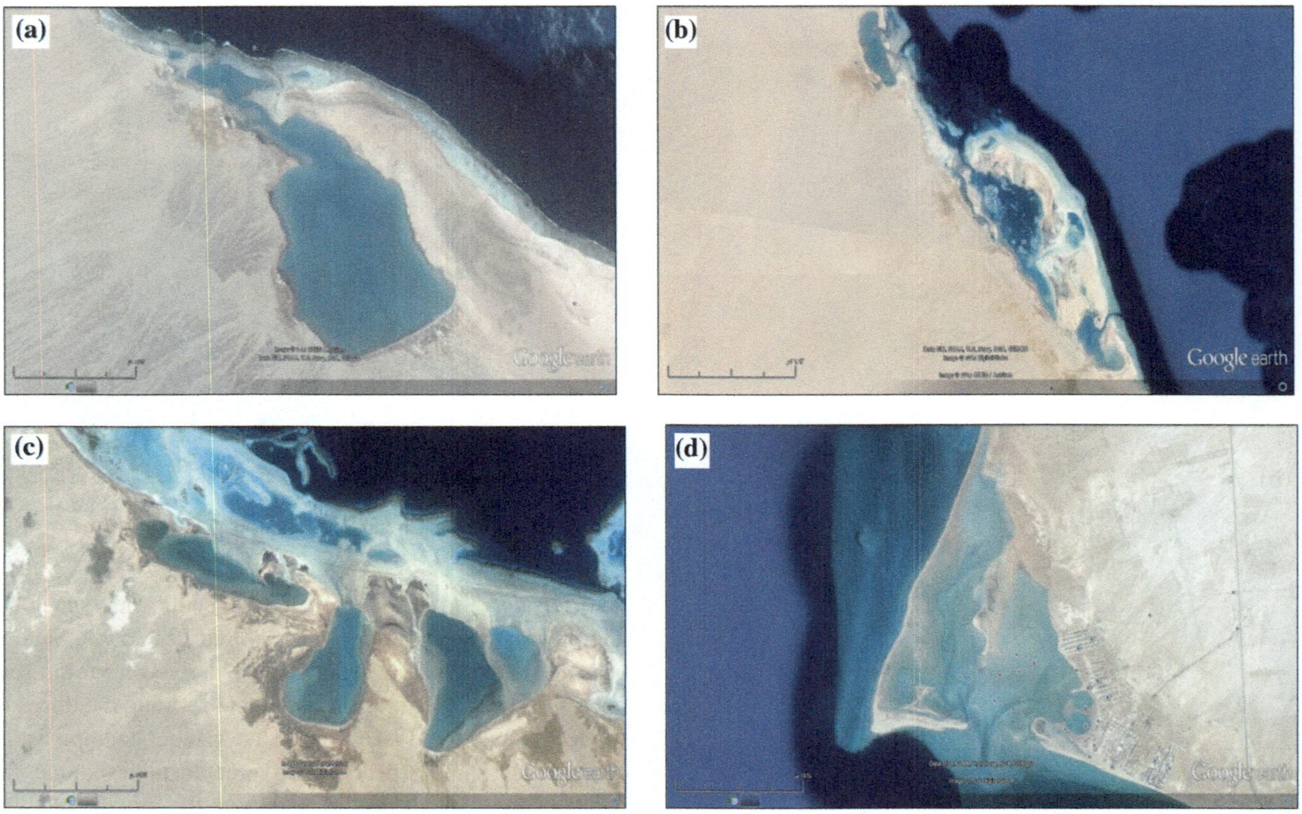

Fig. 21.6 Examples of sand barriers and coastal lagoons developed at the coasts of Red Sea and Sinai. **a** Sand barriers and a compound lagoon, south Ras Banas. **b** Sand barriers of various sizes and a complex of coastal lagoons at Mersa Shaab in the southern part of the Red Sea coast. **c** Compound sand Barriers with a complex of lagoons, 15 km to the south of Mersa Shaab. **d** Lagoons and barriers, 30 km N of Ras Matarmah at the coast of Gulf of Suez. *Source* Google Earth 2014

direction of coastal currents. Mersa Shaab hook is the best example for barriers pointing northward (Fig. 21.6b).

Sand building the barriers is transported first by waves from the wadi mouths and the eroded coral reefs to the coastal platforms, then by coastal currents, which redistribute the sands as barriers parallel to coastline. Barriers size, growth, and connection with the coastline depend on the local conditions. However, the small size of the barriers is an indicator of the limited sand supply available for their development.

21.5 Forms Resulting from Sea Level Changes

This group of forms reflects the relationship between land and sea. Owing to this relationship, new forms either developed or disappeared below seawater. This happens due to several factors, the most effective of which are tectonics and climatic changes. Forms discussed in this section are divided into two groups: submerged forms and emerged forms.

21.5.1 Submerged Forms

Characteristics of submerged forms of the Red Sea coasts are: (**Sharm**/Mersa), submerged deltas, coastal lagoons, submarine canyons, and submarine terraces.

21.5.1.1 Estuaries

Estuaries and their environs were the subject of several geologic and Quaternary studies. The first characteristic revealed by these studies is that the considerable variability of Quaternary marine sedimentation makes difficult correlations from one section to another (Plaziat et al. 1995). This variability is due to the diverse wadi settings and alluvial fans/deltas. This is exemplified by some estuaries of the Gulf of Aqaba, which resemble "Fiords" because of the steep slopes of the lower sections of the wadis reaching the Gulf of Aqaba (Fig. 21.7a). On the contrary, most of estuaries at the Red Sea proper are shallow with low sides, because the lower sections of the wadis cut through depositional forms (deltas/bahadas) (Fig. 21.7b). The second characteristic of the "Sharm"/Mersa is that the Holocene feature of the present Red Sea coasts. This form resulted from the rapid sea

Fig. 21.7 Field photos for two examples of some estuaries of different topographies. **a** Steep-sided estuary on the northern coast of the Gulf of Aqaba. **b** A shallow estuary with Mangrove trees community, Red Sea coast

level rise during the Flandrian transgression 12,000–10,000 years ago (Kidson 1982). Only Sharm El-Bahari developed during the Pleistocene sea level changes (Plaziat et al. 1995). Remains of sediments in the environs of sharms indicated that some micro features developed due to low magnitude changes in land/sea relationship. The presence of gypsum remains at the margins of some sharms and wadi sides—e.g., Wadi Nakara, Quseir El-Qadim, south Marsa Alam, and north Wadi Samadai—indicates that salinas developed there. The age of some of these gypsum remains was found to be 1134 ± 64 years in Quseir El-Qadim (Orszag-Sperber et al. 2001). The presence of mangrove colonies in some lagoons implies that these lagoons are shallow and can be transformed into salinas in a short time under high evaporation rates.

According to Moawad (2013), the submerged lower courses of the wadis exhibit minor small-scale canyon-like shapes, which mainly marked the continental shelf, and to a lesser extent, the continental slope. The canyons are relatively shallow as their depth varies from a few meters to tens of meters (−2 to −70 m bsl). They are characterized by symmetrical winding that is V-shaped and/or U-shaped. The resemblance of submarine canyons to river-cut canyons on land and the juxtaposition of many land and sea canyons have always provided a strong argument that at least the heads of the submarine canyons were originally cut by rivers. Morphometry of such shallow canyons reveals that they might have been developed during the Pleistocene lower sea level some 90–130 m below the present one (Moawad 2013 after Reaches et al. 1987).

Google Earth images show that along some sections of the Red Sea and Gulf of Suez, a narrow submerged platform (100–200 m wide) extends parallel to the coastline and is dissected by drowned sharms. Such findings suggest that the drowned wadis (Sharms) had cut the platforms before the Flandrian sea rise.

21.5.1.2 Submerged Deltas

Embabi (2004) suggested the presence of some **submerged deltas** along the Egyptian coastline of the Red Sea. This suggestion was based on the general configurations of the bathymetric lines of the topographic map (scale 1:500,000). These lines take fan-like shapes and are located seaward of the present-day drainage system. Therefore, it was assumed that they represent submerged deltas of an ancient Red Sea drainage system in the Eastern Desert (Embabi 2004; Moawad 2008). Using remote sensing techniques, Moawad (2008, 2013) contended that the toes of the deltas plunged down into the deep sea for ~500 m below the present-day sea level. In some cases, lateral boundaries overlap indicating the coalescence of some deltas. Images of Google Earth show the presence of some drowned deltas along the southern coast of Ras Banas Peninsula (Fig. 21.3). These submerged deltas were developed at the prolongation of the wadis, which dissect the igneous hills and overlook the sea in the southern area of the Peninsula.

21.5.1.3 Coastal Lagoons

Lagoons are permanent water bodies. Their number cannot be exactly counted along the Red Sea coasts because there are many rather small ones that appear only on high-resolution space images. Small lagoons developed at the low-lying margins of the deltas and along the coastline behind sand barriers. Lagoons can be divided into two categories based on their form. The first type is the simple lagoons that are composed of a single basin, and the second type includes the compound lagoons, which are composed of several lakes of different sizes connected to each other via channels. Examples of the lagoons are illustrated in Fig. (21.6). Most probably, these lagoons have recently developed during the last Flandrian sea rise. Lagoons can be classified into three groups according to mode of development as follows: sand barrier lagoons, marginal lagoons, and reef barrier lagoons.

Sand barrier lagoons develop when a sand spit/hook is formed parallel to the coastline. This type of coastal lagoons of the Red Sea varies in size from rather small ones extending for several tens of meters to longer ones extending several hundreds of meters and occasionally a few kilometers. Large lagoons are composed of several interconnected/separate lagoons; therefore, they become compound/complex. These aspects are acquired due to new barriers that develop off the original ones, resulting in the development of new lagoons. This is exemplified in the two large lagoons of Mersa Shab on the coast of the Red Sea south of Ras Banas and of Wadi Sudr Delta on the eastern coasts of the Gulf of Suez (Fig. 21.6b, d). All lagoons have openings connecting them to the open sea. These openings allow fresh, "new" water to leak into the lagoons by means of tidal currents and channels. Lagoons are normally shallow, and their floors are filled with tidal flats and are cut by tidal channels. This is why mangrove trees colonize some parts of the lagoons. In some lagoons, the mangrove trees occupy relatively wide areas with high density, so that they look like a forest and are called so by locals. This type of lagoons is the most widespread along the Red Sea coasts since the agent, which is responsible for its development covers the coasts of the Red Sea.

Marginal lagoons developed along the coastal margins of the deltas/bahadas, occupying the low-lying areas, which are not fully filled by sediments. Consequently, their shape varies greatly from one locality to another (Fig. 21.6a, c). In some localities, sand barrier lagoons developed alongside these marginal lagoons forming a cluster of coastal lagoons of different origins (Fig. 21.6c). This type of lagoons is surrounded by sabkhas that extend for various distances inland according to local conditions, but in general for short distances.

The occurrence of reef barrier lagoons is very rare. Moreover, it is difficult to recognize these lagoons on high-resolution space images. This is because fringing reefs may form a barrier/platform-like a few meters below sea level, behind which shallow lagoons are developed. Moawad (2008) recognized one case in the Safaga region, where a lagoon developed at the mouth of Wadi Abu Shiqayli behind a reef barrier 1–2 m below sea level.

21.5.1.4 Submarine Canyons

The **submarine canyons** represent a distinctive erosional form marking the outer continental margin along the Egyptian Red Sea coast. The deeply incised canyons are the most common submarine forms that dissect the continental slope. They extend ±500 m below the present-day sea level. Remarkable steep walls characterize submarine canyons, which seem to have been cut in the bedrock. In some cases, head of the canyons may be traced into the upper part of the continental slope at depths 100–150 m bsl. Longitudinal profiles of the canyons are generally concave and sloping strongly seaward from the head of the canyons to the deep sea. Even though the incised canyons sometimes follow the major structural features seaward from the adjacent land, they seem to be a result of a great geological event, which is the Messinian one that influenced the continental margin since they spread along the Egyptian Red Sea coast, mostly without exceptions. According to Moawad (2013) Siroko stated in 2008 that the incised canyons in specific are a result of the Late Miocene Messinian crisis owing to the discontinuity of water supply between the Atlantic Ocean and the Mediterranean Sea. The Mediterranean water level dropped down to 1500 m below its present-day level, which entailed a considerable increase of salinity and intense erosion of all Mediterranean rivers such as the Rhone and Nile canyons. The erosive phase is supposed to have had occurred between 5.59–5.50 Ma and followed by deposition of non-marine sediments between 5.50 and 5.33 Ma. Due to the Messinian crisis, there was a large relief difference between the Red Sea Depression and the surrounding land during the Miocene time. Therefore, the Red Sea Depression received marginal clastic sediments (2–3 km) inter-fingering with evaporites (3–4 km) where the Red Sea Depression continued to subside. It can be deduced that an old drainage system was the major carrier of such sediments, especially during Middle Miocene time when the climate was warm and humid (Kheder 1984).

Some previous work postulated that flash floods that occur in the Red Sea region carry large amounts of sediments into the sea (Fricke 1983; Moawad 2013). It was suggested that the occasional gravity flows fed initially by sediments from aligned Wadis are a major agent in cutting the deep submarine canyons into steep edges of the narrow Red Sea shelves. The close relation between the present-day drainage networks and the bathymetry supports this suggestion.

On the other hand, other studies presented the idea of marine origin of the canyons (Moawad 2013 after Shepard 1972). Nevertheless, this suggestion is improbable in the Red Sea region, as no large active marine erosion has been found throughout the geological evolution of the Red Sea (Moawad 2013). Thence, it is believed that the role of marine erosion might have been restricted to some of forms produced by other processes.

21.5.1.5 Submarine Terraces

Marine terraces are one of the significant features that marked the submerged deltas as flat surfaces of different extents (Moawad 2013). During the Pleistocene and Holocene epochs, the Red Sea experienced dramatic eustatic changes resulting in a sequence of raised coral terraces and alluvial terraces. Moawad (2008) distinguished three submarine terraces at the altitudes of 20, 50, 100 m below the present-day sea level. These submarine terraces are a result of the last glacial maximum (LGM) low sea level ($\sim$100–120 m bsl). He assumed that the terraces are congruent with the generalized curve of sea level rise, since the last ice age. The same interpretation has been proposed by Gornitz

(2007). In this study, notable submarine terraces were found in different locations and altitudes along the coast. These terraces mostly constitute flat narrow surfaces and they differ locally in depth and width. In comparison with the generalized curve of sea level rise since the LGM, it is inferred that terrace 100–120 m bsl may correspond with the last glacial maximum (LGM) in the age range of melt-water pulse (MWP) 1A0 and 1A ($\sim$19–16 ka BP). The sea level rose rapidly 10–15 m in less than 500 yr during the MWP-1A0 and was followed by another increase about 16–24 m during the MWP-1A. Terrace 50–75 m may accord with the melt-water pulse -1B (11.5–11 ka BP) when the sea level might have had a surge by 28 m. Terrace 20–25 m is mostly formed during the fourth melt-water pulse (MWP) 1C (8.2–7.6 ka BP), which relates to Flandrian transgression. It is highly believed that the shallow submarine canyons were probably formed during these intervals of sea level gradual rising.

These terraces conform to the submarine terraces recorded in southern Sinai at 120, 60, 20 m bsl (Gvirtzman 1994). Terraces 100–120 and 50–75 m bsl are in agreement with terraces 90 and 60 m bsl in Sharm El-Sheikh and Ras Burka (Fricke 1983; Reiss and Hottinger 1984) and terraces of Elat (90–50 m bsl) (Reches et al. 1987). Terraces of Elat have likely developed between 70 and 50 ka BP, when the sea level was 100–60 m below the present level (Reches et al. 1987).

21.5.2 Reef Terraces

Reef terraces developed due to drop in sea level; therefore, they are emerged forms. They were found along the coasts of southern Sinai (Gvirtzman et al. 1992; El-Asmar 1997) and along the Red Sea coast (El-Moursi et al. 1994); Manibary 1991). The four studies found that there are several terraces at different levels, though their number is not the same. El-Moursi et al. (1994) recognized eight terraces along the stretch from Hurghada and Marsa Alam, while El-Asmar (1997) recognized five terraces in Ras Mohammed and other four terraces in Um Seed (Sharm El-Sheikh) in southern Sinai. In addition, Gvirtzman et al. (1992) found four terraces in the region of Sharm El-Sheikh along the coasts of Aqaba Gulf in southern Sinai. All the previous studies recorded notches at the lower section of the face of some terraces, indicating the effect of wave action on their development. They also found that these terraces developed between the Middle Pleistocene and the Late Pleistocene. Moawad (2008) summarized the characteristics of reef terraces in the Red Sea according to the findings of El-Moursi et al. (1994). To illustrate, El-Moursi et al. (1994) suggested that the terraces were developed during three cycles of reef formation. These cycles are:

- *The Oldest Cycle* comprises the oldest terrace No. **VIII**, with maximum height 35 m a.s.l. This cycle occurred during the Oxygen Isotope Stage 9 (300–330 ka bp). It was assumed that this cycle corresponds to the Ophira Terrace **I** in the Sharm El-Sheikh region that was dated between 290 and 330 ka bp (Gvirtzman et al. 1992), and in conformity with terrace **I,** recorded in Ras Mohammed and dated 297 ka by Uranium Series (El-Asmar 1997).
- *The Middle Cycle* of reef formation comprises terraces **V**, **VI**, **VII**, which developed during Oxygen Isotope Stage 7 (170–230 ka BP). They were found at heights varying between 22 and 25 m (Terrace **V**) and 30–32 m (Terraces **VI** & **VII**). This cycle corresponds to Mureikha terraces **II** and **III** in southern Sinai, dated 170–215 ka bp (Gvirtzman et al. 1992), and to Terrace **III** in Um seed and Ras Mohammed in southern Sinai that range in age between 202 and 303 ka (El-Asmar 1997).
- *The Youngest Cycle*: During this cycle terraces **I**, **II**, **III**, and **IV** nearly continuous along the Hurghda–Marsa Alam Coastline were developed. Their height ranges between 1.5 and 2 m for terrace I and 9 m a.s.l for the upper surface of this cycle (terrace **IV**), which is equivalent to other terraces recorded along Red Sea coast and the Gulf of Aqaba. According to Moawad (2008), terraces of this cycle were mostly developed during the maximum high sea level stand of the last interglacial cycle (Oxygen Isotope Substage 5e, 120–130 ka BP)

21.6 Conclusion

Red Sea coasts represent an open museum for a large variety of forms of different origins and history. In fact, no other geomorphologic region in Egypt exhibits such a wide variety of forms. This is due to the fact that coastal regions are areas of interaction of marine and continental processes. In addition, the environmental conditions of the Red Sea region played a major role in the development of several forms along its coasts. Meanwhile, climatic changes resulted in a change in the characteristics of environmental conditions. Sea level changes played also a major role in the development of some new forms or the disappearance of some others under seawater.

The great number of small deltas is a result of the steep slopes of the eastern sides of Sinai and Red Sea Mountains. Many of these deltas extend below the present sea level. Is this extension due to normal deposition in the sea or is it due to sea level rise? Previous studies did not provide a clear answer to this question. However, the fan-like shape of the bathymetric lines and the bifurcation of drainage channels on the surfaces of the deltas indicate that these submerged deltas were drowned as a consequence of sea level rise, which occurred during the

Flandrian sea rise. With such rise, another form was developed which is the estuaries (Sharms/Mersas). Nearly all the lower sections of the wadis that reached the coast are drowned at present time.

The drop in sea level exposed the forms that have been developed at previous sea level/or formed underwater. This is exemplified by the reef terraces, which can be seen in several localities along Sinai/Red Sea coasts. The presence of several reef terraces (between four and eight terraces) at different heights, reaching more than 30 m above the present sea level indicates that sea level plunged several times. Such ebbing occurred in intervals during which cliffs with notches and new platforms were formed.

The presence of coral reefs along the Red Sea coasts indicates that environmental conditions are in favor of their formation. The presence of ancient reef terraces at higher levels shows that these favorable environmental conditions prevailed since the formation of these terraces, which was in the Middle Pleistocene (at least since 300 ka ago).

References

Amroon T (2001) Geomorphology of drainage basins and Red Sea coastal belt between Ras El-Esh and the mouth of Wadi Mallaha (in Arabic). Dissertation, Cairo University

Braithwaite C (1987) Geology and palaeogeography of the Red Sea Region. In: Edwards A, Head S (eds) Red Sea. Pergamon Press, Oxford, pp 22–44

El-Asmar H (1997) Quaternary isotope stratigraphy and paleoclimate of coral reef terraces, Gulf of Aqaba, South Sinai, Egypt. Quat Sci Rev 16:911–924

El-Dissouki S (1998) Geomorphology of Wadi Gheweibah Delta (in Arabic). Arab Geogr J 31:299–356

El-Moursi M, Hoang CT, El Fayoumy IF, Hegab O, Faure H (1994) Pleistocene evolution of the Red Sea coastal plain, Egypt: evidence from uranium-series dating of emerged reef terraces. Quatern Sci Rev 13:345–359

Embabi N (1967) A morphological study of the Kharga oasis depression. Western Desert, Egypt. Dissertation, University of Bristol, UK

Embabi N (1976–77) Slope form of barchan dunes in Kharga and Dakhla Depressions. Bulletin de la Societé de Géographie d'Égypte 49–50:13–28

Embabi N (2004) The geomorpholgy of Egypt, vol. I, the Nile Valley and the Western Desert. Special Publication. The Egyptian Geographical Society, Cairo, 447 p

Farag I (1950) Sand dunes. J Egypt Assoc Sci Cult 20:147–178

Fricke H (1983) On the origin of Red Sea submarine canyons. Naturwissenschaften 70:195–197

Google Earth (2014) Egypt Images Latitudes 22-32 N and Longitudes 22-32 E, http://www.google.com

Gornitz V (2007) Sea level, after the ice melted and today. http://www.giss.nasa.gov/re-search/briefs/gornitz_09/. Accessed 17 July 2014

Gvirtzman G (1994) Fluctuations of sea level during the past 400000 years: the record of Sinai, Egypt (Northern Red Sea). Coral Reefs 13:203–214

Gvirtzman G, Kronfeld J, Buchbinder B (1992) Dated coral reefs of southern Sinai (Red Sea) and their implication to late Quaternary sea levels. Mar Geol 10:29–37

Hassan H (2003) Sedimentary nature of El-Aawag playa, south-western Sinai, Egypt. J Sedimentol Soc Egypt 11:29–42

Head S (1987) Corals and coral reefs of the Red Sea. In: Edwards A, Head S (eds) Red Sea. Pergamon Press, Oxford, pp 128–151

Hume W (1925) Distribution of sand deposits in Egypt. In: Hume W (ed) Geology of Egypt, Vol. I: the surface features of Egypt, their determining causes & relation to geological structure. Survey of Egypt, Government Press, Ministry of Finance, Cairo, pp 29–81

Kidson C (1982) Sea level changes in the Holocene. Quat Sci Rev 1:121–151

Kheder E (1984) Sedimentological evolution of the Red Sea continental margin of Egypt and its relationship to sea-level changes. Sed Geol 39:71–86

Kheila E, Khalifa R, El-Haddad A (1989) Holocene carbonate facies model, Ras Shukier hypersaline pool and its surrounding sabkha, west Gulf of Suez. Sed Geol 63:155–169

Manibary N (1991) Some geomorphologic features along the Red Sea coastal plain to the south of Gulf of Suez. Dissertation, Ain Shams University

Moawad M (2008) Application of remote sensing and geographic information systems in geomorphological studies: Safaga-El-Quseir Area, Red Sea Coast, Egypt as an example. Dissertation, Johannes Gutenberg Universität, Mainz

Moawad M (2013) Geomorphological characteristics of the submerged topography along the Egyptian Red Sea Coast: In Scalon L, Ranieri J (eds) Continental shelf: geographical distribution, biota and ecological significance (Earth Sciences in the 21st Century). Nova Science Publishers Inc, New York

Mostafa M (1987) Wadi Feiran Basin: a geomorphologic study. Dissertation, Ain Shams University, Cairo

Orszag-Sperber F, Plaziat J, Baltzer F, Purser B (2001) Gypsum salina-coral reef relationship during the last interglacial (Marine Isotopic Stage 5e) on the Egyptian Red Sea coast: a Quaternary analogue for Neogene marginal evaporates. Sed Geol 140(1–2):61–85

Plaziat J, Baltzer F, Abdelmagib Ch, Cobnchon O, Freytet P, Orszag-Sperber F, Purser B, Raguideau A, Reyss J (1995) Quaternary changes in the Egyptian shoreline of the northwestern Red Sea and Gulf of Suez. Quatern Int 29(30):11–22

Reches Z, Erez J, Garfunkel Z (1987) Sedimentary and tectonic features in the north eastern Gulf of Elat, Israel. Tectonophysics 141:169–180

Reiss Z, Hottinger L (1984) The Gulf of Aqaba: ecological micropaleontology. Ecological Studies 50. Springer, Berlin, 354 p

Shepard F (1972) Submarine canyons. Earth-Science Reviewes. Elsevier, Amesterdam

Bardawil Lake and the Surrounding Sabkhas 22

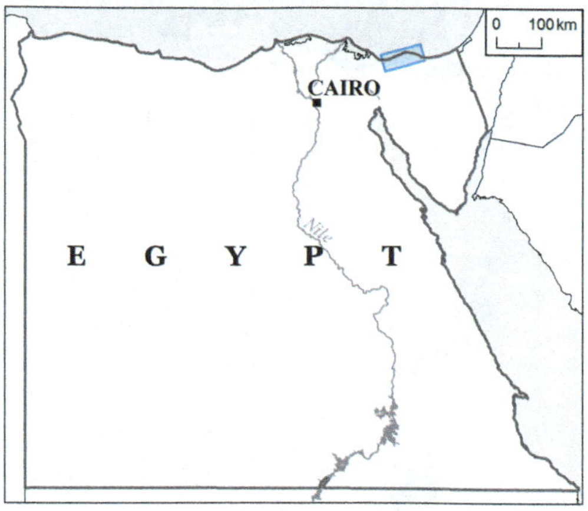

Abstract

Bardawil Lake and the surrounding sabkhas are located centrally along the coastal front of the Mediterranean coast of northern Sinai. The lake is also known as sabkha, due to intermittent connection with the Mediterranean before 1953, when two artificial openings (Bogazes I & II) were dug to allow permanent water flow from the sea. Its maximum length is ~85 km and maximum width is ~20 km, covering an area of ~600 km^2. It is a shallow lake, with a mean depth of ~1.5 m. The lake originated as a tectonic trough, with a depth of ~50 m. Afterward, it has become filled with various sediments. Bardawil is a compound lake, since it is composed of two main basins and several other internal smaller ones. The lake accommodates several minor forms, of which sand barriers are the most widespread. Other forms are drowned dunes and inter-dune low areas—tidal flats and channels. Surrounding sabkhas represents a characteristic feature along the southern shores of the lake.

Keywords

Bardawil Lake • North sinai coastal front • Coastal sabkhas • Sand barriers • Drowned dunes

22.1 General Characteristics of Bardawil Lake

Bardawil Lake lies centrally along the Mediterranean coastal front of North Sinai (Fig. 22.1). Its coastal barrier extends 90 km from east to west, occupying ~50% of the total length of the coast. Although this barrier separates the lake from the Mediterranean, exchange of sea water continues through three openings (two artificial and one natural). Bardawil Lake was also known as "**Sabkhat El-Bardawil**", due to intermittent connection with the Mediterranean Sea. In ancient times, it was known as "**Lac Sirbonis**" (Neev et al. 1987), which seems to be an old Roman naming (Hamdan 1980). In 1953, two artificial openings (Bogaz I and Bogaz II) have been dug, connecting the sabkha with the Mediterranean Sea in order to allow permanent connection, decrease the salinity of the lagoon, and allow the natural immigration of fish into it (Klein 1986; Khalil and Shaltout 2006). Bardawil Lake protrudes into the Mediterranean with a smooth convex sand barrier that extends for ~85 km from east to west. On the contrary, the southern shores are irregular due to the effect of pre-lake topography, of which the sand dunes of North Sinai Sand Sea are most notable (Fig. 22.2). The lake has a maximum width of ~20 km along longitude 33°10′E, covering an area of ~600 km^2 (Klein 1986). It is a very shallow lake, with a mean depth around 1.5 m and a maximum of ~2.75 m.

N.S. Embabi, *Landscapes and Landforms of Egypt*, World Geomorphological Landscapes,
https://doi.org/10.1007/978-3-319-65661-8_22

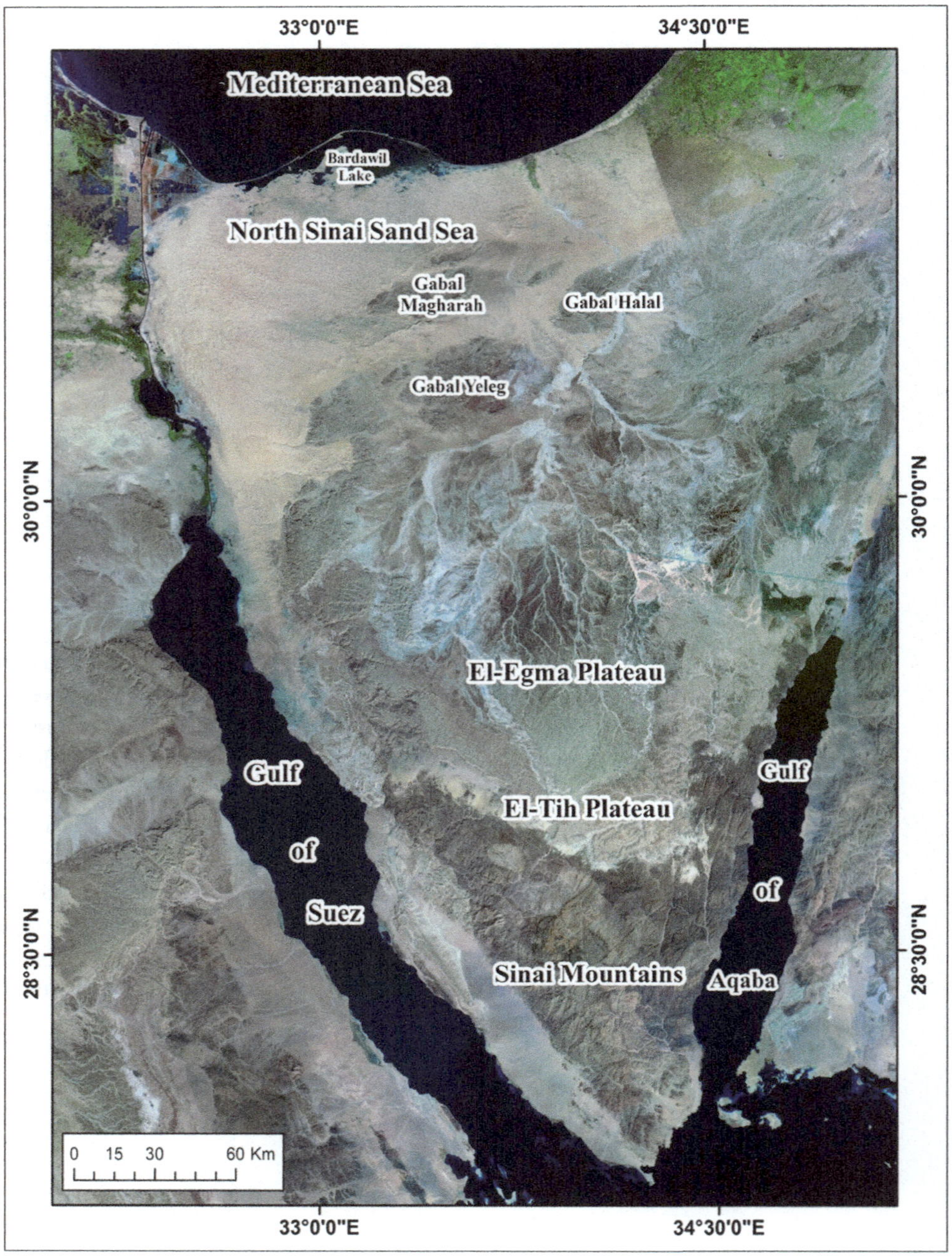

Fig. 22.1 TM-Landsat image of Sinai showing the location of Bardawil Lake

However, maximum depth in Bogaz II is 5.75 m due to dredging. Areas located below 1 m occupy ~20% of the lagoon area, whereas areas with depths between 1 and 1.5 m constitute 65% of the lagoon area, and those deeper than 1.5 m cover 15% only of the lagoon. The results of a survey of depths of the lake showed that the shallowest parts lie in the extreme eastern and western parts, and along the southern shores of the lagoon basin (The Maritime French Company for the Survey of Bardawil Lake, 1982). Islands are a characteristic feature in the lagoon. There are about 50 islands, covering an area of ~2.1% of the total area.

Water salinity of Bardawil Lagoon is much higher than that of the adjacent eastern Mediterranean Sea, making it a hyper-saline lagoon (Kramgatz et al. 1980). In this respect, it is unique, since salinity in other Mediterranean lagoons of the Nile Delta is either lower or much lower than that of the seawater. This is because other lagoons are fed partially by the less-saline water of the Nile through the channels which drain the excess irrigation waters. According to two surveys of water salinity during the period 1967–1971 (Kramgatz et al. 1980) and during the period June 20,1982– July 20, 1982 (The Maritime French Company for the Survey of Bardawil Lake, 1982),salinity increases southward from less than 40–45% in the north to more than 75% in the south and that the southern parts are the most saline. This increase of salinity in the southward direction

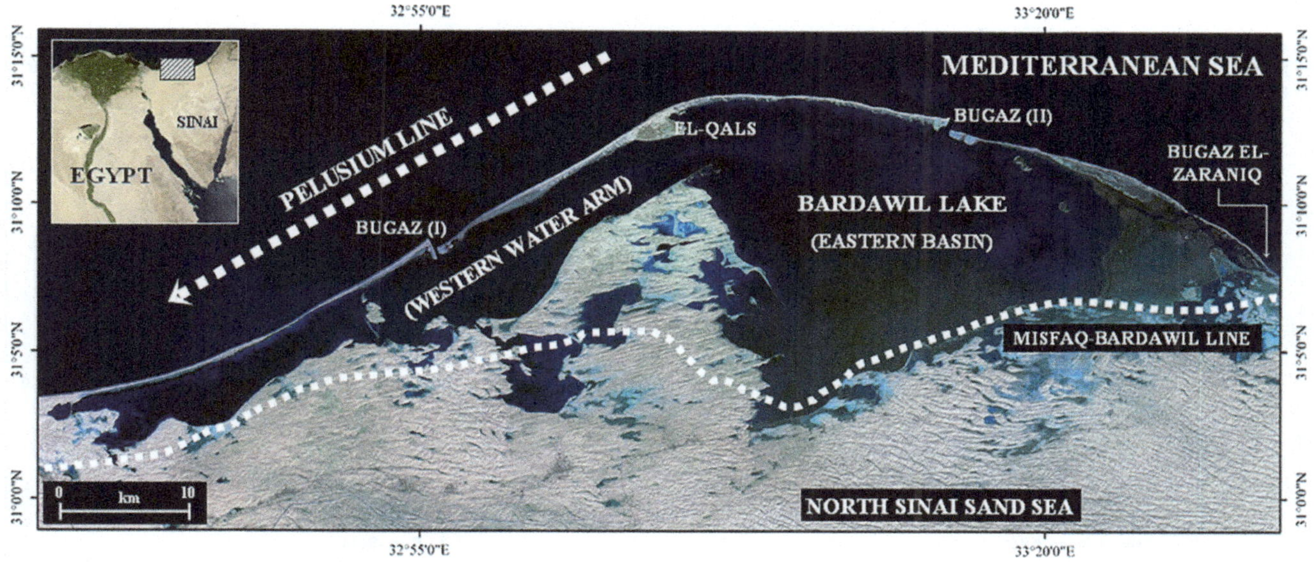

Fig. 22.2 Enhanced ETM + mosaic shows location of Bardawil Lake and the general structure. *Dashed lines* approximately show major lineaments after Neev et al. (1987) and Abdel Aal et al. (2000)

is due to the fact that seawater that feeds the lake enters through the inlets of **Bogazes I, II** and **Bogaz El-Zaranik** (Fig. 22.2) cutting across the coastal barrier in the north, and that water is exposed to high rates of evaporation (evaporation rates: 2190 mm/yr for Port Said, 1542.5 mm/yr for El Arish) while moving southward into the lake basin.

22.2 Tectonic Framework

Figure 22.2 shows that regardless of minor irregularities due to the effect of local landforms, three sections of the lake coasts are straight. This characteristic has been interpreted in some previous studies as caused by the effect of tectonic lineaments that characterize this part of northern Sinai, of which the **Pelusium Line** is the most prominent. This line runs parallel to the western part of the coastal barrier. Figure 22.3 shows a geologic cross section along the eastern part of the Lagoon (Neev et al. 1987). This figure illustrates that since the **Flandrian Transgression**, the Bardawil Basin was filled with several types of sediments, whose thickness exceeds 50 ms. This indicates—the original depth of the basin and its origin as a tectonic trough.

It seems that tectonics in the area of Bardawil Basin continued to exist over a considerable span of time (Neev et al. (1987). A gradual regional tectonic uplift occurred at **Mount Casius** (Fig. 22.3) during the transition from the Pleistocene to Middle Holocene, while folding and uplifting of the **Peninsula of Mount Casius** happened during Late Holocene. Tectonic subsidence took place during the Early to Middle Iron Age (about 3500–2500 y BP), as indicated by rapid accumulation of high-energy beach sand that contains abundant shell lenses. Evidence found along the southern coastline of the Peninsula of Mount Casius indicated another tectonic phase during Post-Persian or even Post-Roman time.

22.3 Local Sediments

According to Levy (1974), the predominant type of sediments on the floor of Bardawil Lake (to a depth of about 1.0 m) is sand, followed in abundance by clayey sand, silt, clay, and sand-silt-clay. Gypsum is precipitated as separate layers between other sediments. The fine sediments concentrate in the central part of the lagoon, while coarser materials are found along the margins. This distribution of sediments in the lagoon basin is explained by the fact that Bardawil receives water and sediments from the Mediterranean through the inlets I, II, and III, and no streams flow into the lagoon at present. In areas of relatively high energy (near the inlets) coarse sediments are laid down, whereas the decreasing energies of water flowing to the central parts lead to the precipitation of fine fractions (silt and clay). Also, wind-blown sand contributes considerably to the sediments in the lagoon, especially along the southern and western areas.

Levy (1974) concluded that the examined lagoon sediments (85 cm) were laid down on an old quartz sand area separated from the Mediterranean by a sand barrier. However, the connection between the Mediterranean and lagoon changed as shown by the composition of the quartz sand. The presence of marine diatoms and pelecypods in the

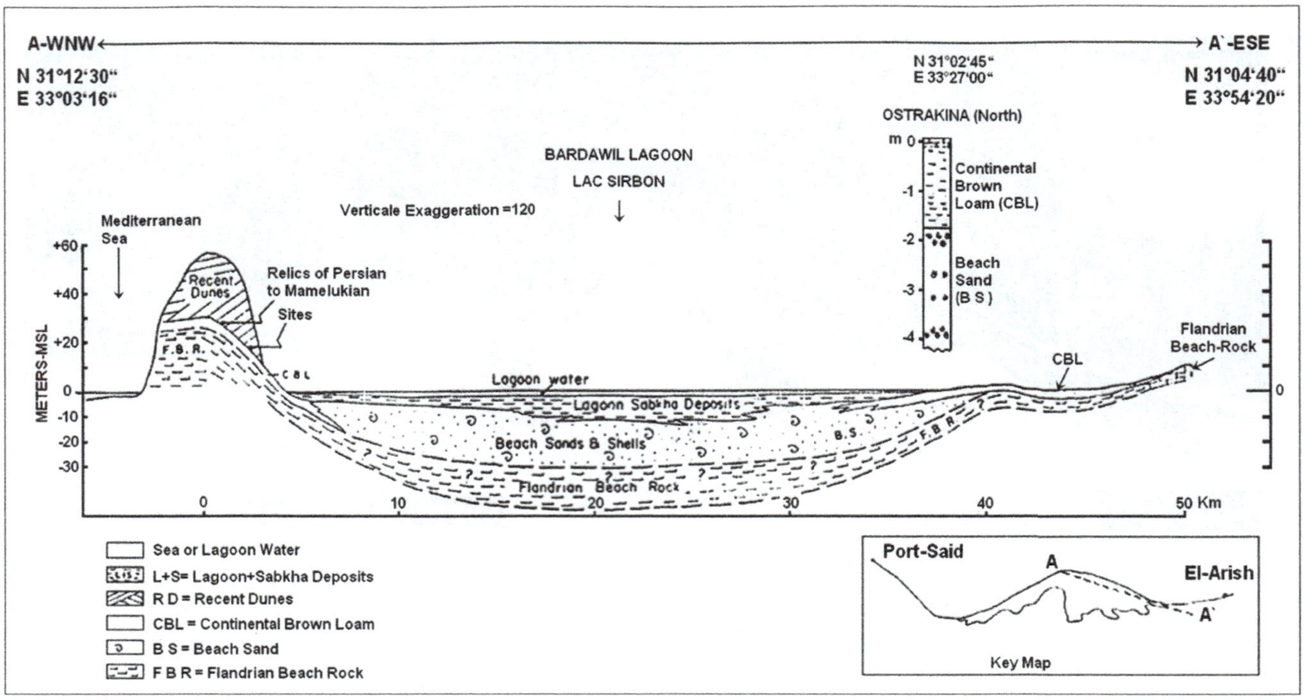

Fig. 22.3 West-northwest geologic cross section across the eastern fringes of Bardawil Lake from El-Meidan to Mount Casius (*Source* Neev et al. 1987)

quartz sand indicates that relatively high-energy conditions prevailed in the earliest periods (Neev et al. 1987). When the connection with the Mediterranean was more restricted at later periods and low-energy conditions prevailed, clayey sediments with gypsum and lagoon diatoms were deposited (Levy 1974). According to Neev et al. (1987), the top layer of the lagoon sediments is composed of sabkha deposits, and bottom of the lake is covered by a layer of Flandrian beach rock (Fig. 22.3).

22.4 Coastal Processes and Hydrodynamics

Frihy et al. (2002) divided the wave climate at the lake into three seasons based on wind intensity and direction. During summer, wave heights rarely exceed 1.5 m and the dominant wind direction is WNW. Waves are much higher during winter, coming primarily from N, NNW, and NW. The dominant northwestern winds are responsible for generating longshore currents in an easterly direction. During spring, the wind direction is reversed coming from N, NNE, and NE, creating a longshore current toward the west. Throughout the year, the average height and period of the wave are 0.5 m and 6.3 s, respectively.

The predominant tide along the North Sinai Coast is semi-diurnal giving rise to two high and low tides per day each 12 h long. The study area represents a micro-tidal environment as the mean spring tide range is less than 2 m. Inman and Jenkins (1984) demonstrated that tides together with wind and evaporation control water exchanges through the inlets of the lagoon in southeastern Mediterranean. They estimated the tidal range at 30 cm. Frihy et al. (2002) revealed that the mean high and low water levels at El Arish, east of Bardawil Lake, were 20.22 and 11.01 cm, respectively, with a tidal range of 31.23 cm. Linnersund and Mårtensson (2008) found that tide amplitude is significantly lower in the lake. The maximum observed value was about 16 cm at the east inlet (**Bogaz II**), but it is a few centimeters lower at the west inlet (**Bogaz I**).

Although tidal amplitude is small and its effect gradually diminishes in the lagoon with the distance from the inlets, tiny fluctuations in the water level of the lake lead to flood/ebb in the adjoining low-lying areas (0–1 m height). Moreover, water invades the inter-dune Depressions through narrow inlets or by capillarity to constitute ephemeral lagoons and/or salt pans since the evaporation rate is high.

The Mediterranean west–east littoral current is responsible for transferring material eroded from the Nile Delta to the eastward direction. Inman and Jenkins (1984) assumed that this eddy current is a result of the interaction between the east Mediterranean gyre and water of the Damietta promontory. The general sediment transport pattern on the north Sinai coast is divided into two littoral sub-cells, namely Port Said and the Bardawil sub-cells. Each sub-cell consists of a sediment source, an area where sand is eroded, and a stretch of coast where sand is accreted (Frihy and

Lotfy 1997). This resulted in silting and closing of the main inlets of the lake that influence its hydrologic system, since it is not fed by any other water supply. Klein (1986) revealed that the azimuth of the east side of El-Bardawil barrier is N105E and therefore the long shore current is much more frequent than that along the west side of the barrier, which has an azimuth of N60E. However, he affirmed that the total amount of entrapment for 10 years (1970–1979) shows an average of 300×10^3 ($m^3\ yr^{-1}$) and that rates of sedimentations at **Bogaz (II)** were higher leading to its complete closure in 1970, after three years of no maintenance work; whereas **Bogaz (I)** remained open. Emanuelsson and Mirchi (2007) calculated the volume of net transport rate at 40×10^3 and 258×10^3 ($m^3\ yr^{-1}$) along the western and eastern sides of the lake barrier, respectively. The central bulge of the Bardawil barrier, in between the inlets, is subjected to erosion, which continues east of the eastern inlet. The Bardawil sub-cell ends with accretion of sediment at El Arish embayment (Frihy and Lotfy 1997).

Bogaz I and II in particular are features through which the lake is opened to material and energy exchange. The balance of the lake's ecosystem depends on the regional and local influences on these inlets. Patterns of outflow and inflow through the inlets were observed by Linnersund and Mårtensson (2008). They concluded that water flow through the east inlet influences a larger body of water inside the lagoon, independent of flow direction. The estimated speed of the inflow and outflow currents through the inlets is up to $0.70\ m\ s^{-1}$. The current speed is constantly higher at Bogaz (II) resulting in larger changes as the waves run almost straight into it, whereas they are perpendicular to Bogaz (I).

However, the main two basins of the lake are characterized by particular hydrodynamics (internal circulation). Linnersund and Mårtensson (2008) modeled internal circulation patterns and magnitude of flow in the lake for one day. Their findings showed that the direction of the currents alternates—several times and mainly due to tidal force. Areas in close proximity to the inlets experience direct effects of either ebb or flow, whereas areas located further away are less affected by—changes at the inlets and more affected by the wind. They estimated internal speed as much as $0.20\ m\ s^{-1}$.

In the western basin, a major current trends SW-NE and generally transports a large amount of sand sediments. These sediments exhibit a long neck of 22 km and an average of 1.2 km width that ends with a fan-shaped toe. This path is fed by other secondary currents of which the inflow current of Boghaz (I) is the most conspicuous. Direction of these circulations differs from the direction of the predominant wind, which is NW-SE. Therefore, it mostly occurs under the influence of gravity since the western basin is generally shallower than the eastern basin. In the east, water seems to be trapped in a semi-closed basin, thus becoming forced to rotate, despite sand deposits that can be traced in some localities (Embabi and Moawad 2014).

Along the internal shores of the lake, there are many secondary currents moving very slowly along the lakeshores. These currents are the result of an angular approach of the waves to the lakeshores, which creates shore-parallel currents. Although they are weak, they contribute to redistributing sands and silts along the lakeshores leading to various small-scale forms such as sand barriers and hooks. During ebb time (low tide), tidal channels are exposed all over the internal lakeshores and corridors of sands, and silt sediments can be clearly observed in situ at shallow water, pointing to directions of the internal currents. Each inland lagoon is a self-contained hydrodynamic regime, although the water movement is very slow.

22.5 Geomorphology of the Lake

A previous study (Embabi and Moawad 2014) revealed that Bardawil, though small as a coastal lagoon, is characterized by several geomorphic aspects and exhibits several forms of different origins reflecting changing environmental conditions during its development (Fig. 22.2). The lake is divided morphologically into two main basins and several other small local basins.

22.5.1 Basins of the Lake

There are two main basins: one in the east and will be called the eastern basin, and the other one is in the west, called the western basin. The eastern basin is larger than the western one occupying nearly two-thirds of the total area of the lake. It takes a triangular shape due to acute changes in the trend of the southern coastline, whereas the western basin takes the form of a water arm parallel to the western **coastal barrier** (Figs. 22.4 and 22.5). This water arm extends for about 40 km in SW direction, with a varying width between 4 and 10 km. The Bardawil Barrier forms the northern shores of this western basin, whereas sand dunes line its southern shores. Along the southern shores, several small local barriers (spits) developed leading to the development of small local lagoons. The eastern basin is bound from the north by the eastern part of the coastal barrier, which extends from El-Qals Peninsula in the west to El-Zaranik in the east and by sand dunes from the south and west. The length of the triangle base from El-Qals in the north to Misfaq in the south is about 20 km, while the length of the other two arms is more or less equal (±45 km).

Figures 22.4 and 22.5 also show the presence of some local basins along the southern peripheries of Bardawil. Most are grouped into interconnected compound basins.

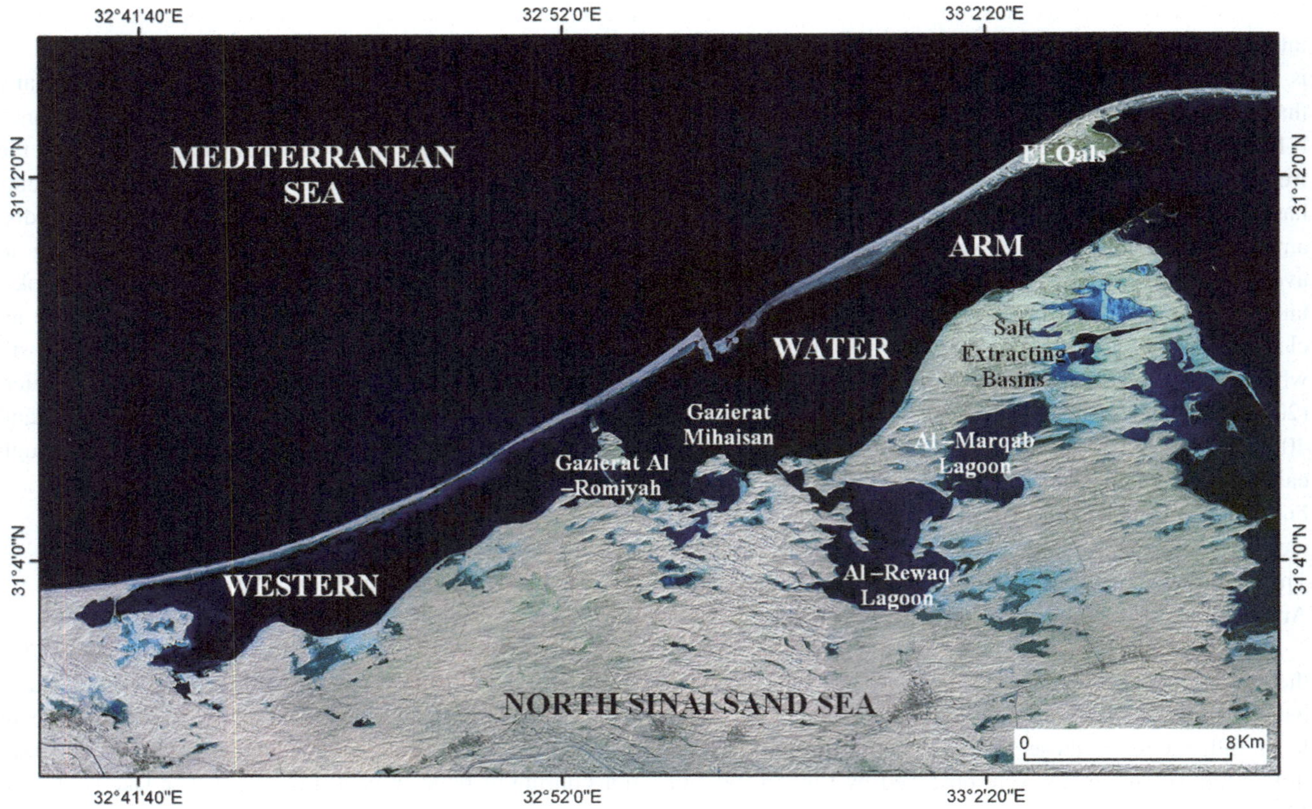

Fig. 22.4 Landsat ETM image showing the morphological aspects of the western part of Bardawil Lake

There are three groups of local basins. The largest group is composed of two interconnected micro-lagoons (El-Rewaq and El-Marqab) and a waterway (El-Misqah) connecting them to the western basin of Bardawil Lake (Fig. 22.4). Both basins are connected by a narrow, small opening. They extend in an N-S direction. The straightness of the northern and southern coasts bears the effect of tectonics on the development of these two basins. The second group of local lagoons exists just to the west of the first, but is much smaller. It is also composed of two micro-lagoons (1.0–1.5 km long), connected with a narrow waterway and connected to the western arm of the main lake with a relatively wide opening.

The third group of basins is present on the southern coast of the main basin of Bardawil, just to the west of El-Telool Jetty (Fig. 22.5). It is composed of a main lagoon that takes an E-W orientation (2.5–1.0 km) and a micro-lagoon in the inner eastern part of the main lagoon due to the development of N-S sand spit in this part. The straightness of the coastlines bears the effect of tectonics on the development of the main lagoon in this group. The fourth group of basins also lies along the southern coasts of the main basin of Bardawil to the east of the third group. It is composed of three small basins, two of them are interconnected, and the third is separated by a minor promontory. This group is separated from the main body of Bardawil Lake by some sand barrier. The largest basin extends for 2.0 km from north to south and 1.5 km in E-W direction, with a relatively large island (El-Artah) in its center (Fig. 22.5).

22.5.2 The Coastal Barriers

There are two types of barriers according to size, location, and age. The first is the large sand barrier, which separates the lake from the open sea, and it will be called "the **Bardawil Barrier**", whereas the second type is represented by the small barriers that developed in the lake basins, and they will be called "the local barriers."

The Bardawil Barrier separates the lagoon from the Mediterranean Sea, with a total length of about 90 km from El-Zaranik in the east to the tip of the lagoon water arm in the west, ranging in width between 300 and 1000 m, and being a few meters high. However, there are some points, which rise several tens of meters above sea level, of which the famous coastal dune called El-Qals (Casius Mont) is the highest (56 m asl). This high dune lies nearly in the center of this barrier. The western section of the barrier is straighter than the eastern one. This characteristic is interpreted in previous studies as due to the effect of the tectonic Pelusium lineament (see Sect. (22.2). Also, it can be seen on the satellite images that the farthest eastern section of the barrier

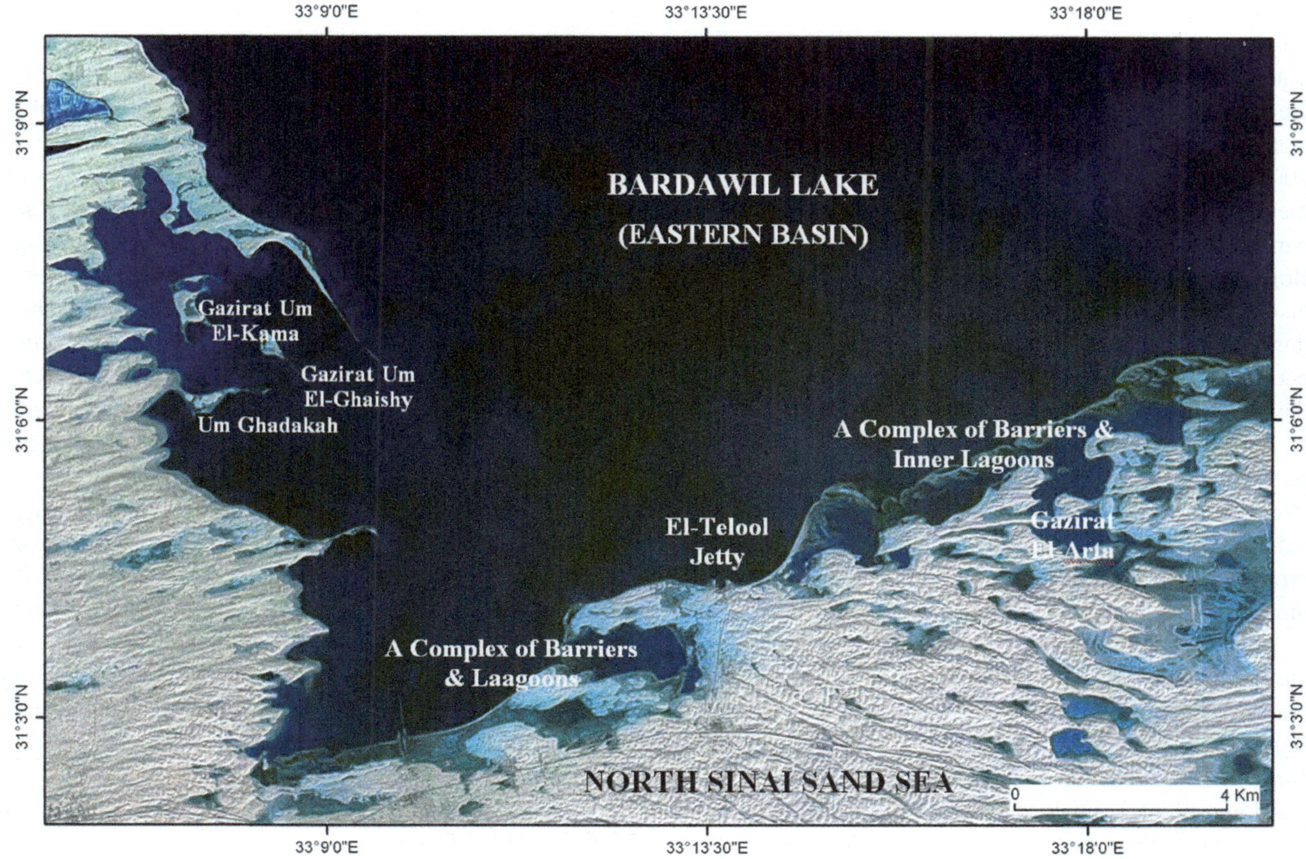

Fig. 22.5 Landsat ETM + image showing micro-lagoons and barriers in the Eastern Basin of Bardawil Lake

(just before Bogaz El-Zaranik) is composed of several small barriers. Most probable, this is due to shallowness of water in this section of the barrier.

There are two types of openings in this barrier. The first type comprises two artificial inlets (**Bogaz I and Bogaz II**) to keep flow of water from and to the Mediterranean going on. Bogaz I opens at the western basin, while the Bogaz II opens at the eastern basin. These two inlets are exposed to morphologic changes due to erosion and sedimentation under the effect of near-shore sea water movements (waves and longshore current). A previous study (Klein 1986) showed that sediments transported by longshore current are deposited in the inlet entrance and are later moved into the inlets by waves entering the lagoon, sedimentation occurs mainly in summer, and barriers appear on the inlet sides and the sides of the approaching channels.

The second type of openings is natural and comprises narrower openings. According to previous studies, there is only one opening at El-Zaranik in the extreme eastern part of the barrier (Assal 1999), but recent space images, Sept. 2009 (Google Earth, 2011), show that there are other two natural openings in this eastern part of the barrier. Some old maps show that there were four natural openings (Survey of Egypt 1935, Scale 1:100,000), and all of them carry names (Bogaz El-Zaranik, Bogaz Geran, Bogaz El-Huseini, and Bogaz Asy) and lie in the extreme eastern part of the barrier (El-Zaranik area). Variations in the number of natural openings from time to time are mostly due to the geomorphologic characteristics of this part of the barrier and lake, where the lake is very shallow, the barrier is very narrow and sandy. Therefore, an opening can be readily developed or silted up in this extreme eastern part of the lake. Four local landforms characterize this outer large barrier. These forms are active coastal sand dunes, sabkhas, submerged lee dunes, fans, and coastal micro-lagoons.

Active dunes are concentrated on the seaward section in general and in particular on the central part of the barrier around the peninsula of El-Qals (Fig. 22.3). El-Qals coastal dune is the highest point on the Bardawil barrier (56 m asl), while other surrounding dunes are much lower than this (10–20 m asl). The extraordinary height of El-Qals dune is due to a fact that it developed on a ridge of beach rock formed during the Flandrian transgression (Neev et al. 1987). Most probably, the sea was the main sand source for such coastal dunes, which extend parallel to the shoreline.

Sabkhas cover the southern part of the Bardawil Barrier as a narrow disconnected strip. Some of these sabkhas were indicated on the maps as dry. However, the field check in the study of Matook (1999) indicated that there are some wet sabkhas on the Bardawil Barrier. Their surface does not rise much above sea level (<2 m). Mostly, these coastal sabkhas are fed by seawater.

Submerged lee dunes: Space images taken in September 2009 (Google Earth, 2016) show that there are some submerged lee dunes formed along the southern shores of the barrier in the section between El-Qals Peninsula and Bogaz I, extending southward for about 50–100 m into the lake basin. Since dunes cannot develop underwater, these lee dunes developed during periods of low stand, before the permanent submergence of the lake basin by seawater, after the two artificial inlets were dug in 1953 (Bogaz I and II).

Fans: These wash-over fans were developed by flows of seawater that pile against the barrier during heavy storms (Neev et al. 1987). Such fans are common along the eastern half of the barrier compared to the western section, most probably due to the abundance of sediments on the shores of the eastern section of the barrier.

Micro-lagoons developed along the southern coasts of the Bardawil Barrier due to the formation of separate minor sand barriers at various localities parallel to the coastline of Bardawil. In one case, **submerged dunes** parallel to the barrier in the extreme eastern part of the laggoon led to development of a small lagoon, which is called “Zaranik Lake” according to 1935 topographic maps (scale 1:100,000). Length of these micro-lagoons ranges between less than several hundreds of meters to a few kilometers, whereas width does not exceed a few hundreds of meters.

The second type of barriers comprises—barriers (barriers, spits, and hooks), which are formed in the lagoon basin along its shores and the inner shores of the main Bardawil Barrier. Their length ranges between 200 and 1700 m, width between 10 and 20 m, and height between zero and a few meters above sea level. Small barriers are simple, whereas the larger ones are compound and complex. The largest compound barriers are those extending parallel to the southern shores of the main basin. This group is composed of several barriers, spits, and hooks all extending along the same line and is connected from both ends to the mainland as two spits. Spits are detected in several localities along the shores of the lake, especially along the shores trending N-S of the main basin. There are two general trends in the direction of barriers: The dominant is E-W to ENE-WSW trend and N-S trend. The first trend is present along the southern shores of the lake, except for the shores, which extends in an N-S direction where barriers take the same trend. All barriers take the trend of the neighboring shores.

In several localities, the barriers take the form of hooks, the largest of which occurs in three points. The first lies at the tip of the southern shore when it changes its direction from SW-NE to N-S, the second is at the southern tip of El-Qals Peninsula, and the third is at the southern shores of the eastern main basin. The first place contains the largest composite hook in the lake and is called locally “Mascat Iblis/Mat Iblis” (Fig. 22.6). It is composed of nearly 12 hooks, the largest of which is about 1.7 km long. These hooks—as they are bending southward—indicate that the lake water moves anticlockwise, coming from the western basin in a general W-E direction parallel to the southern shores of the lake and forming some small spits along these shores. The water current changes its direction toward the south as the shores of the lake change its trend to the south. At the point of change, sand spits develop eastward, but under the effect NW wind, the offshore current changes its direction southward transforming sand spits to sand hooks. As it continues moving southward, the offshore current formed new spits pointing in the same direction of the current.

After a distance of about 15 km, the N-S shore changes its trend toward the east, and also the direction of the offshore current becomes eastward. Along this section of the shores of the eastern basin of the lake, the longest compound barrier developed parallel to the shoreline. In fact, this compound barrier is composed of several barriers closing several small inlets that developed in the inter-dunes low-lying areas, transforming them into small local lagoons.

22.5.3 Tidal Forms

Tidal forms are recognized in the extreme eastern corner of Bardawil Lake. These forms are tidal flats, tidal channels, and tidal inlets.

Tidal flats cover a wide area in this region to the south of the Bardawil Barrier, but they are separated by a wide tidal channel. As indicted on topographic maps (scale 1:50,000) (Mazar Sheet, No. NH 36. O-1b, Military Survey of Egypt 1986), these tidal flats rise only 1 m above sea level after withdrawal of water from its surface. On the surface of the tidal flats, crests of **submerged dunes** rise more than that (2–18 m asl).

Tidal channels develop during the withdrawal of water from the lagoon. During this process, water is concentrated along certain paths which become deeper by time and are transformed into channels. Small channels unite together or join larger ones, and finally, with the highest order channels which end at the two natural inlets present in the eastern part of Bardawil.

Tidal inlets: There are two natural openings in the eastern corner of Bardawil Barrier as appear on space images taken on Nov. 2005 (Google, 2015, but four natural inlets (Bogaz El-Zaranik, Bogaz Geran, Bogaz El-Huseini, and Bogaz Asy) appear on topographic maps (scale 1:100,000) (Survey of Egypt 1935). This indicates that tide is an active agent in the development and disappearance of openings along this eastern extreme section of Bardawil Barrier. Tide

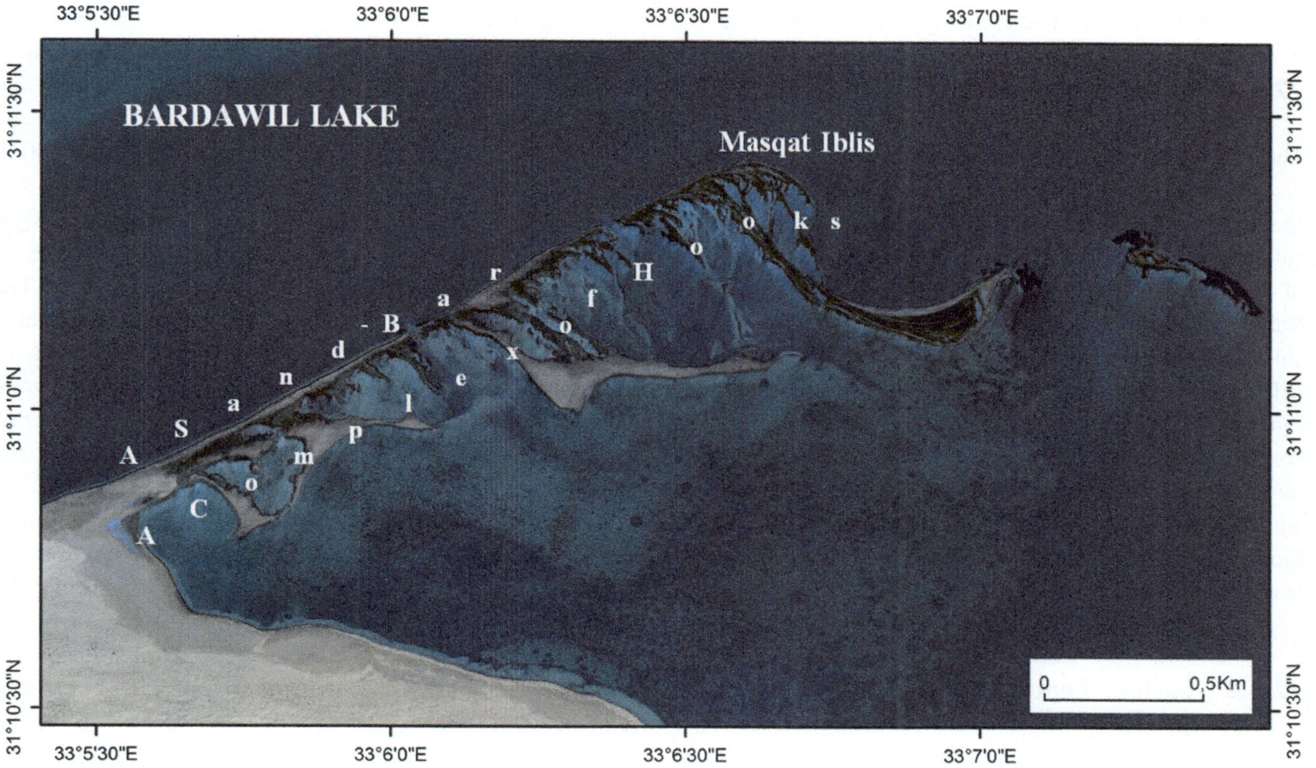

Fig. 22.6 Complex of Hooks of Mascat Iblis (*Source* Google Earth, 2014)

may break the barrier, when it is strong and the barrier is weak, making an opening, or may silt up an opening, when tide is weak and water is shallow, and close it. Therefore, the natural openings of Bardawil Barrier are in fact tidal inlets.

22.5.4 Submerged Forms

These forms developed due to the rise of sea level in the last few thousand years during the Flandrian sea rise. The pre-topography determined types and shapes of drowned forms, of which sand dunes, coastal inlets, and inner lagoons are the most significant.

Sand dunes: The North Sinai Sand Sea developed about 18,000–20,000 years ago (Sarnthein 1978; Pye and Tsoar 1990). At that period, sea level was almost 30 m lower compared to present-day levels. According to topographic maps (scale 1:5000) and space images, many of the islands have a crescentic outline indicating that they are most probably drowned dunes. These dune islands were relatively high dunes—lower section is now covered by water and the upper parts formed the islands. On the other hand, low dunes were submerged completely below the water level of the Bardawil. It can be concluded from the presence of some submerged dunes near the Bardawil Barrier that sand dunes covered the whole surface area of the present-day Bardawil Lagoon. This conclusion conforms to the findings of Levy (1974) that eolian sand is one of the sources of sediments laid down on the floor of Bardawil.

Coastal inlets developed at the low-lying inter-dune (linear dunes) areas. These inlets were formed only along the western coasts of the main basin of Bardawil Lake from Mat Iblis in the north to Misfaq in the south. This section of the coasts takes a nearly N-S trend, whereas the orientation of linear dunes is W-E. This is why sea water intruded into inter-dune areas transforming them into coastal inlets similar to the drowned wadis along the Red Sea coasts. High linear dunes protrude into the lake as headlands. This is why coastal inlets developed only along this N-S section of coasts of the lake, and this is why the coastline takes a zigzag pattern.

Inner lagoons: The rise of sea level resulted in the submergence of nearby low-lying basins, forming very small local lagoons. Of these local lagoons, "El-Rewaq" and "Mallahat El-Merqab" are the largest. They extend a few kilometers in the E-W and N-S directions. Both lagoons exist to the south of the lake arm (the western basin) and are connected to the main water body by a narrow waterway, which is locally called "El-Misqah" (i.e., a feeding channel). Small local lagoons also developed along the southern shores of the eastern basin due to the closure of coastal inlets by small sand barriers. Some of these inner lagoons were used for salt extraction works.

22.6 Sabkhas Surrounding Bardawil Lake

22.6.1 General Aspects

The **Sabkhas of North Sinai** were studied by several researchers. Levy (1974, 1977a, b) studied the sabkhas of Madha, Nahal Ya, Hawash, Al-Mustabik, and another one called by Levi Hafiordim. Assal (1999) studied the sabkhas of Misfaq, El-Zaranik, Rommana, west El-Roda, El-Fateh, Rabaa Salmana, and 6th of October. Attia et al. (1998) studied four sabkhas: two of them are coastal sabkhas (El-Mustabik and Sbeika), and the other two are the inland sabkhas of Sheikh Zuweid and El-Fateh. The following section will discuss the relevant characteristics of sabkhas of northern Sinai according to the findings of previous studies and the data extracted from topographic maps of northern Sinai. Characteristics of sabkhas include two aspects: surface features, and sediments and stratigraphy.

22.6.2 Surface Features

Nearly, all sabkhas have a planar surface, ranging in elevation from sea level to a few meters above. However, some parts of some sabkhas lie 1–2 m below sea level, such as Sabkhat Tolan and Sabkhat Hawash. It was concluded from previous studies of five sabkhas (Sbeika and El-Mustabik as coastal sabkhas, El-Sheikh Zuweid and El-Fateh as inland sabkhas by Attia et al. (1998), and Sabkhat El-Madha by Levy (1977a) that the surfaces of sabkhas could be divided into three zones. These zones are the high outer dry zone, intermediate wet zone, and basin center zone—on the basis of elevation variations, sediment distribution, and sedimentary structures. Coastal sabkhas are fed by lagoon water, but may be fed by underground water which seeps from the base of neighboring dunes. When evaporated, halite is precipitated on the surface, and due to the frequency of this process, thickness of salt increases, forming a salt crust. This crust is found on the surface of most studied sabkhas. The salt crust is mined in some sabkhas for the purposes of industry as well as human consumption.

The surface features will be presented for a few sabkhas considered as examples. These sabkhas are the following: (1) El-Madha (Topographic maps, scale 1:50,000), which is the Misfaq sabkha, according to the study of Assal (1999) and is also called Nahal Yam (Levy 1977a); (2) Sabkhat Hawash; (3) Sabkhat Hafiordim (south of El-Maraqab lagoon); (4) Sabkhat El-Mustabik.

Sabkhet El-Madha lies to the south of Bardawil Lagoon in the vicinity of Misfaq town. It extends for about 8.0 km from east to west, and 0.25–1.0 km from north to south. It is separated from Bardawil by a sand bar in its eastern part and is bordered by low (1–2 m) sand dunes along its southern margins. The surface of Sabkhat El-Madha is divided into three zones (Levy 1977a) on the basis of flooding by lagoon/underground water: (1) Intertidal zone, 100–200 m wide, covered by algal mats, and regularly flooded by tide water. Large gypsum crystals are present within the algal mat layer. (2) An outer transition zone, 100 m wide, is flooded occasionally by lagoon/underground water. The algal mat in this zone is covered by detritus quartz sand. Finally, (3) the high supratidal zone lies at the southern extremity of the sabkhas and is about 500 m wide. This zone is not flooded by lagoon water, but groundwater flows to this area from the south after winter rainfall. Gypsum is the major evaporitic mineral in this sabkha.

Sabkhat Hawash lies to the south of the eastern basin of Bardawil Lagoon, with a maximum E-W extension of about 7 km and a maximum N-S breadth of 5 km. Except for the sand dunes with heights varying between 9 and 31 m asl, which scatter on the surface of the sabkha, its surface rises one meter only asl, with a central area of 2 × 2 km lying 1 m below sea level (Topographic maps, scale 1:50,000, Sheet NH 36 O1b). This sabkha is an example of a sabkha receiving highly saline brine from a restricted lagoon environment and negligible underground water, where gypsum and halite with large crystals cover most of the surface, with the predominance of halite in the southern part (Levy 1977a). These characteristics reflect the restricted hydrographic conditions of the sabkha, due to its location to the south of the eastern basin of Bardawil where maximum salinities prevail and the separation from the eastern basin of Bardawil by sand dunes and bars.

Sabkhat Hafiordim lies to the south of El-Marqab lagoon. It is an example of an inland sabkha that has been transformed from a lagoon beach sabkha in an earlier cycle, due to disconnection from lagoon (Levy 1977a). It is small and narrow, with a maximum width of about 300 m and a maximum length of about 2 km. Most of its surface is covered by a flat layer of halite, which reaches a maximum depth of 1 m in the central part of the sabkha, but the thickness of halite decreases toward the peripheries. The halite bed is intercalated by 1–2 mm laminae of quartz sand, separating layers of pure halite of about 1 cm thick.

Sabkhat El-Mustabik represents another example of inland sabkhas. It lies about 1 km to the south of Sabkhat Hawash and is separated from it by sand dunes. It is a small sabkha, covering an area of about 7 km^2. Its surface is flat, covered by a thin layer of halite, and is encircled by sand dunes. The water table is shallow (10–50 cm) below the surface most of the year (Attia et al. 1998). After winter rainy months, the groundwater level rises up, the sabkha is

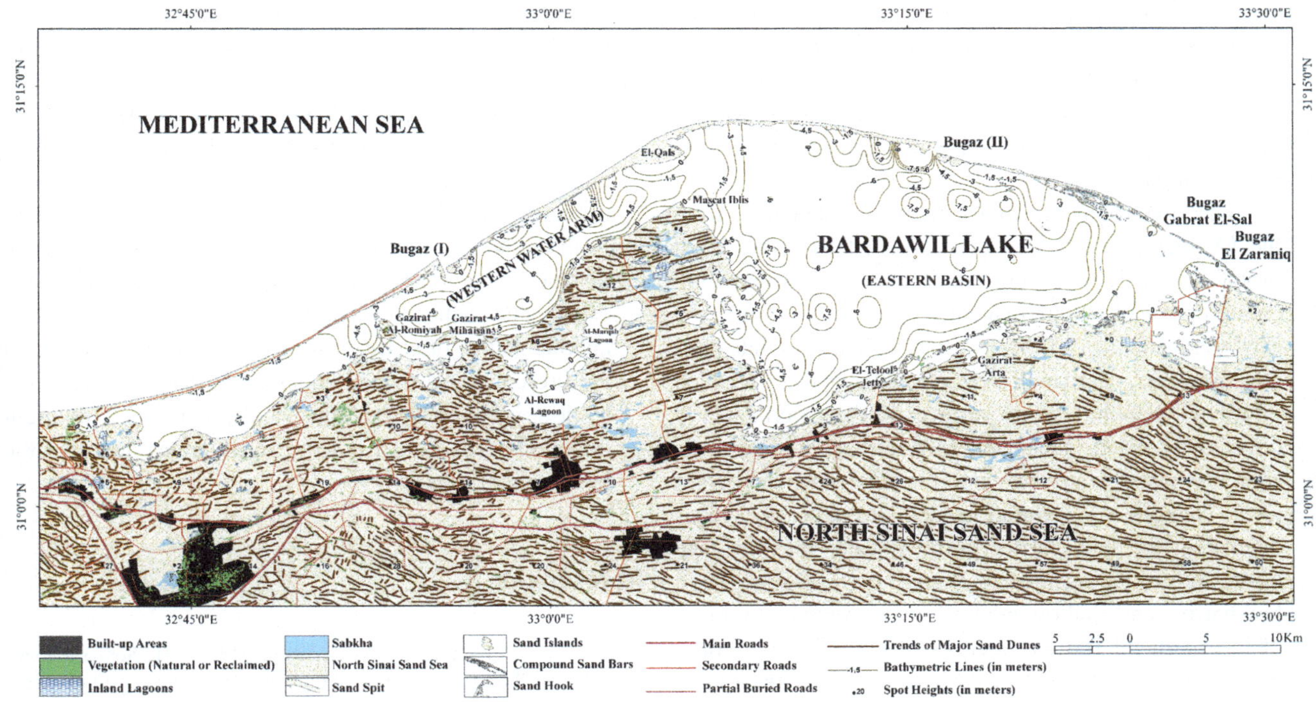

Fig. 22.7 Geomorphologic map of Bardawil Lake and its environs (*Source* Embabi and Moawad 2014)

flooded, and halite is dissolved in most of the sabkha. Due to high rates of evaporation during summer and autumn, water level is lowered and halite is re-crystallized a few weeks after each flooding.

22.6.3 Sediments and Stratigraphy

Sediments of North Sinai sabkhas are mainly composed of sand, gypsum, and algal mats, with the predominance of sand. This sedimentary composition reflects environmental conditions of the coastal area of North Sinai, which is dominated by sand dunes, availability of saline water, and arid climatic conditions (Fig. 22.7). However, there are some local differences in the stratigraphical succession and sediment characteristics. Nearly, Halite forms the top layer in all sabkhas, with a thickness varying between 1 and 10 cm. However, thickness of halite reaches a meter in the central parts of the inland Hafiordim Sabkha and decreases toward the margins. The halite in this sabkha forms well-crystallized cubes. The halite bed is intercalated by 1–2 mm layers of quartz grains, which separate layers of pure halite about 1 cm thick (Levy 1977a). It seems that the salt crust in some sabkhas did not disappear completely during flooding times by water, therefore, its thickness increases by time and becomes a permanent crust. Such sabkhas are called locally Mallaha which means "Salina," such as Mallahat Mikhzan in the east and Tolan in the west.

The section of the inland Sabkhet Mustabik shows a different sedimentological composition than that of other sabkhas. It is composed of dolomite and calcite. According to Levy (1977a), there is dolomite in the sedimentological composition of other inland sabkhas. Its age is identified as 35,000 years as determined by radiocarbon dating.

In Sabkhat Hafiordim, sediments of an earlier cycle (older than 30,000 years) outcrop as low hills protruding from the sabkha surface (Levy 1977a). The thickest sediment is a coquina bed (1–2 m), which is composed of shells cemented by carbonate material. This bed overlies either bedded quartz sand or a layer of shale. Large masses of eroded gypsum occur at the base of the section. Since these sediments exist at a higher elevation than the water table, no interaction occurs between them and the brines, which attain salinities of about ten times that of normal seawater.

22.7 Conclusion

Bardawil Lake is unique among the Mediterranean coastal lagoons in two aspects. The first is that it is fed by sea water only. The second aspect is that the Bardawil Basins are of tectonic origin. The original depth of the lake basin was several times greater compared to present day.

Is Bardawil a lake or a sabkha? Bardawil was a lake during certain periods of its history, such as the Roman Period, and at present as well since the opening Bogaz (I) and Bogaz II in 1953.

These two inlets allowed permanent water flow from the Mediterranean Sea to the basins of the lake. Most probably, sea level during the Roman Period was high enough to allow permanent water flow from the Mediterranean Sea to the lake basin. During other periods, however, Bardawil was a sabkha and was called "Bardawil Sabkha" as its surface was not permanently covered by water. Since the Mediterranean is the only source of water, water flow into the lake depended on sea level changes, and duration and strength of storms overriding the lake barrier pushing water into the lake basin. Before the opening of Bogaz (I) and Bogaz (II), a few narrow and shallow inlets in the extreme eastern part of the Bardawil Barrier, such as El-Zaranik Inlet, could not provide enough water year round to develop a lake, and sabkhas developed instead. Nowadays, the extreme eastern section of the lake is a tidal flat and is a sabkha throughout most of the year. Figure (22.3) shows that sabkha deposits represent the top layer on the floor of the lake basin with a thickness reaching more than 10 m in its central section. This indicates that sabkhas dominated the surface of the Bardawil Basin before the permanent submergence since the opening of Bogaz (I) and Bogaz (II) in 1953.

Although there is no study dealing with the internal movement of the water, the minor forms of Bardawil indicate the presence of such a movement. Apart from forms developed due to the submergence of preexisting topography, the predominant forms are depositional rather than erosional. These are coastal sand barriers, among which hooks and spits are the most widespread. Small spits are developed on the southern shores of the western basin (arm) of Bardawil, pointing eastward and indicating an eastern outward water movement in this western arm. At the end of the southern shores of this arm, a complex of hooks (2 km long) developed (Mascat Iblis/Mat Iblis). Although pointing eastward, hook bends are pointing southward indicating a southern water movement, which continues parallel to the eastern shores of the main basin of Bardawil. Along these shores, several spits and hooks developed pointing southward, confirming the movement of water in the southern direction. Lines of sand barriers, which run parallel to the southern shores of the main basin of Bardawil, indicate an eastward water movement. Therefore, it can be concluded that there is an anticlockwise water movement in this part of the lake.

The sediments in Bardawil Lagoon are derived from various sources, of which the Nile sediments are the main contributor. These sediments are brought to the coasts of northern Sinai by the easterly directed offshore current from the Nile sediments discharged in the Mediterranean Sea at the mouths of the Nile branches and drainage canals and outlets of coastal lagoons. Eolian sand is another source of sediments for Bardawil Lake. Although the presence of drowned dunes in the lake indicates that its floor is covered by aeolian sands, winds from different directions drift sand from the surrounding dunes to the lake body at present.

References

Abdel Aal A, El Barkooky A, Gerrits M, Meyer H, Schwander M, Zaki H (2000) Tectonic evolution of the eastern mediterranean basin and its significance for hydrocarbon prospectively in the ultra deepwater of the nile delta. Lead Edge 19:1086–1102

Assal E (1999) Sedimentological studies on the Quaternary sand dunes and sabkhas, Northern Sinai, Egypt. Dissertation, University of Mansoura

Attia O, Aref M, El-Banna M (1998) Sedimentological significance of modern evaporate settings, Northern Sinai, Egypt. Egypt J Geol 41 (2A):89–118

Emanuelsson D, Mirchi A (2007) Impact of coastal erosion and sedimentation along the northern coast of Sinai Peninsula. Dissertation Lund Institute of Technology, Lund University

Embabi N, Moawad B (2014) A semi-automated approach for mapping geomorphology of El-Bardawil Lake, Northern Sinai, Egypt, Using integrated remote sensing and GIS techniques. Egypt J Remote Sens Space Sci 17:41–60

Google Earth (2014) Egypt Images Latitudes 22-32 N and Longitudes 22-32 E, http://www.google.com

Frihy O, Badr A, El Sayed W, Selim M (2002) Environmental impacts of El Arish power plant on the mediterranean coast of Sinai, Egypt. Environ Geol 42:604–611

Frihy O, Lotfy M (1997) Shoreline changes and beach-sand sorting along the northern Sinai coast of Egypt. Geo-Mar Lett 17:140–146

Hamdan G (1980) The personality of Egypt, vol I: The physical characteristics. Dar El-Hilal, Cairo

Inman D, Jenkins S (1984) The Nile littoral cell and man's impact on the coastal zone of the south eastern Mediterranean. Coastal Engineering Proceedings, North America

Khalil M, Shaltout D (2006) Lake Bardawil and Zaranik Protected Area. Publication of National Biodiversity Unit No. 15, MedWet-Coast Project, Department of Nature Protection, Ministry of State for Environmental Affairs, Cairo

Klein M (1986) Morphological changes of the artificial inlets of the Bardawil Lagoon. Estuar Coast Shelf Sci 22:487–493

Kramgatz B, Hornung H, Oren O (1980) The study of a natural hyper salinelagoon in a desert area (The Bardawil lagoon in Northern Sinai). Estuar Coast Shelf Sci 10: 403–415

Levy Y (1974) Sedimentary reflection of depositional environment in the Bardawil Lagoon, Northern Sinai. J Sed Petrology 44:219–227
Levy Y (1977a) Description and mode of formation of the supratidal evaporate facies in northern Sinai coastal plain. J Sed Petrol 47:463–474
Levy Y (1977b) The origin and evolution of brine in coastal sabkhas, Northern Sinai. J Sed Petrol 47:451–462
Linnersund J, Mårtensson E (2008) Hydrodynamic modelling and estimation of exchange rates for Bardawil Lagoon, Egypt. MSc. TVVR 08/5010, Division of Water Resources Engineering, Department of Building and Environmental Technology, Lund University, Sweden
Matook A (1999) The bardawil barrier: a geomorphologic study. Bulletin of Humanities, Faculty of Arts, Damanhour, University of Alexandria 3:125–193
Neev D, Baker N, Emery K (1987) Mediterranean coasts of Israel and Sinai. Taylor & Francis, New York
Pye K, Tsoar H (1990) Aeolian sand and sand dunes. Unwin Hyman, London
Sarnthein M (1978) Sand deserts during glacial maximum and climatic optimum. Nature 272:143–146

Part VIII
Applied Geomorphology

Geo-Parks and Features of Geomorphic and Geologic Significance

23

Abstract

This chapter deals with various areas and features/forms of geomorphic and geologic significance in the Egyptian landscape. They are not treated in any other part of this book. These areas/features are: (1) A small peneplained plateau in SW Western Desert, (2) Buried (Radar) wadis and channels, (3) The White Desert, (4) Domes of North Sinai, (5) Inverted Wadis, and (6) Djara and Sannur Caves.

Keywords

Peneplained plateau • Buried wadis • Domes of north sinai • Inverted wadis

23.1 Introduction

Features, forms, and areas analyzed in this chapter have not been addressed in any other part in this book. They are selected from various localities, because of their geomorphic and geologic significance. Among these selected areas, there are three **Geo-Parks**. Since 1983, the Egyptian Ministry of Environment declared more than 30 areas as Geo-Parks (Natural protectorates) to conserve their features from deterioration for touristic and scientific purposes. The peneplained igneous plateau, an elevated old erosion surface SW the Western Desert, represents a new type of plateaus that could have been formed by erosion. **Buried wadis**, discovered by radar imaging in the 1980s, under the Selima Sand sheet in the southern part of the Western Desert, represent remnants of an old drainage system. Domes of the Syrian Arcs in the northern region of Sinai are a unique landscape. The clustering of such a large number of anticlines of different sizes gives this region several geomorphic aspects, some of which are textbook examples. Inverted wadis are of a special significance, not only as specific landforms on the Eocene plateau and in some other localities, but also for their geologic aspects as evidence of the Prenile drainage systems. The Djara Cave in the Western Desert and the Sannur Cave in the Eastern Desert are among the selected Geo-Parks, because they are the only complete caves with typical speleothems.

23.2 The Igneous and Metamorphic Peneplained Plateau

Topographic maps (scale 1:250,000), geologic maps (scale 1:500,000), and air photomosaics (scale 1:100,000) show that a rugged plateau lies to the north of the two hills known as Peter and Paul, approximately between Latitudes 22°25′ and 22°50′ N, and Longitudes 25°15′ and 25°30′ E. Prevailing heights range between 850 m and 950 m a.s.l, and the highest peak rises up to 1114 m a.s.l. It has a maximum N–S length of about 47 km, and a maximum E–W width of 25 km (**Fig. 23.1**). The Geological Map (scale 1:500,000) (Sheet No. NF 35 SW) describes this plateau as a **peneplain plateau**. Reference to this plateau was made in three previous publications only (Clayton 1933; Richter and Schandelmeier 1990; Embabi 2004). This plateau was discovered during the exploration of the Gilf Kebir region in April 1932 by the Clayton–Almasy Expedition (Robert Clayton East Clayton, Count L. E. de Almásy, P.A. Clayton and H. W. G. J. Penderel). In 1933, Clayton reported *"north of the Peter and Paul peaks lies an extensive plateau, about 900 m a.s.l,*

N.S. Embabi, *Landscapes and Landforms of Egypt*, World Geomorphological Landscapes,
https://doi.org/10.1007/978-3-319-65661-8_23

with a single peak rising to 1114 m a.s.l. The plateau is mostly granite, and is nearly cut into two divisions by a large wadi." Since then, this plateau has been completely forgotten, and became unknown, due to two reasons: It lies in a very remote and inaccessible area, and necessary tools were lacking such as detailed maps and air photos for its more through research. Recently, Richter and Schandelmeier (1990) considered this plateau, in their description the geology and petrography of the Gabal Uweinat region, as one of its four geographic sub-areas. They described it as an elevated plateau topped by a flat denudation surface and cut off by wadis. However, the publication of topographic maps and air photo mosaics (scale 1:100,000) made it possible—with the aid of geologic maps—to describe the main geologic and geomorphic aspects of this plateau.

23.2.1 Geology

The Geological Map (scale 1:500,000) (Conoco Coral and the Egyptian General Petroleum Corporation 1986–1987, Sheet No. NF 35 SW) shows that this plateau is composed of the following metamorphic and igneous rock units:

(1) Migmatic gneiss, restites, quartizite, marble, calc-cilicate rocks, and amphibolite. These rocks were grouped under one formation called Anatexite Formation and represent the Pre-Pan African old continental crust. These oldest rocks form most of the plateau surface.

(2) Gray green granitiods cover several plots, mainly along the eastern margins.

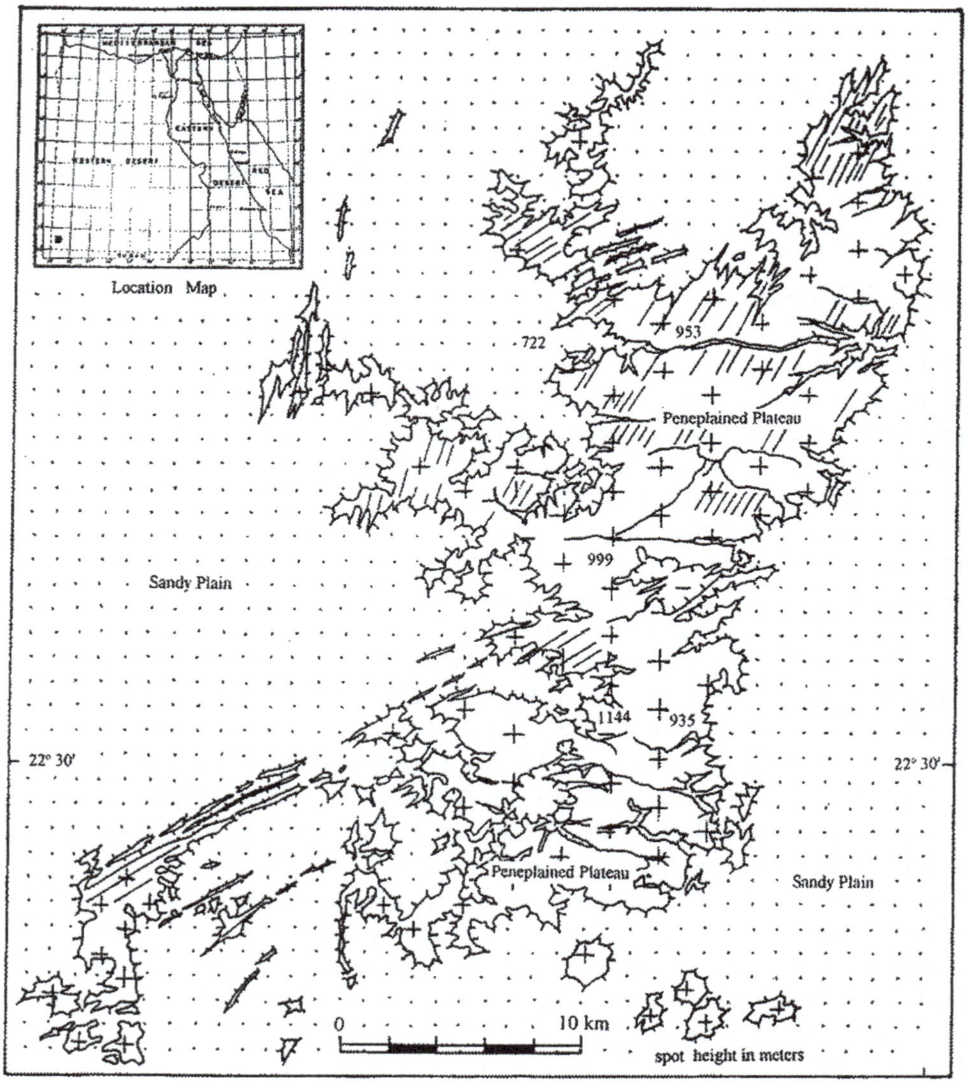

Fig. 23.1 A morphologic map of the Igneous Peneplained Plateau in SW Egypt

(3) Red granite is exposed along some of the western promontories and along a narrow strip along the southern margins of the plateau.
(4) Minor exposures of acid volcanic rocks (quartz porphyry).
All these rock units belong to the Pre-Cambrian Era.
(5) Sandstone of fluviatile origin, possibly marginal marine with occasional marine influence, appears as very minor exposures at several sites on the gneiss surface of the plateau. This group of sandstone is classified as Gilf Kebir Formation, which belongs to Lower Cretaceous.
The plateau is highly dissected by joints and faults, which controls the orientation of several features. The main trends of these tectonic features are W–E, N–S, and NE–SW.

23.2.2 Geomorphology

This plateau is the only plateau originated by erosion not only in the Western Desert, but also nationwide. It seems that its surface was exposed to degradational processes during a long span of time; hence, it acquired the appearance of a peneplain. The presence of some outliers of the sandstone belonging to the Gilf Kebir Formation atop the peneplained metamorphic and igneous rocks indicates that peneplanation took place prior the deposition of the sedimentary cover. Since the Gilf Kebir Formation belongs to Lower Cretaceous, it must have taken place prior to the time of deposition of this formation. The age of the original peneplained surface is pre-Cretaceous.

This surface is similar to that one found on the basement rocks in **Sinai Mountains** and the Eastern Desert, and topped by a sandstone succession of Cretaceous age. The peneplained surface of Sinai Mountains remained unexposed until the Oligocene, when the horst of Sinai developed during the formation of the Gulf of Aqaba and Red Sea graben. By analogy, it can be inferred that the peneplained metamorphic and igneous surface in the Western Desert was covered by a sedimentary succession during the Lower Cretaceous and remained unexposed until the whole block was elevated during one of the tectonic movements which occurred in post-Lower Cretaceous times. Since then, the sedimentary cover was eroded, gradually exposing the original peneplained surface. Since the block is elevated, it takes the appearance of a plateau.

The surface of the plateau is dissected into small parallel ridges (100–300 m), which change their orientation from NNE–SSW to N–S to NE–SW according to the trend of fracture and fault systems. Inter-ridges low-lying grooves are occupied occasionally by the minor tributaries of the main wadis. Along the NE and SW margins of the plateau, some of the inter-ridges corridors are widened, leaving some of the ridges as an independent form separated by relatively wide areas, varying between one and five km (Fig. 23.1).

The surface of the plateau is also dissected along main faults and fractures—by wadis, which take different trends (Fig. 23.1). Due to this difference, those wadis of the western side look like embayments, filled with sand. Wadis of the eastern side are relatively narrow (50–200 m wide in the upper and middle reaches). It seems that the wadis of the eastern side are more controlled by faults and fractures than the western side. Sides of the wadis look like a jigsaw due to the effect of the perpendicular fracture system on the plateau surface.

As a result of the dissection of both the western and eastern sides of the plateau, several parts of different sizes have been separated and become as outliers around the plateau edges. Figure (23.1) illustrates that the outliers of the western side are irregular and take linear trends—in general. It seems that the western side was exposed to dissection and erosion for a longer period than the eastern side. The irregular shape and linear trend of the outliers are due to the effect of the fault and fracture system. As in the Gilf Kebir and Abu Ras Plateaus, the outliers indicate that the present plateau is a remnant of a wider one, which covered an area—most probably—twice that of the present plateau.

23.3 Buried Wadis and Channels

In November 1981, during the journey of the Space Shuttle Columbia, the Space-Borne Imaging Radar (SIR/A) discovered buried forms beneath the sand sheet of Selima Sand Sea. The radar penetrated large areas of the Eastern Sahara where the sand sheet is no more than several tens of centimeters thick. It revealed a web of wadis and smaller channels beneath the desert sands. Since then, several studies have been carried out on the buried features. In this subsection, the author does not wish to make a full review of previous literature, but reference will be made to those studies relevant to this analysis. These buried fluvial forms have been identified in the previous studies by several terms such as "**Radar Rivers**" and "**Subsurface Valleys**" (McCauley et al. 1982; McCauley et al. 1986b; Ghoneim et al. 2007), "**Paleorivers**" (McHugh et al. 1988), "**Paleo-drainages**" (McCauley et al. 1986a; Schaber et al. 1997) and "Paleochannels" (El-Baz and Robinson 1997).

The presence of paleo-drainages buried beneath aeolian sands in the southern part of the Western Desert such as the Great Sand Sea was inferred from topographic maps and space images (Embabi 1998). In the meantime, prior to SIR Mission, fluvial erosion was suspected in the Southern Pediplain depending on regional observations around this vast region. SIR/A mission was followed by another two missions by US space shuttles: SIR/B (1984) and SIR/C (1994). In addition to these SIR missions, the Radarsat of the Canadian Space Agency provided radar images for the southern part of the Western Desert of Egypt and the adjacent areas of northern Sudan. Supported by field studies, all these missions provided valuable information on the paleo-drainage patterns not only in the southern part of Egypt, but also in the whole Eastern Sahara. Three types of drainage forms were recognized in the SIR coverage of southern Egypt (McCauley et al. 1986a). These forms are as follows:

23.3.1 Wide Wadis

These wadis are 10–30 km wide, and hundred kilometers long. These dry valleys are strikingly similar in scale and overall appearance on the radar images to the Nile Valley (McCauley et al. 1986a). These valleys were given the names of local features, such as Wadi Safsaf (after Bir Safsaf), or names describing their major morphologic characteristic, such as Wadi Arid (Wide) and Wadi Mokhtafi (Hidden). Tested by field pits and trenches, these broad valleys are now almost completely filled with fluvio-clastic sediments that are unconformably overlain by aeolian sand sheet. The wadis are separated by low interfluves, which have been greatly reduced by wind erosion as postulated by McCauley et al. (1986a) and are now partly mantled by sand. Some of these wadis have flood plains (**Fig. 23.2**).

23.3.2 Anastomosing Channels

These channels are groups of narrow (0.5–2.0 km) channels separated by islands. Such channel complexes developed as insets in the flood plains or the wadi fill of the above-mentioned wide wadis. The islands separating the channels are formed of the valley fill deposits, which are composed of sand and fine pebbles, cemented by secondary calcium carbonates (McCugh et al. 1989; Szabo et al. 1989; El-Baz and Robinson 1997). Figure (23.2) shows the distribution of the braided Palao-drainage channels in the flood plains of broad wide wadis, such as Wadi Safsaf or Wadi Arid. These **anastomosing channels** and floodplain morphology are very much similar to those of Cooper Creek Central Australia. This river flows in a complex assemblage of waterholes, anastomosing and braided channels, back swamps, and aeolian dune fields before reaching its terminus in Lake Eyre (Nanson et al. 1986). Both valleys in the Southern Pediplain and Australia have floodplains several tens of kilometers wide, containing a complex of channels (0.5 km or less) that are inset into the alluvium covering the valley floor. The only difference between both valleys is the thin to discontinuous aeolian sand sheets that overlie the alluvial valley fill in the southern part of the Western Desert and mask the channels from the human eye or from conventional photography imaging. The mean annual rainfall ranges between 280 mm in the lower section of the basin of Cooper creek and <120 mm at Lake Eyre (Beman et al. 1997). It can be inferred from the previous analysis that the anastomosing and braided channels in the radar wide wadis of the Southern Pediplain developed under climatic condition similar to those arid/semi-arid conditions prevailing at present in the lower section of Cooper Creek.

23.3.3 Regional Reconstruction of Palaeo-Drainage System

Regional reconstruction of these previously unknown palaeodrainage forms, and their systems still need a more extensive radar coverage and field tests. However, McCauley et al. (1982) suggested that the large valleys are relics of a Tertiary system that drained the Eastern Sahara before the integration of the Nile and before the onset of general aridity in the early Quaternary. Misfit tributaries and narrow channels within buried floodplains show that parts of these large Tertiary wadis were used again and again by intermittent running water during the Quaternary pluvial episodes. These

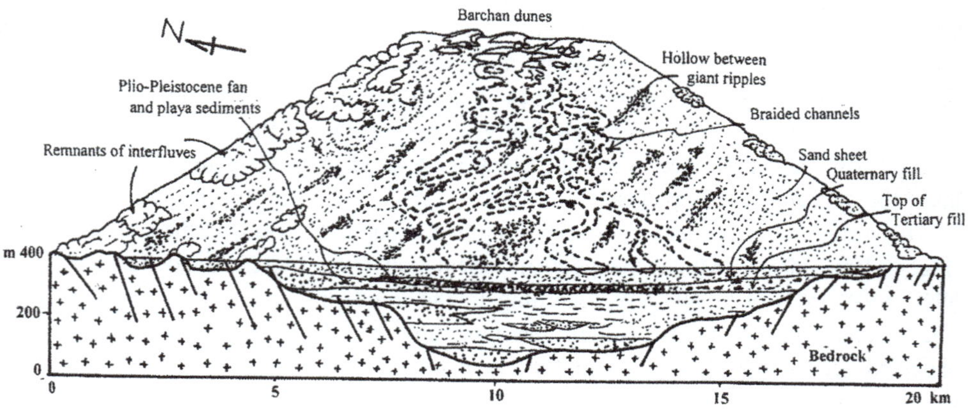

Fig. 23.2 A block diagram showing a wide-buried wadi filled with sediments topped by braided channels and some surface feature (*Source* McCauley et al. 1986a)

suggestions pose a key question: Where were the sources of these palaeodrainage rivers? Or from where did the trunk streams come, and where did they go? Issawi (1983), McCauley et al. (1986b) and Issawi and McCauley (1993) postulated that the Red Sea Mountains are the main source for these valleys. Issawi (1983) and Issawi and McCauley (1993) concluded from the analysis of the characteristics of some major valleys in the Eastern Desert of Egypt such as Wadi Qena and Wadi Allaqi that these wadis were the upper courses of an old westward and southward flowing consequent stream prior to the development of the Nile System (some 7.0–6.0 million years ago). Issawi and McCauley (1993) also concluded that three major palaeodrainage systems developed in response to tectonic uplift and sea level changes from Late Eocene (40 million years ago) to Late Pleistocene (0.25 million years ago). These systems were named the Gilf system, the Qena system, and the Nile system. On the other hand, El-Baz and Robinson (1997) concluded from the study of the palaeo-channels revealed by SIR-C data in the southern part of the Western Desert and Northern Sudan that four major drainage lines drained internally in a basin into the Selima Sand Sheet. Therefore, they called this system "The Selima Drainage System." Beside this system, they found a smaller system that leads into the drainage system of the River Nile.

The youngest major system of the Nile valley (about 7.0–6.0 million ago) represents a landmark in the orientation of drainage systems. This powerful river carved a deep valley by headward erosion, because the level of the Mediterranean Sea dropped, reaching nearly the latitude of Aswan. As this river extended southward, it captured the upper reaches of the previous Qena system and other major drainage systems. This capture left the lower reaches to the west of the Nile Valley as wide-beheaded valleys with misfit channels. Since then, deposition has dominated in these valleys so that they were filled first by fluvial deposits during Pliocene and Pleistocene pluvials, and then by aeolian sediments with the onset of arid conditions in the Holocene.

The destination of these palaeo-drainage systems represents another problem, since the regional picture is still incomplete. McCauley et al. (1986b) hypothesized that these valleys were a part of a regional Tertiary system and are called "the Trans-African Drainage System." This hypothesized system is comparable to a similar Tertiary drainage system (Wadi Ar-Rimah, Wadi El-Batin, Wadi ad-Dawasir, and Wadi Sahba) crossing the Arabian Peninsula from the Hegaz and Asir Mountains in the west, to the Arabian Gulf in the east (Anton 1984). Issawi and McCauley (1993) postulated that the Gilf palaeo-drainage system was a north-flowing system at the end of the Oligocene, whereas that of the Qena system (16 million years ago) was probably running southwest into Chad Basin. They also considered the Tarfa River part of the Qena system, but it fanned on the Eocene plateau at the outskirts of the Kharga Depression. They believe that the thick sand and Pre-cambrian igneous and metamorphic gravel on the surface of the Eocene plateau west of the Nile Valley between Minia and Sohag are deposits of the Tarfa River. Other palaeodrainage rivers drained internally in the extreme south of this pediplain along the Egyptian-Sudanese boundaries, suggesting the development of a lake that is now covered by sand (El-Baz et al. 1998).

Ephemeral watercourses formed at the floors of the trunk wadis when the favorable conditions of the Quaternary pluvials were utilized by successive groups of early humans (McCauley et al. 1986a). Evidence (implements) shows that these palaeo-wadis were inhabited by Paleolithic man since 300,000 BP. Geo-archeological evidences from Wadi Arid and Wadi Safsaf indicate that there were four phases of human occupations. The first is middle Late Acheulean (212 ka), the second is Late Acheulean (141 ka), the third is Middle Paleolithic (45 ka), and the fourth is early terminal Paleolithic–Neolithic since about 15 ka (Szabo et al. 1989).

23.4 The White Desert of Farafra Geo-Park: Product of Karst, Fluvial, and Aeolian Processes

23.4.1 Introduction

The **White Desert** is one of the national Geo-Parks of Egypt and is located in the northern part of Fararfra Depression, 48 km north of Qasr El-Farafra, the capital of the Oases of Farafra. The present-day climate of all Farafra is hyper-arid in which rain is rare, evaporation is high, and temperature is high during summer becoming mild in winter. This desert is composed of a succession of formations belonging to the Mesozoic and Cenozoic, of which those dominated by limestone and chalk are the most widespread. Here, the Eocene limestone caps the surrounding plateaus, while the floor of the Depression is cut into snow-white chalk of relatively smaller shallow basins. In general, the outer basin rises only a few tens of meters above the sea level. According to the contour map, this basin is surrounded by the 40 m a.s.l contour line. The lowest point is 15 m a.s.l (Mostafa 2007). This basin covers an area of ~ 300 km^2, which is about 2.0% of the total area of the **Farafra Depression**.

This small desert is significant for three reasons. Its floor is furnished with snow-white Paleocene chalk, and consequently all forms developed are white—hence the naming, "The White Desert." Moreover, three types of forms of different origins have developed side by side: karst forms, fluvial forms, and aeolian forms, and some forms are modeled by more than one process. It also possesses a scenic

Fig. 23.3 Tower Karst of various shapes and heights in the White Desert

landscape, where the forms are sculptured in a way making them monumental and attractive. The combination of these various forms of the White Desert made the landscape unique not only in Farafra, but also in other Egyptian deserts. Past humid climates, geologic formations, local topography, and present-day arid conditions all contributed to this unique landscape.

23.4.2 Forms

Karst and fluvial forms are inherited from past humid climates and have been exposed to modifications during its long history of development.

The most spectacular karst forms are **rock-towers**, cone karst, and mushroom-shaped rocks (Mostafa 2007). Fields of low-lying **cone karst hillocks** are the most widespread karst form assemblage. The towers are the tallest (>10 m), some of which are smoothed and thin, while others are irregular and huge **(Fig. 23.3)**. Some of the towers are scoured and undermined by wind-driven sand blasting, mostly from one side which faces the wind direction. **Mushroom** rocks are found in various forms and sizes **(Fig. 23.4)**. In between karst forms, fluvial forms have developed. **Playas** with **yardangs** of different shapes and sizes can be seen next to other forms. They occupy small shallow basins, mostly originated by solution. Sediments of the playas have been transported by running water from the surrounding higher areas. Due to the presence of the high percentage of carbonates, the yardangs and playa materials are snow-white **(Fig. 23.5)**, adding more whitish surfaces to the white color of the chalky landscape. This whitish color is kept always fresh by wind scour of Paleocene chalk. Due to the softness of this chalk, sand blasts are very effective as an erosion process **(Fig. 23.6)**.

Fig. 23.4 A mushroom-shaped rock in the White Desert

23.5 Domes of Northern Sinai

23.5.1 Introduction

North Sinai is characterized by the abundance of northeast to east-southeast oriented doubly plunged anticlines. They are known in geologic literature as "the Syrian Arcs" (Said 1990). As can be seen from Fig. (23.7), they are not

Fig. 23.5 A field of white yardangs in an ancient playa in the White Desert

Fig. 23.6 Wind-scoured white chalk, developing a small sharp-edged yardangs in the White Desert (Note the cone karst hills at the background)

restricted to Sinai, but extend eastward into the Negev region. They form a distinctive tectonic and geomorphologic region in North Sinai. Although there are similar forms in various areas in Egypt, not only in North Sinai, domes are clustered to form a specific geomorphological entity. They were divided—according to size—into three categories as large (tens of kilometers long, such as Gabal El-Maghara, G. Yelleq, and G. El-Halal), intermediate (several kilometers long, e.g., Gabal El-Minsherah, G. Kherim, and G. Araif El-Naqa), and small folds (less than 2 km long, such as Gabal Libni, G. Um Makhasa, G. El-Bruk, G. Risan Aneiza, and G. El-Amrar (Mostafa and Khalil 1990). The large and intermediate ones are asymmetric, where the northwest flank dips 5°–20°, while the southeast flanks are steeper and may be vertical or overturned. This southeast flank is affected by thrust faults in a number of cases. Characteristic forms of these structurally controlled hills and mountains can be classified into structural and fluvial forms.

23.5.2 Structural Forms

Three forms were recognized on all domes: hogback ridges, fault escarpments, and inter-mountain basins.

According to Fairbridge (1968), **hogbacks** are those long ridges which developed on sedimentary structures with alternating weak and hard beds, dipping steeply in one direction. This type of ridges represents the remains of the folded beds after eroding the core of the domes. Several parallel ridges of hogbacks may develop according to the number of alternating weak and hard beds.

In Sinai, hogbacks developed on all anticlinal hills and mountains since they have been composed of sedimentary structure with alternating weak and hard beds, where most of the beds' dip is between <10 and >70 degrees as recorded by Mostafa and Khalil (1990). Visual interpretation of high-resolution images provided by Google Earth revealed the following geomorphologic characteristics:

A. Most of the ridges are circular because the anticlinal structures are plunging in two directions.
B. Hogback ridges are repeated and are parallel to each other where alternating weak and hard beds occur on one side or another, such as those developed on the southern part of Gabal El Maghara (Fig. 23.8).
C. With steep dips (30° − +70°), nearly symmetrical hogbacks developed. This occurs on the SE sides of Gabal El Maghara, G. Halal, G. Yelleq, G. e Minsherah, G. Kheirm, on the NW side of G. Araif el Naga, and on most of the sides of Gabal Hamyar. On gentler dips, the ridges often exhibit a dip slope and an escarpment which cuts across them at right angles.
D. Fluvial dissection and differential erosion of hogback ridges tend to develop triangular faces (dip slopes) on the spurs, known as *flatirons* (Lobeck 1939). This form is best developed in the northern section of G. El Maghara.

Various types of faults occur on the **domes of North Sinai**, leading to the development of fault-generated escarpments. Most of the folds are dissected by NW-oriented faults, perpendicular to the strike lines of the folds and that the down-thrown side of most of them is on the SW side. Size and extension of the escarpment is proportional to fault size and type.

The cores of domes and folds represent the weakest parts, least resistant to erosion, and this is why they are eroded at higher rates than the outer margins. These cores are hollowed and become relatively lower compared to the

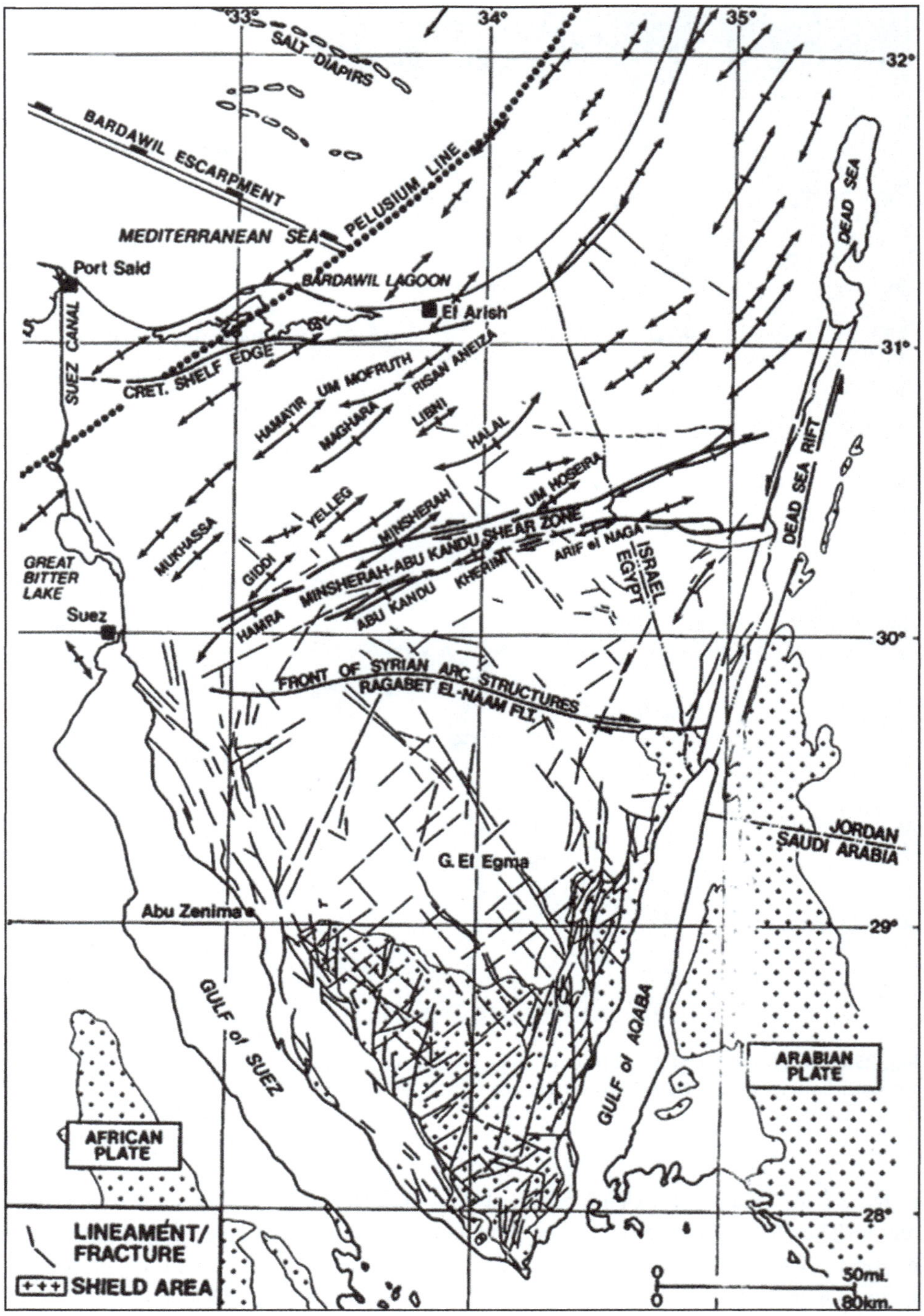

Fig. 23.7 Main structural feature of Sinai

surrounding peripheries, representing internal basins, and are classified as reversed topography. This situation occurs on G. Halal, G. Magharah, and some other smaller domes. In the meantime, basins differ greatly in their dimensions (length, width, and depth) from one dome to another, depending mainly on the size of the dome. The large basins developed on large domes such as El Maghara. On the other hand, structural variations between domes are not less important in determining the size of basins. This is clear on G. El Halal and G. Yelleq, where basins are not proportional to the size of the dome.

23.5.3 Fluvial Forms

These forms have developed by water action during the pluvial periods. They are divided into two groups: (1) wadis and drainage lines; and (2) alluvial fans, bahadas, and playas.

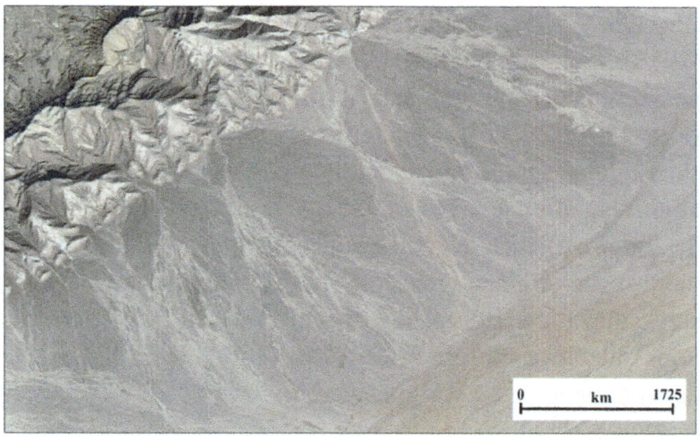

Fig. 23.8 Fans, bahadas, and hogbacks at the southern side of Gabal El Maghara (*Source* Goole Earth 2014)

The dome-like hills and mountains are dissected by drainage nets which are almost independent of each other, because every topographic unit is composed of a separate anticline. Wadis and drainage network are characterized by several geomorphic and geologic aspects. The first aspect is that most of the drainage lines are organized in a radial pattern. The second aspect is that some of the drainage lines unite with the tributaries of Wadi El Arish and drain externally, whereas those dissecting the northern slopes of the northern units disappear under the dunes of the North Sinai Sand Sea, and drain internally. This area is originally a Depression (Kusky and El-Baz 2000), and this is why wadis were not able to reach the Mediterranean coast, except for Wadi El Arish. The third aspect is that some wadis cut across some domes as antecedent valleys. Of these wadis, the one cutting across the eastern part of Gabal Halal (Wadi El Arish) is the most well-known. Another case of the antecedent valley is the one cutting across Gabal Talaat El Badan also by Wadi El Arish, lying to the SW Gabal Halal.

Fans, Bahadas and Playas developed nearly at all the foot slopes of domes. Due to the large number of wadis dissecting the sides of the hogbacks, and due to a long history of fluvial work, fans grew larger by time and coalesced, resulting in the formation of a bahada around each block. At the peripheries of bahadas, some playas developed in low-lying areas. Sediments of the fans and bahadas are composed mostly of gravels and fine materials of sand and silt. Fluvial wadis bifurcate in the foothills into shallow braided channels cutting across the fans and bahadas (Fig. 23.8).

23.6 Inverted Wadis

Inverted wadis are linear, sinuous, and at times branching gravel ridges, which stand out above the surrounding rocky surfaces **(Fig. 23. 9a)**. First, they were recognized by Sandford and Arkel (1929) along the Nile-Fayum divide. They are composed of Pliocene rounded gravelly deposits, capping some hills (G. El-Naaloon and G. El-Rus) on this divide, and was interpreted as representing the remnants of previous beds of wadis, which carried flow toward the Pliocene Nile Gulf.

The Nubian Desert on the western side of the Nile Valley is a second locality where inverted wadis have been recorded. According to Said (1990) and Zaki and Giegengack (2015), they represent ancient beds of an old drainage system, which seems to have had some relationship with the ancient Nile system. Although dendritic drainage patterns represented by inverted wadis were recognized at many sites, the presence of some patterns cross each other led

(a)

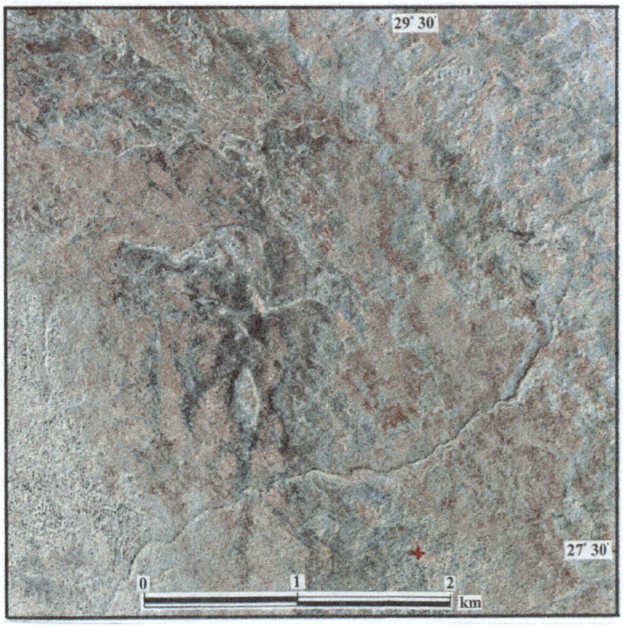

(b)

Fig. 23.9 **a** A Landsat-TM image of an inverted wadi to the west of Ghard Abu Muharik. **b**. A section in the sediments of one of the ridges of inverted wadi on the Eocene Plateau, the Western Desert

Giegengack to the conclusion that multiple episodes of wadi inversion have occurred throughout the region (Zaki and Giegengack 2015). These inverted wadis are indurated wadi deposits of poorly sorted gravel and sand. They include rolled Acheulian implements and are assigned an age of Middle Pleistocene (Zaki and Giegengack 2015).

The central Eocene Plateau is the third locality where inverted wadis have been recorded and described as gravel mounds (El-Hinnawi et al. 1978). The author recorded these inverted wadis during a trip to the Djara Cave in 1995, crossing this central part of the plateau, but as gravel ridges (Fig. 23. 9a). As mentioned in the study of El-Hinnawy et al. (1978), the sediments are constituted of alternating beds of poorly bedded gravel and thin coarse sand-pebble beds **(Fig. 23. 9b)**. The gravel is poorly sorted, sub-rounded to sub-angular, and is embedded in a pale brown sandy matrix. High-resolution images of Google Earth (2014) show that the ridges of the inverted wadis act as a dividing line for more recent drainage lines on the Eocene plateau. They also show that these ridges have been dissected into small sections due to exposure to erosion during later times after development. The original wadis were considered by

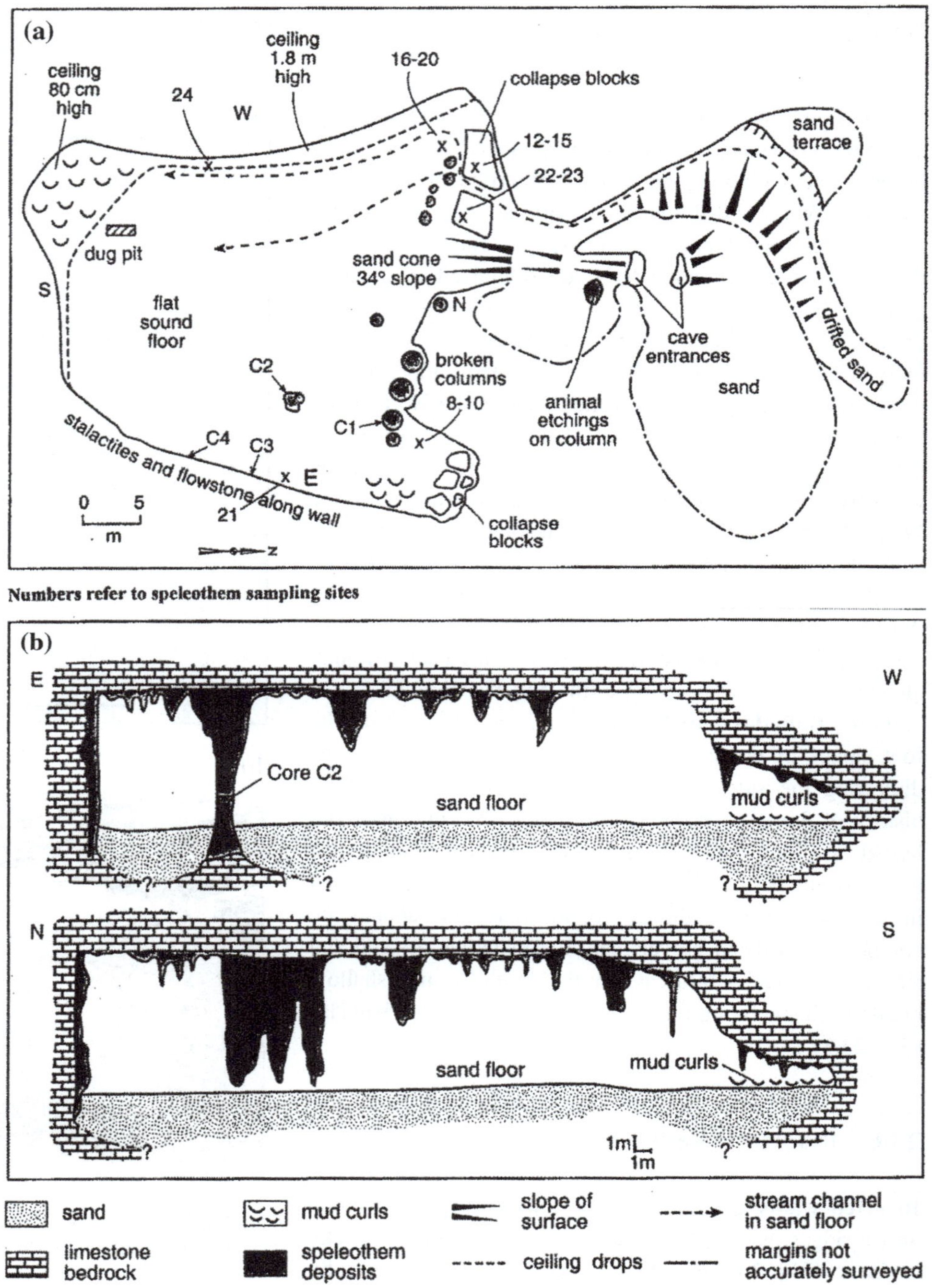

Fig. 23.10 Two sections in El-Djara Cave showing its main morphologic features (*Source* Brook et al. 2002)

geologists to belong to the Oligocene Period (Said 1990). Most probably, they represent relics of the tributaries of the Oligocene drainage system, which started during the Late Eocene and continued to exist until Middle Miocene, and formed a huge delta to the north of Fayum Depression. These inverted wadis need further investigation and field study to reveal their detailed geomorphic and geologic characteristics.

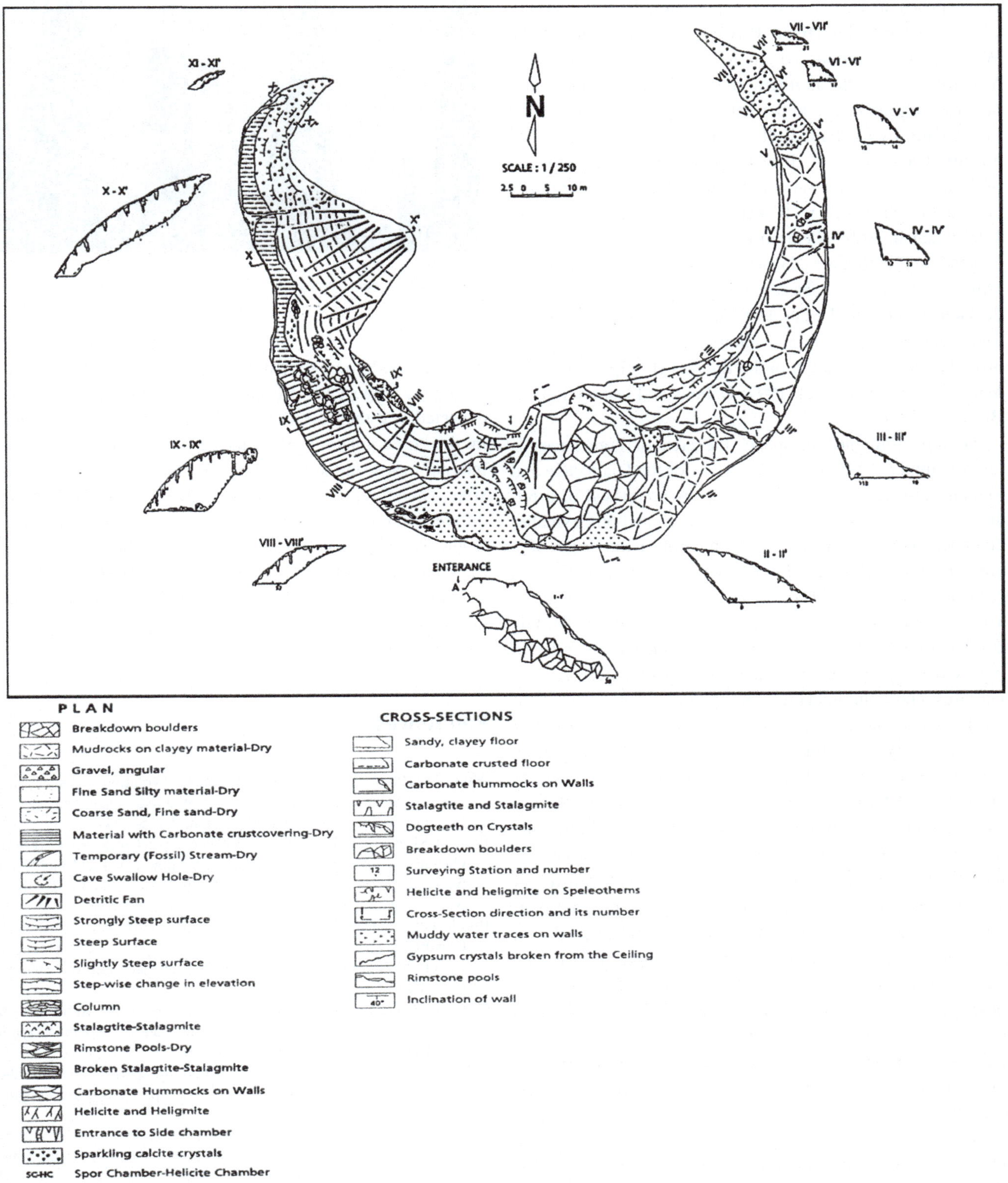

Fig. 23.11 A plan diagram showing major internal forms in Sannur Cave (*Source* Günay et al. 1997)

23.7 Djara and Sannur Caves Geo-Parks

These two caves are not only the largest known caves in Egypt, but they also accommodate their internal speleothems and flowstones in their original condition, because they were not exposed to degradation by quarrying (Embabi 2004). This is because the **Djara Cave** lies in the heart of the Western Desert, about 10 km to the west of Ghard Abu Moharik, and on the northern side of the camel track connecting Farafra Oasis and the Nile Valley. It was discovered incidentally by Gerhard Rolfs on December 24, 1873. The **Sannur Cave** lies in the Eastern Desert on the northern side of a tributary of Wadi Sannur, about 70 km SE of Beni Suef city in the Nile Valley. It was discovered accidentally in 1991 when an artificial entrance to the cave was created after a blast in a carbonate alabaster quarry, locally known as "Mohamed Ali's Alabaster Quarry." Both caves were declared as Geo-Parks due to their geological, geomorphologic, and touristic value. Although few, previous studies provide good information on both caves (Philip et al. 1991; Kuper 1996; Gunay et al. 1997; Dabous and Osmond 2000; Brook et al. 2002).

Despite difference in shape, the area of both caves is almost similar, 1295 m^2 and 1500 m^2 for the Djara cave and Sannur cave, respectively. The Djara is composed of a relatively large square chamber (37 × 35 m) and other smaller chambers **(Fig. 23. 10)**. The Sannur takes the form of a crescent tunnel **(Fig. 23. 11)** that extends for about 275 m, with a maximum width of 12 m and an average height of 10 m. Both caves developed in the Late Early-Middle Eocene limestone. Most of the speleothems are similar in both caves. They include straw- and carrot-stalactites, stalagmites, and columns **(Fig. 23. 12)**. Flowstones developed on the walls and roofs of the caves. They are composed of alternating layers of fine crystalline with milky to yellow coarse calcite bands. These calcite bands are known as "Egyptian Alabaster," quarried in the environs of the Sannur cave. In spite of similarity in cave forms, there are some differences between them. The most characteristic aspect of the Djara Cave is that its floor is covered with a layer of aeolian sand of unknown thickness, since the original bedrock floor has not been reached in six boreholes drilled by a hand auger to a depth of six meters (Kuper 1996). The Sannur cave is characterized by the presence of broken rock blocks collapsed from its walls and roof. They are piled up on the cave floor, concentrated at the cave entrance due to the accumulation of the broken material that resulted from the artificial opening of the entrance while quarrying the alabaster.

Since the two caves have developed in the Eocene limestone, it is expected that they started to form soon after the exposure of the limestone, and the climatic conditions became favorable for their formation. Humid tropical climatic conditions started in Late Eocene and continued during the Oligocene, Miocene, and Pliocene. Therefore, it is expected that the development of these two caves occurred in this time interval. By time, they became wider, and internal forms started to develop and increase in number and size. Therefore, cave forms may have developed during different phases. This is why ages determined for speleothems and flowstones are not similar. They range in age between 140 ± 16 ka and 283 ± 56 ka. However, it was found that some samples from the Djara Cave are beyond the range of the dating methodologies, i.e., older than 500 ka (Brook et al. 2002).

Fig. 23.12 A field photograph of large stalactite columns in the Djara Cave

References

Anton D (1984) Aspects of geomorphologic evolution; palaosols and dunes in Saudi Arabia. In: Gado A, Zotel J (eds) Quaternary period in Saudi Arabia, Springer-Verlag, Vien, 2:275–294

Berman SJ, Nanson GC, Tooth S (1997) Cooper Creek, Southwestern Queensland: A regional perspective, Section 4 in Field Guide for physical environments in Eastern Central Australia. In: Tooth S, Nanson GC (eds) University of Wollongong, 50–64

Brook GA, Embabi NS, Ashour MM, Edwards MM, Cheng H, Cowart JB, Dabous AA (2002) Djara cave in the Western Desert of Egypt: Morphology and evidence of quaternary climatic change. Cave Karst Sci 29:57–66

Clayton P (1933) The Western side of the Gilf Kebir, in A reconnaissance of the Gilf Kebir by the Late Sir Robert Clayton East Clayton. Geogr Jour L31:254–259

Dabous AA, Osmond JK (2000) U/TH study of speleothems from the Wadi Sannur Cavern Eastern Desert of Egypt. Carbonates Evaporites 15:1–16

El-Baz F, Robinson C (1997) Paleo-channels revealed by SIR-C data in the Western Desert of Egypt: implications to sand dunes accumulations. Proceedings of the 20th international conference and workshop on applied geologic remote sensing, Denver, Colorado, 17–19

El-Baz F, Robinson C, Maxwell TA, Hemida IH (1998) Palao-channels of the great selima sand sheet in the Eastern Sahara and implications to ground water. Palaoecology Afr 26

El-Hinnawi M, Abdallah A, Issawi B (1978) Geology of Abu Bayan, Bolaq stretch, Western Desert. Egypt Ann Geological Survey Egypt 8:19–50

Embabi NS (1998) Sand seas of the Western Desert of Egypt. In: Al-Sharhan A, Glennie K, Whittle G, Kendall C (eds) Proceedings of the international conference on quaternary Deserts and climatic change, Al-Ain, Rotterdam, Balkema, 1995

Embabi NS (2004) Geomorphology of Egypt, vol I: The Nile Valley and the Western Desert. The Egypt geogr Soc, Cairo

Fairbridge RW (1968) Encyclopedia of geomorphology. Dowden, Hutchinson & Ross, Inc

Ghoneim E, Robinson C, El-Baz F (2007) Radar topography data reveal drainage relics in the Eastern Sahara. Inter J Remote Sens 28:1759–1772

Google Earth (2014) Egypt Images Latitudes 22-32 N and Longitudes 22-32 E, http://www.google.com

Günay G, Ekmekci M, Bayari CS, Kurttaş T, El-Bedewy F (1997) Sannur Cave: A crescent shaped cave in Egypt. In: Günay G, Johnson AI (eds) Karst: waters & environmental impacts. Balkema, Rotterdam

Issawi B (1983) Ancient rivers of the Eastern Egyptian Desert. Episode 2:3–6

Issawi B, McCauley J (1993) The Cenozoic landscape of Egypt and its river systems. Ann Geol Surv Egypt 19:357–384

Kuper R (1996) Between the Oases and the Nile - Djara: Rohlfs' Cave in the Western Desert, Interregional Contacts in the Later Prehistory of Northeasern Africa: 81–91. Poznan Archaeological Museum, Poznan

Kuski T, El-Baz F (2000) Neotectonics and fluvial geomorphology of the Northern Sinai Peninsula. J. Afr Earth Sci 31:213–235

Lobeck AK (1939) Geomorphology: An introduction to the study of landscapes. McGraw Hill, New York

McCauley J, Schaber G, Breed C, Grolier M, Haynes C, Issawi B, Elachi C, Blom R (1982) Subsurface valleys and geoarcheology of the Eastern Sahara revealed by shuttle radar. Science 218:1004–1020

McCauley J, Breed M, Schaber G, Haynes V, Issawi B, El-Kilani A (1986a) Palaeodrainages of the Eastern Sahara, The radar rivers revisited (SIR-A/B Implications for a Mid-Tertiary Trans-African Drainage System). IEE Trans Geo Sci Remote Sens GE 24:624–648

McCauley J, Breed M, Schaber G (1986b) Megageomorphology of the radar rivers of the Eastern Sahara. Proceedings of the 2nd Space-borne imaging radar symposium (28–30 April), Jet Propulsion Laboratory, Pasadena, California, Publication 86–26

McHugh W, McCauley J, Haynes C, Breed C, Schaber G (1988) Paleorivers and geoarchaeology in the Southern Egyptian Sahara. Geoarchaeology 3:1–40

McHugh W, Schaber G, Breed C, McCauley J (1989) Neolithic adaptation & the holocene functioning of tertiary palaeodrainges in Southern Egypt & Northern Sudan. Antiquity 63:320–336

Mostafa A (2007) The geomorphology of karst forms in Farafra depression, the Western Desert, Egypt. Dissertation, Ain Shams University

Mostafa A, Khalil M (1990) Structural characteristics and tectonic evolution of north Sinai fold belts. In: Said R (ed) The geology of Egypt. Balkema, Rotterdam, pp 381–389

Nanson G, Rust B, Taylor G (1986) Coexistent of mud braids & anastomosing in an arid zone river: Cooper Creak, Central Australia. Geology 14:175–178

Philip G, El Aref MM, Darwish M, Ewais E (1991) Paleoerosion surfaces and karst manifestations including " Egyptian Alabaster" in Gabal Homret Schaibun- Gabal Sannur area, east of the Nile valley, Egypt. Egypt J Geol 34:41–79

Richter A, Schandelmeier H (1990) Precambrian basement inliers of Western Desert: geology, petrology and structural evolution. In: Said R (ed) The geology of Egypt. Balkema, Rotterdam

Said R (1990) Quaternary. In Said (ed) The geology of Egypt, Balkema, Rotterdam

Sandford K, Arkel W (1929) Paleolithic man and the Nile-Faiyum divide: A study of the region during Pliocene and Pleistocene times, vol X. Orient Institute, Chicago University

Schaber G, McCauley J, Breed C (1997) The use of maltifrequency & polarimetric SIR-C/X-SAR data in the geologic studies of Bir Safsaf. Egypt Remote Sens Environ 59:337–363

Szabo B, McHugh W, Schaber G, Haynes C, Breed C (1989) Uranium-Series dated authigenic carbonates and Acheulien sites in Southern Egypt. Science 243:1053–1056

Zaki AS, Giegengack R (2015) Inverted topography in the Southeastern part of the Western Desert of Egypt. J Afr Earth Sc 121:56–61

Geomorphologic Hazards in Egypt

24

Abstract

Geomorphologic hazards in Egypt could be divided into two groups: Earthquakes represent the first group, whereas fluctuations in Nile discharge, coastal erosion, flash floods, sand drift, storms, and dune movement represent the second group of hazards. Although annual fluctuations in Nile discharge and flood risk ceased after the construction of the Aswan High Dam, new hazards came into being due to the deposition of the Nile load in the basin of Lake Nasser. Among these are coastal erosion and retreat, and deprivation of the soils of the floodplain and the Delta from annual rejuvenation by the additions brought in by Nile floods. Flash floods and sand drift, storms, and dune movement are the most recurrent hazards in Egypt. Earthquakes are rather infrequent hazards in Egypt, but when they do occur some can be disastrous.

Keywords

Geomorphic hazards • Nile fluctuations • Lake Nasser • Coastal erosion • Flash floods Sand storms • Dune encroachment • Earthquakes

24.1 Introduction

Before humans appeared on Earth, only geophysical processes shaped the Earth's surface. With the presence of humans, the physical factors became **natural hazards**, and since then, interaction between them and natural agents started. Hence, natural hazards are threatening events, capable of producing damage to the physical and social space where they take place, not only at the moment of their occurrence, but also on a long-term basis due to their associated consequences. When these consequences have a major impact on society and/or infrastructure, they become natural disasters. Hazards are strongly related to geomorphology since they are important ingredients of the dynamics of the Earth's surface. Hazards are the result of unexpected changes in long-term behavior caused by minute/huge changes in the initial conditions. In this sense, geomorphic hazards can be categorized as endogenous (volcanic and tectonics) and exogenous (floods, karst collapse, snow avalanche, channel erosion, sedimentation, mass movement, tsunamis, coastal erosion, sand drift and storms, and dune movement).

Natural disasters are a global issue as they occur everywhere in the world. Even though they may have a considerable impact on countries such as Japan, the USA, France, or Switzerland, their significance in countries such as Egypt, Bangladesh, India, China, Guatemala, Colombia, or Mexico is by far greater. The global death toll due to natural disasters is concentrated in developing countries, and it can be as high as 95% of the total toll (Alca´ntara-Ayala 2002).

In Egypt, all types of geomorphic hazards (endogenous and exogenous) occur, but with different intensities. Previous studies show that the Egyptians have suffered from fluctuations in Nile waters since the Pharaonic times (Said 1993). The burial of Ancient Egyptian monuments under desert sands is evidence of sand encroachment hazards. Earthquakes occur from time to time with various degrees of intensity. Flash floods have become a normal annual incident. Sand storms and drift, and dune encroachment on highways and other infrastructures are widely spread with

N.S. Embabi, *Landscapes and Landforms of Egypt*, World Geomorphological Landscapes,
https://doi.org/10.1007/978-3-319-65661-8_24

the expansion of land reclamation in desert areas. Previous studies have covered nearly all types of hazards in different localities in Egypt. The geomorphologic hazards in Egypt can be classified as follows:

1. Endogenous hazards, such as earthquakes.
2. Exogenous hazards, such as fluvial floods, sand drift and dune encroachment, and coastal erosion.

In the following analysis, exogenous hazards will be dealt with first, because they occur more frequently than earthquakes, and because the resulting damage happens all year round.

24.2 Fluvial Hazards

Fluvial hazards in Egypt are of two types: **nile discharge fluctuations** and **flash floods**. Nile fluctuations are very important to Egypt, due to the fact that most Egyptians are entirely dependent on Nile water. Although they have not occurred since the construction of the **Aswan High Dam** in 1970, the Nile fluctuations will be summarized first to show how they used to threaten the life of Egyptians in the Nile Valley, and how the construction of the High Dam resulted in the creation of new hazards, or the increase in intensity of the existing ones.

24.2.1 Nile Discharge Fluctuations

The main purpose of the construction of the **Aswan High Dam** was to control the discharge of Nile water, which reaches Egypt from the Ethiopian and Equatorial sources, not only during the year, but also from year to year. This is because not only was there a surplus in the Nile water during some years, but there was also a deficiency in the water in other years when water arriving to Egypt through Aswan did not satisfy the consumption demands of the Egyptians. The capacity of the Lake Nasser basin allows storage of excess water which could be used not only in the following year, but also during any forthcoming year with water deficiency for almost a hundred years. This is why the storage of water in Lake Nasser basin is known as century storage. **Nile water fluctuations** and their impact on the population of the Nile Valley were profound, and some cases will be presented to show how they were a serious hazard to Egyptians from the beginning of Pharaonic times.

During the period 2890–2686 BC (Dynasty II), there were 21 records of low Nile levels. Those were years that left a negative impact on the nation, confirmed by records showing of persistent civil unrest and a weakened economy. A return to a more stable Nile discharge is recorded during the period 2686–2345 BC, i.e., dynasties III-V (Said 1993). Available records for the period from 500 BC to 600 AD indicate that the Nile was high, to the extent that Herodotus (ca 450 BC) described the floodplain at the time of floods as a sea. During the Fatimid rule, Egypt was struck by an episode of low Niles, of which the years 1052–1090 AD were the period of the lowest level ever. This episode is known as the years of Al-Shiddah Al-Mustansiriya (Al-Mustansiriya Hardship), or the severe drought during the times of the Fatimid Caliph Al-Mustansir who ruled Egypt at that time. An eyewitness of this period describes food shortage, mass immigration, starvation, death from hunger, spread of disease, and plague. The period also saw a significant decrease in the population of Egypt from 2.4 million at the beginning of the tenth century to 1.5 million at the end of it. The period 1351–1467 AD witnessed very high floods, where the height of water exceeded that of any other period, causing damage. The floods were exceptionally high in the years 1359–1360. According to Al-Maqrizi, the people went out to the desert because they were afraid of being over-flooded (Said 1993). Some of the early years of the nineteenth century (1818–1820) also seen high floods. In his vivid description of the 1818 flood, Belzoni stated that on September 18, 1818, "the Nile rose 3.5 feet above the previous inundation, with uncommon rapidity, and carried off several villages and some hundreds of their inhabitants" (Said 1993). El-Gabarty's chronicles mention the 1818 flood that the Nile rose to such an extraordinary level in Cairo that the Island of Roda was completely submerged, boats crossed the Nile over it, many villages were destroyed, and a considerable number of inhabitants and their animals were drowned. According to El-Gabarty, more disastrous floods occurred in the following year (Said 1993). If we move to the end of the nineteenth century and the twentieth century (1871–1986), the analysis of the records of the Egyptian Ministry of Irrigation showed that it can be divided into three periods: The first thirty years (1871–1900) witnessed high to very high floods, with an annual average of 110 billion m^3; the second period (1901–1966) was characterized by intermediate floods with an annual average of 86.5 billion m^3, but with some disastrous years of high floods ($\sim$100–110 billion m^3) or of low (less than 50 billion m^3); and the third period (1967–1986) witnessed a decreasing trend in the Nile floods from 75 billion m^3 (1968) to 32 billion m^3(1986). This last year (1986) was the fourth consecutive year during which the Nile flood failed to cover the national water needs, because it was less than 55 billion m^3 in any of these years (Embabi 1993). The High Dam and Lake Nasser saved Egypt from a disastrous starvation during those four years of very low floods.

Although the **Aswan High Dam** has protected Egypt from the negative impacts of high and low floods, it has equally been responsible for the development of some negative geomorphologic aspects threatening the life of humans.

Two significant features are the result of water storage in the basin of the High Dam. The first is the deposition of Nile silt in the Lake Nasser basin, and the second is the development of an imbalance between deposition and erosion along the Mediterranean coasts of the Nile Delta. The second phenomenon is a reaction to the first one, because the deposition of the Nile silt in the lake bed deprived the coasts of sediments which would compensate those removed by waves and offshore currents (see also Chap. 6). The deposition of Nile silt in the Lake Nasser basin has also deprived the soils of the floodplain in the Nile Valley, of the annual renewal of soil fertility, due to additions of silt during the flooding season. Therefore, soil fertility has deteriorated, but the use of fertilizers has compensated for the deprived elements. Furthermore, the deposition of Nile silt in the Lake Nasser basin has deprived the building industry of the raw material (silt) used for manufacturing bricks that are used in construction. It was therefore obligatory to discover areas in the Egyptian deserts, where natural materials (shale, clay) could be used for brick manufacturing. Fortunately, Egypt is rich in this type of materials in localities that spread all over the country.

24.2.2 Flash Floods

A **flash flood** is a rapid flooding of low-lying areas. It is characterized by its sudden nature and fast-moving water. In an arid region such as Egypt, it is caused by heavy rains, associated with a severe thunderstorm. They are related to special weather conditions that occur periodically and in unconnected localities. The analysis of rain data in Egypt (Chap. 3) shows that although rainfall is low or very low all over the country, there are a few days during the year (mainly in the winter season) in which rain increases abnormally in a very short time. Therefore, large amounts of water reach the ground and cause floods to develop in a very short time. This is because there is little water loss through evaporation and soil absorption. The number of hazardous events caused by flash floods has increased during the past 40–50 years, due to increase of human activities in the Egyptian deserts for the purposes of land reclamation development. In general, flash floods do not cover the whole country, rather they are localized in nature. It is rare that weather conditions might lead to flash floods that cover most of the country; nevertheless, such conditions did occur in November 1994 (El-Disouki 1995). This case and another two of disastrous flash floods in **Wadi El Arish** in Sinai will be analyzed in the following section. The case of November 1994 has been chosen, because it was rare and does not occur frequently like localized flash floods. Furthermore, the cases of Wadi El Arish have been taken as an example of localized floods, since the lower reaches of this wadi are the most populated region in Sinai and accommodate the capital of North Sinai Governorate, El Arish City. Also, large areas at Wadi El Arish are under cultivation.

24.2.2.1 Flash Floods of 1994

Weather maps of October 30 and 31, 1994, showed that Egypt was under the effect of a large tropical, warm, and humid air mass due to the presence of low pressure to the north of the Red Sea, known as the Sudanese Low (El-Disouki 1995). This was accompanied by a rise in temperature above normal at this time of the year in all stations. Additionally, there was an atmospheric high over North Africa and the Mediterranean, and high pressure in the upper layers of the atmosphere over Central Europe. These conditions indicate that the center of cold air was concentrated in the north, and the center of atmospheric heating was concentrated in tropical and Equatorial latitudes.

As space images showed, clouds extended from Central Africa in the south to the northern and eastern parts of the Red Sea on October 31, 1994, but no rain was reported that day. By the end of that day and the beginning of November 1, 1994, the high atmospheric layers started to cool down and a cold front developed over the country. At the same time, air was still rising, forming some cumulative clouds of high density by the early hours of 2 November as shown on space images. This was followed by the fall of heavy rains accompanied by thunder and lightning. These atmospheric conditions prevailed until 6 November in nearly all stations (El-Disouky 1995). Rainfall varied from one locality to another. During 2 November only, there was 32 mm of rainfall in Alexandria, 13 mm in Asyut, and 18 mm in El-Tor in Sinai.

During the few days of this abnormal meteorological condition (from November 1 to 4, 1994), rainwater accumulated on the slopes of the Red Sea and Sinai, flowing at a high speed and carrying out large quantities of sediments along the dry wadis to the low-lying areas and to the Nile Valley, resulting in the destruction of some villages and highways, and the flooding of cultivated fields (Fig. 24.1). In one area only, the village of Drunka, lying a few kilometers to the southwest of Asyut City, 399 persons lost their lives, a large number of cattle drowned by water that flooded their sheds, and houses were damaged and collapsed. Infrastructure was also damaged, and cars were buried by sediments (Ashour 1992).

24.2.2.2 Flash Floods of 1975 and 2010 in Wadi El Arish

Wadi El Arish covers an area of approximately 20,000 km^2 and drains the northern and central parts of Sinai, extending in an almost SN direction for about 365 km (see Chap. 19). It is the second largest wadi in the Egyptian Desert after

Fig. 24.1 Inundation of cultivated land by water flash flood of November 1994

Wadi Al-Allaqi. It is composed of six major basins. Its catchment area lies in Sinai, except for some small tributaries which come from the Negev Desert joining its lower section. It discharges more than two-thirds of Sinai's waters, which is equal to about 160 million m^3. Although it is dry most of the year, it might carry flow during the winter season. During flood, which follows heavy rainstorms, Wadi El Arish looks like a real river, presenting a hazard to all human activities in its lower section (Hamdan 1980). Two previous studies have recorded and analyzed the two great floods of 1975 (Klein 2000) and of 2010 (Moawad 2013). These were the highest and most extraordinary floods since their discharge was 117 and 119 billion m^3, respectively, resulting in the strongest floods in a period of 85 years from 1925 to 2010 (Moawad 2013).

Flash Flood of 1975: In February 1975, a unique rain event occurred in the El Arish drainage basin which brought about an extreme flood flow. The rain started during the evening of February 19, 1975, spreading throughout the Sinai Peninsula except for its northern part. The rain intensified during the early morning hours of 20 February and continued throughout the day. The storm advanced from southern Sinai to the northeast toward Eilat. During the rainstorm, the following amounts of rain were measured by rain gauges in the area: St Catherine Monastery (south of the basin), 73 mm; Eilat (east of the basin), 68 mm; Nekhel and Qusayima (central part of the basin), 48 and 38 mm, respectively; and only 8 mm in El Arish in the northern part of the basin (Klein 2000 after Gillad 1975). The amount of rain measured in Eilat in 24 h was 64.3 mm, of which 20 mm h^{-1} was measured in the morning. It is estimated that the average rain over the basin was 40–50 mm. As the northern part of the basin did not contribute to the flow, the total water volume on the basin was 800×10^6 m^3 (Klein 2000).

The flow was observed in the middle part of the basin on February 20, 1975. The main source of hydrological information is the data recorded at the hydrological station in the Ruaffa Dam (station no. 25180) operated at the time by the Hydrological Service of Israel. The **Ruaffa Dam** is located about 50 km from the wadi outlet. The flow started at 08:30 Egypt local time on 21 February. The measured peak flow was 1.65 m^3 s^{-1} during the late afternoon hours. The flow terminated at 05:30 on 26 February. The total flow volume was 120×10^6 m^3, which constituted about 15% of the total rainfall volume. The flow reached El Arish at 03:00 h on 22 February. The duration of the flow in the city was 75 h. Large areas were flooded, bridges over the wadi were swept away, and many lives were lost. The average width of the flooded channel was 500 m, and the water depth in the middle of the flow was 4 m. The Ruaffa Dam and the lake behind it constituted a sediment trap for all the coarse material from the drainage basin. Only the fine materials were transported to the north of the dam. The average flow velocity from the dam to the sea was 1 m s^{-1} (Klein 2000).

On January 17 and 18, 2010, severe **flash floods** hit the **Sinai Peninsula**, owing to heavy rains of which the flood of Wadi El Arish was the worst. The flood led to the death of six persons. Dozens of people were wounded, and many went missing. It fully destroyed 592 houses and partially damaged 1487 houses as the water level reached 2 m above the surface of the ground. The water washed away cars, trucks, trees, roads, and electric wires and water pipelines (Moawad 2013). The wadi received a total rain volume of 665.46×10^6 m^3. Most of the rain fell on 17 January with an intensity ranging between 4–8 mm per hour. The estimated total surface runoff was 123.3×10^6 m^3, and the total runoff volume was 124×10^6 m^3 that constitutes 18.8% of the total rain volume (Moawad 2013).

On 18 January, the flowing water reached El Arish City at 16:30 GMT (at 18:30 Egypt local time). The time elapsed since the beginning of the strengthened rainstorms on January 17, 2010, at 12:00 and 18:00 GMT and the arrival time of flow to El Arish City on 18 January was 18:30 and 22:00, respectively. This affirms that the central part of the basin was responsible for most of the flow ($\sim$229–183.7 km south from the pour point). The average flow velocity of the water at the dam was 1.6 m/s. It was assumed that the flow velocity was constant along the last 50 km of the wadi channel until its pour point. In that case, the water passed the dam, most likely, on 18 January at 7:49 GMT (09:49 local time) once the lake had stored some 5.5×10^6 m^3 of water. The average height of the water was 10 cm over the dam and 45 cm over the wadi floor near El Arish City (Moawad 2013).

On 19 January, the height of the water increased considerably to 50 cm over the dam. In El Arish City, the water accumulated at Ezbet Maliem behind the major road for over 4 m high. It inundated the surrounding areas such as villages of Atif, Abu Thaqal, and Al-Badawi to some 2 m high. Therefore, the flowing water completely destroyed the main road connecting the eastern and western sides of El Arish City and hampered rescue operations for many hours. On the other hand, the water flooded vast areas of the northern lowlands because the outlet of the wadi was blocked by the Olympic Sports Centre. This required extraordinary efforts to enable the water reaching the sea. The flow terminated at 08:00 GMT (10:00 local time) on 19 January (Moawad 2013).

24.3 Sand Drift and Dune Encroachment

Sand movement is a natural phenomenon in deserts, as far as sand supply is available and wind speed is capable to move the sands. Sand movement in itself is not a problem, but to people who live in deserts, it becomes a threat and a battle has to be waged against moving sands and dunes. Although the threat affects desert settlers in the Egyptian Oases in the Western Desert from ancient times, its rate has increased since the beginning of the twentieth century when several attempts were carried out to reclaim and cultivate new areas, and railways and highways were constructed to connect the oases with the Nile Valley or to connect the individual oases with each other. Sand is drifted on desert surfaces in three ways: creep, saltation, and suspension. Sand also moves on dune surfaces resulting in dune movement. Each of these movements represents a hazard to humans and their activities when they cross each other.

Fig. 24.2 a Sand-filled irrigation canal in a newly reclaimed land to the east of the Suez Canal (Photo by Prof. Mamdouh Akl of Alex. Univer.) b Sand storm on the central highway in Sinai

24.3.1 Sand Drift Hazards

Hazards resulting from **sand drift** in Egypt are of three types: **sand storms**, decrease in the fertility of arable land, and decrease in the storage capacity of **Lake Nasser**. In a sand storm, huge quantities of sand are driven by strong winds so that sand is carried up in the air making a sandy cloud, resulting in poor, or in some cases nil, visibility (Fig. 24.2a). This poor visibility represents a serious threat to life conditions. Airplane flights, trains, cars, and trucks are the most affected by sand storms. The number of days with sand storms occurring in Egypt is limited. The main season of sandstorms is spring to early summer, which is called the **Khamasin**, or fifty days' season. During this season, hot, dry wind with sand and sometimes dust blows intermittently. Also, sand drift fills new irrigation canals in newly reclaimed land in Sinai to the east of the Suez Canal (Fig. 24.2b).

The fertility of arable land along the western side of the Nile Valley in the governorate of El-Minya is threatened by additions of blown sand from the neighboring desert. Along this western stretch of the floodplain, there is a belt of sand dunes in the desert running from north to south, from which the NW wind carries sand toward the floodplain of the Nile, spreading it on the fertile soils of the Nile silt. The study carried out by Kishk (1990) revealed that sand percentage in the floodplain soils next to the arable lands on the desert is 90%, whereas this percentage beside the River Nile in the east, which is several kilometers from the desert, is 17% only. The same trend is observed in organic content, which is 0.2% and 1.2% next to the desert and beside the Nile, respectively. This indicates that sand drift is a permanent hazard to the arable land along the western fringes of the Nile floodplain by decreasing the fertility of soils.

Sand drift also represents a permanent hazard to the water storage capacity of **Lake Nasser**. This topic is treated in detail in Chap. 5; however, in this chapter on geomorphologic hazards, it is also important to mention that the Aswan High Dam created new hazards in Egypt. Here, the hazard of drift sand reducing Lake Nasser's storage capacity is a direct result of constructing the dam. Before that, sand was drifted into the Nubian Nile, but it was removed by the running water of the Nile, so there was no threat to the Nile channel. However, at present, water in Lake Nasser does not move; therefore, any sand reaching the lake is accumulated on its bed, resulting in a decrease in its capacity.

24.3.2 Hazards of Moving Dunes

Moving dunes have long represented a permanent threat to the humans living and working in the Egyptian deserts. With the expansion in reclaiming desert lands, the threat rate has increased, since the moving dunes cross railways and highways (Fig. 24.3) and encroach on cultivated areas and buildings. This happens in all regions of Egypt, where moving dunes are present. Until the present moment, there have been no measures to stop dunes from moving, but there are several ways and techniques to minimize or avoid the paths of moving dunes. To do this, some studies were carried out on dune movement in all areas, where the hazards are immediate. The first study on dune movement was done in the early years of the twentieth century (Beadnell 1910). Since then, research on dune movement has continued in various localities in Egypt (Tsoar 1974, 1983; Embabi 1979, 1986–1987; Embabi et al. 2012; Philip et al. 1992). The studies concluded that local wind environment is the major controlling factor with regard to local variations of dune movement.

24.4 Erosion Along the Coasts of the Nile Delta

Although sediment removal and accretion along the Nile Delta coasts has been known and evidenced a long time ago, concerns about the negative consequences increased significantly with the completion of the Aswan High Dam in 1964. Ever since, nearly no Nile sediments have reached the Mediterranean Sea, because such sediments are trapped in the basin of Lake Nasser, forming a new delta at the southern section of the Lake (see Chap. 5, Sect. 5.3.1). Deposition of the sediments in Lake Nasser was catastrophic to the coasts of the Nile Delta, because there was no compensation for the eroded sediments from the shore of the Delta. Before the High Dam era, not only a good balance existed between removal by waves and currents and deposition by Nile waters, but also sediments brought by the Nile exceeded those removed by marine processes, resulting in the growth of these two promontories. During the nineteenth century, the two promontories advanced due to deposition of the Nile sediments, and from 1909 to 1973, their coasts retreated due to erosion. This erosion was the result of the construction of the first Aswan Dam 1902 and its heightening in 1912, and then the High Dam in 1964, which controlled the discharge of Nile water and sediments, resulting in a decrease in both. A study of the Nile water discharge during the nineteenth and twentieth centuries indicated that the annual average decreased from 108×10^9 m^3 during the period 1871–1900, to 86.5×10^9 m^3 during the period 1901–1966, until it reached 61.5×10^9 m^3 during the period 1968–1986 (Embabi 1993).

Rates of erosion have accelerated since 1964 along the projections of the Nile Delta, Rosetta, Ras El-Borollos, and Damietta, which are currently undergoing changes due to both natural and anthropogenic pressures. The highest rate of erosion occurs along the outer margins of these promontories. As indicated, this erosion resulted from both the discontinued discharge of the River Nile sediments due to the Aswan High Dam and from the prevailing coastal processes. Erosion at Ras El-Borollos and the tip of the Damietta and Rosetta projections has adversely affected homes and condominiums at Baltim and Ras El Bar (Fig. 24.4), and has destroyed the old coastal road from Damietta to Port Said

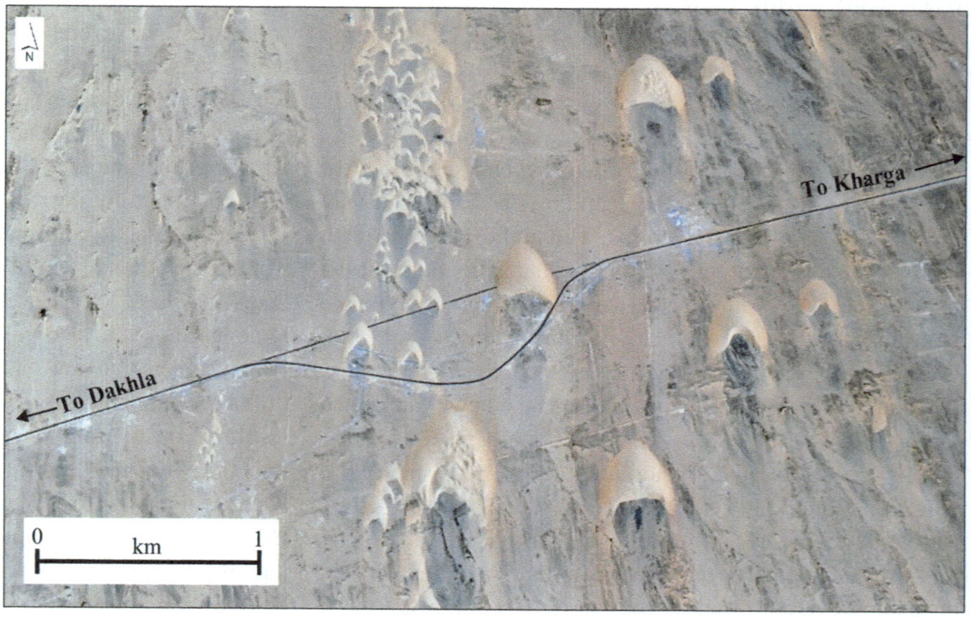

Fig. 24.3 Dune encroachment on the Kharga-Dakhla highway (*Source* Google Earth 2014)

Fig. 24.4 Destruction of houses under the attack of waves in a summer resort at Baltim coastal town, at the center of Delta coast (Photograph taken by M Khidr)

Fig. 24.5 Dolorosa protective concrete blocks along the Nile Delta coasts

and the lighthouses west of the river. However, several protective structures—described in detail by Fanos et al. (1995)—have been constructed along these two projections to protect from erosion (Fig. 24.5)

24.5 Earthquakes

Egypt is considered one of the few countries in the world, where evidence of historical **earthquake activity** has been recorded. Recording covers a period of ~5000 years, from 2800 BC to 1984 AD (Kebeasy 1990). Kebeasy (1990) found that 83 events occurred during this period in and around Egypt, causing damage in various localities. Damage caused by various earthquakes could be direct or indirect and variable in different regions as could be seen from the following review of some cases which occurred during historic times (Kebeasy 1990).

The 2800 BC earthquake in Sharqiya Province, east of the Nile Delta, was a severe one and caused a deep fissure and soil cracks in Tell Basta. The 1210 BC earthquake caused cracks in the temple of Ramses II in Abu Simbel, Upper Egypt. In 221 BC, an earthquake occurred in Siwa Depression and caused destruction in approximately 100 locations in Libya. An earthquake occurred in 27 BC in Thebes, Upper Egypt, and caused great damage leaving only four villages undestroyed in that region. Recent earthquake activity, occurring during the twentieth century (1900–1995) has been recorded in various localities in Egypt, and has caused various degrees of damage. Examples of these earthquakes are as follows: In 1955, an earthquake occurred on 12 September off the shore of Alexandria. It was a strong one, of a magnitude of 6.1 on the Richter scale and was felt in the entire region of the East Mediterranean. In Egypt, it was felt in various localities between Alexandria and Cairo, leaving behind several hundred destroyed houses and several tens of human deaths. The 1969 earthquake occurred on 31 March in Shadwan Island in the Red Sea. There is also the 1978 earthquake which occurred in the unpopulated region of Gilf El-Kebir in southwest Egypt. The 1981 earthquake occurred on 14 November and had a magnitude of 5.5 on the Richter scale. It is of significance, because of its possible relationship with Lake Nasser. Furthermore, two earthquakes occurred in 1992 in Dahshour to the north of Fayum and in 1995 in the Gulf of Aqaba. Both of them resulted in human losses and caused great damage in various localities in Egypt.

All the above-mentioned earthquakes caused direct damage to humans and buildings. Indirect damage has been caused by other earthquakes that occurred in historic times at the floor of the Mediterranean Sea, resulting in the development of **tsunami** sea waves that have affected the Egyptian coasts several times with varying intensities. The largest tsunami was the one resulting from the earthquake of August 8, 1303. The advance of the sea accompanying this event has resulted in the submergence of half the city of Alexandria. Other minor tsunami occurred in 24–20 BC, 1202 AD, and 1908 AD (Kebeasy 1990).

Figure (24.6) shows the distribution of all earthquakes that have occurred in Egypt, the focal mechanisms of some principal earthquakes, and the location of seismographic stations. The analysis of the distribution of epicenters by Kebeasy (1990) suggested that activity tends to occur along three main active lines The first is the northern Red Sea–Gulf of Suez–Cairo–Alexandria clysmic trend. It is the major active trend in Egypt. The second one is the East Mediterranean–Cairo–Fayum Pelusiac trend. The Gilf Kebir event in

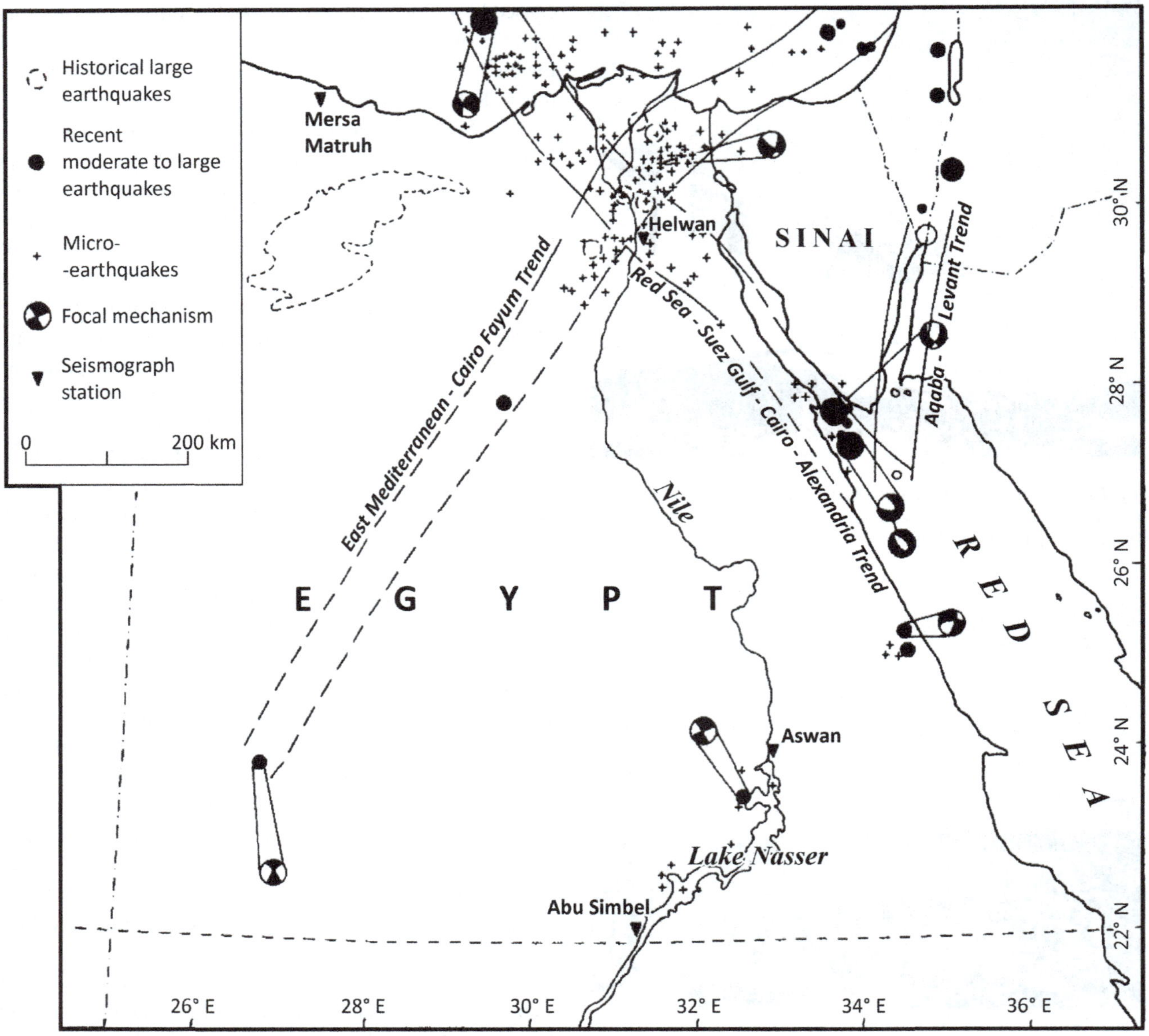

Fig. 24.6 Distribution of epicenters of all earthquakes (*Source* Kebeasy 1990)

1978 may be the SW extension in the Western Desert of this trend. The third one is the Levant–Aqaba trend, which is a continuation of the Levant active fault, and extends along the Gulf of Aqaba and southwest in the Red Sea, where it bisects the first trend.

24.6 Conclusion

Due to the uniqueness of the Egyptian landscape and the aridity of climate since the Mid-Holocene period, Egyptians have concentrated in the Nile Valley because it is the only region with permanent water and fertile soils. This is why fluctuations in Nile discharge were the first and the main hazard threatening the Egyptian people. This situation remained for centuries, until the twentieth century when new technology became available to control Nile discharge. Two dams were constructed on the Nile at Aswan in Upper Egypt: The first is known as the Aswan Dam and was built at the beginning of the twentieth century to control annual fluctuations. However, it did not have the capacity to control extraordinary discharges, so a second dam, the Aswan High Dam, was considered in order to store high and very high floods. It began operating in 1970.

Although the Aswan High Dam secured Egyptians against flooding during high discharge in the summer and against water shortage during low discharge in the winter, it created three new hazards due to the deposition of Nile silt in the basin of Lake Nasser. During pre-dam times, the Nile silt was carried out to the Mediterranean Sea and laid down along the coasts of the Nile Delta. There was a balance between what was eroded by waves and offshore currents and what was laid down by the Nile. After the High Dam took control of the Nile, the coasts were deprived of the Nile silt and a state of disequilibrium came into being whereby erosion of the coasts exceeded the rate of deposition in areas like Damietta and Rosetta promontories. Measures were taken to minimize or stop the retreat, but only a few succeeded. Other problems created by the construction of the High Dam include the decrease in soil fertility and the decrease in the storage capacity of the basin of Lake Nasser due to silting.

Flash floods and sand drift or storms and dune encroachment throughout the country cannot be stopped due to their physical characteristics. Flash floods occur annually and suddenly, and they occur in various places with a different intensity. So far, no accurate weather forecasting is available and this is why measures taken to control their threat have failed in several areas such as in the cases of Wadi El Arish, for in spite of the building up of dam on the wadi (El-Ruaffa Dam), it failed to control or to stop the flood water in 1975 and 2010. Flash flood hazards increase in time, because human activities extend into the deserts to reclaim new land for the growing population of Egypt, or into hazardous areas without taking any measures for protection against flood flow.

Although sand drift, sand storms, and dune encroachment can be observed in several localities in Egyptian deserts, their threat cannot be avoided because they cannot be controlled. Sand storms and their threats to aviation and highways are unavoidable. The same holds for sand drift which deteriorates soil fertility in areas vulnerable to sand creep, such as the western fringes of the Nile floodplain in El-Minya Governorate. Dune movement leads to the encroachment of dunes on any structures that stand in their way such as highways, railways, arable land, villages, and buildings. Complete prevention from dune encroachment is impossible; only temporary stoppage can be done. This is because moving dunes in the Egyptian sand seas are countless, and they are organized in belts where dunes follow each other. The only solution for this hazard is that humans and dunes live side by side in a state of peaceful coexistence and that techniques should be applied in each locality that would guarantee this state of coexistence.

References

Alca´ntara-Ayala I (2002) Geomorphology, natural hazards, vulnerability and prevention of natural disasters in developing countries. Geomorphology 47:107–124

Ashour M (1992) Flash floods in Egypt (A case study of Drunka Village, Upper Egypt). Bull de la Société de Géographie d'Égypte 75:101–114

Beadnell H (1910) The sand dunes of the Libyan Desert. Geographical J 35:379–395

El-Disouki T (1995) Climatic conditions of the November 1994 Flash floods in Egypt. (In Arabic) The Egyptian Geographical Society, Cairo, pp 35–60

Embabi NS (1993) Water resources of Egypt: the Nile water (in Arabic). Special Publication, Middle East Research Centre, Ain Shams University, Cairo

Embabi NS (1979) Barchan dune movement and its effect on economic development in the Kharga Oases Depression, the Western Desert, Egypt. (In Arabic) J Middle East Research Centre 6:51–84

Embabi NS (1986–1987) Dune movement in the Kharga and Dakhla Oases Depressions, the Western Desert, Egypt. Bulletin de la Société de Géographie d'Égypte 59–60:35–70

Embabi NS, Mostafa A, Mahmoud A, Azab M (2012) Geomorphology of Ghard Abu Moharik, Western Desert. Egypt. Bull de la Société Géographie d'Égypte 85:1–28

Fanos A, Khafagy A, Dean R (1995) Protective works on the Nile Delta Coasts. J Coastal Research 11(2):516–528

Google Earth (2014) Egypt Images Latitudes 22-32 N and Longitudes 22-32 E, http://www.google.com

Hamdan G (1980) The Personality of Egypt. Vol I: The physical characteristics, Dar El-Hilal, Cairo

Kebeasy R (1990) Seismicity. In: Said R (ed) the geology of Egypt. Balkema, Rotterdam

Kishk M (1990) Desert encroachment on the fringes of the Nile Valley, Egypt. In El-Baz F, El-Tayeb I, Hassan M (eds) Proc. Inter. Workshop on Sand Transport & Desertification in Arid Lands, 17–26 Nov. 1985, Khartoum, Sudan. World Scientific, Singapore

Klein M (2000) The formation and disappearance of a delta at Wadi El Arish river mouth. Proc. Jerusalem Conference on the Hydrology-Geomorphology Interface: Rainfall, Floods, Sedimentation and Land Use, IAHS Publication No 261, May, 1999

Moawad M (2013) Analysis of the flash flood occurred on 18th January 2010 in Wadi El Arish, Egypt (A case study). Geomatics 4(3):245–274

Philip G, Labib T, Sharaky A (1992) Sand dunes of the Dakhla Depression, the Western Desert, Egypt. In: Sadek A (ed) Proc of the first conference on the geology of the Arab World, vol II. Cairo University, Cairo

Said R (1993) The river nile: geology, Hydrology and Utilization. Pergamon Press, Oxford

Tsoar H (1974) Desert dunes morphology and dynamics, El Arish, Northern Sinai. Zeit Geomorphologie, Suppl Bd 20:41–61

Tsoar H (1983) Dynamic processes acting on longitudinal (Seif) dune. Sedimentology 30:567–578

Index

N.S. Embabi, *Landscapes and Landforms of Egypt*, World Geomorphological Landscapes,
https://doi.org/10.1007/978-3-319-65661-8

The manufacturer's authorised representative in the EU is Springer Nature Customer Service Centre GmbH, Europaplatz 3, 69115 Heidelberg, Germany. If you have any concerns regarding our products, please contact ProductSafety@springernature.com

Printed and bound by CPI Group (UK) Ltd, Croydon, CR0 4YY
01/10/2026
02225885-0001